Practical Microstrip Design and Applications

For a listing of recent titles in the *Artech House Microwave Series,*
turn to the back of this book.

Practical Microstrip Design and Applications

Günter Kompa

ARTECH
HOUSE

BOSTON | LONDON
artechhouse.com

Library of Congress Cataloging-in-Publication Data
Kompa, Günter
Practical microstrip design and applications/Günter Kompa.
p. cm. —(Artech House microwave library)
Includes bibliographical references and index.
ISBN 1-58053-980-7 (alk. paper)
1. Strip transmission lines. 2. Microwave integrated circuits. I. Title. II. Series.

TK7876.K625 2005
621.381'3—dc22

2005048171

British Library Cataloguing in Publication Data
Kompa, Günter
Practical microstrip design and applications. —(Artech House microwave library)
1. Strip transmission lines 2. Strip transmission lines—Design
I. Title
621.3'813
ISBN-10: 1-58053-980-7

Cover design by Yekaterina Ratner

685 Canton Street
Norwood, MA 02062

International Standard Book Number: 1-58053-980-7

10 9 8 7 6 5 4 3 2 1

To Pirkko Marjatta and Sirpa

Contents

Preface

This book is strictly addressed to microstrip design, to treat the subject both theoretically and practically with high thoroughness. In this sense, it may be an exceptional book with respect to other publications offering a broad palette of topics in the general area of planar technology.

The book starts with the fundamentals in microstrip design, so that previous knowledge in planar circuit design is not needed. The book is therefore very suited for a reader who wants to come into the respective field. Although microstrip design is not always present in higher-education curricula, respective knowledge and experience are expected in many cases by communication companies from graduates in communication engineering.

The book does not exclude complex mathematics where needed. However, the reader who is more interested in practical use of the book may rely on the descriptive parts, presented formulas, and graphs for design purpose. The reading of the theoretical analyses, however, will give thorough understanding of the entire subject. This is also valid for parasitic effects such as radiation, surface wave, and higher-order mode excitation, which are less intensively discussed in other books. In particular, the occurrence of a higher-order mode and its effect on the electrical properties are discussed and demonstrated by design examples. The understanding of excitation and propagation of surface waves on microstrip circuits is also important for the microstrip designer, and therefore exhaustively analyzed.

The book is mainly addressed to interested readers in the field of radio frequency (RF) and microwave engineering. It may be attractive also for students and researchers in other scientific disciplines. Since the microstrip is an open waveguiding structure with large scattering field, this can be used for probing. Contactless sensors are commonly treated by specialists educated in applied or technical physics and measurement and control as well.

Organization of This Book

The book is divided into nine chapters, each of which explores a theme through several sections. Arranged alphabetically, several appendices A to J complete the material under discussion and are related to respective chapters. The book comprises an introductory and theoretical part, followed by an extended part concerning the basic microstrip design elements. The last part is devoted to application and technology.

Chapter 1 outlines the challenges of the microstrip designer and contains a review of the evolution of computer-aided design.

Chapter 2 describes initially the origins of microstrip circuit design and respective early applications. It follows an overview about diverse constructions of planar transmission lines, describing the electromagnetic field properties and emphasizing practical design viewpoints. Appendix C concerns a special topic of pulse shaping using coupled microstrip line and slotline.

Chapter 3 is devoted to microstrip analysis. Known diverse methods are grouped as quasi-static, empirical-based, and full-wave analysis approach. A first part of the chapter treats thoroughly the various electrostatic methods proposed by Wheeler (Section 3.1), Schneider (Section 3.2), and Hammerstad (Section 3.3). This is done chronologically so that the reader gets an orientation with respect to the actuality of proposed device formulas and models. The accuracy of a model is mostly related to the time of its publication. In that way, the reader gets some concept for the quality of a given model. As the electrostatic methods utilize conformal mapping, basics of conformal transformation are described in Appendix A. Appendix F contains a table of characteristic impedance and effective dielectric constant values after Hammerstad.

The chapter is completed with the full-wave analysis method (spectral domain method), which is developed step by step (Section 3.4). Intermediate calculations have been included to make the subject readable for those who are not so familiar with the theoretical background. Appendix D reviews some basics in Maxwell theory concerning the calculation of electromagnetic field waveguiding structures. For the closed microstrip line, the orthogonal series expansion method is presented in Appendix H.

In addition to the given design formulas, design graphs for most relevant substrate materials are available, simplifying the practical microstrip design. Thickness correction for nonzero strip conductors are included. Graphs have been plotted for the inductance and capacitance per unit length of a microstrip line.

Chapter 4 is devoted to the dispersion of the microstrip line. The definition of the characteristic impedance is argued. For various substrate materials, the dispersive characteristic impedance and the effective dielectric constant are plotted over a broad frequency range. Section 4.3 describes various measurement techniques to characterize the dispersion of both the quasi-TEM mode, and higher-order modes as well.

Chapter 5 gives an overview about the different waveguide models for the microstrip line to describe its dispersive properties, or to use it as an equivalent waveguiding structure to solve discontinuity and junction problems under application of mode matching procedure. Various proposed waveguide parameters are discussed and weighted with respect to circuit simulation.

Chapter 6 describes most relevant microstrip discontinuities and junctions commonly characterized by their S parameters. Appendix G describes the fundamentals of low- and high-frequency circuit theory involving the definition of scattering parameters. Particular focus is on excitation and radiation of higher-order modes, and surface waves as well. Lumped elements such as series inductance and shunt capacitance are also treated. Appendix I describes the impedance matrix of step discontinuity.

Chapter 7 is devoted to microstrip conductive and dielectric losses. Conductive loss is caused by the skin effect and calculated applying Wheeler's incremental inductance rule. Some remarks about conductor roughness completes the chapter.

Chapter 8 is focused on applications of microstrip circuit design. The Smith Chart (Appendix B) is included in the design procedure if useful for better understanding. Section 8.1 summarizes filter design concept using filter catalogues. As an example, a lowpass filter is designed in three variants showing the different influence of parasitic higher-order modes on the filter response. Section 8.2 is devoted to the design of impedance matching networks of low-impedance high-power amplifiers for base stations for UMTS application. Section 8.3 treats the analysis of TEM quarter-wavelength transformers. Design rules are presented. The theory is approximately applied to a microstrip directional coupler using coupled microstrip lines. Fundamentals of quasi-static and dispersive coupled line parameters are treated in Appendix J. A design example is presented.

Chapter 9 discusses some technology aspects, describes the material parameters of relevant substrates, and discusses their limitation with respect to upper frequency of operation. Furthermore, it addresses the issues in the integration of chip devices in the microstrip design procedure. Section 9.3 illustrates the various circuit fabrication approaches: Hybrid, monolithic, and quasi-monolithic integration technology.

Who Should Read This Book

The motivation to write this book has come from long teaching experience in an international study program "Electrical Communication Engineering" (ECE) at the University of Kassel. ECE was one of the first eight funded international programs in Germany established in 1997. It is a master's program, and a qualified bachelor degree is required for admission. In a preceding study semester, the international students are prepared for sound knowledge in electrical communication before they start with their specialization during the master's course. The main content of this book is given as an introductory course about microstrip integrated circuits during the preparatory semester, with additional practical training in microstrip design using a computer and commercially available software tools. Thorough knowledge and training in the discussed field are regarded as very important factors for future vocational assignment of graduates as design engineers in research and development. The course is also appropriate to graduate students in advanced semesters.

The book may be also useful for design specialists reading selective topics or using the clearly structured collected material as a reference book.

How to Use This Book

The book addresses several reader categories. The reader who wants to come into the field of microstrip design may read the book from the beginning to end. Since the book attaches much importance to the fundamental understanding of the subject, essential theoretical parts are integrated. This not only refers to the fundamental microstrip mode, but also concerns other parasitic effects such as higher-mode excitation and radiation. Thus, the book is a good basis to acquire both thorough theoretical and practical knowledge in the field. There are several appendices that can be selectively read with the main text according to the knowledge available to the reader.

The book can be used on a different reading level. Readers who are more focused on the practical design aspect can pass over the theoretical chapters; for example, Section 3.4 about full-wave analysis of the microstrip line. A preferred practical view of microstrip design can be followed on the basis of derived design formulas, tables, and design graphs for the dispersive parameters of the microstrip line and most relevant discontinuites and junctions.

The book is appropriate as a course book, primarily for graduate students or for undergraduate students in the last year of their studies. Knowledge of vector algebra and vector analysis, and an introduction into Maxwell theory is necessary to understand the theoretical part of microstrip analysis. Respective material is included in the appendices, to be made available at any time.

Acknowledgments

Many people have contributed to the development of this book. Special gratitude goes to Dr. Abhijit Ghose, who reviewed the English language of the entire book and enhanced the writing style. I wish to thank Jürgen Weide who did a great job of preparing the figure material and helped to keep the computer running. I express my gratitude to Alexander Kricke and Alexander Nuss for supporting the graphical material and Arkadiusz Lewandowski from Warsaw University for writing MATLAB programs. I thank all my staff members offering their services at any time for their contributions to simulations with ADS, MWO, MATLAB, and C++. I would like to thank Dr. Bernd Bunz, Dr. Abdulrhman Ahmed, and Asdesach Zena Markos. I am particularly indebted to Dr. Julie Lancashire from Artech House, who motivated me to start the book project. I thank also Tiina Ruonamaa and Katherine Nolan from Artech House for enormous and sustained support during writing. Finally, I wish to express my gratitude and love to my wife, Pirkko Marjatta, without whose infinite, unfailing, and generous support this book could not have been written.

Chapter 1

Introduction

Microstrip design has become mature within one researcher generation. Various research groups were established at universities and institutes in the 1960s all over the world headed by visionary scientific pioneers, intensively following the idea to establish a computer based circuit design platform for future mass production of devices and circuits in electrical communications, such as satellite television. We can notice the fruits of intensive research work of numerous scientific groups, having delivered valuable contributions to the development of design software tools, which are indispensable in the daily work of any circuit designer.

Retrospectively, it can be seen that today's commercially available design software has undergone natural selection and fusion, so that only a few have survived. However, it is also observed that new software products continually appear on the surface of the hard-fought market.

The expectation of any circuit designer is that he can rely on the given design software. However, as it is often stated, simulation only can be as good as the device models in the model library of respective software package. This aspect should always be pointed out with less trained and less experienced designers to prevent them from naive applications. The writer has to know many groups, in particular in the high-academic area, that work mainly on the simulation level. This may induce young designers to rely entirely on simulation, having in mind that any computer model is in fact only some kind of approximation of the given complex physical subject. Very often after fabrication, there remains a gap between specified data and measured circuit response. Thus, the design software is a worthy tool to support the design process, but it cannot substitute the creativity of a design engineer to find an optimum problem-oriented solution. Therefore, it is necessary to get a thorough professional training in the respective field, to be able to analyze especially unexpected simulation results, to understand the origins of the discrepancy between simulation and measurement, and to continue optimization in a systematic approach between the simulated and measured response.

Table 1.1 gives an overview about design software packages currently commercially available. The listed software is only a selection of a variety of commercial products. They are grouped in schematic circuit, planar 2.5D electromagnetic (EM), and full 3D EM simulators. Regarding the most frequently

Table 1.1
Relevant Design Software

Supplier	*Schematic Circuit Simulation*	*Planar 2.5D (EM) Simulator*	*Full 3D (EM) Simulator*
Agilent [1]	Advanced Design System (ADS)	Momentum	
Applied Wave Research (AWR) [2]	Microwave Office (MWO)	EMSight	
APLAC [3]	APLAC RF Design Tool		APLACs ElectroMagnetic Module
Ansoft [4]	Ansoft Designer	Ensemble	High-frequency Structure Simulator (HFSS)
Eagleware [5]	Genesys		
CST [6]			Microwave Studio (MWS)

used schematic circuit simulator, a complex circuit is segmented in basic circuit elements and equivalently described with the help of elementary computer models. The overall performance of a circuit then results in principle in the calculation of a multiloop cascaded network. Thus, it is obvious that crosstalk effects between adjacent circuit elements are not characterized with the provided standard models.

The quality of circuit simulation relies entirely on the accuracy of the computer models for the respective microstrip devices. The model for a given device may differ in various commercial products. Even in the same software package, different models may be offered for the same device. This may introduce uncertainty with less informed design engineers. Therefore, the designer needs sufficient background knowledge in the field of microstrip circuit design, so that he is able to use and control design software to get an optimum solution for his problem. It is expected that the material in hand will be helpful to identify and understand reoccurring design problems in such a way that the designer will be capable to rectify systematically the circuit layout deficiencies.

Except the technical challenge regarding the specifications of a microstrip circuit under development to be achieved, today's design engineers are confronted with further challenging factors. Evolution of communication systems has dramatically accelerated. Lifetimes of communication products have become shorter. Operating frequencies are continuously increasing, approaching the terahertz region above 100 GHz. The global market has strengthened competition with great pressure on the fabrication cost of communication products, and in particular on its development speed. Minimum engineering time, and reduced design, and fabrication cycles have become key constraints imposed on design engineers.

The design process is currently becoming a complex integration task, not only being confined to the circuit itself. It also covers all other mechanical design aspects in the realization of complete microwave modules, such as mounting the substrates in a housing, taking care for holes for clamping the microwave integrated circuit (MIC) substrates in its housing, optimizing the placing of bond-wire connections, embedding chip devices, and providing dc and RF feed throughs [7].

Thus, beyond the schematic circuit design platform, there has been an accelerated trend in the development of full-wave design software. Planar 2.5D EM simulators are based on the method of moments (MoM) [8] and method of lines (MoL) [9], which implies zero-thickness conductors. Therefore, they are well suited for the analysis of multilayer substrates concerning the interaction between different circuit elements.

Thick conductors can be treated approximately with 2.5D EM simulators. By virtue of the skin effect, the strip current mostly appears at the lower face of the actual strip. A smaller part of the current is also flowing on the upper surface. The thick conductor is now equivalently described by two zero-thickness metal layers at the surface locations with the actual current evenly divided between upper and lower conductors [10]. The simulation programs are primarily useful where vertical currents are negligible. However, extensive simulation time is needed for high spacial resolution and large frequency range analysis. Therefore, simulation is commonly restricted to some particular circuit layout problems within a more complex circuit.

Full 3D electromagnetic field simulators are very helpful for the analysis of electromagnetic field problems with more complex geometry, such as microstrip circuit packaging or chip device integration in a microstrip environment.

Figure 1.1 shows the schematic of a 12-GHz microwave oscillator, and Figure 1.2 a possible realization in microstrip hybrid technology [11]. The example illustrates that the microstrip design includes many design subproblems concerning the reliable modeling of a variety of microstrip elements, discontinuities, and junctions, such as open-ended and short-circuited microstrip lines, tapered lines, radial and line stubs, bends, and chip capacitors needed for dc break. Basically, the circuit comprises sections of uniform microstrip lines of different line width, and open-ended radial lines. The size of the circuit, fabricated on a plastic substrate, reads about 4×4 cm.

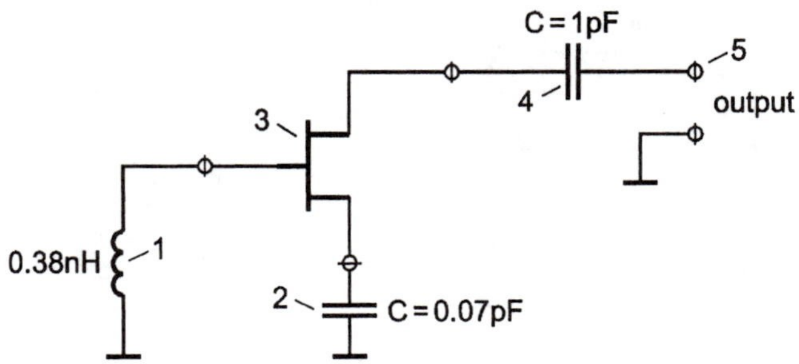

Figure 1.1 Schematic of the designed oscillator circuit shown in Figure 1.2.

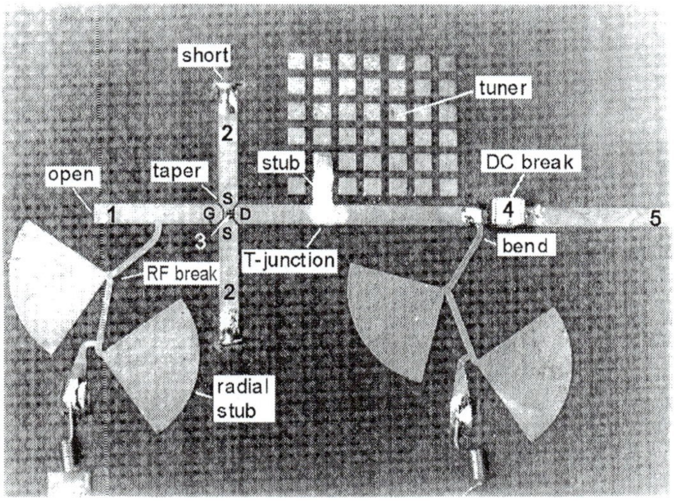

Figure 1.2 12-GHz microstrip oscillator in hybrid technology. Lumped L and C elements, numbered 1 and 2, have been realized by open- and short-circuited lines of lengths greater than λ/4.

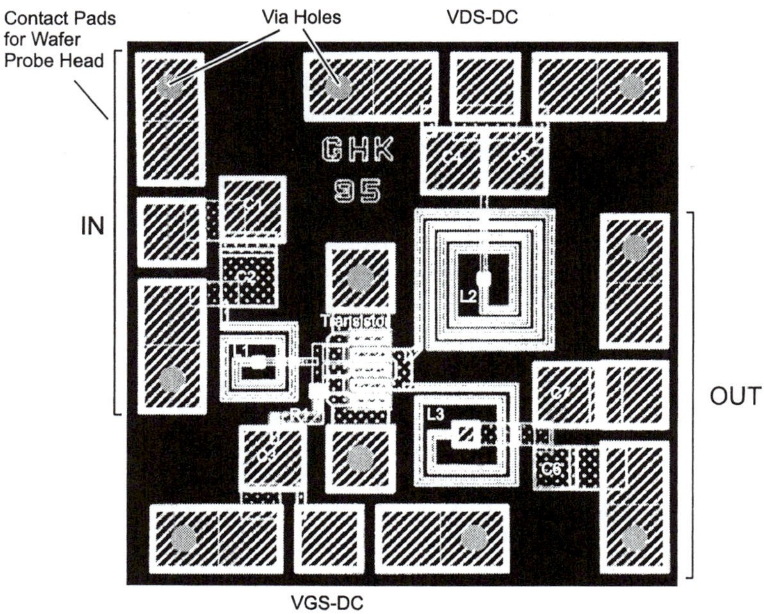

Figure 1.3 Layout of 12-GHz amplifier in MMIC technology.

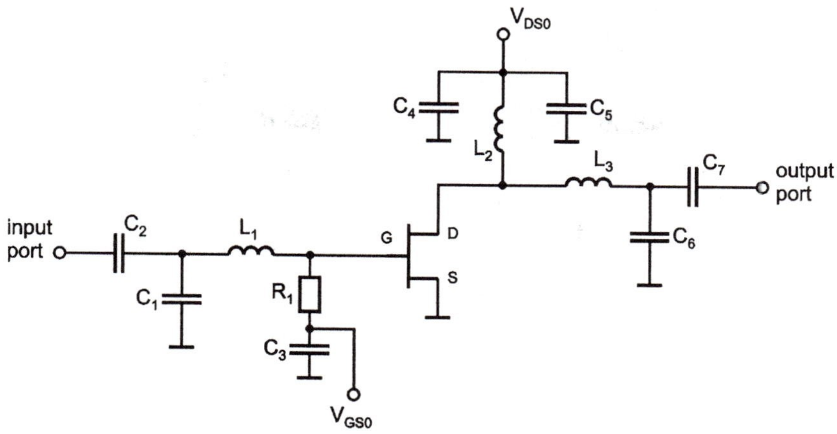

Figure 1.4 Equivalent network of the 12-GHz amplifier in Figure 1.3.

Figure 1.3 shows an example of a 12-GHz microstrip amplifier in monolithic microwave integrated circuit (MMIC) technology on gallium arsenide (GaAs) substrate material [12]. In this case, the chip size is only 1 mm^2. The equivalent network is shown in Figure 1.4. From the notation of circuit elements, it is seen that the lumped equivalent LC-circuit has been implemented using concentrated inductive and capacitive elements. Because monolithic circuits are commonly measured with coplanar wafer probe heads in ground-signal-ground configuration [13], via holes are needed to provide the transition from microstrip ground to respective coplanar grounds on the upper surface. Design tools for creating the layout of the monolithic circuit are provided by the foundry itself, because they are adjusted to the respective running production process.

1.1 Systematics of Microwave Integrated Circuits

Since 1950, planar transmission lines have increasingly displaced conventional hollow waveguides and coaxial transmission lines. Using earlier fabrication techniques, devices were screwed together to set up the desired microwave network. Such systems have the disadvantage of heavy weight, high volume, and high production cost, which make them less suited for mass production.

MICs need no screwing, are compact and light in weight, and can be produced at low cost. Complete receiving units with RF filters, mixers, and IF amplifiers can be fabricated on a single substrate as a chip, thus avoiding any coaxial-planar interconnection (adapter).

Integrated circuit elements are elements that are fabricated simultaneously with the transmission line patterning. Hybrid circuit elements are elements that are inserted after processing the transmission line network in a separate process. As illustrated in Figure 1.5, microwave integrated circuits may be classified with

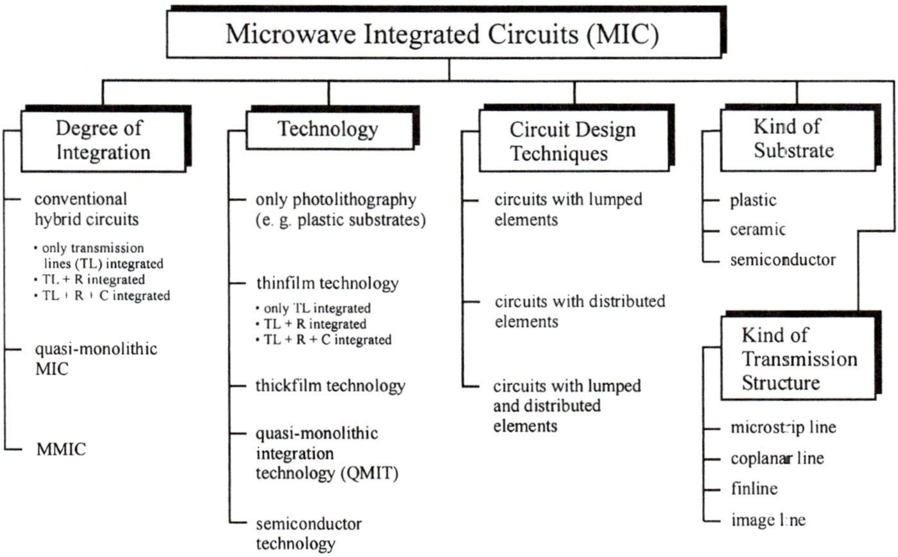

Figure 1.5 Classification of microwave integrated circuits. (After: [14].)

respect to the degree of integration, technology, circuit design technique, kind of substrate, and transmission structure [14]. The design of microwave integrated circuits can comprise either quasi-lumped or distributed elements, or both. In this book, we define a purely lumped network as a circuit consisting of ideal resistors R, capacitors C, and inductors L, with infinitesimal spatial expansion so that voltage and current of a lumped element are only dependent on time. On the other hand, a distributed circuit element has finite expansion and is often realized by a short length of transmission line, so that current and voltage are not only dependent on time, but vary also along the line. We will see in Section 6.6 that a lumped C and L element can be approximated under special conditions by an electrically short length of transmission line. This is the case when the guide wavelength is much larger than the short piece of line. In this case, voltage and current are only weakly nonuniform, so that we may ignore that at the beginning of a microstrip design. Complete information is delivered by EM simulation. We will denote this special group of distributed elements by quasi-lumped elements.

A designed prototype RLC circuit must be transformed into planar structure by equivalently describing the capacitors and inductors in the network by shorted or open-circuited stubs and cascade sections of waveguiding structures, such as microstrip or coplanar lines (Figure 1.6). Hybrid design involves mostly thinfilm resistors, chip capacitors, and inductors.

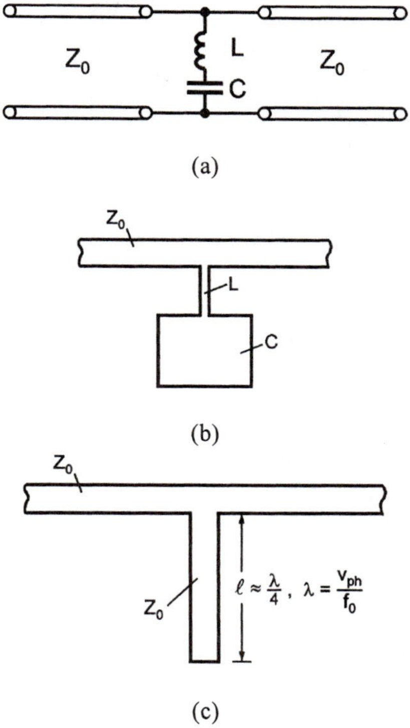

Figure 1.6 Realization of an LC filter in microstrip technique: (a) lumped LC network, (b) quasi-lumped element, (c) distributed microstrip circuit.

1.2 Circuit Design Methods

The development of circuit design techniques relates to the significant advances in the capabilities of computers in the last few decades. In retrospect, we may distinguish between a traditional empirical and an advanced modern computer aided design (CAD) method.

1.2.1 Empirical Design

Until the end of the 1970s, when the personal computer (PC) had not generally been available yet, the circuit design was based mostly on the long-term experience of the circuit designer. At that time, the theoretical treatment of planar devices was not yet well developed, so that microstrip circuit design was heavily dependent on measurements [15]. Design data was commonly based on curve-fitting formulas. A good example is the great effort in measurement and derivation of a closed-form expression for the effective dielectric constant of a microstrip transmission line

[16–18]. The main disadvantage of the experimental approach is the limitation of the derived expressions with respect to the considered microstrip substrate, geometric parameters, and measured frequencies. Extrapolation in any direction principally takes a risk not to meet the circuit specification. Thus, the circuit will not operate, or at least not operate at the right frequency. The remedy involves a long series of cut-and-try experiments, which tend to be time consuming and costly [19].

Figure 1.7 shows a flow diagram of the empirical design process [20, 21], starting with the definition of circuit specifications. An initial circuit layout is then prepared on the basis of hand computation and available empirical design data. The Smith chart, discussed in Appendix B, had proven to be a valuable design tool for solving matching problems. In particular, the layout of a simple network of a filter stub, which is discussed in Section 6.4, needs detailed information about the equivalent circuit of an open-ended microstrip line, the electrical reference plane at which the stub begins, and the frequency-dependent effective dielectric constant. The theoretical response of the considered circuit is analyzed and optimized to get it close as possible to the desired performance. The layout of the circuit was then commonly drawn by hand on a large scale and reduced photographically, or a full-

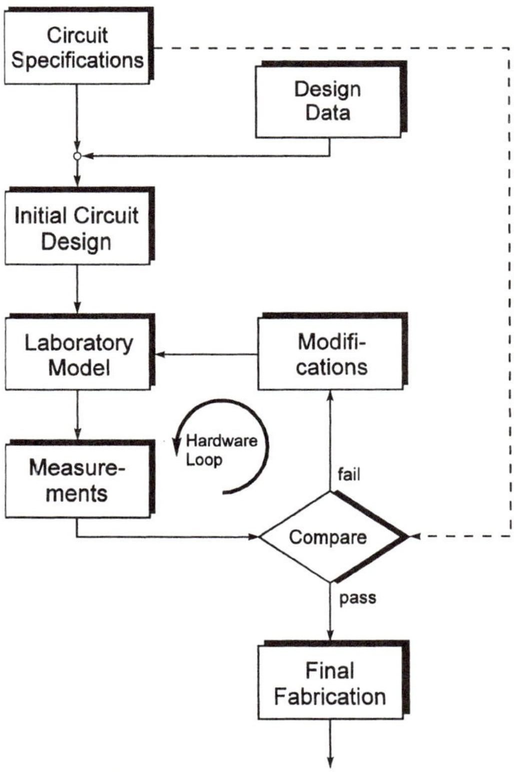

Figure 1.7 Empirical circuit design approach.

scale circuit mask was directly cut from a suitable plastic foil. After the photoetching process, the performance of the finished circuit was measured. Commonly, due to the incomplete design dataset, the desired requirements had not been met, so that the circuit had to be modified intuitively in some way to start the next hardware cycle, hoping the design will work better next time [22]. Change of the circuit parameters was performed with a precision drill or scalpel. Modification and measurement were repeated as long as the specifications were met. Then the circuit was released for fabrication.

In conclusion, the empirical design was a complicated and tedious art [23]. It needed intensive human intervention to refine the design in an interactive mode [24]. Hand formulation limited the size of problems that could be considered [20]. The electrical performance of a circuit is influenced by manifold circuit parasitics (e.g., stray capacitance, cross talk, radiation, loss, and others), which were not accounted for by the simplified circuit models necessary to render the design in a manageable way. The empirical design is less flexible to circuit layout variations. A change in frequency commonly entailed a completely new circuit design.

Further restrictions are given with higher complexity and higher packaging density of integrated circuits, which need, at least for the most critical parts of the circuit, more sophisticated analysis methods, such as an EM 3D field simulator that delivers a numerical solution of the electromagnetic field. Furthermore, since it is extremely difficult and essentially impossible to adjust the circuit characteristics of monolithic circuits once they are fabricated, accurate CAD programs are essential for the design success [25]. In particular, processing tolerances must be taken into account for the design on the basis of a tolerance analysis. Today, design software is challenged by new requirements of circuit design techniques focusing on the synthesis and analysis of highly packaged three-dimensional multilayer planar circuits.

1.2.2 Modern Computer Aided Design

The computer aided design utilizes a computer as a design tool. As seen in Figure 1.8, the design tools for low-frequency analog and digital circuits are known to be well developed. Whether or not to use these tools for microwave circuit design is a frequently asked question. The design features for low-frequency (LF) and microwave and millimeterwave RF circuits are quite different. In a microwave circuit, not only are lumped passive and active devices existent, but also distributed, elements. Furthermore, the placing of transmission lines in the layout, if not carefully done, can cause cross talk between adjacent transmission lines and circuit elements. Therefore, the signal path on a microstrip circuit substrate carrier must be chosen very carefully, because cross talk can alter the circuit performance significantly.

A digital integrated circuit (IC) design can comprise some 10,000 gates or flip-flops that have to be configured and wired. In contrast, a complex microwave circuit can consist of only 100 devices. However, in this case, all devices must have

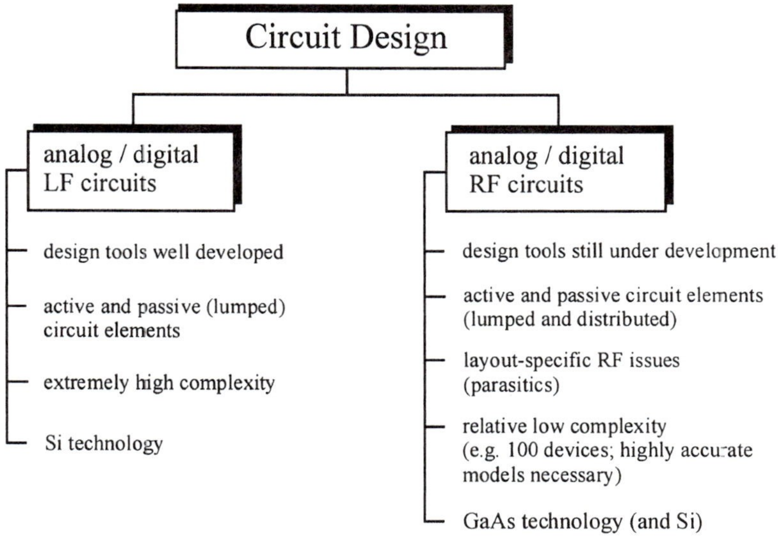

Figure 1.8 Differences in the low-frequency digital and microwave circuit design.

been modeled very accurately on the basis of accurate measurement or theoretical analysis.

The signal flow diagram in Figure 1.9 illustrates the steps of the computer aided circuit design of microwave circuits. The features of the model library building block, the number of stored elements, essentially determine the versatility of the software program. Models may have been derived from rigorous field-theoretical analysis and approximated by curve-fitting, or may have been empirically derived from measurement. Commonly, the specified model accuracy is restricted to limited material, geometry variations, and frequency. Therefore, it is recommended that the circuit designer is familiar with the relevant references of the software in order to have faith in the simulation. Critical usage of software programs and computer models may prevent him from unexpected and questionable design results.

In circuits, manifold passive and active elements are generally present. Therefore, it is expected that the library includes most relevant planar waveguiding structures, like microstrip transmission line, slot line, and coplanar line. Microwave transistors as active devices are stored with standard lumped-element model topologies, both for small signal and large signal operation. However, it should be noticed that the offered models can only be used for initial circuit analysis. Respective simulations are very useful to analyze the influence of circuit parameter variation on the circuit performance. However, the final design needs the implementation of a user-defined model, frequently obtained from measurements.

Furthermore, lumped elements, such as series and shunt capacitances and inductances with a discrete set of element values, are available. In the microwave frequency region, the electrical properties are generally described by their scattering

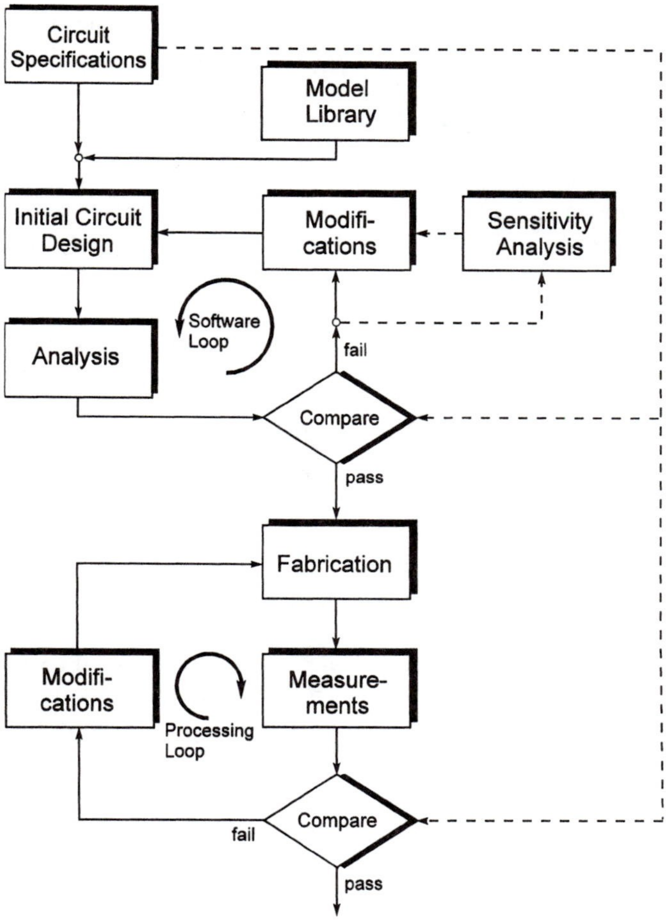

Figure 1.9 Modern computer aided circuit design.

coefficients. Rigorous field-theoretical-based and semiconductor-physics-based characterization of the properties of devices is in general very time consuming, and the numerical analysis often requires very high expenditure of computation. Therefore, with respect to circuit optimization and the respective computing time, the circuit models should be as simple as possible, but sufficiently accurate for the given design purpose.

It can happen that a computer model is unavailable for a device under consideration. In this case, assuming a linear circuit element, the scattering coefficients of the device are measured, and a computer model is derived by curve-fitting. For accurate scattering parameter measurements, vector network analyzers are available up to 110 GHz [26]. The derivation of accurate active device models is a very complex and difficult task, in particular when the model should reflect not only the small signal properties but also the nonlinear behavior of the device. It should be noted that nonlinear device models commonly work with periodical RF signals of arbitrary waveforms. However, broadband device models that are also capable to treat multitone signals as stimulus have drawn attention [27, 28]. These advanced models are urgently needed for the reliable design of circuit components for future broadband digitally modulated UMTS communication systems.

The circuit elements are positioned and interconnected to prepare the layout. The connection leads to different kinds of network topologies, such as cascaded networks with multilevel stub sections or multipath networks [29]. The simplest form of network is a ladder or cascade of circuit elements [24]. Assuming linear networks, the circuit elements can be described by the chain matrices related to the corresponding scattering matrices. The cascaded network is then simply calculated from the product of the respective singular networks. Network analysis programs for complex network topologies are available. A first layout is performed on the basis of available data stored in the computer.

The circuit analysis is focused on a certain response function in terms of the component parameters, circuit topology, and independent excitations [30]. This could be the transfer characteristics of a bandpass filter with defined upper and lower transition frequencies and transition steepness, and insertion loss. In another case, a broadband amplifier with input and output matching networks should be designed and optimized for optimum power gain in a specified frequency range. The objective function [23] may also comprise different goals for simultaneous optimization (e.g., receiving amplifier with low-noise and high linear power gain).

The circuit analysis delivers the response of the circuit, and the result is compared with the design goals. If the agreement is not satisfying, then some relevant circuit parameters have to be changed in an iterative way. This design phase is termed optimization. Optimization describes an iterative change of network parameters, to obtain a specified circuit performance. It focuses on error minimization of actually reached circuit data and the specified objective function. A very general shortcoming of all optimizing methods is the lack of any assurance that the program will converge to the absolute (global) optimum, rather than to the nearest relative local optimum [31]. There is absolutely no guarantee of a "global minima" [23]. The elimination of the local minima problem requires correlated starting parameter values near the global minimum. A respective sophisticated approach has been developed for consistent model parameter extraction of distributed microwave device models [32, 33]. Generally, two types of minimization methods are applied: direct search methods and gradient methods. The application of the latter approach requires the partial derivations of the target function with respect to all variable parameters. Direct search methods do not need any partial derivative, and are therefore very fast and robust in approaching the minimum. A frequently used direct search method is the Simplex algorithm [31]. It is a nonlinear optimization process using simple linear operations and is easy to program. It has also proved to be effective in circuit and systems optimization. Some difficulties appear when the vertices converge too closely or degenerate into a lower-dimensional subspace. The experience has shown that the Simplex algorithm is superior to the classical steepest descent methods [34, 35].

The iterative procedure of circuit analysis, specification comparison, and optimization is continued as long as the specifications have been met. Then the circuit is fabricated and measured. If the measurement result is not satisfying, then further modifications must be introduced. This should be expected, since the models and/or the analysis algorithms are commonly not perfect. Generally, the specifications should have been fulfilled after the third process run.

The final step in most circuit design is a tolerance analysis (i.e., a study of the effect of parameter variation on circuit behavior) [20]. This function makes it possible to investigate the influence of a given circuit parameter change on the overall circuit performance. Such investigations are important to analyze the influence of processing tolerances on the circuit properties.

References

[1] Agilent Technologies, *Advanced Design System (ADS)*, Version 2003(C), http://eesof.tm.agilent.com.

[2] Applied Wave Research (AWR), Inc., *Microwave Office*, http://www.microwaveoffice.com/products/.

[3] APLAC Solutions Corp., *Aplac RF Design Tool*, http://www.aplac.com.

[4] Ansoft, *High-Frequency (HF) Products*, http://www.ansoft.com.

[5] Eagleware Corp., *GENESYS*, http://www.eagleware.com.

[6] CST GmbH, *CST Microwave Studio (CST MWS)*, http://www.cst-world.com.

[7] Hebbar, K., et al., "Plan-Based Evaluation of Designs for Microwave Modules," *Proc. 1996 ASME Int. Design Engineering Technical Conference & Computers and Information in Engineering Conference*, 1996, Irvine, California, pp. 1–12.

[8] Harrington, R. F., *Field Computation by Moment Methods*, IEEE Press, 1993.

[9] Shlepnev, Y. O., "A New Simulator of Passive MIC Elements," Proc. 4th Int. Conf. on Antennas, Radiocommunication Systems and Means (ICARSM'99), Voronezh, Russia, 1999.

[10] Rautio, J. C., "An Investigation of Microstrip Conductor Loss," *Microwave*, December 2000, pp. 60–67.

[11] Werthof, A., *Experimentelle Modellierung aktiver Bauelemente für die Simulation nichtlinearer Mikrowellenschaltungen*, Doctoral Thesis, University of Kassel (1994), published in Fortschritt-Berichte VDI, Reihe 9: Elektronik, Nr. 200, Düsseldorf: VDI Verlag, 1995.

[12] Schmale, I., F. Torres-Molina, and G. Kompa, "In-Depth Analysis of Monolithic 12 GHz MESFET Amplifier," *Int. IEEE Workshop on Experimentally Based FET Device Modelling & Related Nonlinear Circuit Design*, Kassel, July 17–18, 1997, paper 39.1–8.

[13] CASCADE MICROTECH, *Microwave Probe Station Model 42-Instruction Manual*, 1987.

[14] Hoffmann, R. K., *Integrierte Mikrowellenschaltungen*, Berlin: Springer-Verlag, 1983.

[15] Troughton, P., "Design of Complex Microstrip Circuits by Measurement and Computer Modelling," *Proc. IEE*, Vol. 118, No. 3/4, March/April, 1971, pp. 469–474.

[16] Getsinger, W. J., "Microstrip Dispersion Model," *IEEE Trans.*, Vol. MTT-21, 1973, pp. 34–39.

[17] Kompa, G., "S-Matrix Computation of Microstrip Discontinuities with a Planar Waveguide Model," *Arch. Elektr. Übertr.*, Vol. 30, 1976, pp. 58–64.

[18] Yamashita, E., K. Atsuki, and T. Ueda, "An Approximate Dispersion Formula of Microstrip Lines for Computer-Aided Design of Microwave Integrated Circuits," *IEEE Trans.*, Vol. MTT-27, 1979, pp. 1036–1038.

[19] Schmitt, H. J., and K. H. Sarges, "Wave Propagation in Microstrip," *Nachrichtentechnische Zeitschrift*, Vol. 24, 1971, pp. 260–264.

[20] Calahan, D. A., *Computer-Aided Network Design*, New York: McGraw-Hill, Inc., 1972.

[21] Gupta, K. C., R. Gary, and R. Chadha, *Computer-Aided Design of Microwave Circuits*, Norwood, MA: Artech House, Inc., 1981.

[22] Gardiol, F. E., "Microstrip Computer-Aided Design in Europe," *IEEE Trans.*, Vol. MTT-34, 1986, pp. 1271–1275.

[23] Gelnovatch, V. G., and I. L. Chase, "Demon—An Optimal Seeking Computer Program for the Design of Microwave Circuits," *IEEE J.*, Vol. SC-5, 1970, pp. 303–309.

[24] Green, P. E., "General Purpose Programs for the Frequency Domain Analysis of Microwave Circuits," *IEEE Trans.*, Vol. MTT-17, 1969, pp. 506–514.

[25] Itoh, T., *Numerical Techniques for Microwave and Millimeter-Wave Passive Structures*, New York: John Wiley & Sons, 1989.

[26] Datasheet: *8510 Network Analyzer Systems, 45 MHz to 110 GHz*, http://www. agilent.com/ find/8510.

[27] Constantini, A., et al., "Accurate Prediction of PHEMT Intermodulation Distortion Using the Nonlinear Discrete Convolution Model," 2002 IEEE MTT-S Int. Microwave Symp Digest, 2002, pp. 857–860.

[28] Ouarch, Z., et al., "Measurement Based Nonlinear Electrothermal Modeling of GaAs FET with Dynamical Trapping Effects," *1998 IEEE MTT-S Digest*, 1998, pp. 599–602.

[29] Parker, W. N., "DIPNET: A General Distributed Parameter Network Analysis Program," *IEEE Trans.*, Vol. MTT-17, 1969, pp. 495–505.

[30] Monaco, V. A., and P. Tiberio, "Computer-Aided Analysis of Microwave Circuits," *IEEE Trans.*, Vol. MTT-22, 1974, pp. 249–263.

[31] Temes, G. C., and D. A. Calahan, "Computer-Aided Network Optimization—The State-of-the-Art, *Proc. IEEE*, Vol. 55, 1967, pp. 1832–1863.

[32] Mwema, W. N., *A Reliable Optimisation-Based Model Parameter Extraction*, Doctoral Thesis, University of Kassel (2002), published by Kassel University Press, 2002.

[33] Jarndal, A., and G. Kompa, "A New Small Signal Model Parameter Extraction Method Applied to GaN Devices," 2005 IEEE MTT-S Int. Microwave Symp. Digest, 2005, Paper TH1D-2.

[34] Bandler, J. W., "Optimization Methods for Computer-Aided Design," *IEEE Trans.*, Vol. MTT-17, 1969, pp. 533–552.

[35] Bandler, J. W., "Current Trends in Network Optimization," *IEEE Trans.*, Vol. MTT-18, 1970, pp. 1159–1170.

Chapter 2

Overview of Strip Transmission Lines

The historical roots of the planar integrated circuit go back to the 1940s, when V. H. Rumsey and H. W. Jamieson developed an antenna power divider (Figure 2.1) using a planar flat strip coaxial transmission line (Figure 2.2) [1]. Early in 1949, R. M. Barrett had the idea to employ the new transmission line not only to carry energy from point to point, but also to use it to make all types of microwave components, such as filters (Figure 2.3), directional couplers, hybrids, and matched loads (Figure 2.4). His results had first been published in 1951 [2, 3]. In 1952, D. D. Grieg and H. F. Engelmann announced another form of strip transmission line [4], now commonly known as microstrip, which was an adaption to the planar geometry of the two-wire line (Figure 2.5). In 1955 D. D. King announced the dielectric image line (Figure 2.6), which is an adaption of the dielectric waveguide to planar geometry [5]. Using this transmission line configuration, some circuit constructions for millimeter wave applications had become known in the 1970s. However, the dielectric image line did not find any broad application. In fact, the

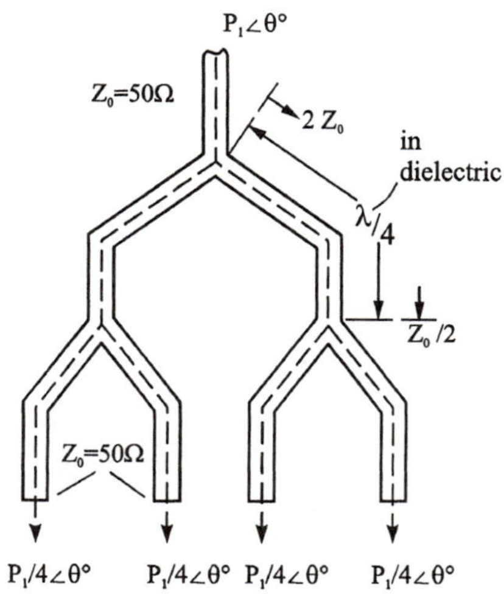

Figure 2.1 Power divider after Rumsey and Jamieson. (From: [2], © 1955 IEEE. Reprinted with permission.)

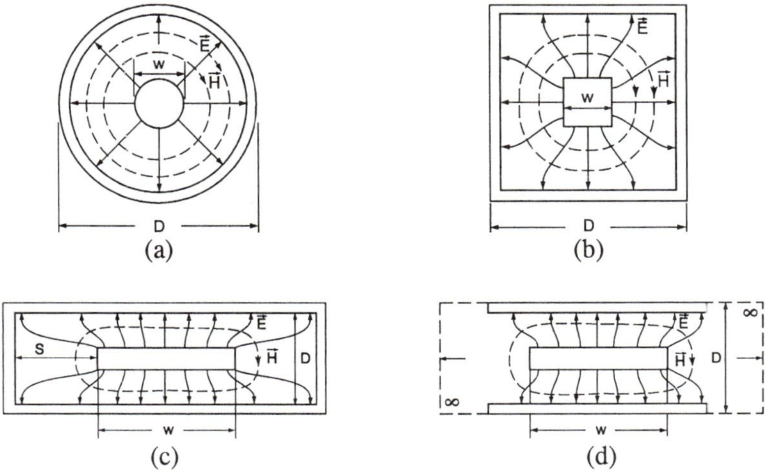

Figure 2.2 Evolution of flat strip transmission line (triplate) from the coaxial transmission system. (From: [2], © 1955 IEEE. Reprinted with permission.)

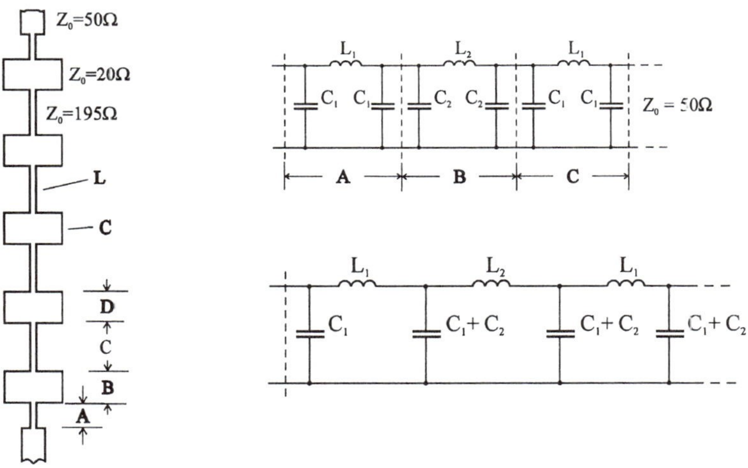

Figure 2.3 Lowpass filter with lumped L and C elements and realization in microstrip technique. (From: [2], © 1955 IEEE. Reprinted with permission.)

dielectric waveguides have become important in integrated optics [6]. Today, the microstrip line and the coplanar waveguide are considered the standard planar waveguiding structures in the microwave and millimeter wave region (see Section 2.1).

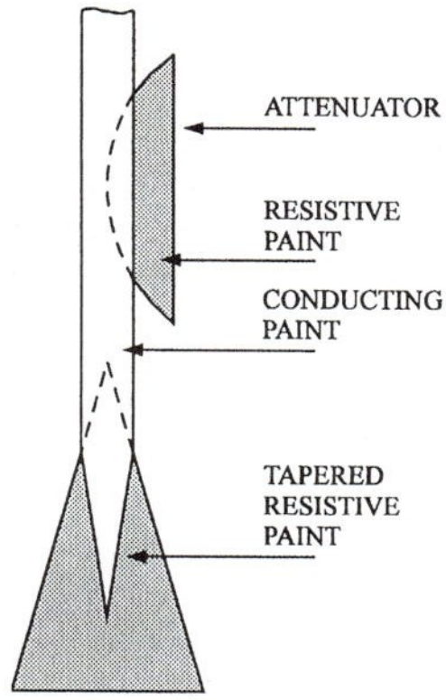

Figure 2.4 Matched load and attenuator. (From: [2], © 1955 IEEE. Reprinted with permission.)

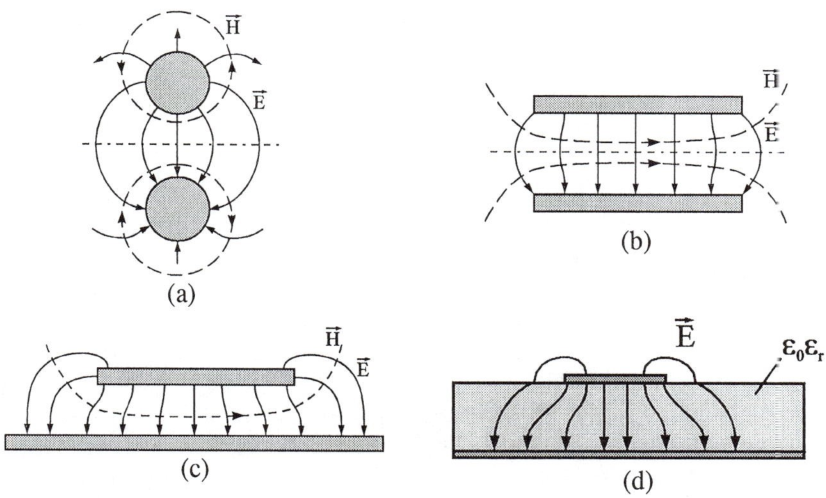

Figure 2.5 Evolution of microstrip transmission line from the two-wire line.

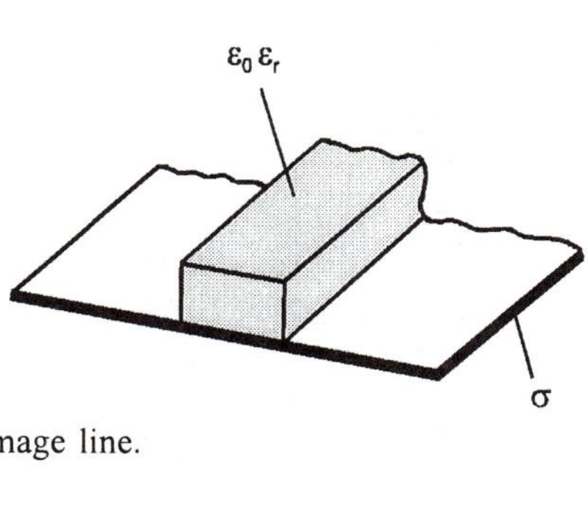

Figure 2.6 Dielectric image line.

The triplate can be regarded as an evolution of the coaxial transmission system illustrated in Figure 2.2. The "flat strip" transmission system is derived as follows. First, the coaxial line (a) is deformed in such a manner that both the center and outer conductors are square (b) or rectangular in cross section (c). Then, the side walls of the rectangular coaxial system are extended to infinity (d). Almost all of the field is concentrated in the region around the strip so that the metallic side walls can be neglected. Figure 2.5 illustrates in a similar way the evolution of the microstrip transmission line with substrate carrier [Figure 2.5(d)] from the two-wire line [Figure 2.5(a)]. The parallel plate transmission line in Figure 2.5(b) is obtained from the two-wire line by deforming the round electrodes to thin conductive strips. Since their plane of symmetry describes a zero-potential surface, it can be replaced by a metal plate. Figure 2.5(c) shows the resulting waveguiding structure which denotes an air-filled microstrip line.

Between 1964 and 1968, Texas Instruments developed an active planar phased array antenna in microstrip technique called *Molecular Electronics for Radar Applications* (MERA) [7, 8]. All of the microwave components were fabricated using chip components on alumina microstrip circuits. A pulsed 2.25 GHz signal (S-band: 2–4 GHz) was supplied to an array of modules. Each module through a solid-state amplifying (silicon power transistor with 2W output and a gain exceeding 3 dB [8]) and ·4 multiplying chain (·4 multiplier using planar varactor diodes) supplied power to the antenna elements at a frequency of 9 GHz. An X-band transmit/receive (T/R) switch using PIN diodes was used to separate the transmitter and receiver functions. Furthermore, an S-band mixer using a Schottky-barrier diode and a 500-MHz IF amplifier were incorporated in each module. Digital phase shifters were used at S-band to provide beam steering capabilities. The array was composed of 604 modules with an average power output of 600 mW.

The integrated circuit technique made further progress between 1960 and 1980. During these years, the idea of MMIC had been coming up regarding both analog and digital functions on a single chip. At an early stage in the MIC/MMIC development, high-resistive silicon of p-type (Boron) was used as substrate and material for devices. From the beginning, GaAs was in great competition. GaAs material has the advantage of higher electron drift velocity, which results in higher transit frequencies. When, in the 1970s, the GaAs field-effect transistor was available for both low-noise and high power applications and the production had reached high maturity, the MMIC development accelerated. Since the 1990s, the coplanar technique has been increasingly important.

In the following, we will define and classify the most relevant strip transmission line structures, their geometries, and related electromagnetic fields. We will discuss the particularities of the waveguiding structures with respect to their theoretical analysis, emphasize their advantages, and will throw light on the drawbacks of their application. It should be noted that beyond the electrical properties of the discussed strip transmission lines, the overall cost for circuit realization is very important for the respective choice. Therefore, manufacturing aspects are also considered. Each type has its advantages with respect to the others.

A strip transmission line is defined as a planar transmission structure used to transmit microwave signals ($f > 300$ MHz). It comprises at least one flat conductive strip of small thickness, and one or two infinite conductive ground plates as counterelectrode (two-wire or multiwire system). For mechanical stability, the strip and metal ground planes are deposited on a dielectric substrate that serves as a carrier for the electrodes [9]. The relative magnetic permeability of the substrate material is commonly $\mu_r = 1$. If the relative dielectric permittivity is $\epsilon_r = 1$, then the transmission system is termed "air-filled" (microstrip air line).

For design purposes, it is necessary to know how the characteristic impedance, phase velocity, and attenuation coefficient of the dominant mode depend on geometrical factors and on the frequency.

2.1 Microstrip Line

The microstrip line comprises a single upper conductor above an infinite ground plate with a dielectric substrate as carrier (Figure 2.7).

Electromagnetic Properties

The microstrip transmission line is an open waveguiding structure, and its electromagnetic field is theoretically defined in an infinite space. This makes the analysis very difficult in general. Therefore, metallic shielding or at least a metal coverplate of infinite dimensions above the microstrip line is used to define well-determined boundary conditions for the electromagnetic field [6]. For the analysis of microstrip properties, the upper conductor is often assumed to be infinitely thin and free of losses.

The microstrip transmission line is a two-electrode guiding system, and therefore the dominant mode has a zero cutoff frequency. Due to the mixed air-dielectric material, however, the microstrip line cannot propagate a pure transverse electromagnetic (TEM) wave. This is only possible in the special case of air-filled microstrip line. This case is not relevant from a practical point of view, because the strip needs to be suspended in some way.

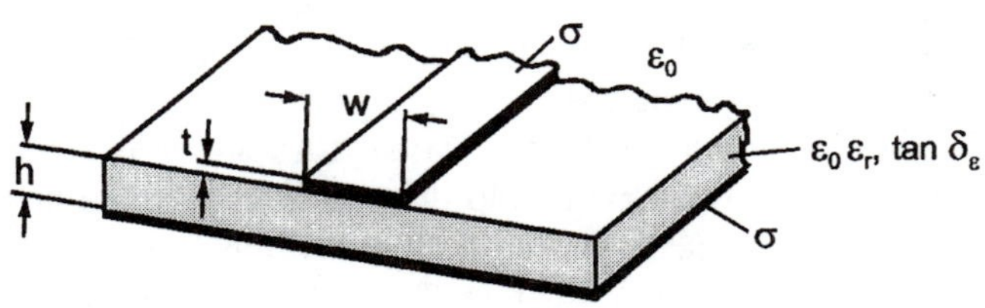

Figure 2.7 Microstrip transmission line with material and geometrical parameters.

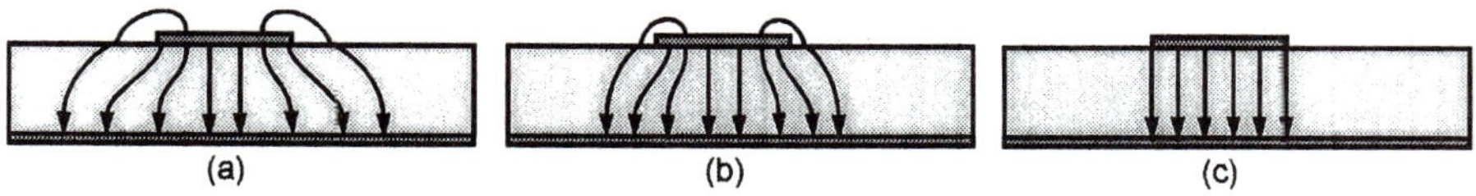

Figure 2.8 Approximate electric field distribution versus frequency: (a) very low frequency, (b) medium frequency, and (c) infinite frequency.

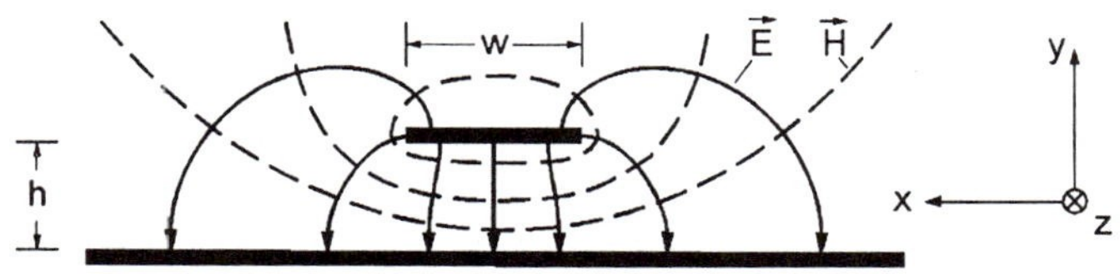

Figure 2.9 Electromagnetic field of air-filled microstrip.

As will be shown in Section 3.4.3, no pure transverse electric (TE) or transverse magnetic (TM) mode can exist. TE and TM modes are also denoted as E and H modes, respectively, in the literature, emphasizing the existence of the longitudinal electric and magnetic field component, respectively. In fact, both kinds of modes are necessary to satisfy the boundary conditions of a microstrip line involving transverse inhomogeneity in the dielectric and the presence of a strip. Thus, the lowest-order mode is a hybrid mode with nonzero longitudinal electrical and magnetic field. It becomes TEM only at zero frequency.

At very low frequencies, we may expect a TEM-like behavior of the fundamental mode propagating with a phase velocity that lies between the free-space velocity c_0 and the dielectric material velocity $c_0/\sqrt{\epsilon_r}$. With respect to the mixed dielectric material, we may define an effective dielectric constant ϵ_{eff}, which is expected to be greater than the dielectric constant of air and less than that of the dielectric substrate. Experience has shown that the electromagnetic field of a mode in a mixed waveguiding structure increasingly penetrates the medium with the highest dielectric constant with growing frequency. In the theoretically limiting case of infinite frequency, the wave is totally "absorbed" by the dielectric material with highest dielectric constant (Figure 2.8). In the case of the microstrip transmission line, it is supposed that the electromagnetic field is completely concentrated beneath the strip at infinite frequency, as illustrated in Figure 2.8(c). Because of the frequency-dependent phase velocity of the waves, the microstrip transmission line is a dispersive waveguiding structure.

Figure 2.9 shows the electric and magnetic field distributions of the fundamental microstrip mode on an air-filled microstrip line that is assumed to be completely lossless. The transversal electric field corresponds to the electrostatic field of a parallel plate capacitor of the same structure. The electromagnetic field propagates with the phase velocity $v = c_0$, where c_0 is the velocity of light. The dominant mode on the transmission line is a pure TEM mode, and therefore has no electric and magnetic longitudinal field components ($E_z = 0,\ H_z = 0$). Therefore, voltage and current on the line are uniquely defined. The voltage between the two conductors is given by the path integral

$$V = \int \vec{E} \cdot \vec{ds} \tag{2.1}$$

where $\vec{E}$ is the electric field of the propagating TEM wave. The integration may be carried from the upper strip to any point on the ground plate due to path independency of the integral. The current carried by the conductor is evaluated by Ampère's Law

$$I = \oint \vec{H} \cdot \vec{ds} \tag{2.2}$$

where $\vec{H}$ is the magnetic field, and the path of integration may be around the surface of the conductor in a transverse plane. In this case, the characteristic impedance $Z_0\,(\epsilon_r\!=\!1)$ of the microstrip air line can be uniquely defined,

$$Z_0\,(\epsilon_r = 1) = \frac{V}{I} = \frac{\int \vec{E} \cdot \vec{ds}}{\oint \vec{H} \cdot \vec{ds}} \tag{2.3}$$

The electrostatic field problem of a transmission line in a homogeneous dielectric material can be solved by several methods described in Section 3.1. Widely used is the conformal mapping procedure, which images the complicated electrode configuration of the microstrip line onto an ideal parallel plate transmission line structure with a homogeneous transversal electromagnetic field, where the integrals of (2.1) and (2.2) become simple line integrals with constant electric and magnetic fields.

With the inductance per unit length L' and capacitance per unit length C' of the transmission line, the characteristic impedance can be written as

$$Z_0(\epsilon_r = 1) = \sqrt{\frac{L'}{C'}} = \frac{\sqrt{L'C'}}{C'} = \frac{1}{c_0 C'} \tag{2.4}$$

where

$$c_0 = \frac{1}{\sqrt{\epsilon_0 \mu_0}} = \frac{1}{\sqrt{L'C'}} \tag{2.5}$$

Equation (2.4) indicates that the analysis of the characteristic impedance can be reduced to the determination of the capacitance per unit length C'. The capacitance per unit length of an ideal air-filled parallel plate transmission line of effective width w_{eff} and height h reads

$$C' = \epsilon_0 \frac{w_{eff}}{h} \tag{2.6}$$

Substituting C' in (2.4) gives

$$Z_0(\epsilon_r = 1) = \sqrt{\epsilon_0 \mu_0}\ \frac{h}{\epsilon_0 w_{eff}} = \sqrt{\frac{\mu_0}{\epsilon_0}}\ \frac{h}{w_{eff}} \tag{2.7}$$

with $\sqrt{\mu_0/\epsilon_0} = 120\pi\,\Omega \approx 377\,\Omega$. Applying conformal mapping transformation and introducing diverse approximations, Wheeler [10] has obtained closed-form analytical expressions for the effective strip width w_{eff} and the characteristic impedance Z_0, both being dependent on the shape ratio w/h (strip-width over separation) (see Section 3.1). Schneider [9] derived an exact solution for the characteristic impedance of the homogeneously filled transmission line, which comprises elliptic integrals (see Section 3.2).

Figure 2.10 shows the approximate field distribution of the microstrip line with substrate carrier at low frequency. As can be seen, the magnetic field lines are continuous at the air-dielectric interface due to $\mu_r = 1$ in the whole field region. Regarding the electric field, the tangential electrical field $\vec{E}_{\tan}$ and normal displacement field $\vec{D}_n$ must be continuous. Therefore, the tangential and normal electric field components at the air-dielectric interface must fulfill the conditions $\vec{E}_{\tan,air} = \vec{E}_{\tan,diel}$ and $\vec{E}_{n,air} = \epsilon_r \vec{E}_{n,diel}$. The degree of discontinuity in the electric field increases with higher dielectric constant.

If the line is partially filled with a dielectric, a pure transversal electromagnetic field can only exist at zero frequency. At frequencies $f > 0$, no pure TEM mode can propagate, which can be verified in a simple way. The electromagnetic field is distributed both in air and dielectric regions. In air, the phase velocity v of a TEM wave is equal to light velocity in vacuum ($v = c_0$). In the dielectric region ($\epsilon_r > 1$), a TEM wave would propagate with a phase velocity $v = c_0/\sqrt{\epsilon_r}$. The phase velocities of TEM modes ($E_z = 0$, $H_z = 0$) are different in both regions, so that

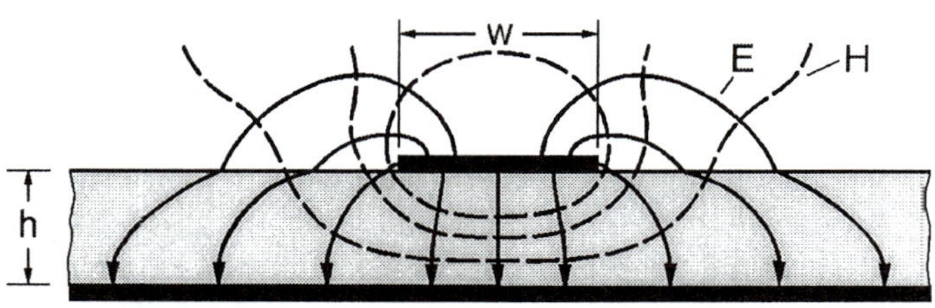

Figure 2.10 Approximate transverse field distribution on a microstrip line with substrate carrier.

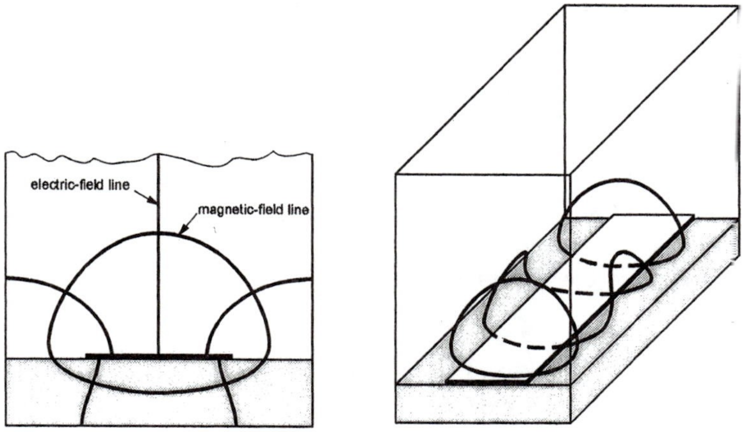

Figure 2.11 Calculated distribution of the electric and magnetic field lines of a shielded microstrip transmission line. (From: [11], © 1970 IEE. Reprinted with permission.)

—strictly speaking—a pure TEM mode cannot exist as a transmission line mode. The characteristic impedance of a microstrip line may be approximately calculated by assuming that the electromagnetic field in the line has a quasi-TEM nature. It will be shown that the quasi-static approximation can be used up to approximately 1 GHz, depending on the substrate and transmission line geometry.

In the early 1970s, the fundamental microstrip mode was analyzed by several authors. Grünberger et al. [11, 12] had theoretically and experimentally investigated the wave properties of the dominant mode in a shielded microstrip transmission line. They observed increasing discrepancy of the actual field with the ideal conception of a TEM wave as the frequency increased. Electrical and magnetic field strengths had been measured with a probe in a very large scale model. Notice, the microstrip line had a strip width of 10 cm, height of 4.5 cm, and length of 1.8m. The substrate had a dielectric constant $\epsilon_r = 2.89$. Measurements were performed in the range from 100 to 500 MHz. The field was probed through holes in the shielded box. In particular, field components of considerable magnitude had been found in the direction of propagation. The longitudinal components increase with increasing frequency, but are theoretically already existent at very low frequencies. They are responsible for the dependence of the phase velocity on frequency, a fact which had also been verified by measurement. The phase velocity decreases with increasing frequency, since the field is then increasingly concentrated in the dielectric layer. It was found that the electric field resembles closely the static field even at high frequencies. The longitudinal electric component can be neglected in the most interesting range of frequencies (i.e., between zero and the cutoff frequency of the first higher-order mode). The magnetic field, however, varies considerably with increasing frequency. The typical field pattern of the dominant mode is shown in Figure 2.11.

It is generally assumed that the electromagnetic field of the dominant mode has six field components: E_x, E_y, E_z, H_x, H_y, and H_z. Electromagnetic waves with electric and magnetic field components in the direction of propagation are commonly termed hybrid modes. H_z field component can be neglected in a first order approximation, since it is weak for $h/\lambda \ll 1$, which is fulfilled below 1 GHz in any case. The fundamental hybrid mode on the microstrip transmission line is termed quasi-TEM mode. With increasing frequency in the gigahertz region, the electromagnetic wave increasingly concentrates in the dielectric. This is the reason why the phase velocity of the mode becomes frequency dependent (dispersion).

Practical Design Viewpoints

As discussed earlier, microstrip line is an open guide structure with a fringing electromagnetic field at the side edges. Closely positioned lines couple with each other. This may be desired, for example, to build up microstrip couplers. In general, however, the crosstalk reduces isolation. Therefore, a rule of thumb is to keep a distance of at least 5 strip widths between parallel transmission lines (line-to-line coupling [13]). This will, however, limit the maximum degree of packaging density. Crosstalk also can be caused by surface waves, propagating on the dielectric-coated ground plate. They have a zero cutoff frequency and can be excited by the microstrip mode. However, coupling is significant only in the presence of phase velocity synchronism [13, 14].

Discontinuities and junctions may cause radiation. High-efficient radiation is desired in the subject of antenna design. However, regarding common applications, radiation is a parasitic effect, and contributes to transmission loss and coupling with other lines [13].

The characteristic impedance range of a microstrip line is typically 20Ω to 120Ω. The upper limit is set by technological constraints on the minimum line width that can be realized and the production tolerances, while the lower limit is essentially set by the appearance of higher-order modes (see Chapter 5, Figure 5.7 and Figure 5.8).

The fabrication of microstrip circuits is relatively simple and can be done on the basis of a low-cost technology. The dimensions of microstrip transmission lines on standard substrates are relatively large, so that the demand for highly precise photolithography is less stringent.

Compared to dielectric-embedded strip transmission lines such as triplate and sandwich lines, the microstrip line has the advantage that semiconductor and passive chip components can be easily incorporated to make active hybrid circuits. Circuit adjustment can be easily done from the top. Series connected beam-leaded active and passive devices can be bonded directly to metal strips on the dielectric substrate. A shunted hole in the dielectric is required for mounting a semiconductor device between the strip and ground. Devices and components incorporated in hybrid integrated circuits are accessible for probing and circuit measurements [15].

2.2 Shielded Microstrip Line

Shielded microstrip line comprises a standard microstrip line that is covered with metal plates on the top and on the sides (Figure 2.12).

Electromagnetic Properties
The transmission line is a closed waveguiding structure without any radiation loss. The shielded form of the microstrip transmission line makes the theoretical treatment less difficult because the field region is confined to the inner of the metal box. Analysis can be performed on the basis of the orthogonal series expansion method (Appendix H), which is straightforward but very time consuming.

Practical Design Viewpoints
The main purposes of the housing or package are to provide mechanical strength, EM shielding to eliminate radiation losses, hermetization, and heat sinking in the case of high-power applications. Packaging is utilized to protect the circuitry from moisture, dust, salt spray, and other environmental contaminants. In order to package the circuit, certain methods of sealing can be used, such as conductive epoxy, solder, gasket materials, and metallization tape.

A microwave integrated circuit mounted into a housing may be considered to be a dielectrically loaded cavity resonator. Therefore, the dimensions of the housing or the sheet conductivity of the shielding material have to be chosen in such a way that excitation of transverse electric modes, transverse magnetic modes, and box resonances are suppressed [9]. The top and side covers essentially redistribute the field of the more theoretical microstrip and understandably have an influence on the effective dielectric constant. When the ratio of the distance from the top cover to the dielectric substrate and the substrate thickness $(b-h)/h$ is greater than 10, the enclosure effects can be considered negligible [16]. The housing also can be covered with absorbing material to suppress parasitic waveguide modes.

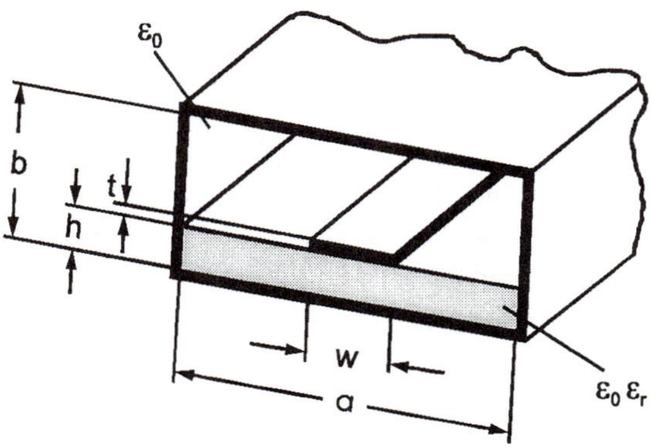

Figure 2.12 Shielded microstrip line.

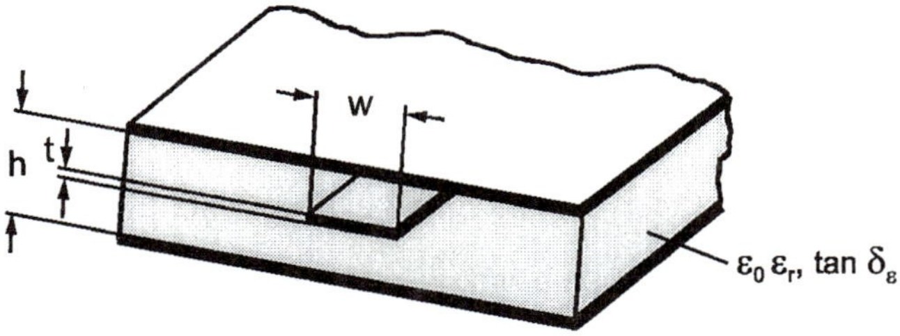

Figure 2.13 Triplate—geometry and material parameters.

2.3 Triplate

The triplate comprises infinite ground plates below and above the strip and is embedded in homogeneous dielectric material (Figure 2.13).

Electromagnetic Properties
As shown in Figure 2.2, the triplate can be regarded as a modified version of the coaxial transmission line, and therefore has similar electrical properties. If the size of the ground plates is infinite, then the waveguiding structure is completely closed, so that no radiation loss and EM interference occur. For finite plates, this is still valid to a large extent. The dielectric of the triplate line is homogeneous. Therefore, the fundamental mode is a pure TEM mode (if the conductor and dielectric are lossless), even at very high frequencies.

Practical Design Viewpoints
The application is commonly restricted to passive components, such as planar filters. Directional couplers with very high directivity can be realized using two coupled lines, because the fundamental even and odd modes (see Section 2.8) have the same phase velocity. Adjustment of the circuit cannot be done easily in comparison with the adjustment of the microstrip structure.

Serious problems can occur with circuit fabrication. The final assembly of the triplate circuit requires that the upper metal cladded substrate is pressed onto the lower substrate half, which carries the planar structure. Because the strip metallization has finite thickness, the presence of trapped air near the strip edge can cause dispersion effects (see Section 2.5). The circuit halves can be simply riveted to plastic substrates. The coaxial adapter comprises a flat and thin strip center conductor to be contacted with the triplate. Caution is needed to prevent any break of the contact flag during assembly.

2.4 Suspended Stripline

The suspended stripline is a triplate system in which the strip conductor is suspended symmetrically between two parallel ground planes and supported upon a thin dielectric substrate (Figure 2.14).

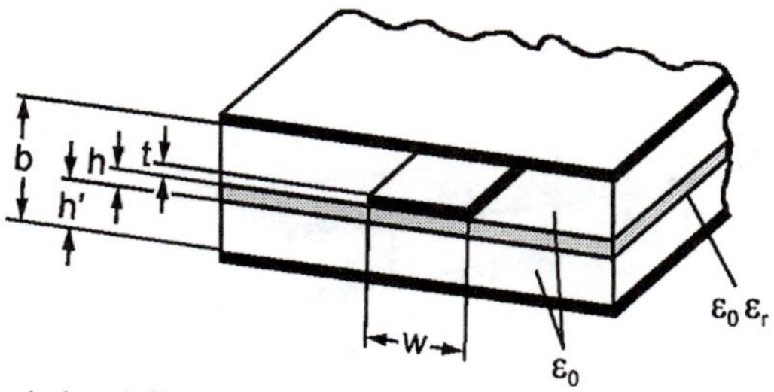

Figure 2.14 Suspended stripline.

Electromagnetic Properties
Theoretically, no pure TEM mode can propagate at frequencies greater than 0 due to the different phase velocity in the dielectric layer and air region. However, the field distribution of the fundamental mode resembles a TEM mode at low frequencies.

The suspended stripline is a low-loss version of the standard triplate guide system [17]. The dielectric area of the line is drastically reduced, which gives rise to substantially smaller dielectric loss. All of the electrical parameters of the high-Q-triplate are close to the electrical parameters of the air-filled stripline.

Practical Design Viewpoints
Due to the low effective dielectric constant (approximately 1) for air, the strip width is wider for a given impedance value compared with the standard triplate. This means that high impedance lines can be realized easier [18]. In practice, the suspended stripline will be totally enclosed by a metal box. Sufficient sidewall spacing must be provided to prevent the striplines from influencing the fundamental mode propagation [19]. The mechanical construction is complex. The thin substrate carrier is clamped between the two halves of a conductive housing, or is fixed by mechanical fasteners. In particular, the appearance of vibration and shock can affect the electrical performance or destroy the circuit.

Balanced mixers, bandpass filters, and several other components, namely SPST switch, hybrid race coupler, PIN attenuator, branch line coupler, and backward wave coupler, have been realized at operating frequencies up to X-band [20].

2.5 Sandwich Line

The sandwich line comprises a standard microstrip line that is covered by a dielectric plate (Figure 2.15).

Electromagnetic Properties
The placement of an additional upper dielectric layer on the top of the strip reduces dispersion due to the larger amount of electromagnetic field confined to the dielectric material [16]. The overlay dielectric must be sufficiently thick so that the

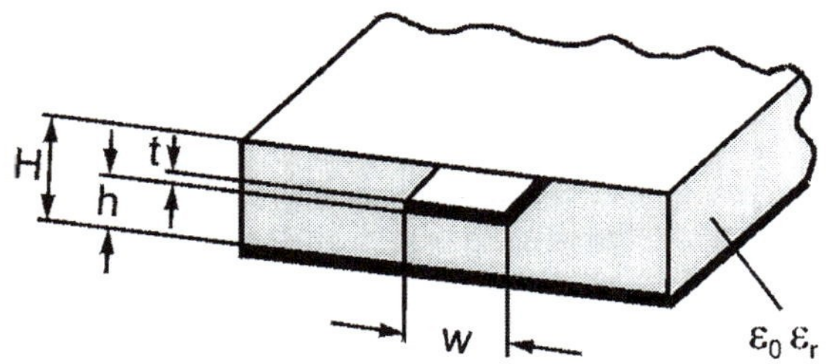

Figure 2.15 Sandwich line.

residual penetrating field becomes negligible. The higher the value of the dielectric constant, the faster the decay of the electromagnetic field with increasing distance from the line surface. The sandwich line has less dispersion compared with the microstrip line. However, the dielectric loss becomes larger. Impedance level is lowered upon overlay placement.

Practical Design Viewpoints
The dielectric overlay is commonly confined locally and is used to enhance the directivity of directional coupler with coupled microstrip lines [21–23]. It also can be used for fine tuning of planar circuits [24]. Deposition of thin dielectric films is applied for passivation of semiconductor surfaces to protect against atmospheric contaminants [9].

Problems may occur with the assembly of the line pressing the dielectric cover plate closely onto the strip carrying the bottom part. It is of utmost importance that air gaps be eliminated during overlay placement. The technique used in [21] was to soften the overlay with suitable solvent. Dielectric type Hi K 707L-10, manufactured by Custom Material Inc., has been used as overlay material, which softens easily when using acetone as a solvent. The overlay was clamped in place, after which an adhesive may be used on the overlay edges to ensure firm placement.

The sliding load configuration treated in Chapter 4 also may be considered as a sandwich line of finite length. In this case, lossy and high-dielectric material is placed on the microstrip line to absorb the energy of the propagating quasi-TEM mode, thus representing a matched termination.

2.6 Slot Line

The slot line consists of a slot in a conducting plane on one side of a dielectric sheet (Figure 2.16).

Electromagnetic Properties
The slot line was first proposed by S. B. Cohn in 1969 [24–26], and considered as a planar waveguiding structure alternative to microstrip line. It has been regarded as a guideline dual to microstrip, since it can be derived from the latter by

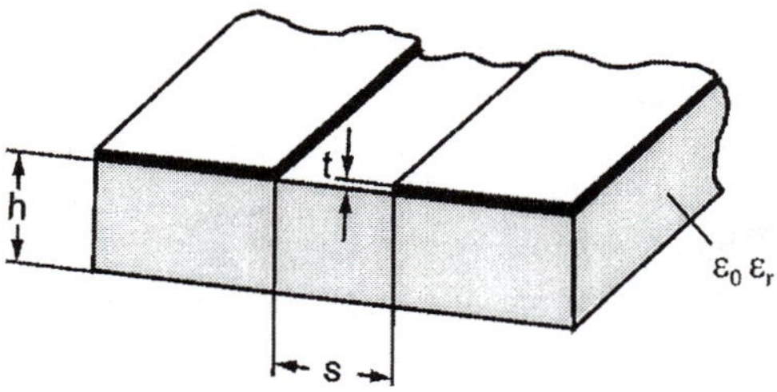

Figure 2.16 Slot line.

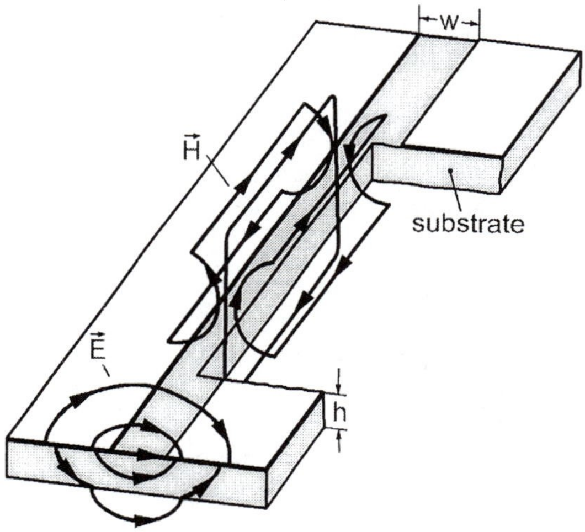

Figure 2.17 Electric and magnetic field distribution of slot line in three-dimensional view. (From: [27], © 1971 IEEE. Reprinted with permission.)

replacing metal-dielectric interfaces with dielectric-dielectric interfaces, and vice versa [18].

The slot line is an open waveguiding structure. To prevent noticeable radiation, high permittivity substrate with dielectric constant $\epsilon_r > 10$ must be used. The slot line is a two-conductor system and therefore the fundamental mode has a zero cutoff frequency. However, the fields will be closely confined near the slot with increasing frequency. As pointed out in [26], using a substrate with a dielectric constant $\epsilon_r = 16$, the electrical field at a frequency of 3 GHz and at a distance of 1 in (25.4 mm) from the slot center compared to the field amplitude directly across the slot is down 28.4 dB, while at 1.3 in, it is down 38.6 dB. Typical parameters of slot lines investigated experimentally in [26] are: (1) $\epsilon_r = 20.3$, $h = 3.38$ mm, $s = 0.6$ mm; and (2) $\epsilon_r = 14.5$, $h = 1.75$ mm, $s = 0.6$ mm.

Figure 2.17 illustrates the slot-mode field in three-dimensional view. Transverse electric and magnetic fields actually are in-phase (notice, the transverse fields in Figure 2.17 are shifted for clearness), resulting in a Poynting vector in propagation direction, and thus the power flow, along the line. A voltage difference exists between the slot edges. The electric field extends across the slot. It can be seen that the magnetic field lines that are perpendicular to the slot curve on both sides and return to the slot at half-wavelength intervals. Due to the strong longitudinal magnetic field component, the fundamental mode in the slot line is approximately an *H* (*TE*) mode. A propagating wave has a magnetic field with elliptically polarized regions that can be usefully applied in creating certain ferrite components [27].

Because of the non-*TEM* nature of the slot wave, definition of characteristic impedance is somewhat arbitrary (see Section 4.2). The definition chosen in [26] is the power-voltage definition (P-I definition, see Chapter 4). Voltage is peak voltage amplitude across the slot and power is the average power flow of the wave. As pointed out in [26], this definition does not necessarily yield the best match in a transition to coaxial or microstrip line. In fact, experimental results indicate that a 50Ω coaxial or microstrip line requires about 75Ω slot impedance.

Practical Design Viewpoints

Stimulation of the slotwave can be simply arranged using a miniature semirigid line with the inner and outer conductor attached to the two metal plates near the gap by means of solder. Filters, couplers, ferrite devices, and circuits containing semiconducter elements have been realized.

Fields of slot line and microstrip line can be coupled in an easy way [9]. A combination is frequently advantageous with the microstrip line on one side and the slot line on the opposite side of a dielectric substrate [28]. This has been utilized, for example, to generate fast sampling pulses for balanced sampling circuits [29, 30] (Appendix C).

The fabrication of the slot line is simple and is formed by photoetching a narrow slot along the length of one face of the substrate while the copper is removed from all other surfaces [26]. It is well suited to shunt mounted elements such as diodes, resistors, and capacitors, since both conductors are on one side of the substrate. The need for holes through the substrate for shunt mounting of elements is eliminated [27].

2.7 Coplanar Line

A coplanar waveguide consists of a strip of thin metallic film deposited on the surface of a dielectric slab with two ground electrodes running adjacent and parallel to the strip on the same surface (Figure 2.18).

Electromagnetic Properties

The coplanar line was proposed by C. P. Wen in 1969 [31]. Beside the microstrip line, the coplanar line is the most frequently used planar transmission line in integrated circuits. The coplanar line can be regarded as two coupled slot lines.

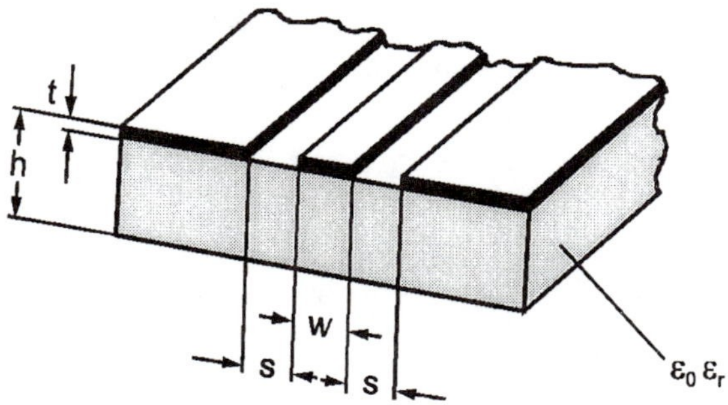

Figure 2.18 Coplanar waveguide.

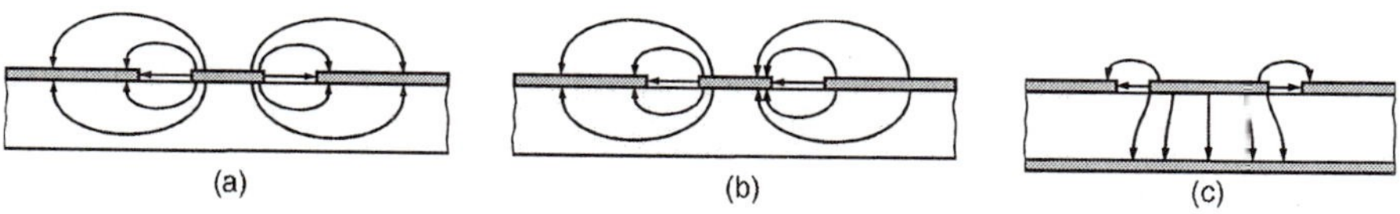

Figure 2.19 Schematic electrical field distribution in coplanar waveguide: (a) desired even mode, (b) parasitic odd mode, and (c) parasitic parallel plate mode.

Therefore, similar properties of a slot line may be expected. The coplanar line consists of three conductors with the exterior ones used as ground plates. These need not necessarily have same potential. As is known from transmission line theory of a three-wire system, even and odd mode solutions exist as illustrated in Figure 2.19. The desired even mode, also termed coplanar mode [Figure 2.19(a)], has ground electrodes at both sides of the centered strip, whereas the parasitic odd mode, also termed slot line mode, has opposite electrode potentials. When the substrate is also metallized on its bottom side, an additional parasitic parallel plate mode with zero cutoff frequency can exist [Figure 2.26(c)] [32]. When a coplanar wave impinges on an asymmetric discontinuity such as a bend, parasitic slot line modes can be excited [33]. To avoid these modes, bondwires or airbridges are connected to the ground plates to force equal potential [13].

Figure 2.20 shows the electric and magnetic field distribution of the even mode at low frequencies, which is TEM-like. At higher frequencies, the fundamental mode evolves itself approximately as a TE mode (H mode) with elliptical polarization of the magnetic field in the slots. This readily lends itself to nonreciprocal magnetic device applications because of the built-in circularly polarized magnetic vector at the air-dielectric boundary between the conductors. Practical applications have been experimentally demonstrated by measurements on resonant isolators and differential phase shifters fabricated on low-loss dielectric substrates with high dielectric constants.

A coplanar line has low dispersion. As reported in [34], coplanar lines on GaAs and sapphire substrate were stimulated with a 100 fs laser, creating 0.6 ps electrical

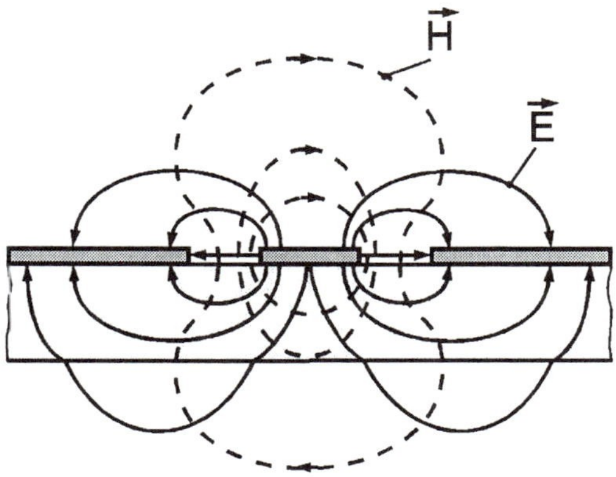

Figure 2.20 Transversal electromagnetic field of even coplanar mode at low frequency.

pulses, which means a 3 dB bandwidth of about 500 GHz. In this case, dispersion becomes insignificant until 300 GHz and above.

Practical Design Viewpoints

The coplanar line is a single plane waveguiding structure. The circuit realization is done on one side in one plane, which is convenient for mounting both shunted and series devices. The dimensions of the coplanar lines such as slot and strip width are relatively small, which may hamper the realization of hybrid circuits. Because the coplanar line is surrounded by conductors with ground potential, decoupling of neighboring transmission lines is much more efficient. This permits higher packaging density and smaller chip size. Viaholes that are required with microstrip lines are not necessary, which results in low-inductive ground connection. Due to the small strip cross sectional area, power transfer is limited [13]. This also causes higher attenuation.

As can be seen from Figure 2.20, when the dielectric substrate is not sufficiently thick, the field leaks from the backside. Because the coplanar circuit is commonly attached to a conductive package or located on a wafer-prober station for measurement, this can lead to some change of the electrical properties of the line. The thickness of the dielectric substrate becomes less critical with higher relative dielectric constants. In practice, the thickness of the substrate should be one or two times the width of the slots. It is obvious that the finite thickness of the substrate will influence the dispersion characteristics of the transmission line. Radiation problems may occur when the total distance from one ground to another approaches one-half wavelength. This will pose a limit on the dielectric constant of the substrate employed at a given frequency and the transverse dimension of the transmission line.

The characteristic impedance of the coplanar transmission line can be approximately derived applying conformal mapping [26]:

$$Z_o = \frac{30\pi}{\sqrt{\epsilon_{eff}}} \frac{K'(k)}{K(k)} \ \Omega \tag{2.8}$$

$K(k)$ and $K'(k)$ are the complete elliptic integrals of first kind and its complement, respectively. They are related to each other through the relation $K'(k) = K(k')$, with $k' = \sqrt{1 - k^2}$. The argument of the elliptic integrals is related to the coplanar line parameters, $k = s/(s + 2w)$.

For the ratio K/K', accurate and simple expressions are known from the literature [35]:

$$\frac{K(k)}{K'(k)} = \pi / \ln\left(2\,\frac{1 + \sqrt{k'}}{1 - \sqrt{k'}}\right) \quad \text{for} \quad 0 \le k \le 1/\sqrt{2} \tag{2.9}$$

$$\frac{K(k)}{K'(k)} = \frac{1}{\pi} \ln\left(2\,\frac{1 + \sqrt{k}}{1 - \sqrt{k}}\right) \quad \text{for} \quad 1/\sqrt{2} \le k \le 1 \tag{2.10}$$

Applying curve fitting procedure to field-theoretical results, the following expression was derived for the effective dielectric constant [36]:

$$\varepsilon_{eff} = \frac{\varepsilon_r + 1}{2}\left\{ \tanh\left[1.785 \log\left(\frac{h}{w}\right) + 1.75\right] + \frac{kw}{h}\left[0.04 - 0.7k + 0.01(1 - 0.1\,\varepsilon_r)(0.25 + k)\right]\right\} \tag{2.11}$$

The accuracy is supposed to be better than 1.5% for $h/w \ge 1$.

Coplanar lines are widely used as the transmitting medium in MMICs on GaAs material basis. GaAs wafer material is very expensive and therefore not acceptable for hybrid circuit fabrication. Alternatively, high-resistivity silicon (HRS) substrate material with a comparable high-dielectric constant (see Section 9.3.3) has become available at moderate cost. Figure 2.21(a) shows a fabricated 50Ω coplanar line on HRS substrate using a less elaborate photoetching technology. On-wafer measurements can be easily performed with coplanar probes that need convenient contacting pads [Figure 2.21(b)]. A loss of 1 dB/mm was obtained at 25 GHz and about 1.3 dB/mm at 40 GHz [37].

Figure 2.22 shows a modification of the coplanar waveguide on HRS substrate. Using micromachining, dielectric material between the metallization pattern is removed by dry etching. This results in a lower effective dielectric constant, and therefore lower dielectric loss.

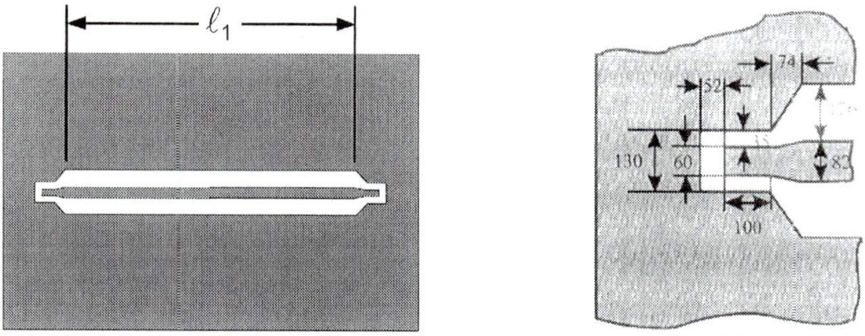

(a) (b)

Figure 2.21 50 Ω coplanar waveguide realized on high resistivity silicon (HRS) substrate material (strip width 126 μm, slot width 82 μm, width of ground plane 834 μm, substrate height 375 μm).

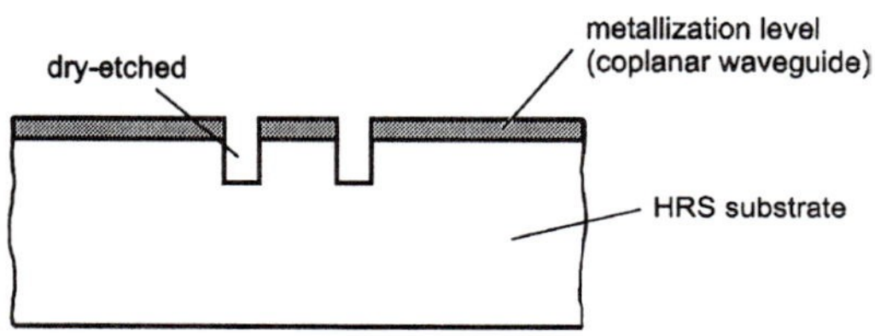

Figure 2.22 Micromachined high-Q coplanar waveguiding structure.

2.8 Coupled Microstrip Line

Coupled microstrip line consists of two closely spaced planar conductors on the surface of a dielectric carrier with an infinite ground plate at the backside (Figure 2.23).

Electromagnetic Properties

Regarding two microstrip lines in parallel as shown in Figure 2.24, and assuming that the distance s between them is sufficiently large, the electromagnetic fields of both lines will not interfere. In this case, the microstrip lines are decoupled, and their electromagnetic fields correspond to that of a single microstrip mode. This changes when the spacing of the lines becomes closer. As illustrated in Figure 2.25, at very close spacing ($s \ll 5w$), two different modes, even and odd mode, exist on the coupled microstrip line [38].

The even mode is excited by a source with both strips having the same potential to ground. As illustrated in Figure 2.25(a), the even mode comprises a symmetric electric field with respect to the plane of symmetry. The smaller the gap between the two metal strips, the stronger is the influence of the adjacent conductor upon the stray field, which is located in the symmetry plane between the conductors. As

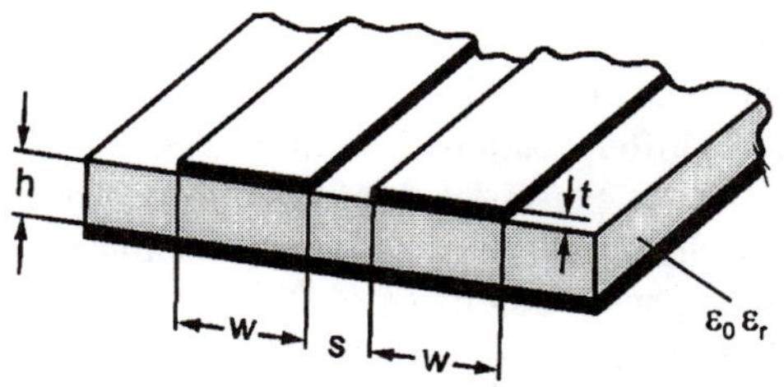

Figure 2.23 Coupled microstrip line.

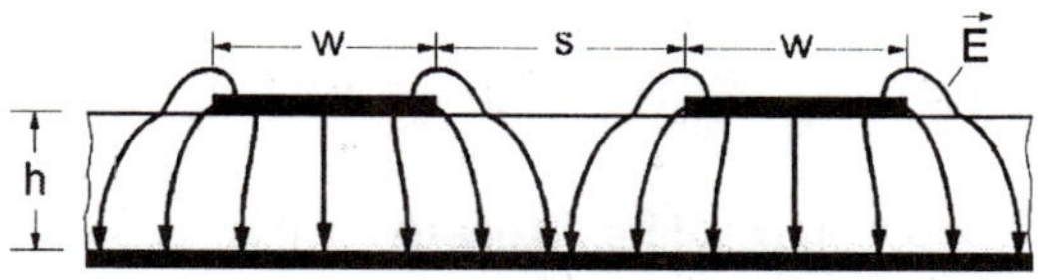

Figure 2.24 Transversal electrical field distribution of two adjacent microstrip lines not yet coupled.

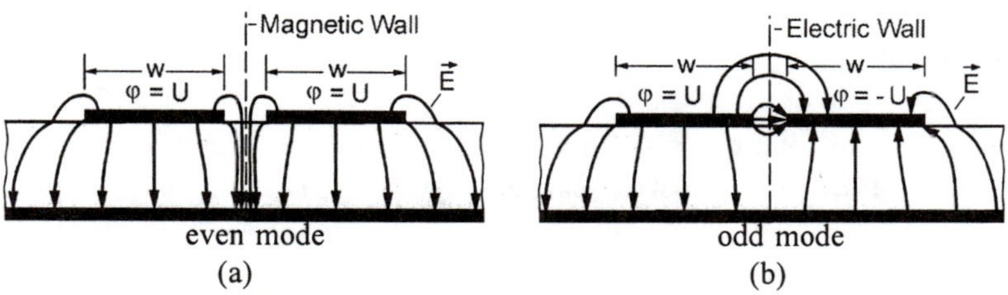

Figure 2.25 Fundamental modes in coupled microstrip transmission line: (a) even mode, and (b) odd mode.

can be seen from the curved electrical field lines, only a tangential electric field strength exists in the symmetry plane. We call such property an ideal electric open circuit or a magnetic wall. Upon a magnetic wall, the magnetic field strength is always perpendicular. For reason of the symmetric electrical field distribution, this propagation mode is named even mode.

The odd mode in Figure 2.25(b) can be stimulated by a source with its electrodes connected to the strips. The strips have an opposite potential that differs from ground. In the case of an asymmetric excitation, the electric field lines are vertical upon the symmetry plane. Therefore, this symmetry plane is a so-called electric wall, and can be substituted by an electric short circuit. Because a significant part of the field lines of the odd mode is running through the air region, its effective permittivity is lower than in the case of the even mode excitation. Therefore, the phase velocity of the odd mode is larger than that of the even mode. The general excitation field is a composition of both an even and an odd mode. Dielectric overlays have been used to adjust the effective dielectric constants of the even and odd mode [21].

Practical Design Viewpoints

Coupled microstrip lines are utilized extensively as basic elements for directional couplers, filters, phase shifters, baluns, matching networks, and a variety of other useful circuits [39]. The successful development of broadband directional couplers has been hampered by directivity degradation attributable to a difference between the even and odd mode phase velocities.

In [22], a directional coupler has been realized on alumina substrate (h = 0.64 mm, $\epsilon_r \approx 10$). Mode velocity equalization has been obtained by an overlay, cut from the same substrate material, visually placed, and fixed to the circuit by Stycast HiK cement ($\epsilon_r \approx 10$). When placing the dielectric overlays on the circuit, it is important to fill all the voids in the coupling gap, which otherwise can significantly affect the coupler's performance. The width of the dielectric overlay was chosen to only span the distance across the coupled conductors. However, as pointed out in [21], because of the strong edge field, any trapped air bubbles or other adhesive inhomogeneities would have a great effect on coupler isolation. Therefore, it is recommended to choose the overlay sufficiently wide, so that the bonding agent (which may have arbitrary permittivity) has less influence on the electrical performance.

2.9 Antipodal Slot Line

The antipodal slot line comprises two overlapping semi-infinite metal plates on both sides of a dielectric substrate carrier (Figure 2.26).

Electromagnetic Properties

The antipodal slot line is a two-conductor system. Therefore, at very low frequencies, a TEM-like mode with zero cutoff frequency can propagate. Practically, the waveguiding structure only can be operated at higher frequencies, at which the electromagnetic field is concentrated near the overlapping region. Notice that when the two half-planes are grounded by mounting the slot line in a conductive housing, then wave propagation is only possible above a finite cutoff frequency.

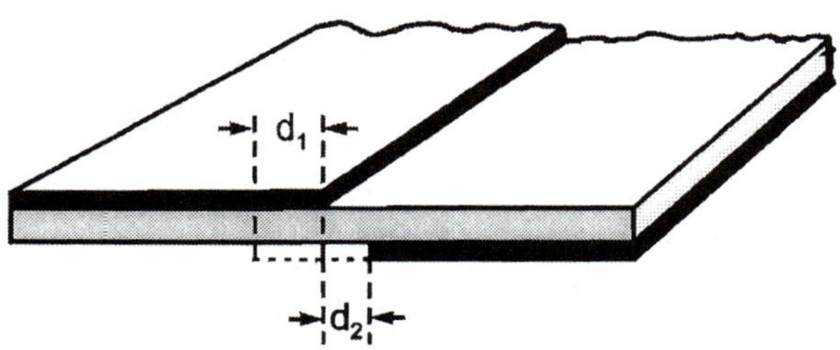

Figure 2.26 Antipodal slot line [40].

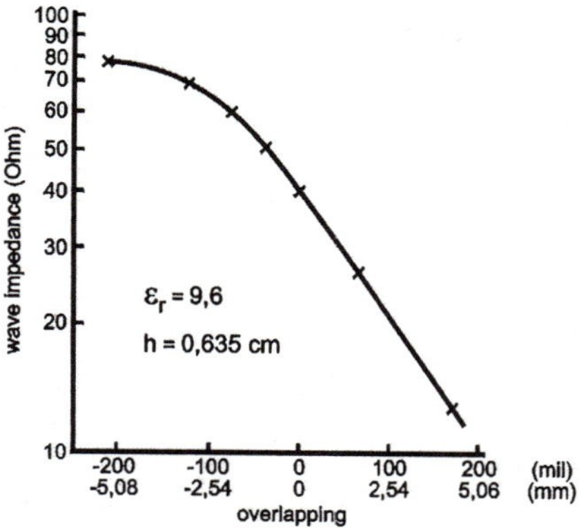

Figure 2.27 Impedance as function of metallization overlapping [40].

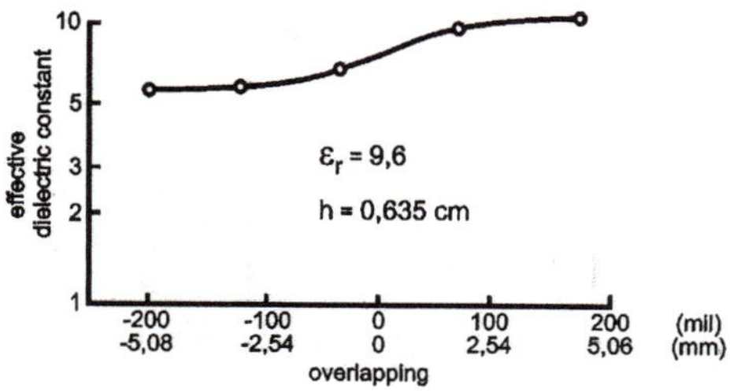

Figure 2.28 Effective dielectric constant as function of metallization overlapping [40].

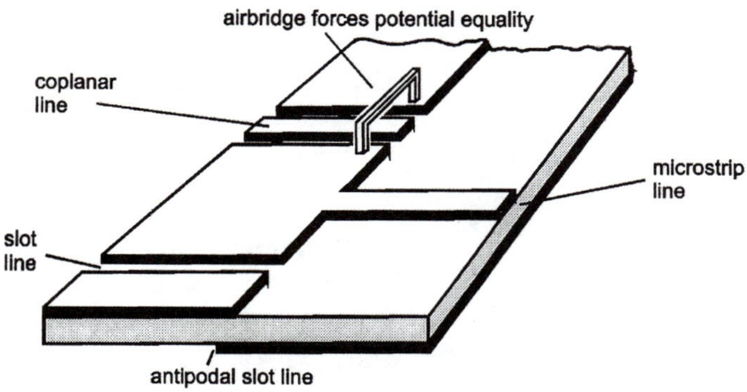

Figure 2.29 Representation of various strip transmission lines in a circuit [40].

Practical Design Viewpoints

As shown in Figure 2.27 the characteristic impedance decreases with increasing overlapping of the metal plates. For the pictured case of alumina ($\epsilon_r = 9.6$, $h = 0.635$ mm), an impedance level of about 10Ω is attained. See Figure 2.27.

Combination of Different Planar Transmission Structures

As the characteristic impedance range differs with the planar transmission structures, it may be convenient to combine some configurations for optimum circuit performance. An illustrative example is given in Figure 2.29, where microstrip transmission lines, slot lines, and antipodal strip lines have been combined on one single substrate [40]. This leads to the widest possible choice of circuits to be built. Especially for millimeter wave applications, the integration of multifunction circuits is likely to require the use of more than one type of transmission structure to achieve the desired level of performance [23].

2.10 Finline

The finline comprises a slot pattern on one side (or on both sides) of a thin substrate carrier in the *E*-plane of a rectangular waveguide.

Electromagnetic Properties

The finline technique combines the advantages of the planar transmission structure (e.g., low-cost fabrication, system integration) with the advantages of the properties of a hollow waveguide (e.g., no radiation). Figures 2.30 to 2.32 show the three-dimensional view of three commonly used finlines. In contrast to the unilateral finline (Figure 2.32), conductors of the bilateral line are deposited on both sides of the substrate (Figure 2.31). The structures may be regarded as a ridged waveguide with a zero-thickness ridge backed by a dielectric slab in the E-plane. The suspended fins concentrate the field energy in the gap region, leading to a capacitive loading for the dominant mode of propagation in the slab-loaded waveguide. This effectively lowers the cutoff frequency of the fundamental mode to a large extent, and that of the next higher order mode to a small extent, thus

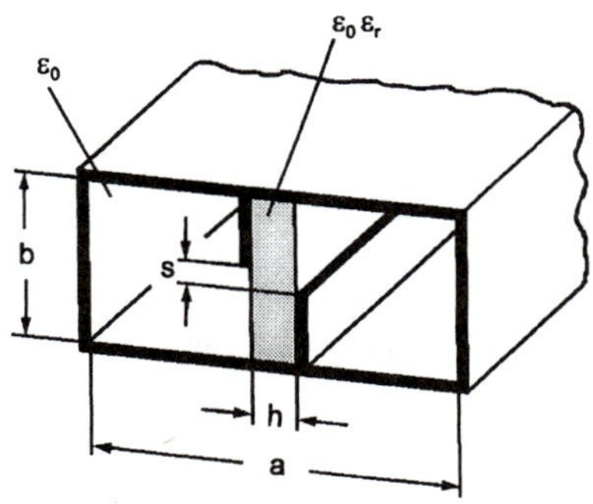

Figure 2.30 Antipodal finline ($Z_0 \geq 10\Omega$).

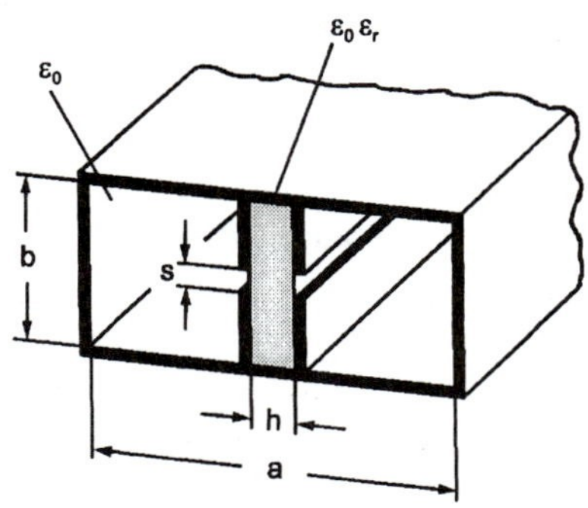

Figure 2.31 Bilateral finline.

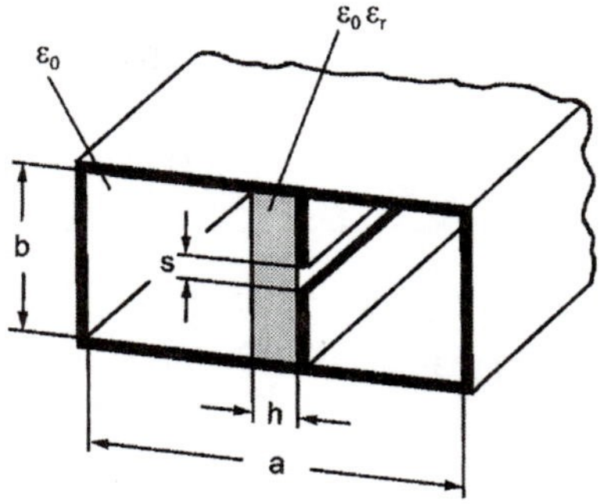

Figure 2.32 Unilateral finline ($Z_0 \geq 100\Omega$).

leading to a larger fundamental mode bandwidth [41]. For an exhaustive account of finline characteristics and circuits, the reader is referred to [42].

Practical Design Viewpoints

Low production tolerances of the waveguide is of less importance, because the electromagnetic field is concentrated near the slot. Fiber-strengthened teflon (17-μm Cu-plate with 3-μm Au-sheet for passivation) has been used as substrate, as well as quartz with an Au-sheet of 3 to 5 μm. The finline technique has been utilized to realize filters with a center frequency of 150 GHz. Radar receivers have been developed at frequencies approximately equal to 35 GHz.

Comparison of Various Planar Waveguiding Structures

Table 2.1 summarizes some important properties of planar waveguiding structures [28, 43]. The total Q-factor of a transmission line is related to the phase coefficient β and total attenuation coefficient α,

$$Q = \frac{\beta}{2\alpha} \tag{2.12}$$

α includes conduction and dielectric losses α_c and α_d. Radiation losses α_{rad} occur at discontinuities. For weakly lossy waveguiding structures, (2.12) reads

$$\frac{1}{Q} = \frac{2\alpha}{\beta} = \frac{2}{\beta}(\alpha_c + \alpha_d + \alpha_{rad}) = \frac{1}{Q_c} + \frac{1}{Q_d} + \frac{1}{Q_{rad}} \tag{2.13}$$

Regarding the open waveguiding structures such as microstrip line, slot line, and coplanar waveguide, the slot line exhibits highest loss, radiation, and dispersion. Low-impedance values less than 60Ω are difficult to realize in slot line technique. However, series and shunt elements can be easily implemented. The coplanar waveguide permits a large impedance range from 20Ω to 250Ω. As with slot lines,

Table 2.1
A Comparison of Various Planar Waveguiding Structures

Transmission Line	*Unloaded Q*	*Radiation*	*Dispersion*	*Impedance Range*	*Chip Mounting*
Microstrip	100–200	Low	Low	20–120	Difficult for shunt, easy for series
Stripline	400	Low	None	35–250	Poor
Suspended Stripline	500–1500	Low	None	40–150	Fair
Slot Line	100	Medium	High	55–300	Easy for shunt, difficult for series
Coplanar Line	150	Medium/ High	Low	25–155	Easy for series and shunt
Finline	500	None	Low	10–400	Fair

the mounting of series and shunt elements is easy. Dispersion is low. However, because of the small strip area, the power handling capabilities are unsatisfactory. This restriction is not valid for the microstrip line that is preferably used in power amplifier design. Series chip mounting is easy. However, installation of shunt devices or the realization of shorts needs a feed through to ground, which is accomplished by drilling or cutting in case of soft substrate, and etching or laser beam processing with hard substrates. Characteristic impedance values lower than 20Ω can be designed by simply increasing the width of the strip conductor, but this results in a risk of parasitic transverse modes. At low-to-moderate frequencies, microstrip impedance values down to slightly below 10Ω are thus achievable [41]. We will see, in stub filter design, that low impedances can be realized by shunting two stubs with moderate characteristic impedance values. The upper limit to characteristic impedance is set by the manufacturing tolerances of very narrow strip.

The closed waveguding structures, such as standard stripline, suspended stripline, and finline, exhibit higher Q-factors. This results in less radiation, and less dielectric losses, particularly with finline and suspended stripline. The impedance value range is larger than in the group of open structures. This is particularly valid for finline covering a large impedance region of 10Ω to 400Ω.

References

[1] Ayer, D. R., and C. A. Wheeler, "The Evolution of Strip Transmission Line," *The Microwave Journal*, May 1969, pp. 31–40.

[2] Barrett, R. M., "Microwave Printed Circuits—A Historical Survey," *IEEE Trans.*, Vol. MTT-3, 1955, pp. 1–9.

[3] Howe, H., "Microwave Integrated Circuits—An Historical Perspective," *IEEE Trans.*, Vol. MTT-32, 1984, pp. 991–996.

[4] Grieg, D. D., and H. F. Engelmann, "Microstrip—A New Transmission Technique for the Kilomegacycle Range," *Proc. I.R.E.*, 40 (1952), pp. 1644–1650.

[5] King, D. D., "Properties of Dielectric Image Lines," *IEEE Trans.*, Vol. MTT-3, 1955, pp. 75–81.

[6] Itoh, T., *Numerical Techniques for Microwave and Millimeter-Wave Passive Structures*, New York: John Wiley & Sons, 1989.

[7] National Museum of American History: http://smithsonianchips.si.edu/texas/t_280.htm.

[8] McQuiddy, D. N., et al., "Monolithic Microwave Integrated Circuits: An Historical Perspective," *IEEE Trans.*, Vol. MTT-32, 1984, pp. 997–1008.

[9] Schneider, M. V., "Microstrip Lines for Microwave Integrated Circuits," *Bell System Technical Journal*, Vol. 48, 1969, pp. 1421–1444.

[10] Wheeler, H. A., "Transmission-Line Properties of Parallel Wide Strips by a Conformal Mapping Approximation," *IEEE Trans.*, Vol. MTT-12, 1964, pp. 280–289.

[11] Grünberger, G. K., V. Keine, and H. H. Meinke, "Longitudinal Field Components and Frequency-Dependent Phase Velocity in the Microstrip Transmission Line," *Electronics Letters 6*, 1970, pp. 683–685.

[12] Grünberger, G. K., and H. H. Meinke, "Experimenteller und theoretischer Nachweis der Längsfeldstärken in der Grundwelle der Mikrowellen-Streifenleitung," *Nachrichtentechnische Zeitschrift*, Vol. 24, 1971, pp. 364–368.

[13] Goyal, R., *Monolithic Microwave Integrated Circuits*, Norwood, MA: Artech House, Inc., 1989.

[14] Louisell, W. H., *Coupled Mode and Parametric Electronics*, New York: John Wiley & Sons, 1960.

[15] Mao, S., S. Jones, and G. D. Vendelin, "Millimeter-Wave Integrated Circuits," *IEEE Trans.*, Vol. ED-15, 1968, pp. 517–523.

[16] Krage, M. K., and G. I. Haddad, "Frequency-Dependent Characteristics of Microstrip Transmission Lines," *IEEE Trans.*, Vol. MTT-20, 1972, pp. 678–688.

[17] Bochtler, U., and F. M. Landstorfer, "Suspended-Substrate Technique," in *Handbook of Microwave Technology* (T. K. Ishii, ed.), Vol. I, Components and Devices, Academic Press, 1995.

[18] Gunston, M. A. R., *Microwave Transmission-Line Impedance Data*, London: Van Nostrand Reinhold Company, 1972.

[19] Brenner, H. E., "Numerical Solution of TEM-Line Problems Involving Inhomogeneous Media," *IEEE Trans.*, Vol. MTT-15, 1967, pp. 485–487.

[20] Koul, S. K., B. Bhat, and M. C. Krishnaswamy, "Passive Components in Inverted Microstrip and Suspended Microstrip Configurations," *IEEE MTT-S*, Boston, 1983, pp. 412–414.

[21] Paolino, D. D., "MIC Overlay Coupler Design Using Spectral Domain Techniques," *IEEE Trans.*, Vol. MTT-26, 1978, pp. 646–649.

[22] Sheleg, B., and B. E. Spielman, "Broad-Band Directional Couplers Using Microstrip with Dielectric Overlays," *IEEE Trans.*, Vol. MTT-22, 1974, pp. 1216–1220.

[23] Spielman, B. E., "Integrated-Circuit Media for Millimeter-Wave Applications," in *Millimeter and Submillimeter Wave Propagation and Circuits*, AGARD Conference Proceedings, Vol. 245, E. Spitz and G. Cachier (eds.), 1979, pp. 20-1–20-8.

[24] Cohn, S. B., "Slot Line—An Alternative Transmission Medium for Integrated Circuits," IEEE G-MTT Intern. Microwave Symposium Digest, 1968, pp. 104–109.

[25] Cohn, S. B., "Slot Line on a Dielectric Substrate," *IEEE Trans.*, Vol. MTT-17, 1969, pp. 768–778.

[26] Mariani, E. A., et al., "Slot Line Characteristics," *IEEE Trans.*, Vol. MTT-17, 1969, pp. 1091–1096.

[27] Harrison, G. R., et al., "Ferrimagnetic Parts for Microwave Integrated Circuits," *IEEE Trans.*, Vol. MTT-19, 1971, pp. 577–588.

[28] Kneppo, I. et al., *Microwave Intgrated Circuits*, Chapman & Hall, 1994.

[29] Merkkelo, J., and R. D. Hall, "Broad-Band Thin-Film Signal Sampler," *IEEE Trans.*, Vol. SC-7, 1972, pp. 50–54.

[30] Kompa, G., "Entfernungsmeßgerät nach dem Impulslaufzeitverfahren," European Patent, No. 0015566, granted 1984.

[31] Wen, C. P., "Coplanar Waveguide: A Surface Strip Transmission Line Suitable for Nonreciprocal Gyromagnetic Device Applications," *IEEE Trans.*, Vol. MTT-17, 1969, pp. 1087–1090.

[32] Klingbeil, H., *Studien zu passiven MMIC-Strukturen der Höchstfrequenztechnik*, Fortschritt-Bericht VDI, Reihe 9: Elektronik, Nr. 260, Düsseldorf: VDI Verlag, 1997.

[33] Klingbeil, H., K. Beilenhoff, and H. L. Hartnagel, "Finite-Difference Analysis of Structures Consisting of Roundly and Rectangularly Shaped Domains," *IEEE Microwave and Guided Wave Letters*, Vol. 6, No. 8, 1996, pp. 295–297.

[34] Frankel, M. Y., et al., "Terahertz Attenuation and Dispersion Characteristics of Coplanar Transmission Lines," *IEEE Trans.*, Vol. MTT-39, 1991, pp. 910–916.

[35] Hilberg, W., "From Approximations to Exact Relations for Characteristic Impedances," *IEEE Trans.*, Vol. MTT-17, 1969, pp. 259–265.

[36] Gupta, K. C., R. Garg, and I. J. Bahl, *Microstrip Lines and Slotlines*, Norwood, MA: Artech House, Inc., 1979.

[37] Cheng, Y., *Study on Partitioning Design Approach for K-Band Oscillator Design*, Doctoral Thesis, University of Kassel, 2004.

[38] Wolff, I., *Einführung in die Mikrostrip-Leitungstechnik*, Aachen: Verlag H. Wolff, 2. Auflage, 1978.

[39] Bahl, I., and P. Bhartia, *Microwave Solid State Circuit Design*, New York: John Wiley & Sons, 1988.

[40] RCA, "Kreuzglied für Mikrowellenfrequenzen," Patent, DE 22 64 978.

[41] Ishii, T. K., *Handbook of Microwave Technology*, Vol. I, Components and Devices, Academic Press, 1995.

[42] Bhartia, P., and P. Pramanick, "Finline Characteristics and Circuits," in *Topics in Millimeter-Wave Technology*, K. J. Button, (ed.), Vol 1, San Diego: Academic Press, 1988.

[43] Edwards, T., *Foundations for Microstrip Circuit Design*, 2nd ed., New York: John Wiley & Sons, 1992.

Chapter 3

Methods of Microstrip Analysis

The ever increasing demand for microstrip circuits being operated at higher and higher frequencies during the last decades has required an improvement in the methods of microstrip analysis. This trend was associated with the growing commercial exploitation of the microstrip technique in monolithic microwave integrated circuits, where hardware tuning after circuit fabrication is very difficult, and only known in some particular cases. In the following, we will discuss the key features of major published analysis methods, as shown in Figure 3.1 [1, 2].

At the beginning, electrostatic methods prevailed, using the method of conformal transformation [3–5], the finite difference method [6–8], integral equation [9], and variational method [10, 11]. Conformal mapping is frequently applied to planar transmission lines to analyze the propagation of waves at lower frequencies where dispersion effects are negligible. Therefore, this method will be treated in more detail in Section 3.1. Regarding the finite difference method, the solution of the Laplace equation in the cross-sectional area is approximated by the simultaneous solution of a set of linear equations. Following this method, the field region under analysis is subdivided into a net of small squares with assigned discretized electrostatic potentials. In order to develop the set of equations, the potential at a given point is measured and the potential of all directly adjacent mesh

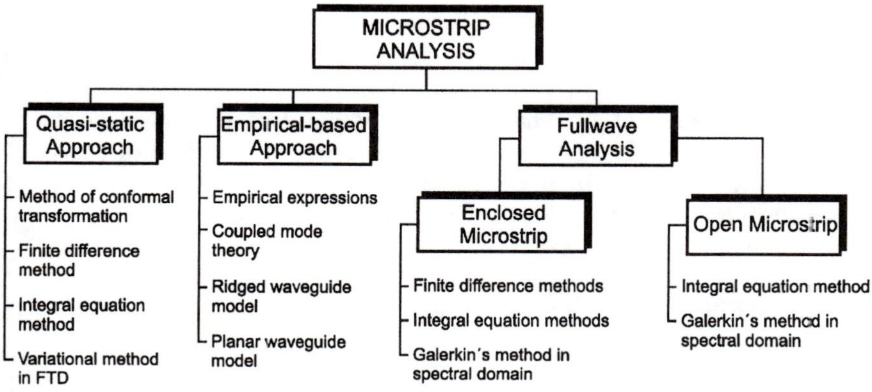

Figure 3.1 Overview of microstrip analysis methods. (After: [1].)

points are expanded in Taylor's series about the considered point. Neglecting higher-order expansion terms, a simple algebraic relation is derived between the centered and adjacent potentials.

The integral equation method describes the total electrostatic potential at a given point, with the integrand containing the distribution of the surface charges on the conductors and the Green's function. As the Green's function describes the potential distribution due to a unit line charge, integration must be carried out over all conductor surfaces to obtain the potential at an arbitrary point. The Green's function is calculated on the basis of the method of images [9].

The variational method permits the calculation of the capacitance per unit length of a microstrip line from merely approximate knowledge of the charge distribution over all the conductor surfaces. The calculation of the capacitance is based on a variational integral equation [12] comprising the potential on the conductors and distributed charge. The variational expression has the significant feature that if the charge distribution deviates by a differential amount from the actual one, then there will be no influence on the calculation of the capacitance value. That also means an approximate trial function for the charge distribution will give good results for the electrostatic capacitance under consideration.

Dispersion is generally known as the frequency dependence of the phase velocity of a wave and occurs in any waveguiding structure with mixed dielectric material in the cross-sectional area. In the microstrip line, the upper electrode is placed on a dielectric substrate carrier, so that the electromagnetic field exists partly in air and partly in dielectric material. As the frequency is growing, the energy of the wave is increasingly concentrated in the material with the highest dielectric constant, and the limiting phase velocity for infinite frequency corresponds to the phase velocity in the respective dielectric medium. Experimental investigation of wave propagation along a microstrip transmission line was first performed with the standing wave detector [13]. Then, with the advent of vector network analyzers [14], measurement became more efficient. Numerous microstrip lines with different geometric parameters and materials have been analyzed, and the measured frequency dependence of the phase velocity of the dominant mode described empirically. The dispersion has commonly been described by closed-form analytical expressions found by curvefitting procedure [15–18]. Dispersion phenomenon also was described by the interaction of a TEM mode and surface waves on the basis of coupled mode theory [19–21]. It is known from multimode waveguiding structures that modes will become coupled when the phase velocities of waves are approximately the same. The coupled mode theory commonly supposes weak coupling, so that the field distribution of the original waves is only slightly perturbed. Assuming a hypothetical TEM wave on the microstrip line, which really cannot exist, with the phase coefficient $\beta_{TEM} = \omega(\epsilon_{eff_1}{}^{1/2})/c_0$, and dielectric constant $\epsilon_{eff_1} = (\beta_{TEM}\, c_0/\omega)^2$, and defining an effective dielectric constant $\epsilon_{eff_2} = (\beta_{TM_0} c_0/\omega)^2$ for the fundamental TM_0 surface wave, the coupled mode theory delivers the following relation for the frequency-dependent effective dielectric constant:

$$\epsilon_{eff}(f) = \frac{\sqrt{\epsilon_{eff_1}} + \sqrt{\epsilon_{eff_2}}}{2} + \sqrt{\left(\frac{\sqrt{\epsilon_{eff_1}} - \sqrt{\epsilon_{eff_2}}}{2}\right)^2 + \xi} \tag{3.1}$$

where ξ denotes the coupling coefficient between the modes. If the velocities of waves are different for very low and high frequencies, no coupling is predicted by theory, meaning that the coupling coefficient ξ becomes zero. Then from (3.1) it reads $\epsilon_{eff}(f = 0, \infty) = \epsilon_{eff_1}$ which does not agree with the dispersion behavior at very high frequencies. Therefore, a frequency-dependent coupling coefficient was proposed. As can be simply verified, the coupling coefficient should vary with frequency from zero to $\xi(f = \infty) = \left(\sqrt{\epsilon_{eff_2}} + \sqrt{\epsilon_{eff_1}}\right)/2 + \sqrt{\epsilon_{eff_1}}\big)/2 + \left[\left(\sqrt{\epsilon_{eff_2}} - \sqrt{\epsilon_{eff_1}}\right)/2\right]^2$ at infinite frequency.

It has been shown that the model of a dielectrically loaded ridged waveguide is very useful to describe the principal frequency dependence of the phase velocity on microstrip lines. Values of some formal (nonphysical) parameters must be determined by curvefitting procedure in the case under consideration [22]. A further approach is known, which describes the dispersion of the fundamental microstrip mode, and that of the higher-order modes as well, with a unified guide model. This planar waveguide model [23–25] represents an equivalent waveguiding structure for the open microstrip line. It comprises frequency-dependent effective dielectric constants, and frequency-dependent effective widths as well. It will be treated in Section 5.

The full-wave analysis methods are regarded as the most advanced methods, because they consider the hybrid mode characteristics. Both enclosed and open microstrip transmission lines have been analyzed using finite difference [26], integral equation [27, 28], and Galerkin's method in spectral domain [29, 30].

Applying the finite difference method, the differential equations for the electrical and magnetic scalar potentials (see Section 3.4) are discretized. For that, the cross-sectional field region of the enclosed microstrip line is subdivided into small squares. The second order partial derivatives of the potential in a given grid point are expressed by potential differences to the adjacent points, and substituted in the scalar differential equation for the electric and magnetic potentials. After applying the boundary and interface conditions, a complete set of finite difference equations is obtained in the form of a matrix eigenvalue equation, which can be solved by several methods [26].

Regarding the grounded dielectric slab configuration, the natural propagating modes are surface waves. However, the inclusion of the metal strip on the dielectric surface with transverse and longitudinal surface current flow disposes the surface modes, but generates a hybrid mode, which can be represented as a linear combination of TM and TE surface modes. The surface modes form a complete set of eigenfunctions of which an arbitrary field can be presented. This implies that the surface modes exhibit the property of orthogonality [12]. For dispersion analysis of the fundamental mode, the field components of surface modes are analytically described so that they fulfill the dielectric-air interface and the boundary conditions on the ground, and at infinity in the air region. The inclusion of both a longitudinal and a transverse component of current on the strip leads to the introduction of potential functions that represent Fourier transforms with respect to the x-coordinate. With respect to the y-coordinate, the potential functions fulfill the boundary at ground plate and the radiation condition at infinity; that is, all field components decay exponentially with growing distance from the line. The continuity of the longitudinal and transverse electric field and the step in the respective magnetic field according to the longitudinal and transverse surface current lead to a coupled pair of integral equations (in the space domain). Simplification is introduced by neglecting the transverse current and satisfying the strip interface conditions only at the center of the strip. Thus, a single integral equation is obtained for the unknown ξ, which denotes the frequency dependent effective dielectric constant:

$$\int_0^\infty \frac{\beta_1 Z\, I_z(\gamma)}{(YH)^2 + \dfrac{\epsilon_r Z}{\xi}\lambda_0^2\beta_1^2\left(Q\cot\beta_1 d - \dfrac{\beta}{\epsilon_r\beta_1}\right)}\, d\gamma = 0 \tag{3.2}$$

where

$$\gamma = \alpha w;\; H = (\lambda_0/w)\gamma,\; Q = (\xi - 1)/(\mu_r K - \xi),\; I_z(\gamma) = J_0(\gamma/2),$$

$$Y = (\mu_r K - 1)/(\mu_r K - \xi),\; Z = Q\,\mu_r \tan\beta_1 d + \beta_2/\beta_1$$

ϵ_r denotes the dielectric constant of the substrate material, w the width of the strip, λ_0 the free-space wavelength, I_z the longitudinal strip current, J_0 the Bessel function of zero order, and β_1 and β_2 the separation parameters in the y direction in dielectric and air medium.

The Galerkin method [29, 30] uses basically the same formulation as given in [27]. The essential difference is that the two final boundary conditions on the strip are now applied in the Fourier transform or spectral domain. This results in a set of two algebraic equations, in contrast to the coupled integrals in [27], which is solved in the spectral domain for the calculation of the phase coefficient for the fundamental microstrip mode (see Section 3.4.3)

In the following sections, we will deal more closely with the selected approaches listed in Figure 3.1; that is, the method of conformal transformation, the planar waveguide model, and the Galerkin's method in spectral domain for the laterally open microstrip transmission line.

3.1 Conformal Transformation Method After Wheeler

The characteristic impedance of a waveguiding structure is one of the most relevant electrical parameters that is required for circuit design. In 1964, H. A. Wheeler first published closed-form expressions for the characteristic impedance of a microstrip line [4]. He analyzed the properties of parallel wide strips in air using an approximate conformal mapping transformation. In 1965, he extended his analysis to parallel strips separated by a dielectric sheet [5]. The transmission structure was assumed to be lossless. As explained by Wheeler, his intention was to derive simple explicit design formulas for practical use with "slide-rule" accuracy. He has stated that his approach is a departure from the straightforward exact analysis leading to solutions in implicit form in terms of elliptic functions.

The analysis was primarily applied to the symmetrical parallel plate transmission line shown in Figure 3.2(a). The width of the conductors is $2a$ and the thickness of the dielectric substrate carrier $2b$. Regarding the fundamental mode, the horizontal plane of symmetry describes an equipotential plane that can be replaced by a metal sheet, leading to the microstrip transmission line shown in Figure 3.2(b) with the parameter notation commonly used, strip width w and substrate height h. The dimensions of Wheeler's parallel-plane line apply to the microstrip transmission line with $2a = w$ and $b = h$. The characteristic impedance of a TEM transmission line is calculated as $Z_0 = \sqrt{L'/C'}$, where L' is the inductance per unit length and C' the capacitance per unit length. It is obvious that, because of the upper and lower strip, the inductance of guide (a) is twice the inductance value of the microstrip line whereas, due to the twofold distance of the electrodes, the capacitance of guide (a) is only one-half the capacitance value of the microstrip line. Therefore, taking index 1 for guide (a) and index 2 for the microstrip line, it reads

$$Z_{0_2} = \sqrt{\frac{L_2'}{C_2'}} = \sqrt{\frac{\frac{1}{2}L_1'}{2\,C_1'}} = \frac{1}{2}Z_{0_1} \tag{3.3}$$

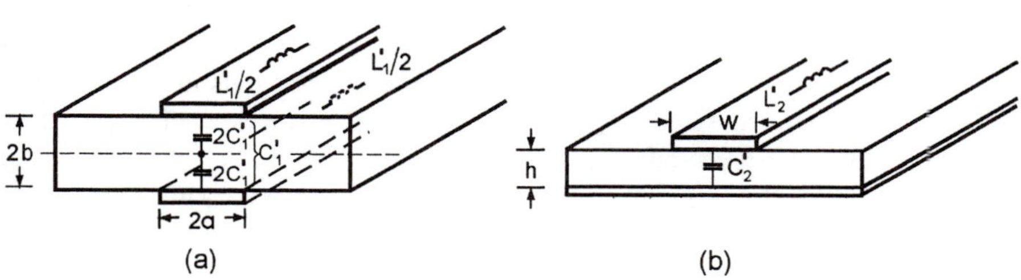

Figure 3.2 Parallel-plane guide originally analyzed by (a) Wheeler and (b) related microstrip transmission line.

Thus, the impedance of the microstrip is one-half the impedance of the parallel guide [31]. Notice that the wavelength of the fundamental mode is equal on both guides.

Wheeler has analyzed the air-filled microstrip transmission line first. Under application of the conformal mapping procedure, the complicated microstrip electrode configuration with inhomogeneous field is mapped on an ideal parallel-plate transmission line with homogeneous electric and magnetic field. As is known, the exact mapping transformation includes elliptic functions (see Section 3.2.1). The most remarkable achievement of Wheeler's work is that, although having introduced a lot of approximations throughout his analysis, he got a rather accurate closed-form expression for the characteristic impedance of the air-filled microstrip transmission line. The influence of the dielectric is investigated in a second step, applying the derived mapping transformation onto the dielectric areas of the actual microstrip line. The analysis has been performed separately for two ranges of shape ratio w/h, characterized by wide strips with $w/h \geq 2$ or narrow strips with $w/h < 2$.

3.1.1 Air-Filled Microstrip Line

Figure 3.3(a) shows the cross section of an air-filled microstrip line with physical primed parameters. The strip is assumed to be a perfect conductor of zero thickness. It is spaced from a parallel ground plane by the separation h. The shape ratio w/h is the significant parameter for the characteristic impedance. Because of uniform free-space medium ($\epsilon_r = 1$), a pure TEM mode exists. The two-dimensional electrical field distribution can be determined by application of conformal mapping (Appendix A). The characteristic impedance of the air-filled microstrip line is

$$Z_0 = \sqrt{\frac{L'}{C'}} = \frac{1}{c_0 C'} \tag{3.4}$$

where L' is the inductance and C' the capacitance per unit length. The velocity of light c_0 reads

$$c_0 = (L'C')^{-1/2} = (\epsilon_0 \mu_0)^{-1/2} \tag{3.5}$$

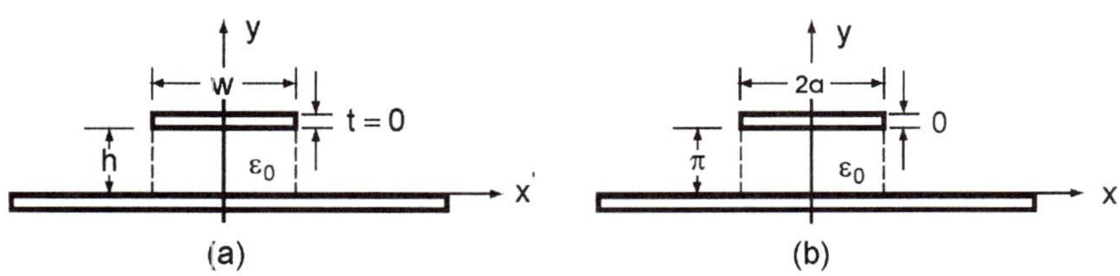

Figure 3.3 Cross section of air-filled microstrip line with (a) physical and (b) normalized geometrical parameters.

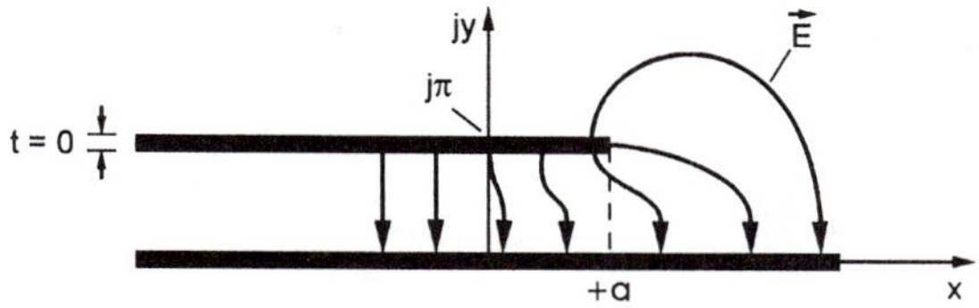

Figure 3.4 Electrical stray field near the edge of the parallel plate capacitor structure.

with $c_0 = 29.98 \times 10^9$ cm/s (or approximately $30\ \mathrm{GHz \cdot cm}$). Thus, the characteristic impedance can be found if the capacitance per unit length is known. For the application of the conformal mapping transformation, all geometric quantities are normalized to an arbitrary quantity l_N. If this is defined as
30 GHz·cm

$$l_N = h/\pi \tag{3.6}$$

then the normalized strip w/l_N reads

$$w/l_N = w\pi/h = 2a \tag{3.7}$$

and the normalized dielectric sheet thickness h/l_N is given as

$$h/l_N = \pi \tag{3.8}$$

Figure 3.3(b) shows the microstrip structure with normalized geometrical parameters in the complex z-plane ($z = x + jy$). In order to derive closed-form analytical expressions for the characteristic impedance, Wheeler made the following assumptions. First, the transmission structure is symmetric with respect to the y-axis. This is also valid for the electric field distribution of the fundamental mode. In this case, the capacitance per unit length can be derived from the analysis of only one-half of transmission structure [e.g., the region $Re(z) \geq 0$]. In the symmetry plane of the microstrip transmission line, the electric field lines are parallel to the y-axis. The magnetic field lines are perpendicular to the electric ones. In the symmetry plane, the normal electric field and the tangential magnetic field vanish. Such boundary condition describes an ideal electric open or magnetic wall. Second, assuming wide strips (i.e., $w/2 > h$ and $a > \pi$, respectively), the electrical field in region $Re(z) \geq 0$ approximately agrees with the edge region of a parallel plate capacitor of infinite extension in x-direction ($x = -\infty$), as illustrated in Figure 3.4. Notice that the electric field line within the capacitor at $x = 0$ is slightly bent, in contrast to the respective field line in the actual microstrip transmission line. Wheeler's theory is based on the Schwarz-Christoffel mapping theorem (see Appendix A.2).

3.1.2 Mapping of the Microstrip Conductor Boundaries into a Parallel Plate Structure

In the following section, we will discuss the mapping of the complicated electrode configuration of a microstrip line with inhomogeneous electrical field distribution onto an ideal parallel plate transmission line with a homogeneous electrical field. It will be shown that twofold Schwarz-Christoffel transformation is needed. As is described in the Appendix A.2, the theorem can be used to map a polygonal curve onto the real axis of an imaged plane. In case of the microstrip configuration with different electrical potentials on the upper strip and ground, the mapped figure consists of two conductive half-planes with an included angle of 180°. Similarly, the Schwarz-Christoffel theorem also can be used to map the ideal parallel plate electrodes onto such single-plane capacitor configurations. Combining both transformations then will lead to a mapping equation, which relates the width of the imaged parallel plate line to the width of the microstrip transmission line. The height of both lines is supposed to be invariant in the mapping procedure.

As already pointed out, Wheeler confined his analysis to one-half of the microstrip line as shown in Figure 3.5. The field area of interest is designated by the points 1 to 8. At the left of the half cross section of the microstrip line, the conductor lines are extended to infinity. The extension is designated by the points 4, 5′, 5′′ at $x = -\infty$, and 6. This region will be excluded in the final mapping. Notice that point 5′ identifies the transition from the positive to the negative potential of the microstrip electrodes. The dashed line between points 3 and 7 describes a field line off the edge of the strip. The electrical field lines inside and outside of this line terminate on the inner and outer face of the strip conductor, respectively. The vertical dashed lines, which coincide with the imaginary axis of the z-plane, represent the bisection boundaries of the electrical field. In the mapping approximation, these lines are taken to be vertical straight lines in the z-plane. Because of stray effects near the edge of the parallel plate capacitor configuration under analysis, the electrical field line in point 4 will not exactly end in point 6, as is approximately assumed here, but a little shifted to the right of point 6 (compare Figure 3.4). If the aspect ratio $w/(2h) = a/\pi$ is sufficiently large, then the error can be neglected.

Figure 3.6 shows the mapping of the z-plane onto the upper half of the w-plane. The excluded area in the z-plane appears inside a region, which approximately describes a semicircle in the w-plane. When the w-plane is completed by imaging at the u-axis, then the cross section of a guide line occurs, which consists of a pair of circular dielectric wires. These dielectric areas are of particular interest because they include relevant transformation parameters. On closer examination, these parameters are nearly equal; a maximum parameter difference $g - d$ is estimated to be less than 0.097. The location of the significant points 1 to 8 are also indicated. In particular, point 2 is placed on $w = 0 + j0$, point 3 on point $w = c$ (and $w = -c$, respectively), and point 5′′ at infinity on point $w = d$ (and $w = -d$, respectively).

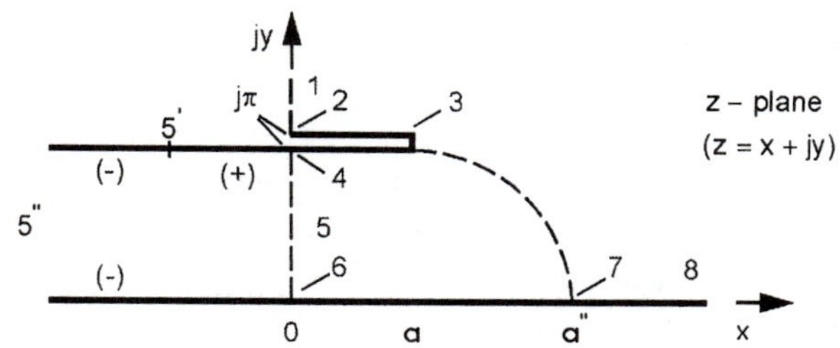

Figure 3.5 Space coordinates in the *z*-plane and critical points for mapping. (From: [4], © 1964 IEEE. Reprinted with permission.)

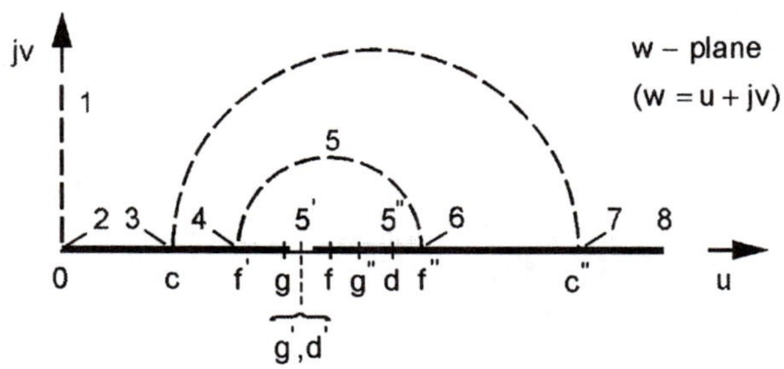

Figure 3.6 Microstrip electrodes in the *z*-plane mapped on two infinite half-planes in the *w*-plane. (From: [4], © 1964 IEEE. Reprinted with permission.)

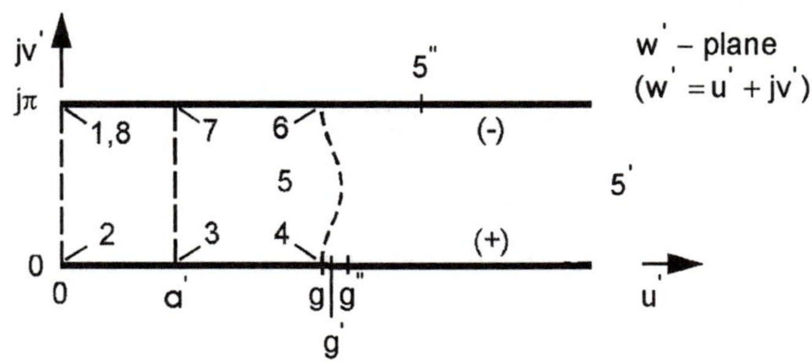

Figure 3.7 Transformation of the *w*-plane onto the *w*′-plane. (From: [4], © 1964 IEEE. Reprinted with permission.)

Figure 3.7 shows the mapped single plane capacitor configuration in the w'-plane. The relevant field region is bounded by the point sequence 1-2-3-4-5-6-7-8. The regions outside of this rectangle (area 4-5′-5′′-6) correspond to areas inside the round dielectric guide lines in the w-plane. The solid lines $jy' = 0$ and $jy' = \pi$ represent electrical equipotential boundaries (conductors) having a separation of $j\pi$, which is equal to the normalized height of the microstrip line. The curved line from point 3 to 7 in the z-plane occurs as a straight line at $u' = a'$. As we have discussed above, the line 5 in the z-plane is accurately not a field line; therefore the transformation results in a curvature as shown in Figure 3.7. The coordinates g and g' signify the minimum and maximum value of the line coordinates. For the calculation of the capacitance per unit length, an average value g' between g and g'' is derived. The width g' can directly be related to the half-width a of the microstrip line. The difference between them is a measure of the electrical stray field at the edge of the line.

First, we are searching for a transformation function that maps the two strip boundaries in the z-plane on a single plane capacitor (w-plane). The Schwarz-Christoffel mapping theorem generally permits the transformation of the polygonal structure of the plate capacitor with $x \to -\infty$, as shown in Figure 3.5, via an intermediate mapping in the w-plane (Figure 3.6) onto an ideal parallel plate capacitor in the w'-plane (Figure 3.7).

The mapping gradient dz/dw in (A.63) is specified as follows. There are two complete reversals of the polygonal contour in the z-plane at points 3 and 5′′. For point 3, the corresponding external angle δ_3 reads $\delta_3 = -\pi = -\mu_3\pi$, so that $\mu_3 = 1$. Similarly, for $P_{5''}$, it follows $\delta_5'' = +\pi = -\mu_5''\pi$, and $\mu_{5''} = -1$. Thus, the transformation equation (A.63) reads with the freely chosen values $u_3 = c$ (point 3) and $u_5'' = d$ (point 5′′)

$$\frac{dz}{dw} = A\,(w - c)\,(w - d)^{-1} \tag{3.9}$$

If the quadrant of the w-plane is completed by imaging with respect to the z-axis, then further mathematical treatment is simplified,

$$\frac{dz}{dw} = A\,(w - c)\,(w - d)^{-1}\,(w + c)\,(w + d)^{-1} \tag{3.10}$$

and

$$\frac{dz}{dw} = A\,\frac{w^2 - c^2}{w^2 - d^2} \tag{3.11}$$

respectively. The correlation of other points of the polygonal contour in the z- and w-planes can be taken from the respective figures. In order to relate the transformation parameters A, c, and d to the known normalized half-width a of the microstrip line, the mapping gradient equation (3.11) is analyzed in the points 2, 4, and 7. An infinitesimal shift dz in the z-plane relates to an infinitesimal shift dw in the w-plane, and the mapping gradient equation identifies the angle between both. Equation (3.11) delivers the following relation for the arguments of the complex mapping parameters dz and dw:

$$\arg(dz) = \arg(dw) + \arg(A) + \arg(w-c) + \arg(w+c) - \arg(w-d) - \arg(w+d) \tag{3.12}$$

As pointed out in Section A.2, the angle of dw is always zero due to the definition of circulation direction. Regarding point 7, $\arg(dz)$ is also zero. Furthermore, because the coordinate value w of point 7 is larger than those of points 3 and 5′′, all other angle terms in (3.12) vanish. Thus, a positive infinitesimal shift of dw results in $\arg(A) = 0$. As the third freely chosen value of parameter of Schwarz-Christoffel transformation, we take $|A| = 1$ for simplicity. This means that the constant A has been chosen to match the far field on both planes, where the mapping gradient approaches unity. Then

$$\frac{dz}{dw} = \frac{w^2 - c^2}{w^2 - d^2} = 1 + \frac{d^2 - c^2}{2d} \frac{2d}{w^2 - d^2} \tag{3.13}$$

Integration of the differential equation (3.13) delivers [32]

$$z = w - \frac{d^2 - c^2}{2d} \ln \frac{1 + d/w}{1 - d/w} = w - \frac{d^2 - c^2}{2d} \ln \left(\frac{1 + w/d}{1 - w/d} e^{-j\pi} \right) \tag{3.14}$$

$$= w - \frac{d^2 - c^2}{2d} \left(\ln \frac{1 + w/d}{1 - w/d} - j\pi \right) \tag{3.15}$$

$$z = w - \frac{d^2 - c^2}{2d} 2 \operatorname{artanh}(d/w) \tag{3.16}$$

We investigate the transformation of point 3 in the z-plane and w-plane to find a relation of the coordinate c to the microstrip parameter a. The point 3 with the coordinates $z = a + j\pi$ is transformed in $w = c$. Substituting the respective coordinates into (3.15) results in

$$a + j\pi = c - \frac{d^2 - c^2}{2d} \left(\ln \frac{1 + c/d}{1 - c/d} - j\pi \right) \tag{3.17}$$

From the imaginary parts, it follows

$$\frac{d^2 - c^2}{2d} = 1 \tag{3.18}$$

which can be written as

$$c = \sqrt{d(d - 2)} = \sqrt{(d - 1)^2 - 1} \tag{3.19}$$

Equation (3.19) implies $d > 2$, because the radicant must be positive, so that c can be approximated as

$$c \approx d - 1 \tag{3.20}$$

From (3.19), we obtain

$$d = 1 + \sqrt{c^2 + 1} \tag{3.21}$$

Regarding the real parts of (3.17) and taking into account (3.18), we obtain

$$a = c - \ln \frac{1 + c/d}{1 - c/d} = c - 2\,\mathrm{artanh}(c/d) \tag{3.22}$$

Equation (3.22) can be written as

$$a = c - \ln\left(\frac{d + c}{d - c}\right) = c - \ln\left(\frac{1 + \sqrt{c^2 + 1} + c}{1 + \sqrt{c^2 + 1} - c}\right) \tag{3.23}$$

$$= c - \ln\left(c + \sqrt{c^2 + 1}\right) \tag{3.24}$$

Equality of (3.23) and (3.24) can be proved simply by comparison of the arguments of the logarithmic functions. Equation (3.24) can be written as hyperbolic function [32],

$$a = c - \mathrm{arsinh}(c) \tag{3.25}$$

Equation (3.25) is an implicit equation for the calculation of c of a. With c known, d also can be calculated using (3.20). Considering (3.18), Equation (3.16) is simplified to

$$z = w + j\pi - \ln \frac{1 + w/d}{1 - w/d} = w + j\pi - 2\,\mathrm{artanh}(w/d) \tag{3.26}$$

In the second step, the mapping gradient between the w'-plane and the w-plane has to be specified. There is only one reversal in the w'-plane; point 5′ at infinity in the w'-plane is transformed to $w = d'$ on the real axis of the w-plane. Applying the Schwarz-Christoffel transformation equation (A.63) yields

$$\frac{dw'}{dw} = A'(w - d')^{-1}(w + d')^{-1} = \frac{A'}{w^2 - d'^2} \tag{3.27}$$

with supposed mirror symmetry to the imaginary axis. Shifting point 7 by dw' and using (3.12) results in $\arg(A') = \pi$. $|A'|$ is chosen as $2d'$ to make the following integration easier. Thus, we get from (3.27)

$$\frac{dw'}{dw} = \frac{2d'}{d'^2 - w^2} \tag{3.28}$$

Integration yields

$$w' = \ln\frac{1 + w/d'}{1 - w/d'} = 2\,\mathrm{artanh}(w/d') \tag{3.29}$$

Resolving for w gives

$$w = d'\tanh\left(w'/2\right) \tag{3.30}$$

With (3.26) and (3.30), the transformation equations are known. It remains to find the relations between the coordinates of the z-plane and w'-plane. The quantities d and d' can be eliminated after long mathematical treatment and diverse approximations by putting the coordinates of points 3, 4, and 6 into the transformation equations. It has been found that the values for the coordinates f, g, g', d', g'', and d are nearly of the same size. Under these approximations, it follows from (3.30)

$$w \approx d\tanh\left(w'/2\right) \tag{3.31}$$

Substituting w' in (3.15) leads to the simple relation between the z-plane and the z'-plane

$$z \approx w + j\pi - w' = d\tanh(w'/2) + j\pi - w' \tag{3.32}$$

Equation (3.32) can be used to calculate the value of the coordinate a' of point 3 in the w'-plane, which will become important for the calculation of the microstrip line with the dielectric substrate carrier. Substituting the coordinates of point 3 of all three planes ($z = a + j\pi$, $w = c$, $w' = a'$), the following relation is obtained from the real part of (3.20), (3.21), and (3.24),

$$a' \approx c - a = \ln\left(c + \sqrt{c^2 + 1}\right) \approx \ln[2(d - 1)] \tag{3.33}$$

With the approximations $d \approx g$ and $g' \approx g$, the ratio

$$\frac{a'}{g'} \approx \frac{\ln[2(g - 1)]}{g} \tag{3.34}$$

can be calculated. The coordinate g' in the w'-plane is needed as a function of a for the calculation of the characteristic impedance. The coordinates of point 3 are inserted in (3.26). The real part delivers

$$a = c - \ln\left(\frac{d + c}{d - c}\right) \tag{3.35}$$

Recalling the approximation $c \approx d - 1$ in (3.20), we find

$$a \approx d - 1 - \ln(2d - 1) \tag{3.36}$$

After multiplication by 2 and reorganizing, (3.36) reads

$$2d - 1 \approx 2a + 1 + 2\ln(2d - 1) \tag{3.37}$$

It is shown that the argument of the logarithmic function $2d - 1$ is the equation itself. Thus, we write

$$\begin{aligned} 2d - 1 &\approx 2a + 1 + 2\ln[2a + 1 + 2\ln(2d - 1)] \\ &= 2a + 1 + 2\ln\left[2a + 1 + 2\ln(4\pi - 1) + 2\ln\frac{2d - 1}{4\pi - 1}\right] \end{aligned} \tag{3.38}$$

The last term, $2\ln(2d - 1)/(4\pi - 1)$, is estimated as $0.004 \ll 1$, and therefore neglected,

$$2d - 1 \approx 2a + 1 + 2\ln[2a + 1 + 2\ln(4\pi - 1)] \tag{3.39}$$

Then, a' in (3.33) is obtained as

$$a' \approx \ln\{2a + 2\ln[2a + 1 + 2\ln(4\pi - 1)]\} \tag{3.40}$$

Solving for d in (3.39) and recalling $d \approx g'$, provides

$$g' \approx d \approx a + 1 + \ln[2a + 1 + 2\ln(4\pi - 1)] = a + 1 + \ln[2\pi(a/\pi + 0.94)] \tag{3.41}$$

As is pointed out in [4], Wheeler changed the constant 0.94 to 0.92 to make the formula nearly exact for $g' = 2\pi$, so that we get the final result for the normalized width g' in the w'-plane

$$g' \approx a + 1 + \ln\left\{2\pi\left(\frac{a}{\pi} + 0.92\right)\right\} \tag{3.42}$$

Then the denormalized half-effective microstrip width w_{eff} reads with (3.6) and (3.8)

$$\frac{w_{eff}}{2} = g' l_N \approx \frac{h}{\pi}\left\{\frac{w\pi}{2h} + 1 + \ln\left[2\pi\left(\frac{w}{2h} + 0.92\right)\right]\right\} \tag{3.43}$$

and the full-width reads

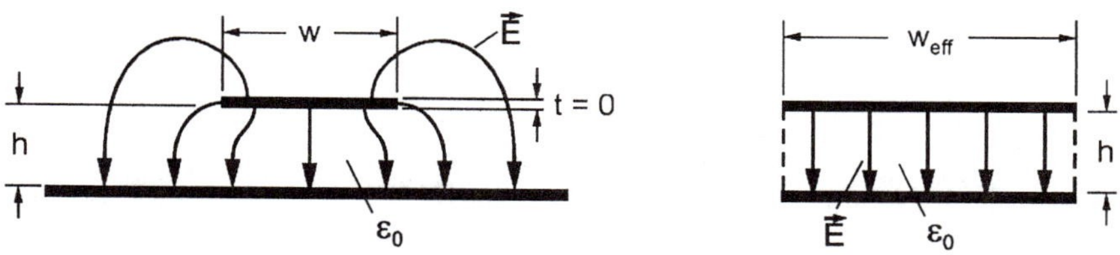

Figure 3.8 Mapping of the air-filled microstrip onto an ideal parallel plate transmission line.

$$w_{eff} \approx w + \frac{2h}{\pi}\left\{\ln\left[2\pi e\left(\frac{w}{2h} + 0.92\right)\right]\right\} \tag{3.44}$$

The capacitance per unit length reads

$$C' = 2\frac{\epsilon_0 g'}{\pi} = \frac{\epsilon_0 w_{eff}}{h} \tag{3.45}$$

Then with (3.4), the characteristic impedance of the air-filled microstrip transmission line is calculated as

$$Z_0 \approx \sqrt{\frac{\mu_0}{\epsilon_0}}\,\frac{0.5}{\frac{w}{2h} + \frac{1}{\pi}\ln\left[2\pi e\left(\frac{w}{2h} + 0.92\right)\right]} \quad (w > 2h) \tag{3.46}$$

Figure 3.8 illustrates the result of the approximate conformal mapping transformation. The characteristic impedance of the air-filled ideal parallel plate transmission line is simply calculated as

$$Z_0 = \frac{U}{I} = \frac{\int \vec{E}d\vec{r}}{\oint \vec{H}d\vec{r}} = \frac{E_y}{H_x}\frac{h}{w_{eff}} = \sqrt{\frac{\epsilon_0}{\mu_0}}\frac{h}{w_{eff}} = 120\pi\frac{h}{w_{eff}} \quad \text{(in ohms)} \tag{3.47}$$

3.1.3 Dielectric-Filled Microstrip Line

The actual microstrip transmission line comprises a dielectric layer to keep the metal strip parallel to ground. For the following analysis, the dielectric material is supposed to be homogeneous, isotropic, and free of dissipation, so that the dielectric properties are completely described by the relative dielectric constant ϵ_r. As is illustrated in Figure 3.9, the free-space electrical field line that terminates on the edge of the strip and is characterized by the points 3 and 7, divides the field region under analysis into a homogeneous dielectric region bounded by the points

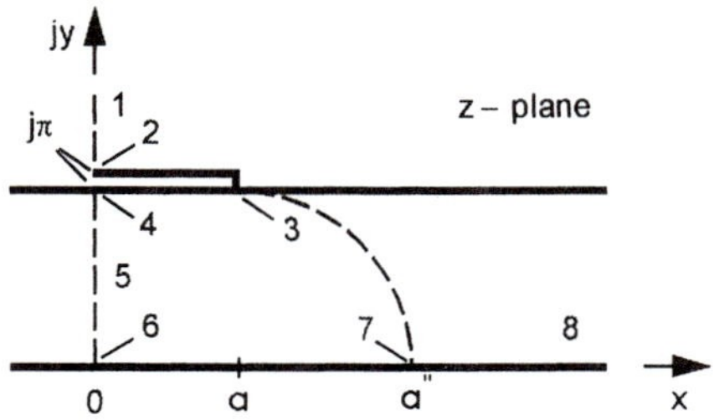

Figure 3.9 Microstrip with dielectric carrier in the *z*-plane. (From: [5], © 1965 IEEE. Reprinted with permission.)

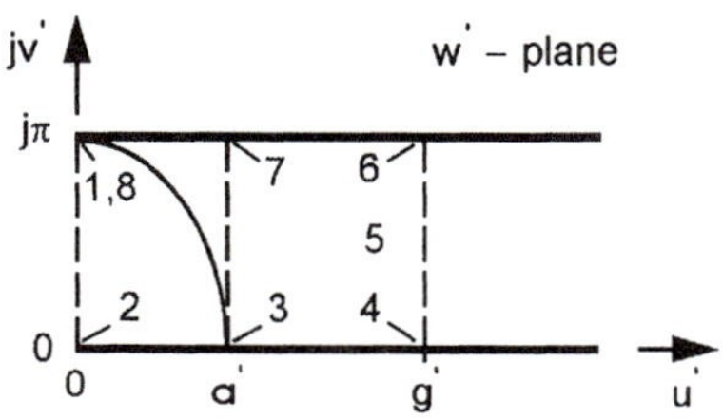

Figure 3.10 Microstrip with dielectric carrier mapped onto the w'-plane. (From: [5], © 1965 IEEE. Reprinted with permission.)

3 to 7, and a mixed dielectric-air region denoted by the point sequence 1-2-3-7-8. As was shown in Section 3.1.2, the homogeneous dielectric region is mapped onto a rectangle in the w'-plane of width $g' - a'$ and height π. The mixed dielectric area in the z-plane leads to an inhomogeneously filled area in the w'-plane, as is illustrated in Figure 3.10. The curve of the mapped dielectric–air interface is found by applying the derived transformation equation (3.32) to the straight boundary of the dielectric layer in the z-plane which is described by the linear equation $z = x + j\pi$. Introducing this relation into (3.32) and evaluating the imaginary part of the relation [32] gives

$$u' = \operatorname{arcosh}\left(d\,\frac{\sin v'}{v'} - \cos v'\right) \tag{3.48}$$

With (3.47), the area A of the dielectric fraction, which is set equal to $\pi s'$, can be calculated by solving the integral

$$A = \pi s' = \int_0^{\pi} (a' - u')\,dv' = -\int_0^{\pi} \ln\left(\frac{\sin v'}{v'}\right) dv' \tag{3.49}$$

The maximum value for s' is estimated as $s'_{max} = 2\ln(\pi/2)$. As shown in Figure 3.11, for approximate calculation the total area A is divided into a rectangular parallel region of area $\pi s''$, which effectively adds s'' to the dielectric area of

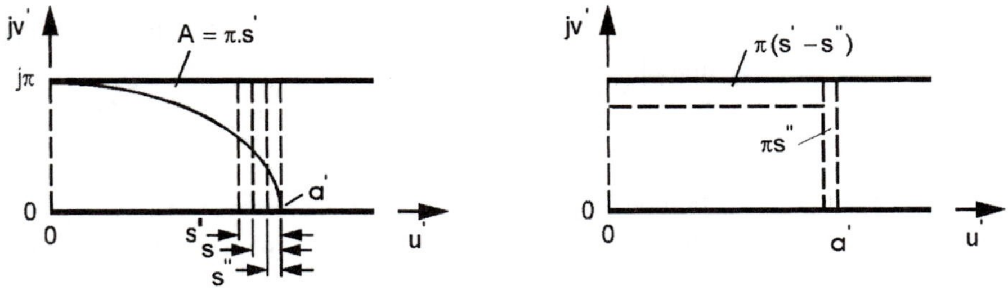

Figure 3.11 Mapped ellipsoidal air-dielectric interface approximated by rectangular limiting curves.

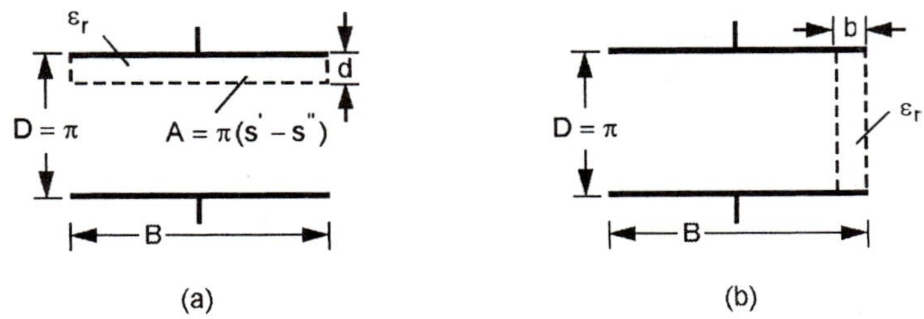

Figure 3.12 Equivalent capacitance per unit length for (a) thin series and (b) shunted dielectric sheet.

width $g' - a'$, and a remaining rectangular part of area $\pi(s' - s'')$, which is effectively in series with the free-space region outside of the dielectric. The latter can be equivalently described by a capacitor configuration with shunted air and dielectric filling, as shown in Figure 3.12. The capacitance C_a' per unit length of Figure 3.12(a) should equal the capacitance C_b' per unit length of Figure 3.12(b). Supposing $d \ll D$ we obtain

$$C_a' = \frac{\epsilon_0 B}{D}\left(1 + \frac{d}{D}\frac{\epsilon_r - 1}{\epsilon_r}\right) \tag{3.50a}$$

and

$$C_b' = \frac{\epsilon_0 B}{D}\left(1 + \frac{b}{B}(\epsilon_r - 1)\right) \tag{3.50b}$$

Thus, for $C_a' = C_b'$, the width b of the shunted dielectric region is calculated as

$$b \approx \frac{dB}{D\epsilon_r} = \frac{\pi(s' - s'')}{\pi\epsilon_r} \tag{3.51}$$

Thus, as indicated in Figure 3.11, the inhomogeneous air–dielectric region of the rectangle 1-2-3-7-8 (Figure 3.10) can be equivalently described by a shunted dielectric width s calculated as

$$s = s'' + \frac{1}{\epsilon_r}(s' - s'') = \frac{1}{\epsilon_r}s' + \frac{\epsilon_r - 1}{\epsilon_r}s'' \qquad (3.52)$$

A filling fraction q is defined as the ratio of dielectric area over total area in the rectangle of field mapping. This can be simply expressed in terms of mapped quantities g', a', and s,

$$q = \frac{g' - a' + s}{g'} = 1 - \frac{a'}{g'} + \frac{1}{\epsilon_r g'}(s' - s'') \qquad (3.53)$$

The air–dielectric interface is approximated by an ellipse, to calculate the area $A = \pi s'$ as a function of a'. Then, s' is obtained as [4]

$$s' = 0.732\,[a' - \operatorname{arcosh}(0.358\cosh a' + 0.953)] \qquad (3.54)$$

It remains to determine s''. This is found by the consideration of a triplate configuration with strip width $w = 2a$ and height $b = 2\pi$ to the upper and bottom ground plate (Figure 3.13). From [33, 34], the effective width $D = 2a''$ of a triplate is exactly known from conformal mapping and reads

$$D = w + \frac{2b}{\pi}\ln 2 = w + 2\Delta a$$

where Δa describes the fringing field at the edge. With $b = 2\pi$, it follows

$$\Delta a = \frac{b}{\pi}\ln 2 = 2\ln 2 = \ln 4 \qquad (3.55)$$

Supposing the triplate is now half-filled with dielectric, then the system is rather similar to that under analysis. This is in particular valid for very high dielectric material. In this case, the filling fraction would be 0.5. Regarding (3.52) for high dielectric constant value, s' vanishes, and thus $s = s''$. Regarding Figures 3.13 and 3.14, a'' can be written as

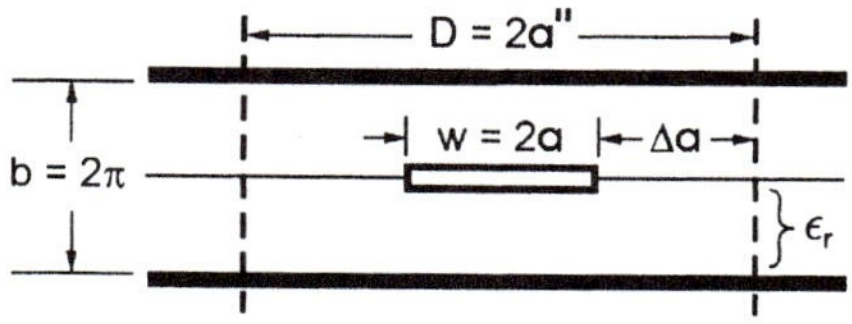

Figure 3.13 Triplate system as reference for s'' estimation. (From: [5], © 1965 IEEE. Reprinted with permission.)

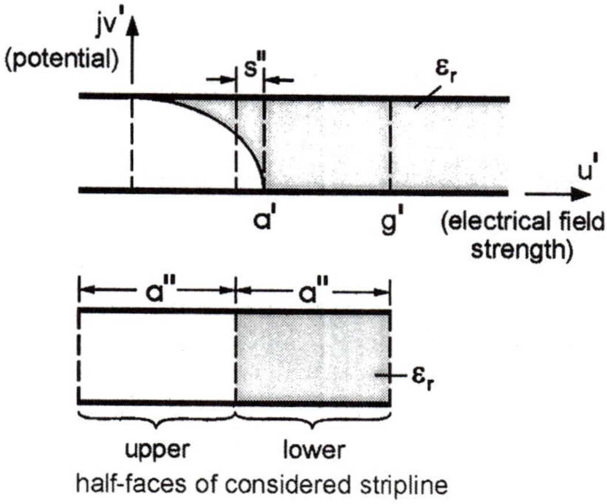

Figure 3.14 Used mappings for finding s''. (From: [5], © 1965 IEEE. Reprinted with permission.)

$$a'' \approx g' - a' + s'' = a + \ln 4 \tag{3.56}$$

Solving for s'', we get

$$s'' \approx a'' - (g' - a') = \ln 4 - (g' - a' - a) \tag{3.57}$$

With (3.19), (3.25), and (3.33) related to the parameters c, d, and a', and regarding $d \approx g'$, the following relation can be derived:

$$s'' \approx \ln 4 - 1 - e^{-a'} \approx \ln 4 - 1 \tag{3.58}$$

for $a' \gg 1$.

The dielectric fraction q in (3.53) signifies the effect of the mixed dielectric on the capacitance per unit length of the microstrip line. It can be used to define an effective dielectric constant ϵ_{eff},

$$\begin{aligned}\epsilon_{eff} &= q\epsilon_r + (1 - q) = 1 + q(\epsilon_r - 1)\\ &= \epsilon_r - \frac{\epsilon_r - 1}{g'}\left(a' - s'' - \frac{s' - s''}{\epsilon_r}\right)\end{aligned} \tag{3.59}$$

or

$$\epsilon_{eff} = \epsilon_r\left[1 - \frac{\epsilon_r - 1}{g'\epsilon_r}\left(a' - s'' - \frac{s' - s''}{\epsilon_r}\right)\right] \tag{3.60}$$

Extracting the root using the approximation $\sqrt{1 + x} \approx 1 + x/2$, (3.60) becomes [35]

$$\sqrt{\epsilon_{eff}} = \sqrt{\epsilon_r} - \frac{\epsilon_r - 1}{2\sqrt{\epsilon_r}\, g'}\left(a' - s'' - \frac{s' - s''}{\epsilon_r} \right) \tag{3.61}$$

With (3.57) and (3.42), $s'' - a'$ is calculated as

$$s'' - a' \approx \ln 4 - (g' - a' - a) - a' \approx \ln 4 - \ln\left[e\left(\frac{a}{\pi} + 0.92 \right) \right] \tag{3.62}$$

The difference $s' - s''$ is approximated by the maximum values of the parameters,

$$s' - s'' \approx s'_{\max} - s''_{\max} = 2\ln\frac{\pi}{2} - \ln 4 + 1 \tag{3.63}$$

Thus, with (3.62) and (3.63), the effective dielectric constant reads

$$\sqrt{\epsilon_{eff}} \approx \sqrt{\epsilon_r} + \frac{\epsilon_r - 1}{2\sqrt{\epsilon_r}\, g'}\left\{ \ln 4 - \ln\left[2\pi e\left(\frac{a}{\pi} + 0.92 \right) \right] \right\} + \frac{\epsilon_r - 1}{2 g' \sqrt{\epsilon_r}} \, \frac{\ln\left(\frac{\pi}{4} \right)^2 + 1}{\epsilon_r} \tag{3.64}$$

With (3.8), the effective dielectric constant for wide strips ($w \geq 2h$) is obtained as a function of the shape ratio w/h and the dielectric constant ϵ_r,

$$\sqrt{\epsilon_{eff}} \approx \sqrt{\epsilon_r} + \frac{(\epsilon_r - 1)\left\{ \ln\left(\frac{\pi}{4} \right)^2 + 1 - \epsilon_r \ln\left[\frac{\pi e}{2}\left(\frac{w}{2h} + 0.94 \right) \right] \right\}}{2\sqrt{\epsilon_r}\, \epsilon_r \left\{ \frac{w\pi}{2h} + \ln\left[2\pi e\left(\frac{w}{2h} + 0.94 \right) \right] \right\}}$$

or with different writing, (3.65)

$$\sqrt{\varepsilon_{eff}} \approx \frac{\sqrt{\varepsilon_r}\left\{ \frac{w\pi}{2h} + \ln 4 + \frac{\varepsilon_r + 1}{2\varepsilon_r} \ln\frac{\pi e}{2}\left(\frac{w}{2h} + 0.94 \right) + \frac{\varepsilon_r - 1}{2\varepsilon_r^2}\left[\ln\left(\frac{\pi}{4} \right)^2 + 1 \right] \right\}}{\pi\left\{ \frac{w}{2h} + \frac{1}{\pi}\ln\left[2\pi e\left(\frac{w}{2h} + 0.94 \right) \right] \right\}} \tag{3.66}$$

The characteristic impedance $Z_0(\epsilon_r)$ with dielectric substrate is calculated as

$$Z_0(\epsilon_r) = \frac{Z_0(\epsilon_r=1)}{\sqrt{\epsilon_{eff}}} \tag{3.67}$$

Substituting $\sqrt{\epsilon_{eff}}$, the characteristic impedance for wide strips ($w \geq 2h$) reads

$$Z_0(\epsilon_r) \approx \sqrt{\frac{\mu_0}{\epsilon_0}} \frac{0.5/\sqrt{\epsilon_r}}{\frac{w}{2h} + \frac{1}{\pi}\ln 4 + \frac{\epsilon_r+1}{2\pi\epsilon_r}\ln\left[\frac{\pi e}{2}\left(\frac{w}{2h}+0.94\right)\right] + \frac{\epsilon_r-1}{2\pi\epsilon_r^2}\left[\ln\left(\frac{\pi}{4}\right)^2+1\right]}$$

or

$$Z_0(\epsilon_r) \approx \sqrt{\frac{\mu_0}{\epsilon_0}} \frac{0.5/\sqrt{\epsilon_r}}{\frac{w}{2h} + 0.441 + \frac{\epsilon_r+1}{2\pi\epsilon_r}\left[\ln\left(\frac{w}{2h}+0.94\right) + 1.451\right] + \frac{\epsilon_r-1}{\epsilon_r^2}0.082} \tag{3.68}$$

Though the effective dielectric constant was derived under the assumption of wide strips ($w \geq 2h$), it is shown that the result also delivers sufficient accuracy for shape ratios $w/h \leq 2$.

As has been shown by Wheeler, formulas for the effective dielectric constant and the characteristic impedance can be derived for narrow strips ($w < 2h$) in a similar way using conformal mapping transformation. Here, we will only present the final relations for the effective dielectric constant and the characteristic impedance:

$$\sqrt{\epsilon_{eff}} = \sqrt{\frac{\epsilon_r+1}{2}}\left[1+\frac{1}{2}\frac{\epsilon_r-1}{\epsilon_r+1}\frac{\ln\left(\frac{\pi}{2}\right)+\frac{1}{\epsilon_r}\ln\left(\frac{4}{\pi}\right)}{\ln\left(\frac{8h}{w}\right)+\frac{1}{32}\left(\frac{w}{h}\right)^2-\frac{1}{2}\frac{\epsilon_r-1}{\epsilon_r+1}\left[\ln\left(\frac{\pi}{2}\right)+\frac{1}{\epsilon_r}\ln\left(\frac{4}{\pi}\right)\right]}\right] \tag{3.69}$$

$$Z_0(\epsilon_r) \approx \sqrt{\frac{\mu_0}{\epsilon_0}}\frac{1}{\pi\sqrt{2(\epsilon_r+1)}}\left[\ln\left(\frac{8h}{w}\right)+\frac{1}{32}\left(\frac{w}{h}\right)^2-\frac{1}{2}\frac{\epsilon_r-1}{\epsilon_r+1}\left(\ln\frac{\pi}{2}+\frac{1}{\epsilon_r}\ln\frac{4}{\pi}\right)\right] \tag{3.70}$$

Wheeler had claimed a relative accuracy of 1% for the derived formulas. The comparison with numerical results based on the integral equation with Green's function [17] has shown the following results.

Largest relative errors for ϵ_{eff} and Z_0 have been found in the transition region $w = 2h$.

Air ($\epsilon_r = 1$):

narrow strip: $\Delta Z_0 = 1.2\%$;
wide strip: $\Delta Z_0 = 0.1\%$.

Glass-reinforced Rexolite 2200 ($\epsilon_r = 2.62$):

narrow strip: $\Delta\epsilon_{eff} = 0.8\%$, $\Delta Z_0 = 1.4\%$;
wide strip: $\Delta\epsilon_{eff} = 2.2\%$, $\Delta Z_0 = -1.0\%$.

Fused silica quartz, SiO_2 ($\epsilon_r = 3.78$):

narrow strip: $\Delta\epsilon_{eff} = 0.8\%$, $\Delta Z_0 = 1.4\%$;
wide strip: $\Delta\epsilon_{eff} = 2.6\%$, $\Delta Z_0 = -1.2\%$.

Alumina, Al_2O_3 ($\epsilon_r = 9.6$):

narrow strip: $\Delta\epsilon_{eff} = 1.1\%$, $\Delta Z_0 = 1.3\%$;
wide strip: $\Delta\epsilon_{eff} = 3.2\%$, $\Delta Z_0 = -1.5\%$.

For $w/h \le 1$, the relative error for the given materials is $Z_0 \le 0.5$, and for $w/h > 5$, the error is <1%. The error increases with increasing ϵ_r value.

The following Figures 3.15 to 3.20 show the square root of the effective dielectric constant ϵ_{eff} and the characteristic impedance Z_0 for various geometries (w/h) and frequently used substrates of different dielectric constants (ϵ_r), such as GaAs, HRS, alumina (Al_2O_3), epoxy laminate, and polyolefin (Polyguide). Substrate data is given in Chapter 9. Notice that the graphs for wide and narrow strips do not match at the transition from narrow to wide strips ($w/h = 2$), and many proposals have been submitted to bridge the gap by defining some average value in the transition region. Here, we have set any modification of the original formulas aside. It can be noticed that the step in the squared dielectric constant and characterisitic impedance is marginal and mostly negligible with primary design attempts. The final circuit layout is based on the optimization using commercial software tools, which include advanced refined computer models for any kind of device in microstrip technique.

Wheeler's comprehensive treatise has provided reasonably simple yet very accurate closed-form equations for the effective dielectric constant ϵ_{eff} and the characteristic impedance $Z_0(\epsilon_r)$ of zero-thickness microstrip line.

The formulas given above are useful for analysis (i.e., determining the impedance from a given strip width). Usually the engineering problem is one of synthesis (i.e., determining the line width for a given impedance). Wheeler has published the following microstrip synthesis formulas [5]:

$$
w/h = \begin{cases} \dfrac{8}{e^{A} - 2e^{-A}} & (w/h \le 2) \\ \dfrac{2}{\pi}\left\{B - 1 - \ln(2B-1) + \dfrac{\epsilon_r - 1}{2\epsilon_r}\left[\ln(B-1) + 0.39 - \dfrac{0.61}{\epsilon_r}\right]\right\} & (w/h \ge 2) \end{cases} \tag{3.71}
$$

where

$$
A = \frac{\pi}{\eta_0}\sqrt{2(\epsilon_r + 1)} \cdot Z_0(\epsilon_r) + \frac{\epsilon_r - 1}{\epsilon_r + 1}\left(0.226 + \frac{0.120}{\epsilon_r}\right)
$$

and

$$
B = \frac{\pi \eta_0}{2\sqrt{\epsilon_r} \cdot Z_0(\epsilon_r)} \qquad (\eta_0 = 376.73\Omega)
$$

The relative error is not more than 1% [17]. This is because the synthesis formulas give better overlap in the transition between wide and narrow strips.

Problem 3.1: Determine the width w and dielectric constant ϵ_{eff} for a 50Ω microstrip transmission line for the following substrate materials: (a) Al_2O_3 ($\epsilon_r = 9.8$, $h = 0.635$ mm); (b) Polyguide ($\epsilon_r = 2.32$, $h = 1.59$ mm); (c) Epoxy laminate ($\epsilon_r = 3.8$, $h = 1.5$ mm).

Solution:
(a) Al_2O_3 ($\epsilon_r = 9.8$): From Figure 3.18 we read a w/h ratio of 0.98 for a characteristic impedance of 50Ω and substrate material with $\epsilon_r = 9.8$. With w/h known, we calculate the microstrip width as $0.98h = 0.62$ mm. From Figure 3.17, we read a value of 2.55 for $\sqrt{\epsilon_{eff}}$. The effective dielectric constant ϵ_{eff} is then calculated as 6.5.

(b) Polyguide ($\epsilon_r = 2.32$): As seen in Figure 3.18, a 50Ω microstrip line with $\epsilon_r = 2.32$ dielectric substrate material yields a w/h ratio of 2.96. Multiplication by the given substrate height delivers a strip width $w = 4.7$ mm. For the determined w/h ratio we evaluate a $\sqrt{\epsilon_{eff}}$ value of 1.425. The effective dielectric constant ϵ_{eff} turns out to be 2.03.

(c) Epoxy laminate ($\epsilon_r = 3.8$, $h = 1.5$ mm): From Figure 3.18 the w/h ratio for a 50Ω line ($\epsilon_r = 3.8$) is obtained as 2.0, resulting in a strip width of $w = 3.0$ mm. The w/h value results in a $\sqrt{\epsilon_{eff}}$ value of 1.75. The effective dielectric constant ϵ_{eff} is then calculated as 3.06.

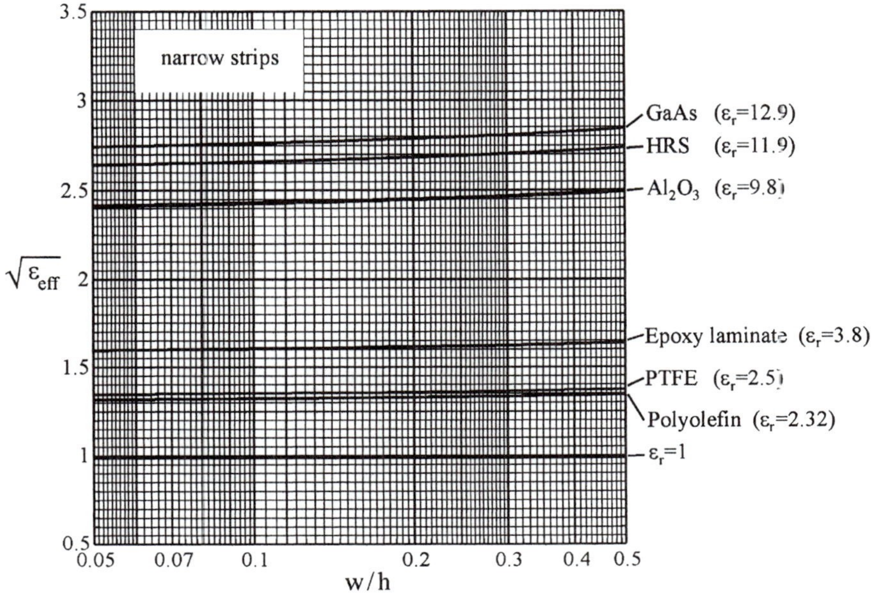

Figure 3.15 Narrow strips—Square root of the effective dielectric constant ε_{eff} as a function of *w/h* with ε_r as parameter.

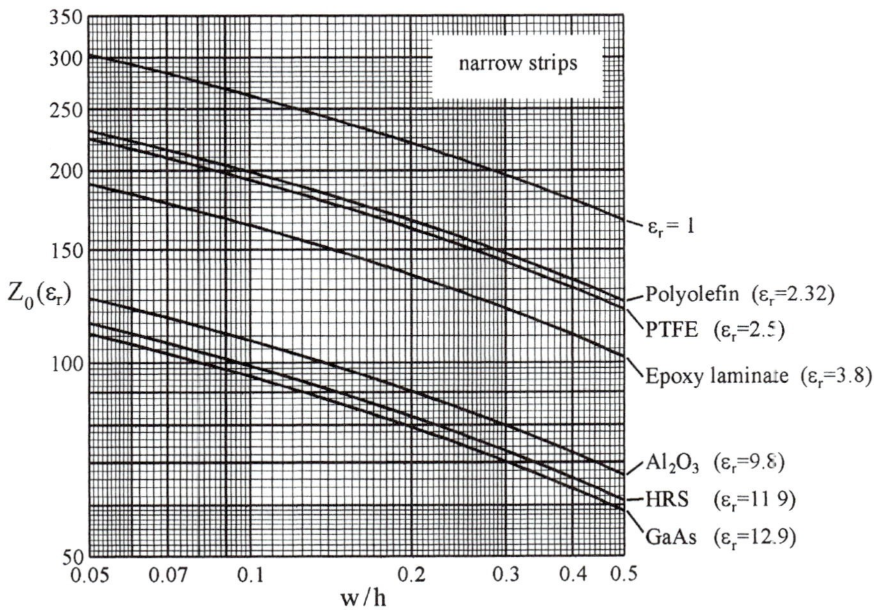

Figure 3.16 Narrow strips—$Z_0(\varepsilon_r)$ as function of *w/h* with ε_r as parameter.

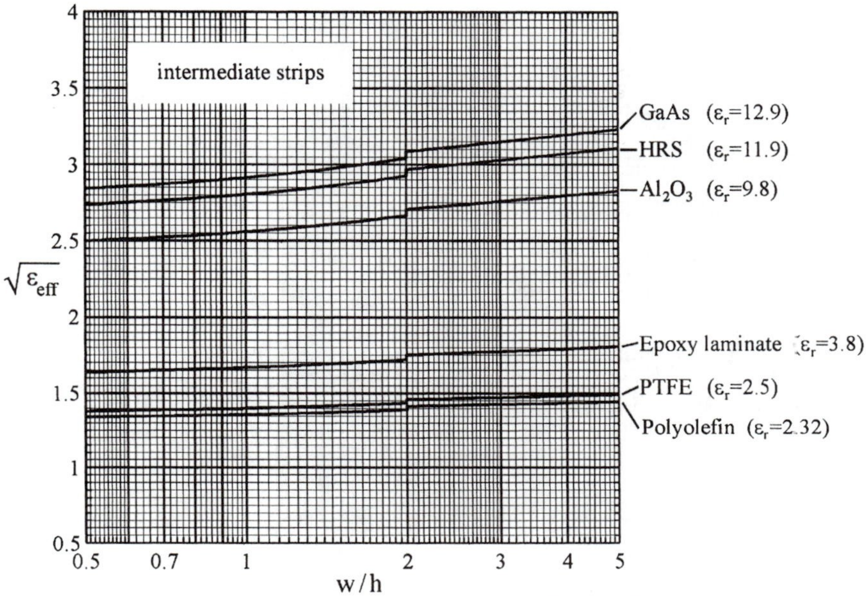

Figure 3.17 Intermediate strips—Square root of the effective dielectric constant ε_{eff} as a function of *w/h* with ε_r as parameter.

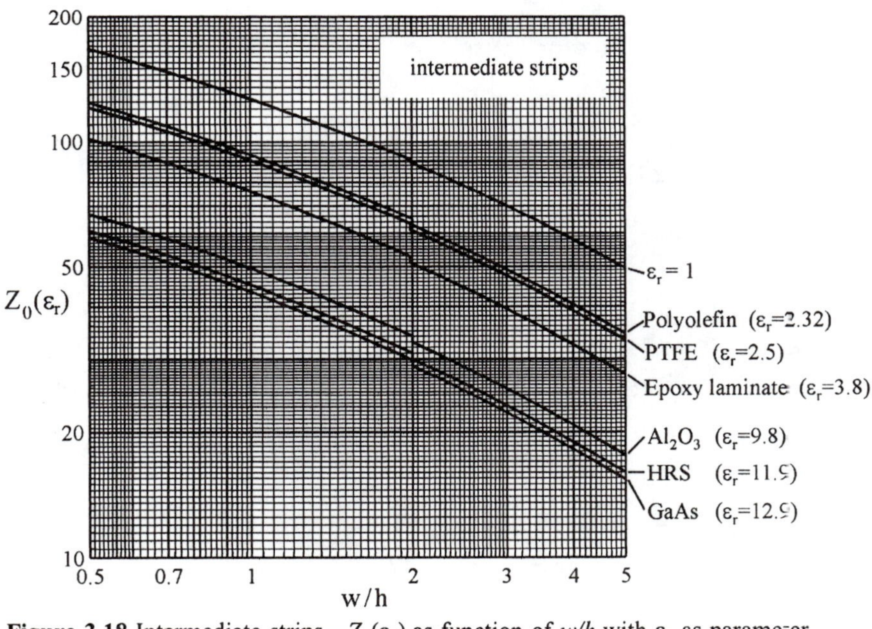

Figure 3.18 Intermediate strips—$Z_0(\varepsilon_r)$ as function of *w/h* with ε_r as parameter.

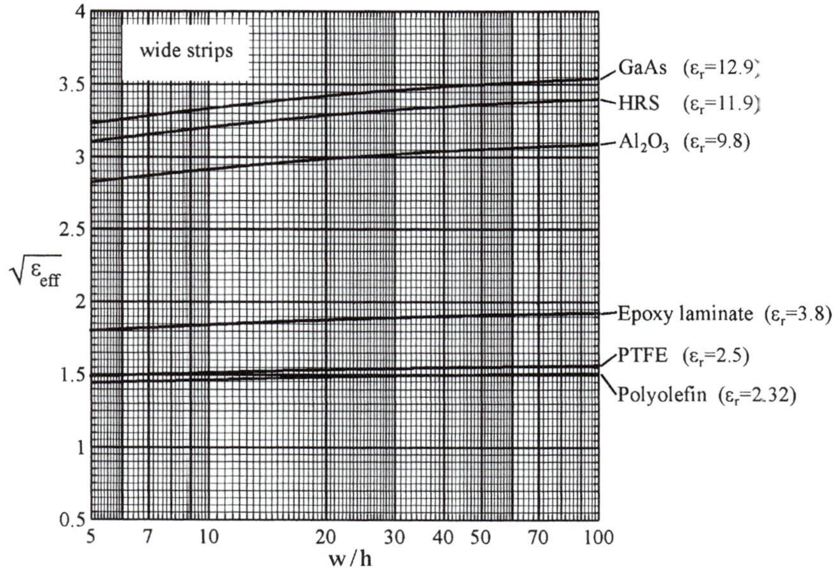

Figure 3.19 Wide strips—Square root of the effective dielectric constant ε_{eff} as a function of *w/h* with ε_r as parameter.

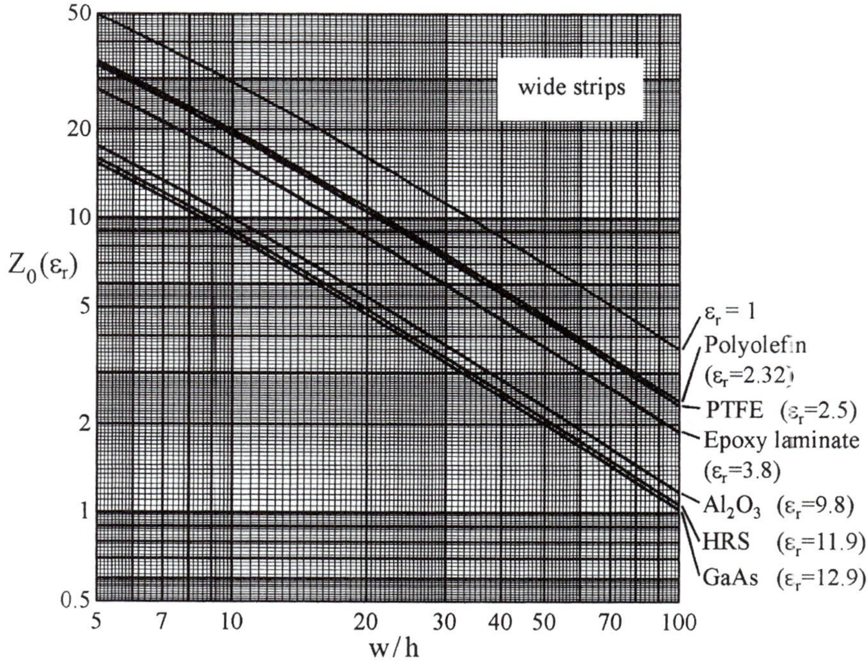

Figure 3.20 Wide strips—$Z_0(\varepsilon_r)$ as function of *w/h* with ε_r as parameter.

3.1.4 Effect of Strip Thickness

The derivation of the microstrip design formulas presented until now are limited to infinitely thin strips ($t = 0$). However, in practice the thickness is finite in the range up to hundreds of microns. The electrical field lines terminating at both sides of the strip cause an additional increase of the capacitance per unit length, and thus decrease the characteristic impedance of the microstrip transmission line. A small thickness can be compensated by a reduction of the width. While retaining the same separation h, the strip is taken to have a rectangular cross section of small thickness t less than the half-width $w/2$, and much less than the separation h. Each edge recedes by a small amount $\Delta w/2$, which is the edge correction to retain the same properties. We define a primed width $w' = w'(z)$,

$$w' = w + \Delta w \tag{3.72}$$

and assume that a microstrip line with the thickness $t = 0$ and primed width w' have the same characteristic impedance as the actual line with thickness $t \neq 0$ and width w (Figure 3.21). Notice that one has to distinguish carefully between the primed width w' and the effective width w_{eff}, after Wheeler. The quantity Δw can be found by comparison of the conformal mapping result of a microstrip line with a strip thickness $t = 0$ and $t \neq 0$.

This edge correction Δw is simply formulated after Wheeler for free-space ($\epsilon_r = 1$) as follows, for wide and narrow strips:

$$\Delta w = \begin{cases} \dfrac{t}{\pi}\left\{1 + \ln\left(\dfrac{2h}{t}\right)\right\} & \text{for} \quad \dfrac{w}{h} \geq \dfrac{1}{2\pi} \\[2ex] \dfrac{t}{\pi}\left\{1 + \ln\left(\dfrac{4\pi w}{t}\right)\right\} & \text{for} \quad \dfrac{w}{h} < \dfrac{1}{2\pi} \end{cases} \tag{3.73}$$

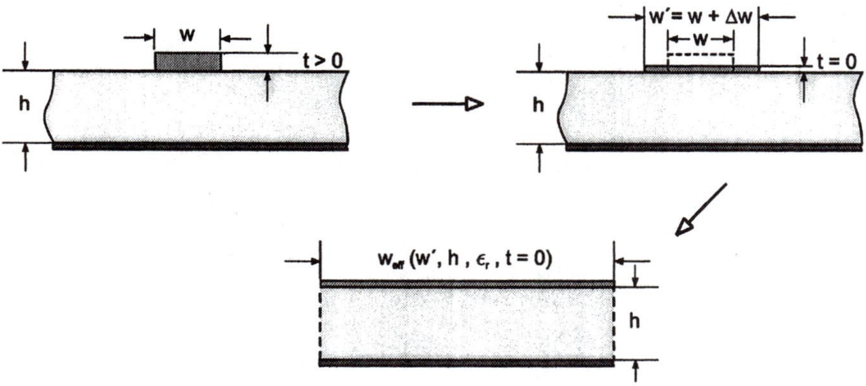

Figure 3.21 Influence of finite strip thickness on the effective width of microstrip line.

in the range of validity of

$$\frac{t}{\Delta w} < 0.75, \quad t < \frac{h}{2\pi}, \quad t < \frac{w}{2} \tag{3.74}$$

It is assumed that the strip thickness t is several times larger than the skin depth. As the loss increases rapidly for conductor thicknesses below three skin depths, it is recommended to maintain a conductor thickness of three to five skin depths [36]. The derived relations for the correction of the physical width of finite strip thicknesses can be used as first order approximations. Regarding microstrip lines on a substrate with a high dielectric constant ϵ_r, the microstrip mode will mainly propagate within the dielectric; that is, only a few electric field lines will terminate on the sides of the strip. In the extreme case of high ϵ_r values, the edge correction approaches zero ($\Delta w|_{\epsilon_r \to \infty} = 0$). Therefore, the equation above should be dependent on the dielectric constant. It is proposed that the edge correction Δw in these formulas is multiplied by $1/\sqrt{\epsilon_r}$, as a sensible interpolation between the extremes [5],

$$\Delta w_\epsilon \approx \Delta w / \sqrt{\epsilon_r} \tag{3.75}$$

In 1977, Wheeler published more acccurate approximations for thickness correction based on conformal mapping [37]. They cover the whole region $0 < w/h < \infty$, under the condition that $t/w < 1$ and $t/h < 1$. For the air-filled case,

$$\Delta w/t = \frac{1}{\pi} \ln \frac{4 \cdot e}{\sqrt{(t/h)^2 + \left(\dfrac{1/\pi}{w/t + 1.10}\right)^2}} \quad \text{(analysis)} \tag{3.76}$$

Using the synthesis formula, Δw can be calculated from the primed width w' and the strip thickness, as follows

$$\Delta w/t = \frac{1}{\pi} \ln \frac{4 \cdot e}{\sqrt{(t/h)^2 + \left(\dfrac{1/\pi}{w'/t - 0.26}\right)^2}} \quad \text{(synthesis)} \tag{3.77}$$

The relations (3.73) and (3.76) are represented in Figure 3.22 and Figure 3.23, respectively.

Problem 3.2: Specify the characteristic impedance $Z_0(\epsilon_r)$ of a microstrip line of width $w = 10$ μm and thickness $t = 5$ μm on lapped GaAs substrate ($\epsilon_r = 12.9$) of height $h = 100$ μm.

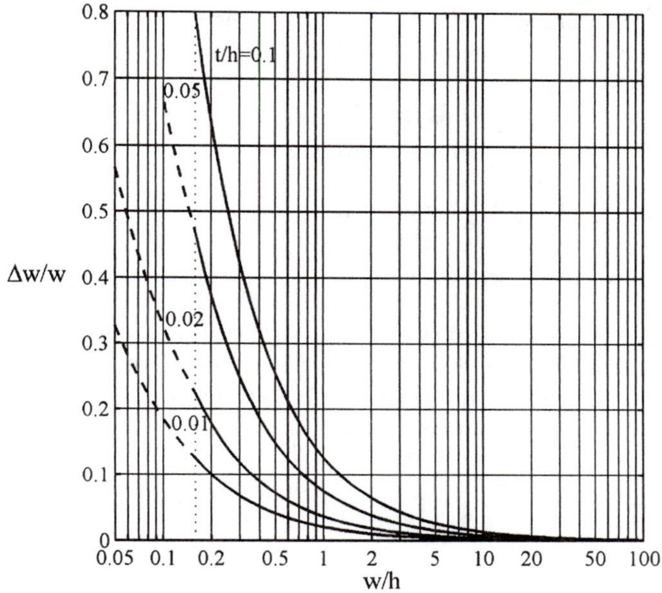

Figure 3.22 Effect of strip thickness versus *w/h* ratio with the parameter *t/h*, according to (3.73).

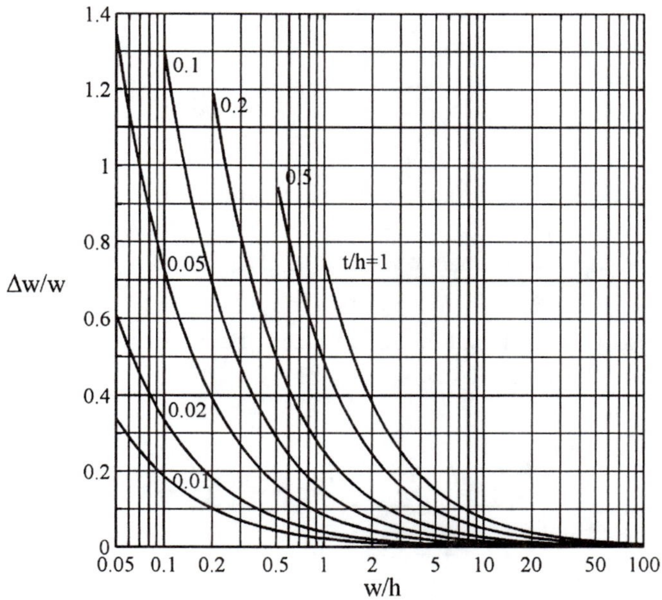

Figure 3.23 Effect of strip thickness versus *w/h* ratio with the parameter *t/h*, according to (3.76).

Solution: We apply the analysis formula (3.76), and first calculate the needed terms $t/h = 0.05$ and $w/t = 2$. Substituting these values in (3.76) results in $\Delta w/t = 1.45$. Then, the edge correction reads $\Delta w = 7.25$ μm. Taking into account the dielectric substrate correction, we get from (3.75) $\Delta w_\epsilon \approx 7.25\,\mu\text{m}/\sqrt{\epsilon_r} = 2\,\mu\text{m}$. The equivalent strip width w' that determines the characteristic impedance is thus calculated as $w' = w + \Delta w = 10\,\mu\text{m} + 2\,\mu\text{m} = 12\,\mu\text{m}$. The shape ratio w'/h then reads 0.12. From Figure 3.14, we obtain from the curve with $\epsilon_r = 12.9$ a characteristic impedance value of 91Ω. Notice, without taking into account the finite thickness, a characteristic value of 95Ω would have been obtained.

Problem 3.3: Specify the strip width of a microstrip transmission line on PTFE substrate (h = 1.0 mm, $\epsilon_r = 2.5$) with a characteristic impedance of 50Ω. The thickness of metallization is t = 100 μm.

Solution: From Wheeler's graph (Figure 3.18), we obtain for $Z_0\,(\epsilon_r = 2.5)$ a shape ratio w/h = 2.85, and $w = 2.85h$, respectively. This would be the convenient strip for zero-thickness of metallization. The finite thickness, however, causes additional fringing field and therefore effective extension of the strip width. In order to maintain the 50Ω, we have to reduce the width w' determined from Wheeler's graph by the edge correction effect. For this, we use the synthesis formula (3.77). With $t/h = 100\,\mu\text{m}/1.0\,\text{mm} = 0.1$, and $w'/t = 2.85\,\text{mm}/100\,\mu\text{m} = 28.5$, we obtain $\Delta w/h$ = 1.49, and $\Delta w = 1.49h \approx 150$ μm, respectively. Considering the substrate influence, we get $\Delta w_\epsilon = \Delta w/\sqrt{2.5} \approx 90$ μm. Then, the width for the characteristic impedance of 50Ω is finally calculated as $w = w' - \Delta w_\epsilon$ = 2.85 mm - 0.09 mm = 2.76 mm.

3.1.5 Microstrip Inductance per Unit Length

Figure 3.24 shows the mapped cross section of a microstrip line with effective width w_{eff}, effective dielectric constant ϵ_{eff}, and height h after Wheeler. The magnetic flux per unit length ϕ' of the mapped microstrip line relates to the strip current I and the magnetic inductance B,

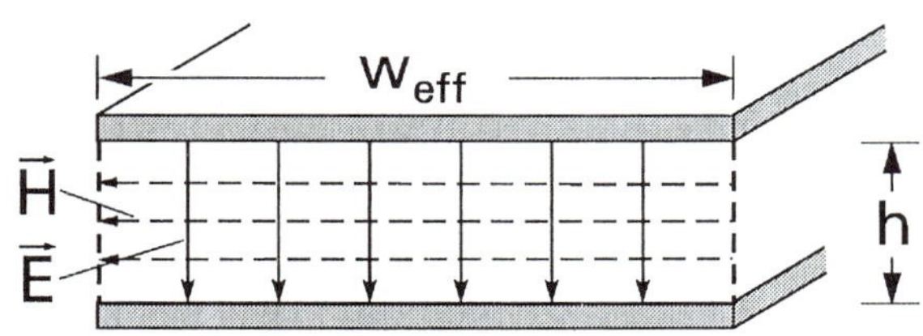

Figure 3.24 Mapped parallel plate transmission line after Wheeler.

$$\phi' = L'I \tag{3.78}$$

and

$$\phi' = B h = \mu_0 H h \tag{3.79}$$

With

$$I = \oint \vec{H} \cdot d\vec{s} = H w_{eff} \tag{3.80}$$

the inductance per unit length L' can be written as

$$L' = \frac{\phi'}{I} = \mu_0 \frac{h}{w_{eff}} \tag{3.81}$$

L' depends only upon the microstrip geometry and is independent of the shape and the dielectric properties of the substrate carrier. With

$$Z_0(\epsilon_r) = \frac{120\pi}{\sqrt{\epsilon_{eff}}} \frac{h}{w_{eff}} \ \Omega \tag{3.82}$$

L' also can be calculated from the air-line characteristic impedance and velocity of light in free space. With

$$Z_0(\epsilon_r) = \sqrt{\frac{L'}{C'}} = \sqrt{\frac{L'L'}{C'L'}} = L' v_{ph} \tag{3.83}$$

it follows

$$L' = \frac{Z_0(\epsilon_r)}{v_{ph}} = \frac{Z_0(\epsilon_r = 1)}{\sqrt{\epsilon_{eff}}} \frac{\sqrt{\epsilon_{eff}}}{c_0} \tag{3.84}$$

$$= \frac{Z_0(\epsilon_r = 1)}{c_0} \tag{3.85}$$

Figure 3.25 shows the calculated inductance per unit length L' for a uniform microstrip transmission line with conductor widths $0 \le w/h \le 10$. In addition, the capacitance C' per unit length calculated from (2.6) is shown in Figure 3.26 for $\epsilon_r = 2.3$ and $\epsilon_r = 9.6$. Dispersion is not included. Calculation was done using the formulas after Hammerstad (Section 3.3).

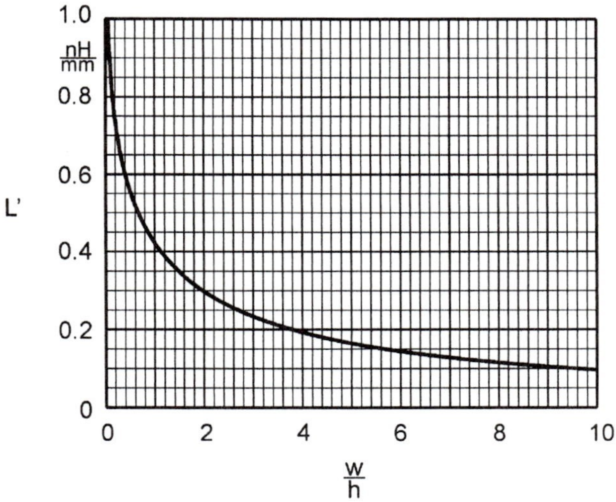

Figure 3.25 Inductance per unit length of a homogeneous microstrip line.

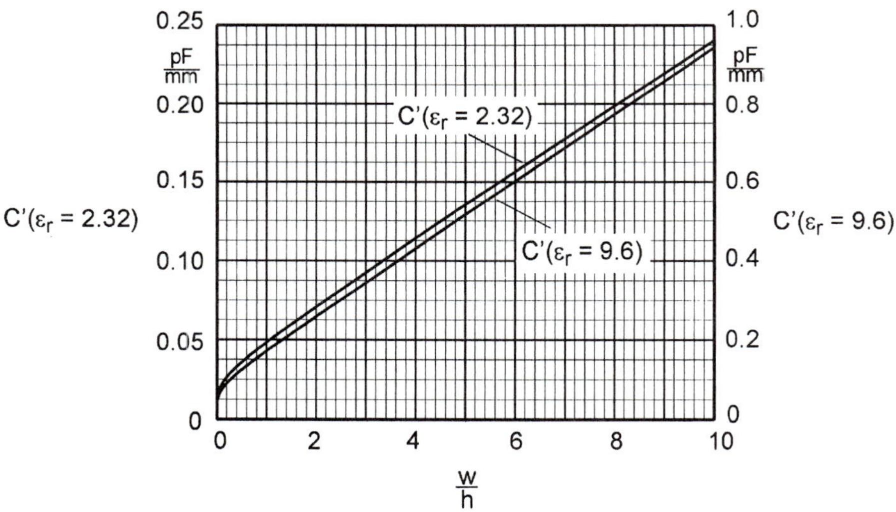

Figure 3.26 Capacitance per unit length of a homogeneous microstrip line.

Problem 3.3: A microstrip transmission line on Polyguide substrate material ($\epsilon_r = 2.32$, $h = 1.56$ mm) has a characteristic impedance of $Z_0 = 35\Omega$. Determine the inductance and capacitance per unit length L' and C', respectively.

Solution: From Figure 3.18, we obtain for a 35Ω line a shape ratio $w/h = 4.95$, and width $w = 4.95h = 7.821$ mm, respectively. The air-filled characteristic impedance value reads 50Ω (Figure 3.18). Applying (3.85), we calculate an inductance per unit length $L' = 50\Omega/(30\text{ GHz cm}) = 0.167$ nH/mm. Alternatively, this value can directly be taken from Figure 3.25 for the given w/h ratio.

The capacitance per unit length C' is calculated from the characteristic impedance and the effective dielectric constant as $C' = 1/(v_{ph} Z_0(\epsilon_r)) = \sqrt{\epsilon_{eff}}/(c_0 Z_0(\epsilon_r))$. For $w/h = 4.95$, we take from Figure 3.15 a value for the effective dielectric constant $\sqrt{\epsilon_{eff}} = 1.4$. Then C' is calculated as $C' = 1.4/(30$ GHz cm $35\Omega) = 1.33$ pF/cm. This value can also approximately be taken directly from Figure 3.26.

3.2 Theory After Schneider

In the special case of a single-dielectric microstrip with zero-thickness strip, a rigorous analytic solution for the characteristic impedance of a microstrip line can be found by conformal mapping transformation [38]. In Section 3.1, we have seen that the Schwarz-Christoffel mapping theorem is very useful for tackling the problem. We have found that the formulation of the mapping gradient is rather simple. However, the integration leads to elliptic functions that can only be evaluated numerically. An exact solution is very useful for checking other methods of microstrip analysis.

3.2.1 Single-Dielectric Microstrip Line

In the following section we suppose a single-dielectric microstrip line with $\epsilon_r = 1$ (microstrip air-line). We will see that the Schwarz-Christoffel transformation described in Appendix A is expressed in terms of theta functions and complete elliptic integrals. Because these mathematical terms are not commonly used, we will summerize some essential features.

The basic theta function is defined by the Fourier series [39, 40]

$$\theta(z,t) = \sum_{n=-\infty}^{+\infty} e^{-j\pi t n^2} e^{j\pi n(2z+1)} \tag{3.86}$$

with a complex variable z and complex parameter t satisfying the condition $Re(t) > 0$. $f(z)$ is periodic with period π $[f(z+\pi) = f(z)]$. In solving mathematical problems, it has been proven convenient to introduce four variants of (3.86), theta functions θ_1, θ_2, θ_3, and θ_4, which are interrelated to each other. These theta functions were introduced by Jacobi, a reason why they are also termed Jacobian theta functions. These functions and their logarithmic derivatives are essential mathematical tools for solving conformal mapping problems. In Schneider's theory, the first and fourth theta functions $\theta_1(z,t)$ and $\theta_4(z,t)$ are of special importance. They are defined as [40]

$$\theta_1(z,t) = j\sum_{n=0}^{+\infty}(-1)^n e^{j\pi t(n-1/2)^2} e^{j\pi(2n-1)z} = 2\sum_{n=0}^{+\infty}(-1)^n q^{(n+1/2)^2}\sin[(2n+1)\pi z] \tag{3.87}$$

$$\theta_4(z,t) = \sum_{n=-\infty}^{+\infty} e^{-j\pi t n^2} e^{j\pi n(2z+1)} = 1 + \sum_{n=1}^{+\infty}(-1)^n 2q^{n^2}\cos(2\pi nz) \tag{3.88}$$

with $q = \exp(j\pi t)$. For most applications, the parameter t is purely imaginary as given for the discussed mapping problem. The logarithm of the fourth theta function, which is also needed in the theory, reads [40]

$$\ln\theta_4(z,t) = \ln C - 2\sum_{n=1}^{+\infty}\frac{q^n}{1-q^{2n}}\frac{\cos(2\pi nz)}{n} \tag{3.89}$$

with Euler's constant $C = 0.577215$. Equation (3.89) converges rapidly [40].

As was already mentioned in Section 3.1, the exact mapping transformation of the microstrip onto a rectangle includes also complete elliptic integrals. The term "complete" denotes the upper integration limit $x = 1$. The complete elliptic integral of the first kind with modulus m is designated as [40–43]

$$K(m) = \int_0^1 [(1-t^2)(1-mt^2)]^{-1/2}\,dt \tag{3.90}$$

and the complete elliptic integral of the second kind with modulus m is defined as

$$E(m) = \int_0^1 [(1-t^2)^{-1/2}(1-mt^2)]^{1/2}\,dt \tag{3.91}$$

It is also defined as [41]

$$K' = K(m') = K(1-m) \tag{3.92}$$

and

$$E' = E(m') = E(1-m) \tag{3.93}$$

with $m' = 1 - m$. Figure 3.27 shows the complete elliptic integrals $K(m)$ and $K'(m)$ versus the modulus m within the interval $0 \le m \le 1$. Similarly, Figure 3.28 shows the complete elliptic integrals $E(m)$ and $E'(m)$.

After having reviewed the most essential relations of elliptic and theta functions, we may now attend to Schneider's theory. Figure 3.29 shows the transformation of a single-dielectric microstrip line with the two strip boundaries in the z-plane onto a rectangle in the w-plane with top and bottom conductive plates. The

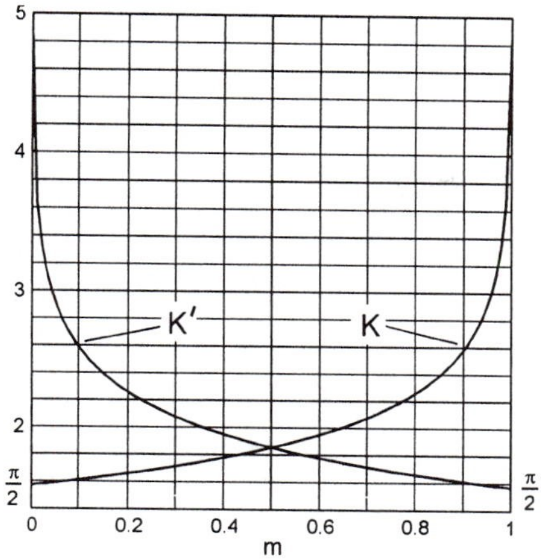

Figure 3.27 Complete elliptic integrals of the first kind $K(m)$ and $K'(m)$ with modulus m.

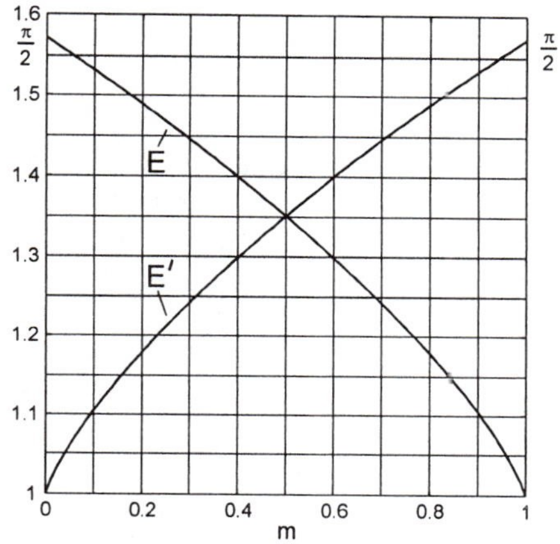

Figure 3.28 Complete elliptic integrals of the second kind $E(m)$ and $E'(m)$ with modulus m.

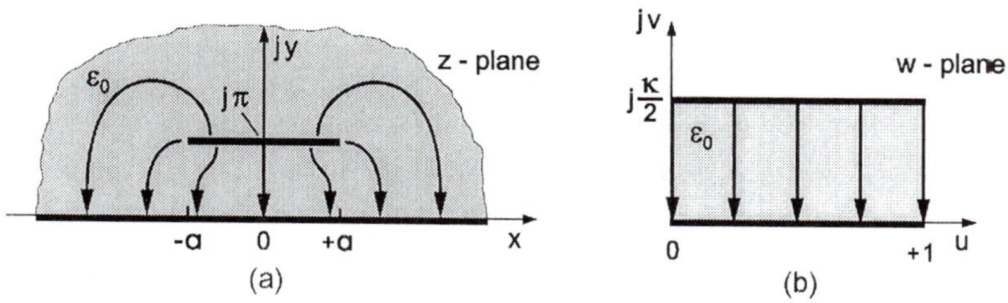

Figure 3.29 Normalized geometric parameters of (a) the microstrip and (b) its mapping onto a parallel plate line.

application of the Schwarz-Christoffel theorem results in the following mapping function

$$z = -2K \frac{\partial}{\partial w} \ln\left[\theta_1(w,\kappa)\right] \tag{3.94}$$

where $\theta_1(w,\kappa)$ is the first theta function of complex variable $w = u + jv$ and the real parameter κ. κ is the ratio of the complete elliptic integrals of the first kind $K(m)$ and $K'(m)$ with modulus m,

$$\kappa = K'/K \tag{3.95}$$

The characteristic impedance of the air-filled microstrip $Z_0(\epsilon_r = 1)$ with width w, height h, and thickness $t = 0$ is obtained by solving the following set of equations:

$$\frac{w}{h} = \frac{2}{\pi} \frac{\partial}{\partial \zeta} \ln \theta_4(\zeta,\kappa) \tag{3.96}$$

$$dn^2(2K\zeta) = E/K \tag{3.97}$$

$$Z_0(\epsilon_r = 1) = \frac{1}{2}\sqrt{\frac{\mu_0}{\epsilon_0}}\frac{K'}{K} = 60\,\pi\,\Omega\,\frac{K'}{K} \tag{3.98}$$

where dn is the Jacobian elliptic function with parameter m. The logarithmic derivative of theta function $\theta_4(\zeta,\kappa)$ in (3.96) is directly obtained from (3.89) and reads

$$\frac{\partial}{\partial\zeta}\ln\theta_4(\zeta,\kappa) = 4\pi\sum_{n=1}^{\infty}\frac{\sin(2n\pi\zeta)}{\exp(n\pi\kappa) - \exp(-n\pi\kappa)} \tag{3.99}$$

Figure 3.30 illustrates the rapid convergence of the series for chosen values $m = 0.1, 0.7, 0.9$. It is shown that with only four terms, the limit values have been attained.

Rigorous Solution for Computing Z_0

Equations (3.95) to (3.99) provide an implicit relationship between the characteristic impedance $Z_0(\epsilon_r = 1)$ and the line parameters w and h. The dependence of the characteristic impedance Z_0 on width w and height h can be found as follows. The procedure is illustrated in Figure 3.31. For a given modulus m, the elliptic integrals K, K', and E are calculated. Then, the ratio $\kappa = K'/K$ is known, and the characteristic impedance as function of m can be calculated.

The intermediate variable ζ of the Jacobian elliptic function in (3.97) can be calculated by the known ratio E/K using look-up tables for the Jacobian elliptic functions. For known ζ and κ, the shape ratio w/h of the microstrip line can be

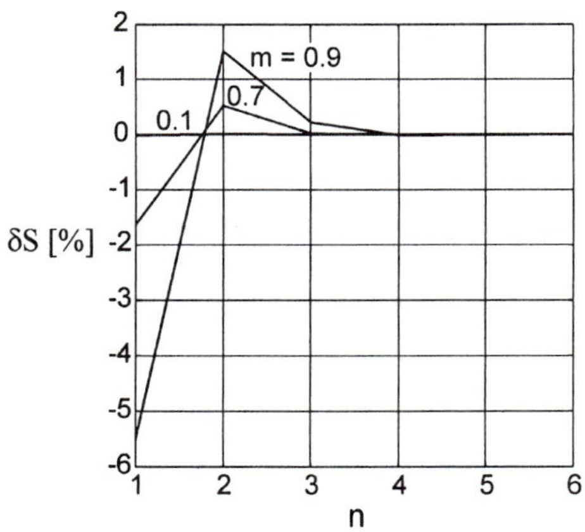

Figure 3.30 Convergence analysis of series expansion of the logarithmic derivative of theta function (3.99). δS denotes the deviation from the limit value.

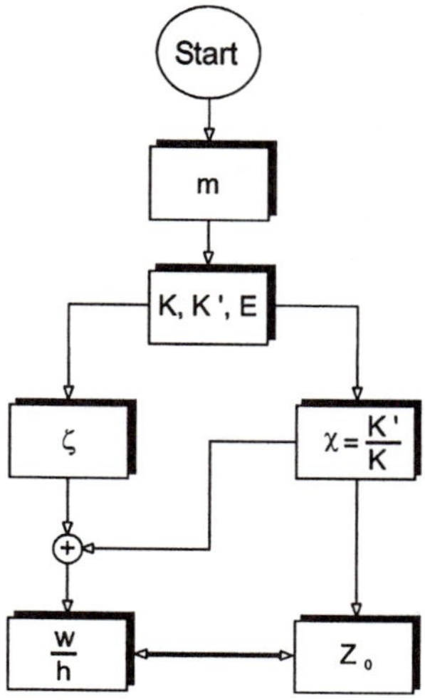

Figure 3.31 Flow graph to calculate $Z_0 = f(w/h)$.

calculated from (3.96) using the approximation (3.99). Thus, for the calculated impedance Z_0 with given modulus m, we have found the corresponding w/h ratio. Multiple repetition of the procedure delivers the function $Z_0 = Z_0(w/h)$ in the form of a look-up table. Intermediate values can be obtained by interpolation. In the past, a solution of the set of equations could be obtained with the help of numerical tables [44]. However, as is shown in the following, Schneider's equations can simply be evaluated using advanced programming like MATLAB [The MathWorks, Inc., http://www.mathworks.com, MATLAB Version 7.0.1 (R14), September 13, 2004], which comprises all elliptic functions needed for impedance calculation. Lowest impedance value is limited to >15.2559Ω corresponding to a modulus value of $m < 1 - 2\exp(-52)$ for double precision format. The program follows the signal flow graph in Figure 3.31:

```
% program_schneider
% exact calculation of Z0 and w/h for m = 0...1, step 0.01,
% on the basis of elliptic functions.
double precision
for n =1:101
   m = (n-1)/100;
   if (m==1)
      m=1-eps;
   end;
   [Z0, w_h]=schneider(m)
end;
% function [Z0, w_h]=schneider(m)
```

```
%     calculation of Z0 and w/h for a given modulus
%     lower limit value of w/h and Z0 due to
%     limited accuracy of MATLAB: m < 1-(2exp-52)
%     i.e. w/h < 21.3187 and Z0 > 15.2559
function [Z0,w_h,K,Kp,E,kappa,dn,u,zeta,sum]=schneider(m)
  % calculating elliptic integrals
  format long
  [K,E]=ellipke(m);
  [Kp,Ep]=ellipke(1-m);
  kappa=Kp./K;
  % calculating characteristic impedance
  Z0=60*pi*kappa;
  % calculating w/h
  dn=sqrt(E./K);
  u=arcdn(dn, m);
  zeta=u./(2*K);
  % calculating the series
  N=30;
  S=zeros(N,length(m));
  S(1,:)=(exp(-pi*kappa).*sin(2*pi*zeta)./ ...
      (1-exp(-2*pi*kappa)));
  for n=2:N
    S(n,:)=S(n-1,:)+(exp(-n*pi*kappa).* ...
        sin(2*n*pi*zeta)./(1-exp(-2*n*pi*kappa)));
  end;
  sum=S(N,:);
  w_h=8*sum;
% endfunction

function u=arcdn(dn_, m)
% function u=arcdn(dn, m)
%     Calculates argument u of the Jacobi elliptic function dn
%     for a given value of the function dn and modulus m
% opt=optimset('Display','iter');
  u=zeros(1,length(dn_));
  [K,E]=ellipke(m);
  for i=1:length(dn_)
    u(i)=fzero(['dn(x, ' num2str(m(i)) ')-' num2str(dn_(i))],[0 K(i)]);
  end
% endfunction

function dn=dn(u,m)
% function dn=dn(u,m)
%      Interface for the function arcdn
```

```
  [sn,cn,dn]=ellipj(u,m);
% endfunction
```

The results are given in Table 3.1.

Table 3.1

Rigorous Calculation of Air-Filled Characteristic Impedance $Z_0(\varepsilon_r=1)$ after Schneider Based or Elliptic Integrals

m	Z_0	w/h	K	K'	E	κ	dn	$arc\ dn$	ζ	S
0.00	∞	0.0000	1.571	∞	1.571	∞	1.000	0.000	0.000	0.000
0.01	442.3643	0.0050	1.575	3.696	1.567	2.347	0.997	0.787	0.250	0.001
0.02	400.4718	0.0101	1.579	3.354	1.563	2.125	0.995	0.789	0.250	0.001
0.03	375.8376	0.0152	1.583	3.156	1.559	1.994	0.992	0.790	0.249	0.002
0.04	358.2679	0.0204	1.587	3.016	1.555	1.901	0.990	0.791	0.249	0.003
0.05	344.5679	0.0256	1.591	2.908	1.551	1.828	0.987	0.792	0.249	0.003
0.06	333.3145	0.0309	1.595	2.821	1.547	1.768	0.985	0.794	0.249	0.004
0.07	323.7486	0.0363	1.599	2.747	1.543	1.718	0.982	0.795	0.249	0.005
0.08	315.4171	0.0417	1.604	2.684	1.539	1.673	0.980	0.797	0.248	0.005
0.09	308.0278	0.0472	1.608	2.628	1.535	1.634	0.977	0.798	0.248	0.006
0.10	301.3808	0.0527	1.612	2.578	1.531	1.599	0.974	0.799	0.248	0.007
0.11	295.3340	0.0583	1.617	2.533	1.527	1.567	0.972	0.801	0.248	0.007
0.12	289.7821	0.0639	1.621	2.493	1.523	1.537	0.969	0.802	0.247	0.008
0.13	284.6452	0.0696	1.626	2.455	1.518	1.510	0.966	0.804	0.247	0.009
0.14	279.8614	0.0754	1.631	2.421	1.514	1.485	0.964	0.806	0.247	0.009
0.15	275.3812	0.0813	1.635	2.389	1.510	1.461	0.961	0.807	0.247	0.010
0.16	271.1651	0.0872	1.640	2.359	1.506	1.439	0.958	0.809	0.247	0.011
0.17	267.1805	0.0932	1.645	2.331	1.502	1.417	0.956	0.810	0.246	0.012
0.18	263.4004	0.0992	1.650	2.305	1.498	1.397	0.953	0.812	0.246	0.012
0.19	259.8024	0.1054	1.655	2.281	1.493	1.378	0.950	0.813	0.246	0.013
0.20	256.3673	0.1116	1.660	2.257	1.489	1.360	0.947	0.815	0.246	0.014
0.21	253.0787	0.1179	1.665	2.235	1.485	1.343	0.944	0.817	0.245	0.015
0.22	249.9227	0.1243	1.670	2.214	1.480	1.326	0.942	0.818	0.245	0.016
0.23	246.8870	0.1307	1.675	2.194	1.476	1.310	0.939	0.820	0.245	0.016
0.24	243.9609	0.1372	1.680	2.175	1.472	1.294	0.936	0.822	0.245	0.017
0.25	241.1351	0.1439	1.686	2.157	1.467	1.279	0.933	0.824	0.244	0.018
0.26	238.4014	0.1506	1.691	2.139	1.463	1.265	0.930	0.825	0.244	0.019
0.27	235.7523	0.1574	1.697	2.122	1.459	1.251	0.927	0.827	0.244	0.020
0.28	233.1814	0.1643	1.702	2.106	1.454	1.237	0.924	0.829	0.243	0.021
0.29	230.6827	0.1713	1.708	2.090	1.450	1.224	0.921	0.831	0.243	0.021
0.30	228.2509	0.1784	1.714	2.075	1.445	1.211	0.918	0.833	0.243	0.022
0.31	225.8812	0.1856	1.720	2.061	1.441	1.198	0.915	0.835	0.243	0.023
0.32	223.5693	0.1929	1.726	2.047	1.436	1.186	0.912	0.836	0.242	0.024
0.33	221.3112	0.2003	1.732	2.033	1.432	1.174	0.909	0.838	0.242	0.025
0.34	219.1033	0.2079	1.738	2.020	1.427	1.162	0.906	0.840	0.242	0.026
0.35	216.9423	0.2155	1.744	2.008	1.423	1.151	0.903	0.842	0.241	0.027

m	Z_0	w/h	K	K'	E	κ	dn	$arc\ dn$	ζ	S
0.36	214.8250	0.2233	1.751	1.995	1.418	1.140	0.900	0.844	0.241	0.028
0.37	212.7487	0.2311	1.757	1.983	1.413	1.129	0.897	0.846	0.241	0.029
0.38	210.7108	0.2392	1.764	1.972	1.409	1.118	0.894	0.848	0.241	0.030
0.39	208.7087	0.2473	1.771	1.961	1.404	1.107	0.890	0.851	0.240	0.031
0.40	206.7403	0.2556	1.778	1.950	1.399	1.097	0.887	0.853	0.240	0.032
0.41	204.8034	0.2640	1.785	1.939	1.395	1.087	0.884	0.855	0.240	0.033
0.42	202.8960	0.2726	1.792	1.929	1.390	1.076	0.881	0.857	0.239	0.034
0.43	201.0163	0.2813	1.799	1.918	1.385	1.066	0.877	0.859	0.239	0.035
0.44	199.1625	0.2902	1.806	1.909	1.380	1.057	0.874	0.862	0.239	0.036
0.45	197.3329	0.2992	1.814	1.899	1.375	1.047	0.871	0.864	0.238	0.037
0.46	195.5259	0.3084	1.822	1.890	1.371	1.037	0.867	0.866	0.238	0.039
0.47	193.7401	0.3178	1.829	1.880	1.366	1.028	0.864	0.869	0.237	0.040
0.48	191.9740	0.3273	1.837	1.871	1.361	1.018	0.861	0.871	0.237	0.041
0.49	190.2263	0.3371	1.846	1.863	1.356	1.009	0.857	0.874	0.237	0.042
0.50	188.4956	0.3470	1.854	1.854	1.351	1.000	0.854	0.876	0.236	0.043
0.51	186.7806	0.3571	1.863	1.846	1.346	0.991	0.850	0.879	0.236	0.045
0.52	185.0801	0.3675	1.871	1.837	1.341	0.982	0.846	0.881	0.235	0.046
0.53	183.3930	0.3781	1.880	1.829	1.335	0.973	0.843	0.884	0.235	0.047
0.54	181.7180	0.3889	1.890	1.822	1.330	0.964	0.839	0.887	0.235	0.049
0.55	180.0540	0.3999	1.899	1.814	1.325	0.955	0.835	0.890	0.234	0.050
0.56	178.4000	0.4112	1.909	1.806	1.320	0.946	0.832	0.892	0.234	0.051
0.57	176.7547	0.4228	1.918	1.799	1.315	0.938	0.828	0.895	0.233	0.053
0.58	175.1172	0.4346	1.929	1.792	1.309	0.929	0.824	0.898	0.233	0.054
0.59	173.4863	0.4467	1.939	1.785	1.304	0.920	0.820	0.901	0.232	0.056
0.60	171.8609	0.4591	1.950	1.778	1.298	0.912	0.816	0.904	0.232	0.057
0.61	170.2400	0.4719	1.961	1.771	1.293	0.903	0.812	0.907	0.231	0.059
0.62	168.6225	0.4849	1.972	1.764	1.287	0.895	0.808	0.911	0.231	0.061
0.63	167.0072	0.4984	1.983	1.757	1.282	0.886	0.804	0.914	0.230	0.062
0.64	165.3931	0.5122	1.995	1.751	1.276	0.877	0.800	0.917	0.230	0.064
0.65	163.7789	0.5264	2.008	1.744	1.271	0.869	0.796	0.921	0.229	0.066
0.66	162.1636	0.5410	2.020	1.738	1.265	0.860	0.791	0.924	0.229	0.068
0.67	160.5458	0.5561	2.033	1.732	1.259	0.852	0.787	0.928	0.228	0.070
0.68	158.9242	0.5716	2.047	1.726	1.253	0.843	0.783	0.932	0.228	0.071
0.69	157.2976	0.5876	2.061	1.720	1.248	0.834	0.778	0.936	0.227	0.073
0.70	155.6646	0.6042	2.075	1.714	1.242	0.826	0.773	0.940	0.226	0.076
0.71	154.0236	0.6213	2.090	1.708	1.236	0.817	0.769	0.944	0.226	0.078
0.72	152.3731	0.6391	2.106	1.702	1.230	0.808	0.764	0.948	0.225	0.080
0.73	150.7115	0.6575	2.122	1.697	1.224	0.800	0.759	0.952	0.224	0.082
0.74	149.0368	0.6766	2.139	1.691	1.217	0.791	0.754	0.957	0.224	0.085
0.75	147.3472	0.6965	2.157	1.686	1.211	0.782	0.749	0.962	0.223	0.087
0.76	145.6405	0.7172	2.175	1.680	1.205	0.773	0.744	0.966	0.222	0.090
0.77	143.9144	0.7388	2.194	1.675	1.198	0.763	0.739	0.971	0.221	0.092
0.78	142.1663	0.7614	2.214	1.670	1.192	0.754	0.734	0.977	0.221	0.095
0.79	140.3934	0.7850	2.235	1.665	1.185	0.745	0.728	0.982	0.220	0.098

m	Z_0	w/h	K	K'	E	κ	dn	arc dn	ζ	S
0.80	138.5925	0.8098	2.257	1.660	1.178	0.735	0.723	0.988	0.219	0.101
0.81	136.7600	0.8360	2.281	1.655	1.172	0.726	0.717	0.994	0.218	0.104
0.82	134.8919	0.8636	2.305	1.650	1.165	0.716	0.711	1.000	0.217	0.108
0.83	132.9834	0.8927	2.331	1.645	1.158	0.705	0.705	1.006	0.216	0.112
0.84	131.0293	0.9237	2.359	1.640	1.151	0.695	0.698	1.013	0.215	0.115
0.85	129.0232	0.9568	2.389	1.635	1.143	0.684	0.692	1.020	0.214	0.120
0.86	126.9578	0.9922	2.421	1.631	1.136	0.674	0.685	1.028	0.212	0.124
0.87	124.8241	1.0302	2.455	1.626	1.128	0.662	0.678	1.036	0.211	0.129
0.88	122.6114	1.0714	2.493	1.621	1.121	0.650	0.671	1.045	0.210	0.134
0.89	120.3064	1.1163	2.533	1.617	1.113	0.638	0.663	1.054	0.208	0.140
0.90	117.8926	1.1655	2.578	1.612	1.105	0.625	0.655	1.064	0.206	0.146
0.91	115.3486	1.2201	2.628	1.608	1.096	0.612	0.646	1.076	0.205	0.153
0.92	112.6463	1.2813	2.684	1.604	1.088	0.598	0.637	1.088	0.203	0.160
0.93	109.7474	1.3508	2.747	1.599	1.079	0.582	0.627	1.101	0.200	0.169
0.94	106.5978	1.4314	2.821	1.595	1.070	0.566	0.616	1.117	0.198	0.179
0.95	103.1163	1.5271	2.908	1.591	1.060	0.547	0.604	1.134	0.195	0.191
0.96	99.1732	1.6449	3.016	1.587	1.051	0.526	0.590	1.155	0.192	0.206
0.97	94.5370	1.7976	3.156	1.583	1.040	0.502	0.574	1.182	0.187	0.225
0.98	88.7218	2.0147	3.354	1.579	1.029	0.471	0.554	1.217	0.181	0.252
0.99	80.3197	2.3904	3.696	1.575	1.016	0.426	0.524	1.273	0.172	0.299
1 - δ	15.2559	21.3187	19.408	1.571	1.000	0.081	0.227	2.163	0.056	2.665

Problem: We calculate the same example as given in [38]. We suppose a modulus $m = 0.86$. Then from Table 3.1, we read: $K = 2.421$, $K' = 1.631$, $E = 1.136$. Thus, κ is calculated as $\kappa = K'/K = 0.674$ so that the impedance yields 126.9578Ω. With E and K known, the Jacobian elliptic function is calculated as $dn = 0.685$. The inverse Jacobian elliptic function delivers $\text{arc}\, dn \sqrt{E/K} = \text{arc}\, dn(0.685) = 1.028$. Then ζ can be determined, $\zeta = 1.028/(2K) = 0.212$. The sum S is calculated as $S = 0.1240$ taking into account the first 30 terms, though only about four terms would be sufficient, as seen in Figure 3.28. Then the w/h is calculated to be 0.9922.

When Schneider proposed the calculation algorithm, the procedure was very laborious and not very convenient for practical design purpose. Therefore, using the exact solution as a reference, Schneider applied curvefitting procedure to obtain well-approximated explicit equations for the characteristic impedance of very wide ($w \gg h$) and very narrow ($w \ll h$) strips. The solution for these limit cases are well known and are expressed as follows [45]:

$$Z_0 \approx 60 \ln\left(\frac{8h}{w}\right) \ \Omega \qquad (w \ll h) \tag{3.100}$$

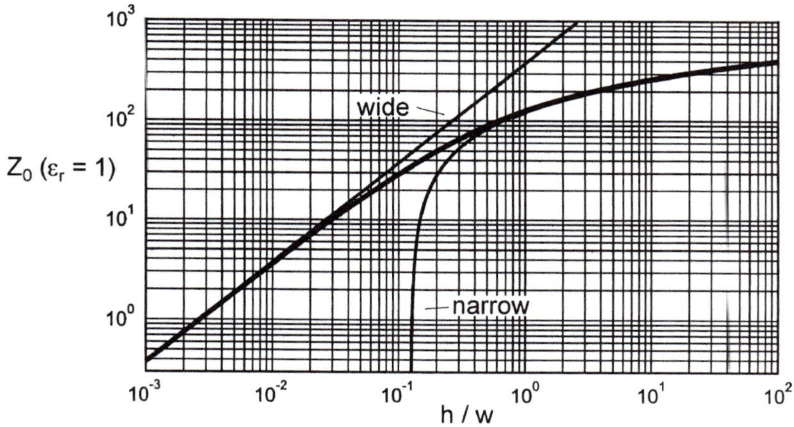

Figure 3.32 Asymptotic approximation for very small and wide strips.

$$Z_0 \approx 120\pi \frac{h}{w} \;\; \Omega \qquad (w \gg h) \tag{3.101}$$

The relations for the regarded asymptotic cases, illustrated in Figure 3.32, are used to describe the characteristic impedance Z_0 for intermediate cases. The following series expansions for narrow and wide strips have been proposed.

$w \leq h$:

$$Z_0 = 60 \ln \sum_{n=1}^{-\infty} a_n \left(\frac{h}{w}\right)^n \Omega = 60 \ln \left\{ a_1 \left(\frac{h}{w}\right) + a_0 + a_{-1}\left(\frac{w}{h}\right) + \dots \right\} \tag{3.102}$$

$w \geq h$:

$$Z_0 = \frac{120\pi}{\sum_{n=1}^{-\infty} b_n \left(\frac{w}{h}\right)^n} \Omega = \frac{120\pi}{b_1\left(\frac{w}{h}\right) + b_0 + b_{-1}\left(\frac{w}{h}\right)^{-1} + b_2\left(\frac{w}{h}\right)^{-2} \dots} \Omega \tag{3.103}$$

The number of terms after which the series is terminated determines the accuracy of the approximations. Since the exact solution, which contains elliptic integrals, is known, the expansion coefficients a_n and b_n can be chosen in such a way that the power series fits the exact solution within desired accuracy [38, 46]. As claimed by Schneider, the following expressions describe the characteristic impedance of the air-filled microstrip with an accuracy of ±0.25% with respect to the elliptic function–based exact solution:

$w \leq h$:

$$Z_0 \approx 60 \ln\left(\frac{8h}{w} + \frac{w}{4h}\right) \ \Omega \tag{3.104}$$

Thus, it is sufficient to consider only two nonzero expansion terms of (3.102), namely, $a_1 = 8$ and $a_{-1} = 1/4$.

$w \geq h$:

$$Z_0 \approx \frac{120\pi}{\frac{w}{h} + 2.42 - 0.44\frac{h}{w} + \left(1 - \frac{h}{w}\right)^6} \ \Omega \tag{3.105}$$

The fitting of the air-filled characteristic impedance for wide strips includes $(w/h)^n$ expansion terms with $-6 \leq n \leq 1$. The accuracy claimed by Schneider is $\pm 1.0\%$ for strip width $w/h > 10$.

3.2.2 Effective Dielectric Constant

Schneider has made the following assumption to derive an expression for the effective dielectric constant. The effective dielectric constant lies between unity and the dielectric constant ϵ_r. If the strip is extremely small ($w/h \ll 1$), then the electric field energy will approximately be equal in the air and dielectric region. Therefore, in this case, the effective dielectric constant may be roughly estimated as a mean value of air and dielectric constant [i.e., $\epsilon_{eff} = (\epsilon_r + 1)/2$]. With increasing w/h, the effective dielectric constant will approach the limit value of ϵ_r. Therefore, over the range $0 \leq w/h < \infty$ and $1 \leq \epsilon_r < \infty$, one can analytically describe the function $\epsilon_{eff}(w, h, \epsilon_r)$ as

$$\epsilon_{eff} = \frac{\epsilon_r + 1}{2} + \frac{\epsilon_r - 1}{2} F\left(\epsilon_r, \frac{w}{h}\right) \tag{3.106}$$

$F(\epsilon_r, w/h)$ is still an unknown function. For any value of ϵ_r, F should become 0 for $w/h > 0$, and 1 for $w/h = \infty$. The following series expansion fulfills these requirements:

$$F\left(\epsilon_r, \frac{w}{h}\right) = \left[1 + \sum_{n=1}^{\infty} c_n \left(\frac{h}{w}\right)^n\right]^m \tag{3.107}$$

with $m < 0$ and $c_n = c_n(\epsilon_r)$. The expansion coefficients c_n are determined by curvefitting of the function $\epsilon_{eff} = \epsilon_{eff}(w/h, \epsilon_r, t=0)$ proposed by Wheeler [5]. See

(3.65) in Section 3.1. It turns out that, to a good approximation, the coefficients are independent of the dielectric constant. This is also confirmed in [43] for $\epsilon_r \leq 16$. With $m = -0.5$ and one single term of the series, an accuracy of $\pm 2\%$ compared to Wheeler's results has been obtained. In [17], an accuracy of $\pm 2.3\%$ was found using numerical results based on the integral equation with Green's function. Thus, the function $F(w/h)$ can be written as

$$F\left(\epsilon_r, \frac{w}{h}\right) = \left(1 + \frac{10h}{w}\right)^{-(1/2)} \tag{3.108}$$

Thus, the effective dielectric constant ϵ_{eff} is

$$\varepsilon_{eff} = \frac{\varepsilon_r + 1}{2} + \frac{\varepsilon_r - 1}{2}\left(1 + \frac{10h}{w}\right)^{-(1/2)} \tag{3.109}$$

The expressions proposed by Wheeler and Schneider were intensively analyzed by Hammerstad (see Section 3.3), by comparing their results with exact numerical static methods.

3.2.3 Analysis and Synthesis Equations After Wheeler

In 1977, Wheeler also applied curvefitting procedures to exact numerical results [37, 47]. He published the following formulas for the calculation of the characteristic impedance of the air-filled microstrip covering the whole strip region $0 < w/h < \infty$ (analysis equation):

$$Z_0 = \frac{42.4}{\sqrt{\epsilon_r + 1}} \ln\left\{1 + \left(\frac{4h}{w}\right)\left[\left(\frac{14 + 8/\epsilon_r}{11}\right)\left(\frac{4h}{w}\right) + \sqrt{\left(\frac{14 + 8/\epsilon_r}{11}\right)\left(\frac{4h}{w}\right)^2 + \frac{1 + 1/\epsilon_r}{2}\pi^2}\right]\right\} \tag{3.110}$$

As stated by Wheeler, the error is less than 2%. "Over most of range" an accuracy better than 1% is obtained. The effective dielectric constant is calculated by the ratio $Z_0(\epsilon_r)/Z_0(\epsilon_r = 1) = 1/\sqrt{\epsilon_{eff}}$. The following expression permits the calculation of w/h of an air-filled microstrip, if the characteristic impedance Z_0 is known (synthesis equation):

$$w/h = \frac{8}{A}\sqrt{\frac{A}{11}\left(7 + 4/\epsilon_r\right) + \frac{1}{0.81}\left(1 + 1/\epsilon_r\right)} \tag{3.111}$$

where

$$A = \exp\left(Z_0 \frac{\sqrt{\epsilon_r + 1}}{42.4}\right) - 1$$

3.3 Impedance Design Formulas After Hammerstad

Hammerstad [48] stated that the formulas for the air-filled characteristic impedance and effective dielectric constant derived by Wheeler and Schneider should be improved for purposes of more accurate computer aided circuit design. Accuracy should be derived from typical variations in substrate parameters and manufacturing tolerances. A maximum relative error of 1% was assumed to be a reasonable goal for both.

The relative error of the characteristic impedance is calculated from the difference of simulated and actual characteristic impedance $Z_{0,sim}$ and $Z_{0,act}$, respectively,

$$\frac{\Delta Z_0(\epsilon_r)}{Z_0(\epsilon_r)} = \frac{Z_{0,sim} - Z_{0,act}}{Z_{0,act}} \tag{3.112}$$

where

$$Z_0(\epsilon_r) = \frac{Z_0(\epsilon_r = 1)}{\sqrt{\epsilon_{eff}}} \tag{3.113}$$

The relative error of a product or ratio is equal to the sum of the relative errors of the factors. With (3.113), we obtain

$$\begin{aligned}\frac{\Delta Z_0(\epsilon_r)}{Z_0(\epsilon_r)} &= \delta_{Z_0(\epsilon_r)} = \delta_{Z_0(\epsilon_r = 1)} + \frac{1}{2}\delta_{\epsilon_{eff}} \\ &= \frac{\Delta Z_0(\epsilon_r = 1)}{Z_0} + \frac{1}{2}\frac{\Delta\epsilon_{eff}}{\epsilon_{eff}} \le 1\%\end{aligned} \tag{3.114}$$

Thus, if the effective dielectric constant is known with an uncertainty of 1%, then the uncertainty of the calculated characteristic impedance should be within 0.5%.

In the microstrip handbook [17] written by Hammerstad in 1975, formulas had become known for the first time which permitted the analysis of the transmission line parameters within an accuracy of less than 1%. The expressions had been found by curvefitting, using numerical results obtained by the integral equation method with Green's function. For the air-filled microstrip ($\epsilon_r = 1$), he used the exact values of the characteristic impedance derived from conformal mapping procedure.

To improve the accuracy of the effective dielectric constant ϵ_{eff}, the formula (3.109) proposed by Schneider [38] was slightly modified [17, 48]:

$$\epsilon_{eff} = \frac{\epsilon_r+1}{2} + \frac{\epsilon_r-1}{2} F(w/h) \tag{3.115}$$

with

$$F(w/h) = \begin{cases} (1 + 12h/w)^{-(1/2)} + 0.04(1 - w/h)^2 & (w/h \le 1) \\ (1 + 12h/w)^{-(1/2)} & (w/h \ge 1) \end{cases} \tag{3.116}$$

It was claimed that the relative error is not larger than 1% for $0.05 \le w/h \le 20$ and $\epsilon_r \le 16$. To obtain a better approximation, it would be necessary to let $F(w/h)$ vary with ϵ_r. The dependence of $F(w/h)$ on ϵ_r decreases with increasing ϵ_r. It was estimated that the error outside this range ($\epsilon_r > 16$) is not larger than 2% when (3.115) is used for $w/h < 0.05$. However, such narrow strips are commonly not used in practice. The wavelength being proportional to the inverse of the square root of the effective dielectric constant is determined with a maximum relative error of less than 0.5%.

Regarding the characteristic impedance, (3.104) for narrow width due to Schneider was found to be less than 0.1%. However, the maximum error for wide strips (3.105) was found to be 3%. Therefore, the two constants 2.24 and 0.44 were slightly changed to 2.64 and 0.49, respectively, and the error could be reduced to 0.4% for $w/h \le 10$, increasing to 2% for larger values of w/h,

$$Z_0(\epsilon_r = 1)/\Omega = \begin{cases} 60 \ln\left(\dfrac{8h}{w} + \dfrac{w}{4h}\right) & (w/h \le 1) \\ 120\pi \left[\dfrac{w}{h} + 2.46 - 0.49 \dfrac{h}{w}\left(1 - \dfrac{h}{w}\right)^6 \right]^{-1} & (w/h \ge 1) \end{cases} \tag{3.117}$$

In [48], Hammerstad proposed revised formulas with higher accuracy. The characteristic impedance Z_0 is calculated due to Schneider (3.104) if $w/h \le 1$. If $w/h \ge 1$, then the function Z_0 is better described by Wheeler (3.69) with modified constants,

$$Z_0(\epsilon_r = 1) = \begin{cases} 60\,\Omega \ln\left(\dfrac{8h}{w} + \dfrac{w}{4h}\right) & (w/h \le 1) \\ 120\pi\,\Omega \left\{ \dfrac{w}{h} + 1.393 + 0.667 \ln\left(\dfrac{w}{h} + 1.444\right)\right\}^{-1} & (w/h \ge 1) \end{cases} \tag{3.118}$$

Maximum relative error is claimed to be 0.4%, and approximately 0.8% in $Z_0(\epsilon_r)$. Although not definitely mentioned in [48], the same range of validity is supposed as given for the effective dielectric constant (3.115). Furthermore, it was found that the synthesis equations after Wheeler (3.71) deliver the same accuracy, after having modified some constants:

$$w/h = \begin{cases} \dfrac{8}{e^{A} - 2\,e^{-A}} & (w/h \le 2) \\ \dfrac{2}{\pi}\left\{B - 1 - \ln(2B-1) + \dfrac{\epsilon_r - 1}{2\epsilon_r}\left[\ln(B-1) + 0.39 - 0.61/\epsilon_r\right]\right\} & (w/h \ge 2) \end{cases} \tag{3.119}$$

where

$$A = \frac{\pi}{\eta_0}\sqrt{2(\epsilon_r + 1)} \cdot Z_0(\epsilon_r) + \frac{\epsilon_r - 1}{\epsilon_r + 1}(0.23 + 0.11/\epsilon_r)$$

and

$$B = \frac{\pi\sqrt{\mu_0/\epsilon_0}}{2\sqrt{\epsilon_r} \cdot Z_0(\epsilon_r)}$$

An accuracy of 1% was obtained by comparison with numerical results. Only for alumina and $w/h = 2$, a worse accuracy of -1.4% was obtained applying the formula for wide strips.

In 1980, Hammerstad and Jensen [49] published the following "very accurate and simple" equations:

$$Z_0(\epsilon_r = 1) = 60\,\Omega\,\ln\left[\frac{F_1}{w/h} + \sqrt{1 + \left(\frac{2}{w/h}\right)^2}\,\right] \tag{3.120}$$

where

$$F_1 = 6 + (2\pi - 6)\exp\left[-(30.666\,h/w)^{0.7528}\right]$$

It was claimed that the accuracy is better than 0.01% for $w/h \le 1$, and 0.03% for $w/h \le 1000$. Kirschning [50] has proven the accuracy of (3.120) on the basis of numerical results gained from Galerkin's method in spectral domain (Section 3.4.3). An examination covering the range of $\epsilon_r = 2(2)20$ for $w/h = 0.1(0.1)1$, and $w/h = 2(1)10$, respectively, resulted in a maximum error of less than 0.3%.

The effective dielectric constant was proposed to be calculated as

$$\epsilon_{eff} = \frac{\epsilon_r + 1}{2} + \frac{\epsilon_r - 1}{2}\left(1 + \frac{10}{w/h}\right)^{-a(w/h)\cdot b(\epsilon_r)} \tag{3.121}$$

where

$$a(w/h) = 1 + \frac{1}{49}\ln\left[\frac{(w/h)^4 + [(w/h)/52]^2}{(w/h)^4 + 0.432}\right] + \frac{1}{18.7}\ln\left[1 + \left(\frac{w/h}{18.1}\right)^3\right]$$

and

$$b(\epsilon_r) = 0.564\left(\frac{\epsilon_r - 0.9}{\epsilon_r + 3}\right)^{0.053}$$

It was claimed that the accuracy is better than 0.2% for at least $\epsilon_r \le 128$ and $0.01 \le w/h \le 100$. Kirschning [50] has verified the accuracy within the range of $\epsilon_r = 2(2)20$ for $w/h = 0.1(0.1)1$, and $w/h = 2(1)10$, respectively.

In conclusion, the approximated formulas (3.120) and (3.121) for the quasi-static microstrip line parameters $\epsilon_{eff}(f = 0)$ and $Z_0(f = 0)$ proposed by Hammerstad are sufficiently accurate for practical CAD purpose.

Nonzero strip thickness was taken into account on the basis of Wheeler's approach [34]. However, Hammerstad introduced some modifications to enhance the accuracy for narrow strips and substrates with a low-dielectric constant [49]:

$\epsilon_r = 1$:

$$\Delta w/h = \frac{t/h}{\pi}\ln\left(1 + \frac{4 \cdot e}{t/h\,\coth^2\sqrt{6.517\,w/h}}\right) \tag{3.122}$$

$\epsilon_r > 1$:

$$\Delta w_\epsilon/h = \left(1 + 1/\cosh\sqrt{\epsilon_r - 1}\right)\Delta w/h \tag{3.123}$$

By defining the thickness-corrected strip widths with $w' = w + \Delta w$, and $w'_\epsilon = w + \Delta w_\epsilon$, respectively, the effect of strip thickness may be induced in the above equations:

$$Z_0(w/h, t, \epsilon_r) = Z_0(w'_\epsilon/h, \epsilon_r = 1)/\sqrt{\epsilon'_{eff}(w'_\epsilon/h, \epsilon_r)}$$

$$\epsilon_{eff}(w/h, t, \epsilon_r) = \epsilon_{eff}(w'_\epsilon/h, \epsilon_r)\left(\frac{Z_0(w/h, \epsilon_r = 1)}{Z_0(w'/h, \epsilon_r = 1)}\right)$$

In Appendix F, the characteristic impedance $Z_0(\epsilon_r)$ (3.120) and the effective dielectric constant ϵ_{eff} (3.121) as a function of the w/h aspect ratio are listed for the range $0.1 \le w/h \le 20$.

Table 3.2
Longitudinal Wavenumber and Type of Mode

k_z	*Propagation Term*	*Type of Mode*
real	$e^{-jk_z z}$	propagating mode
negative imaginary	$e^{-\lvert k_z \rvert z}$	evanescent mode

3.4 Full-Wave Analysis

In the following section a full-wave analysis is presented using the spectral domain method. This method has been frequently used to analyze planar transmission line structures with high accuracy. This method assumes that the conductivity of the strip is infinite and its thickness t is zero. The method was proposed by Itoh and Mittra in 1973, investigating the dispersion of shielded microstrip transmission lines [29, 30]. We first consider modes on stripless structures, which are discussed in Section 3.4.1, subjected to the parallel plate transmission line with dielectric layer and the grounded dielectric slab in Section 3.4.2.

3.4.1 Modes of Inhomogeneous Parallel Plate Structure

Figure 3.33 shows a parallel plate transmission line with grounded dielectric layer. We assume wave propagation in the z direction. We search for a solution of Maxwell equations, taking into account the boundary conditions at the conductive bottom and top plate, as well as the air-dielectric interface.

The propagation term of modes in the z direction is commonly given as $e^{-jk_z z}$. If k_z is real, then the mode denotes a propagating mode, and if k_z is negative imaginary, then the mode denotes an evanescent mode (Table 3.2). An evanescent mode is a local wave that decays exponentially. Stimulated evanescent modes do not propagate. In lossless parallel plate structures, modes only exist with

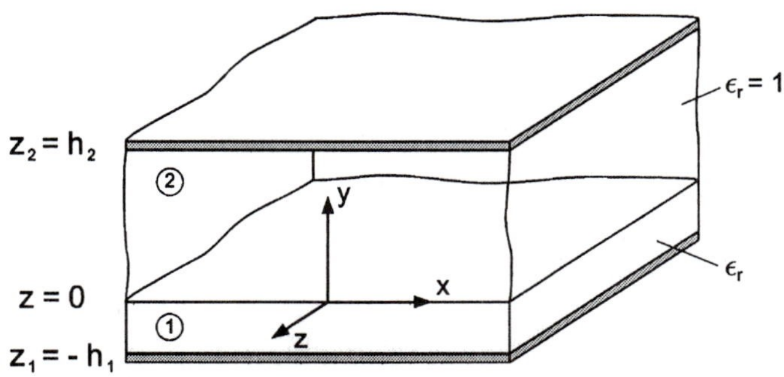

Figure 3.33 Parallel plate transmission line with dielectric layer.

longitudinal wavenumbers which are real or imaginary. The spectrum of modes (system of eigenmodes) comprises both propagating and evanescent modes. The lowest order propagating mode is an E mode (TM_0) and has no low cutoff frequency. All other modes exist as evanescent modes below their cutoff frequency. The longitudinal electric field component E_z fulfills the partial differential equation (Appendix D),

$$\vec{\nabla}_t^{\,2} E_z + k_t^2 E_z = 0 \tag{D.118}$$

where

$$k_t^2 = k^2 - k_z^2 \tag{D.88}$$

and $\vec{\nabla}_t^2 = d^2/dy^2$ for no variation of the electromagnetic field with x. Equation (D.88) reads for the dielectric region (Index 1) and air region (Index 2):

$$k_{t_1}^2 = k_1^2 - k_z^2 \tag{3.124}$$

$$k_{t_2}^2 = k_2^2 - k_z^2 \tag{3.125}$$

The difference of the squared transverse wavenumbers gives

$$k_{t_1}^2 - k_{t_2}^2 = k_1^2 - k_2^2 = (\epsilon_r - 1)\,k_0^2 \tag{3.126}$$

with $k_1 = k_0\sqrt{\epsilon_r}$ and $k_2 = k_0$. The longitudinal field strength E_z in both regions read

$$\frac{d^2E_z}{dy^2} + k_{t_1}^2 E_z = 0 \qquad (-h_1 \le y \le 0) \tag{3.127}$$

$$\frac{d^2E_z}{dy^2} + k_{t_2}^2 E_z = 0 \qquad (0 \le y \le h_2) \tag{3.128}$$

The boundary conditions are:

$$E_z(y) = 0: \qquad y = -h_1, +h_2$$

$$E_z(y) \text{ continuous at } y = 0$$

$$H_x(y) \text{ continuous at } y = 0$$

From (D.126), we obtain for H_x:

$$H_x = \frac{j\omega\epsilon}{k_t^2}\frac{\partial E_z}{\partial y}$$

and applied to the air and dielectric regions,

$$H_{x_1} = \frac{j\omega\epsilon_0\epsilon_r}{k_{t_1}^2}\frac{\partial E_z}{\partial y} \tag{3.129}$$

$$H_{x_2} = \frac{j\omega\epsilon_0}{k_{t_2}^2}\frac{\partial E_z}{\partial y} \tag{3.130}$$

Continuity of the tangential magnetic field $H_{x_1} = H_{x_2}$:

$$\frac{\epsilon_r}{k_{t_1}^2}\frac{\partial E_z}{\partial y} = \frac{1}{k_{t_2}^2}\frac{\partial E_z}{\partial y} \tag{3.131}$$

From (D.126), we calculate E_y of E_z:

$$E_y = -\frac{jk_z}{k_t^2}\frac{\partial E_z}{\partial y} \tag{3.132}$$

In the two regions, this reads

$$E_{y_1} = -\frac{jk_z}{k_{t_1}^2}\frac{\partial E_z}{\partial y} \tag{3.133}$$

$$E_{y_2} = -\frac{jk_z}{k_{t_2}^2}\frac{\partial E_z}{\partial y} \tag{3.134}$$

To fulfill the boundary conditions on the conductive bottom and top plates, we write for $E_z(y)$

$$E_z(y) = C_1 \sin k_{t_1}(y + h_1) \qquad (-h_1 \le y \le 0) \tag{3.135}$$

$$E_z(y) = C_2 \sin k_{t_2}(y - h_2) \qquad (0 \le y \le h_2) \tag{3.136}$$

$E_z(y)$ is continuous at the air–dielectric interface ($y = 0$), which results in

$$C_1 \sin(k_{t_1}h_1) = -C_2 \sin(k_{t_2}h_2) \tag{3.137}$$

Differentiating $E_z(y)$ in (3.135) and (3.136),

$$\frac{dE_z(y)}{dy} = C_1 k_{t_1} \cos k_{t_1}(y + h_1) \qquad (-h_1 \le y \le 0)$$

$$\frac{dE_z(y)}{dy} = C_2 k_{t_2} \cos k_{t_2}(y - h_2) \qquad (0 \le y \le h_2)$$

the boundary condition (3.131) requires

$$\frac{\epsilon_r}{k_{t_1}^2} C_1 k_{t_1} \cos k_{t_1} h_1 = \frac{1}{k_{t_2}^2} C_2 k_{t_2} \cos k_{t_2} h_2 \tag{3.138}$$

Then, dividing (3.137) by (3.138),

$$k_{t_1} \tan(k_{t_1} h_1) = -\epsilon_r k_{t_2} \tan(k_{t_2} h_2) \tag{3.139}$$

Equation (3.139) must be solved simultaneously with (3.126), which results in an infinite number of E modes with longitudinal wavenumber

$$k_z = \sqrt{k_0^2 - k_{t_2}^2} = \sqrt{\epsilon_r k_0^2 - k_{t_1}^2} \tag{3.140}$$

When k_z becomes negative, an evanescent mode occurs. We are interested in wave propagation. In this case, k_z must take a real value between k_0 and k ($k_0 \le k_z \le k$). Thus, we suppose that k_{t_2} is imaginary, and we define $k_{t_2} = jk_{t_{20}}$. Then, with $\tan(jy) = \sin(jy)/\cos(jy) = j\sinh(y)/\cosh(y) = j\tanh(y)$, we get from (3.139)

$$k_{t_{10}} \tan(k_{t_{10}} h_1) = \epsilon_r k_{t_{20}} \tanh(k_{t_{20}} h_2) \tag{3.141}$$

and from (3.126)

$$k_{t_{10}}^2 + k_{t_{20}}^2 = (\epsilon_r - 1) k_0^2 \tag{3.142}$$

We consider solutions for the limit case of very low and very high frequencies [51].

Low-Frequency Solution

Assume a very low frequency (e.g., $f = 1$ MHz), then the free-space wavenumber becomes a small value of $k_0 = 2.09 \times 10^{-4}$ rad/cm. Then, with (3.142), the transversal wavenumbers k_{t_1} and k_{t_2} also become very small values. This is also valid for $k_{t_1} h_1$ and $k_{t_2} h_2$ regarding usual values h_1 and h_2. Then, the tangent function and hyperbolic tangent function can be approximated by their arguments:

$$k_{t_{10}}^2 h_1 = \epsilon_r k_{t_{20}}^2 h_2 \tag{3.143}$$

With (3.142)

$$k_{t_{20}} = \frac{h_1(\epsilon_r - 1)k_0^2}{h_1 + \epsilon_r h_2} \tag{3.144}$$

The solution for k_z is

$$k_z = \sqrt{k_0^2 + k_{t_0}^2} = k_0\sqrt{\frac{\epsilon_r(h_1 + h_2)}{h_1 + \epsilon_r h_2}} = \sqrt{\epsilon_{eff}}\,k_0 \tag{3.145}$$

with

$$\epsilon_{eff} = \frac{\epsilon_r(h_1 + h_2)}{h_1 + \epsilon_r h_2} \tag{3.146}$$

For $h_1 = 0$, the effective dielectric constant is $\epsilon_{eff} = 1$. When $h_2 = 0$, then we get $\epsilon_{eff} = \epsilon_r$.

High-Frequency Solution

With a large number of k_0 at high frequencies, k_{t_1} and k_{t_2} become very large. Thus, the tangent hyperbolic function can be approximated by 1, $\tanh(k_{t_{20}}h_2) \approx 1$. Then (3.141) reads

$$k_{t_{10}}\tan(k_{t_{10}}h_1) = \epsilon_r k_{t_{20}} \tag{3.147}$$

With (3.142)

$$k_{t_{10}}\tan(k_{t_{10}}h_1) = \epsilon_r\sqrt{(\epsilon_r - 1)k_0^2 - k_{t_{10}}} \tag{3.148}$$

Equation (3.148) is independent of the plate separation h_2. The electrical field component E_z can be calculated from (3.135) and (3.136), substituting $k_{t_2} = jk_{t_{20}}$,

$$E_z(y) = C_1 \sin k_{t_{10}}(y + h_1) \qquad (-h_1 \le y \le 0) \tag{3.149}$$

$$E_z(y) = jC_2 \sinh k_{t_{20}}(y - h_2) \qquad (0 \le y \le h_2) \tag{3.150}$$

Then, the boundary condition for E_z at $y = 0$ delivers

$$C_2 = jC_1\frac{\sin k_{t_{10}}h_1}{\sinh k_{t_{20}}h_2} \tag{3.151}$$

Then

$$E_z(y) = C_1 \sin(k_{t_{10}} h_1) \frac{\sinh k_{t_{20}}(h_2 - y)}{\sinh(k_{t_{20}} h_2)} \qquad 0 \le y \le h_2$$

$$\approx C_1 \sin(k_{t_{10}} h_1) \frac{e^{k_{t_{20}}(h_2 - y)}}{e^{k_{t_{20}} h_2}} = C_1 \sin(k_{t_{10}} h_1)\, e^{-k_{t_{20}} y} \tag{3.152}$$

The field decays exponentially in the y direction and does not depend on h_2, as long as $k_{t_{20}} h_2$ is large. This field is bound to the grounded dielectric plate and will exist after removing the upper plate. Modes that are guided by dielectric interfaces are called surface waves. Equation (3.148) also delivers solutions of higher-order modes, which will be treated in the following section regarding the field analysis of grounded dielectric plates.

3.4.2 Surface Waves on Dielectric Plate

Electromagnetic waves can be guided not only with conductor configurations. In the following section, we will see that electromagnetic waves can even propagate along dielectric layers. It is assumed that the dielectric plate waveguide that is depicted in Figure 3.34 has infinite extension in the x and z direction. We further assume that the plate is lossless. Outside of the plate, the free space is filled with air ($\epsilon_r = \mu_r = 1$). We consider such solutions, whose fields are not dependent on the x coordinate ($\partial/\partial x = 0$). Then, the x-component of the transversal wavenumber k_t also vanishes ($k_x = 0$). The propagation of considered modes shall be in the z direction, so that the propagation term is denoted by $f(z) = \exp(-jk_z z)$. First, we search for such solutions, which have a transversal magnetic field—E modes with respect to the z direction.

3.4.2.1 *E* Modes

We intend to calculate the fields from the electric vector potential $\vec{A}^e$ or the respective scalar potential ψ^e, in accordance with Section D.2.2.2. With $\partial/\partial x = 0$, it follows from (D.172)

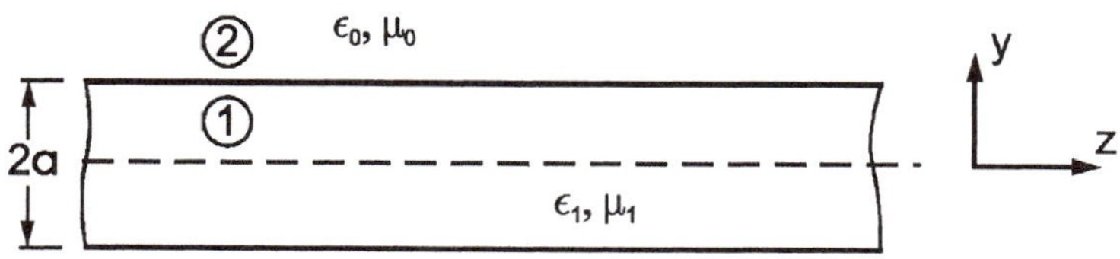

Figure 3.34 Dielectric plate waveguide.

$$E_x = 0 \qquad H_x = \frac{\partial \psi^e}{\partial y}$$
$$E_y = -\frac{k_z}{\omega\epsilon} H_x \qquad H_y = 0 \qquad (3.153)$$
$$E_z = \frac{k_t^2}{j\omega\epsilon} \psi^e \qquad H_z = 0$$

The dielectric plate represents a symmetric waveguide configuration with respect to the plane $y = 0$. In this case, the general solution of the field problem can be split into an even and odd partial solution. The symmetry plane can be described by an electric wall or a magnetic wall, respectively.

Solutions for the Odd Wave Function $\psi^{e,o}$

We first examine a partial solution resulting from a wave function $\psi^{e,o}$ that is odd with respect to the coordinate y. For simplification, we drop the designation (e,m) of the electromagnetic potential in the following. For the potential $\psi^{e,o}$, we choose the following statement for the homogeneous field ranges:

$$\psi_1^{e,o} = A \sin(k_{y_1} y)\, e^{-jk_z \cdot z} \qquad |y| < a$$
$$\psi_2^{e,o} = B\, e^{-jk_{y_2} y}\, e^{-jk_z \cdot z} \qquad y > a \qquad (3.154)$$
$$\psi_2^{e,o} = -\,B\, e^{+jk_{y_2} y}\, e^{-jk_z \cdot z} \qquad y < -a$$

As is illustrated in Figure 3.35, within the dielectric plate, only harmonic functions occur in the y direction, while the fields in the y direction are pure evanescent

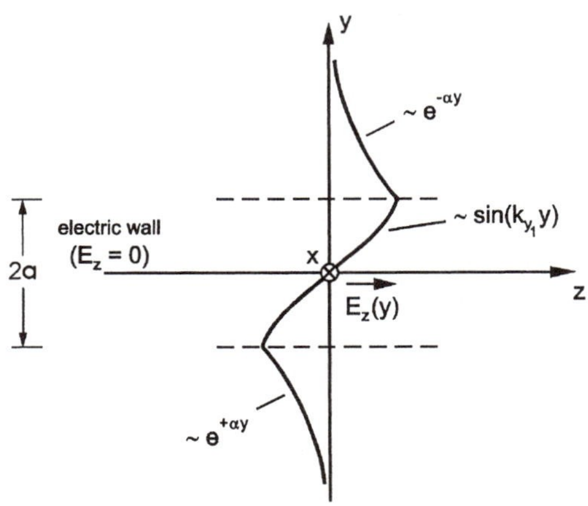

Figure 3.35 Longitudinal electric field distribution from odd scalar potential $\psi^{e\,o}$.

fields in the y direction outside of the plate (i.e., all field components are exponentially decreasing in the y direction). Therefore, k_{y_2} is purely negative imaginary. The boundary conditions require that the continuity of E_z and H_x is fulfilled at the position $y = \pm a$. With $k_{x_i} \equiv 0$ and

$$k_{t_i}^2 = k_{x_i}^2 + k_{y_i}^2 = k_{y_i}^2 \tag{3.155}$$

where $i = 1,2$, it follows from (3.153) that

$$E_{z_i} = \frac{k_{y_i}^2}{j\omega\epsilon_i} \psi_i^{e,o} \tag{3.156}$$

$$H_{x_i} = \frac{\partial \psi_i^{e,o}}{\partial y} \tag{3.157}$$

$$E_{y_i} = -\frac{k_z}{\omega\epsilon_i} H_{x_i} \tag{3.158}$$

with $i = 1,2$ and using (3.154), we find

$$|y| < a: \quad E_{z_1} = A\frac{k_{y_1}^2}{j\omega\epsilon_1}\sin(k_{y_1}y)e^{-jk_z \cdot z}, \qquad H_{x_1} = A\,k_{y_1}\cos(k_{y_1}y)\,e^{-jk_z \cdot z} \tag{3.159}$$

$$y > a: \quad E_{z_2} = B\frac{k_{y_2}^2}{j\omega\epsilon_0}\,e^{-jk_{y_2}y}\,e^{-jk_z \cdot z}, \qquad H_{x_2} = -B\,(jk_{y_2})\,e^{-jk_{y_2}y}\,e^{-jk_z \cdot z} \tag{3.160}$$

$$y < a: \quad E_{z_2} = -B\frac{k_{y_2}^2}{j\omega\epsilon_0}\,e^{+jk_{y_2}y}\,e^{-jk_z \cdot z}, \qquad H_{x_2} = -B\,(jk_{y_2})\,e^{+jk_{y_2}y}\,e^{-jk_z \cdot z} \tag{3.161}$$

The continuity of the tangential field components must be fulfilled at the location $y = \pm a$.

(a) $E_{z_1} = E_{z_2}$ for $y = \pm a$

$$y = a: \qquad A\frac{k_{y_1}^2}{j\omega\epsilon_1}\sin(k_{y_1}a) = B\frac{k_{y_2}^2}{j\omega\epsilon_0}\,e^{-jk_{y_2}a}$$

$$y = -a: \qquad -A\frac{k_{y_1}^2}{j\omega\epsilon_1}\sin(k_{y_1}a) = -B\frac{k_{y_2}^2}{j\omega\epsilon_0}\,e^{-jk_{y_2}a}$$

or

$$y = \pm a: \qquad \frac{A}{\epsilon_1}\, k_{y_1}^{\;2} \sin(k_{y_1} a) = \frac{B}{\epsilon_0}\, k_{y_2}^{\;2}\, e^{-jk_{y_2} a} \tag{3.162}$$

(b) $H_{x_1} = H_{x_2}$ for $y = \pm a$: From (3.159) to (3.161), it turns out that

$$A k_{y_1} \cos(k_{y_1} a) = -B(jk_{y_2}) e^{-jk_{y_2} a} \tag{3.163}$$

Equations (3.162) and (3.163) represent a system of linear simultaneous equations for the unknown coefficients A and B,

$$\begin{aligned} \frac{k_{y_1}^{\;2}}{\epsilon_1} \sin(k_{y_1} a)\, A - \frac{k_{y_2}^{\;2}}{\epsilon_0}\, e^{-jk_{y_2} a}\, B &= 0 \\ k_{y_1} \cos(k_{y_1} a)\, A + jk_{y_2}\, e^{-jk_{y_2} a}\, B &= 0 \end{aligned} \tag{3.164}$$

For a nontrivial solution, it is required that the coefficient determinant of this system vanishes:

$$\begin{vmatrix} \dfrac{k_{y_1}^{\;2}}{\epsilon_1} \sin(k_{y_1} a) & -\dfrac{k_{y_2}^{\;2}}{\epsilon_0}\, e^{-jk_{y_2} a} \\ k_{y_1} \cos(k_{y_1} a) & jk_{y_2}\, e^{-jk_{y_2} a} \end{vmatrix} = 0 \tag{3.165}$$

Therefore, the following relation is valid:

$$\frac{k_{y_1}^{\;2}}{\epsilon_1} (jk_{y_2}) \sin(k_{y_1} a) e^{-jk_{y_2} a} + k_{y_1} \frac{k_{y_2}^{\;2}}{\epsilon_0} \cos(k_{y_1} a) e^{-jk_{y_2} a} = 0 \tag{3.166}$$

or after transformation

$$k_{y_1} a \tan(k_{y_1} a) = j \frac{\epsilon_1}{\epsilon_0} k_{y_2} a \quad \Bigg| \times \frac{\epsilon_0}{\epsilon_1} \tag{3.167}$$

Equation (3.167) is the characteristic equation for the determination of k_z with

$$k_{y_1}^{\;2} + k_z^2 = k_1^2 = \omega^2 \epsilon_1 \mu_1 \tag{3.168}$$

$$k_{y_2}^{\;2} + k_z^2 = k_0^2 = \omega^2 \epsilon_0 \mu_0 \tag{3.169}$$

We want to find solutions of (3.167) in a graphical way. Therefore, we rewrite this equation,

$$jk_{y_2}a = \frac{\epsilon_0}{\epsilon_1}(k_{y_1}a)\tan(k_{y_1}a) \tag{3.170}$$

From (3.168) and (3.169), it follows

$$k_{y_1}^{\ 2} - k_{y_2}^{\ 2} = k_1^2 - k_0^2 = \omega^2(\epsilon_1\mu_1 - \epsilon_0\mu_0)$$

or normalized to the half-plate width a

$$(k_{y_1}a)^2 + (jk_{y_2}a)^2 = (\omega a)^2(\epsilon_1\mu_1 - \epsilon_0\mu_0) \tag{3.171}$$

where the radius R reads

$$R = \omega a\sqrt{(\epsilon_1\mu_1 - \epsilon_0\mu_0)} = \frac{\omega a}{c_0}\sqrt{\epsilon_r - 1} \tag{3.172}$$

As can be seen from Figure 3.36, the solutions of the characteristic equation turn out to be the intersections of the circle of (3.171) with the tangent function of (3.170). If the transversal wavenumbers of the different field regions are known, then the longitudinal wavenumber can be determined as well from (3.168) and (3.169),

$$k_z^2 = (jk_{y_2})^2 + k_0^2 \tag{3.173}$$

$$k_z^2 = -k_{y_1}^{\ 2} + k_1^2 \tag{3.174}$$

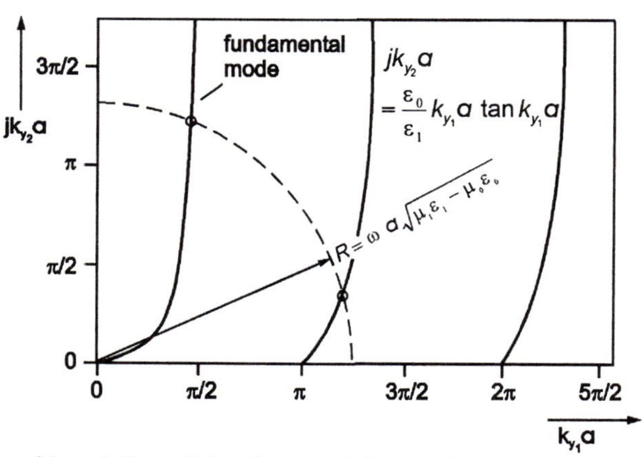

Figure 3.36 Graphic solution of the characteristic equation.

Multiplication of (3.173) with k_1^2, and of (3.174) with k_0^2, and calculating the difference of both equations lead to

$$k_z = \left(\frac{\epsilon_0 \mu_0 k_{y_1}^{\;2} - \epsilon_1 \mu_1 k_{y_2}^{\;2}}{\epsilon_1 \mu_1 - \epsilon_0 \mu_0} \right)^{\frac{1}{2}}$$

$$= \sqrt{\frac{k_{y_1}^2 - \epsilon_r k_{y_2}^2}{\epsilon_r - 1}} \tag{3.175}$$

with k_{y_2} being a negative imaginary value. The lower the frequency, the smaller the radius, and the less the value of $(jk_{y_2} a)$, as can be seen from Figure 3.36. This means that the transversal damping (evanescent) field increases its extension with decreasing frequency. The range of the fundamental mode (dominant mode) is limited to frequencies, for which the condition $R(f) \le R_{max}(f) = \pi/2$ is valid. For higher frequencies, higher odd modes become capable of propagation.

For the fundamental mode, the field configuration of Figure 3.37 results. The cutoff frequencies of the higher-order odd E-modes can be calculated from

$$\frac{\omega_c a}{2} (\epsilon_1 \mu_1 - \epsilon_0 \mu_0)^{1/2} = n\pi \qquad (n = 1, 2 \ldots) \tag{3.176}$$

to be

$$f_{c_n} = \frac{n}{a(\epsilon_1 \mu_1 - \epsilon_0 \mu_0)^{1/2}} \tag{3.177}$$

Regarding the different waveguiding structures, such as hollow waveguide or coaxial transmission line, solution of the wave equation with respective boundary

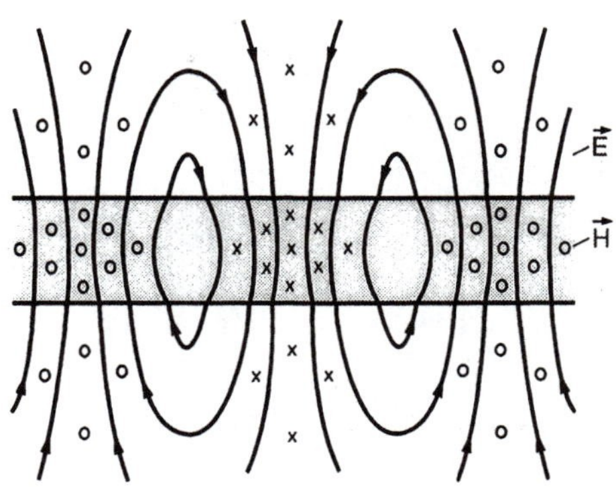

Figure 3.37 Field distribution of the fundamental mode.

conditions exists at any frequency. The cutoff frequency designates the transition from nonpropagating to propagating mode. Both the evanescent and propagation modes represent a complete system of eigenmodes. Any arbitrary field distribution can be described by a superposition of an infinite series of eigen-modes. Surface mode properties differ strongly from waveguide modes because they exist only as waves, above their cutoff frequency.

Problem 3.4: We suppose an infinite dielectric plate as depicted in Figure 3.34. The thickness is $2a = 10$ mm and the dielectric constant $\epsilon_r = 9.6$. Determine graphically with the enclosed diagram $jk_{y_2}a = f(k_{y_1}a)$ the normalized wavenumbers $jk_{y_2}a$ and $k_{y_1}a$ at a frequency $f = 9.16$ GHz.

Solution:
From (3.172), the radius is calculated as $R = 2\pi \cdot 9.1\,\mathrm{GHz} \cdot 0.5\,\mathrm{cm} \cdot \sqrt{8.6}/30\,\mathrm{GHz\,cm} = 2.79$. Crossing of a circle with $R = 2.79$ with the tangent function delivers: $k_{y_1}a = 1.5$ and $|k_{y_2}a| = 2.35$.

Solutions for the Even Scalar Potential $\psi^{e,e}$
In analogy to the preceding section, we make the following statement for the even wave function $\psi^{e,e}$:

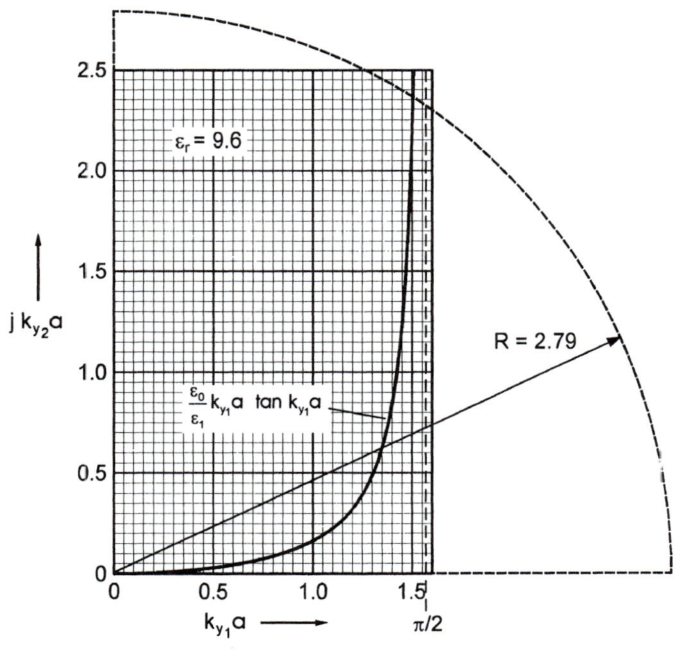

Figure 3.38 Graphic solution.

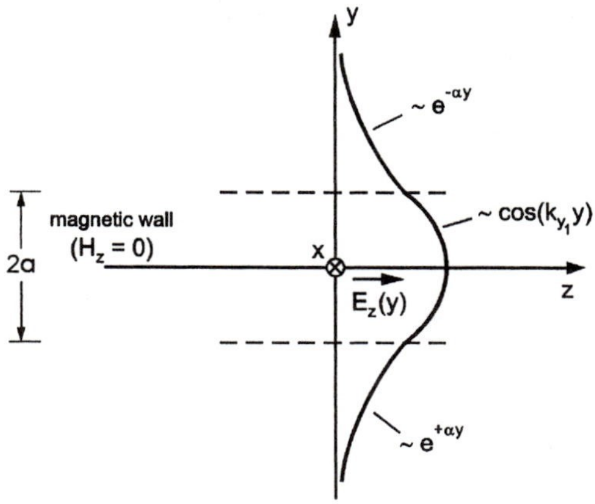

Figure 3.39 Longitudinal electric field distribution from even scalar potential $\psi^{e,e}$.

$$
\begin{aligned}
\psi_1^{e,e} &= A' \cos(k_{y_1} y)\, e^{-jk_z \cdot z} && |y| < a \\
\psi_2^{e,e} &= B'\, e^{-jk_{y_2} y}\, e^{-jk_z \cdot z} && y > a \\
\psi_2^{e,e} &= B'\, e^{+jk_{y_2} y}\, e^{-jk_z \cdot z} && y < -a
\end{aligned}
\tag{3.178}
$$

Figure 3.39 illustrates the field distribution of the longitudinal electrical field inside and outside of the dielectric plate. The alignment of the tangential electric and magnetic field strengths at the location $y = \pm a$ leads us again to a linear system of simultaneous equations for the coefficients A' and B'. The characteristic equation for this case reads

$$-k_{y_1} a \cot(k_{y_1} a) = j \frac{\epsilon_1}{\epsilon_2} k_{y_2} a \tag{3.179}$$

Under consideration of (3.168) and (3.169), we again can determine the longitudinal wavenumber k_z from the above equation.

3.4.2.2 *H* Modes

From (D.178) in Section D.2.2.2, it follows with $\partial/\partial x = 0$ that the electric and magnetic field components read

$$E_x = -\frac{\partial \psi^m}{\partial y} \qquad H_x = 0$$

$$E_y = 0 \qquad H_y = \frac{k_z}{\omega \mu} E_x \tag{3.180}$$

$$E_z = 0 \qquad H_z = \frac{k_t^2}{j \omega \mu} \psi^m$$

We again search for partial even and odd solutions of the boundary value problem under discussion.

Solutions for the Odd Scalar Potential $\psi^{m,o}$

Because the procedure is in analogy to the treatment of the E-modes, we only give the final result for the characteristic equation for H-modes arising from the odd scalar potential $\psi^{m,o}$ here:

$$-k_{y_1} a \tan(k_{y_1} a) = j \frac{\mu_1}{\mu_0} k_{y_2} a \tag{3.181}$$

Solutions for the Even Scalar Potential $\psi^{m,e}$

The solution for the even scalar potential $\psi^{m,e}$ results in the characteristic equation

$$k_{y_1} a \cot(k_{y_1} a) = j \frac{\mu_1}{\mu_0} k_{y_2} a \tag{3.182}$$

Regarding (3.181) and (3.182), it must be noted that k_{y_2} is again negative imaginary.

3.4.2.3 Dielectric Cladded Metal Plate

As can be seen from (3.159) and (3.180), the E-modes, which can be derived from an odd scalar potential, and the H-modes, which arise from an even scalar potential, do not have any tangential electric field strength at the plane of symmetry. Therefore, the symmetry plane represents an electric wall, which can be substituted by an electrically conductive plane, as shown in Figure 3.40. The electric and magnetic field of the modes can be derived from the characteristic equations (3.167) and (3.182) of the dielectric plate waveguide. The cutoff frequency of the fundamental mode, shown in Figure 3.41, is $f_c = 0$ Hz. At low

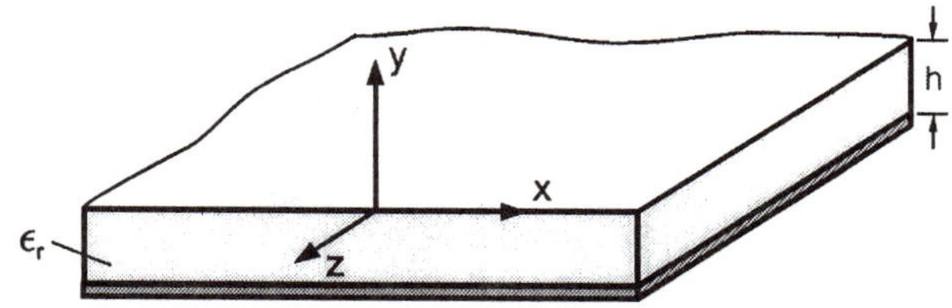

Figure 3.40 Grounded dielectric plate.

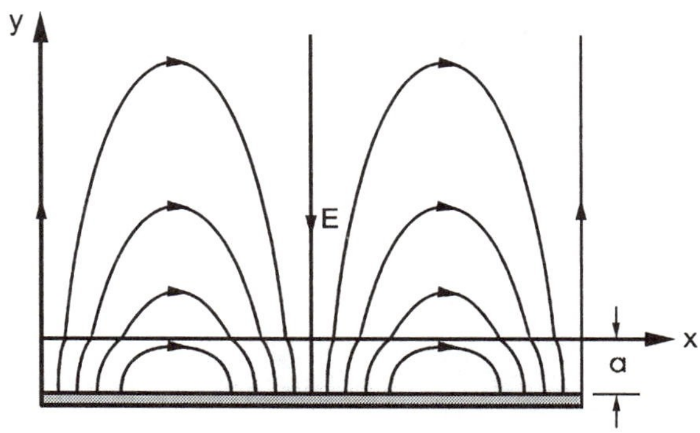

Figure 3.41 Fundamental TM_0 surface wave.

frequency, the electromagnetic field extends very far in the transversal direction. The wave is only loosely bound to the given waveguide configuration. This is also valid for higher-order modes at frequencies that are slightly above the respective cutoff frequencies.

Problem 3.5: The dielectric cladded metal surface shown in Figure 3.40 has a substrate height $h = 1.6$ mm and dielectric constant $\epsilon_r = 2.32$. Determine the wavelength of the lowest-order surface wave at a frequency of 26.0 GHz (Hint: Use the enclosed graph in Figure 3.42).

Solution:

The radius R is obtained as $R = 2\pi f h/c_0 \sqrt{\epsilon_r - 1} = 2\pi \cdot 26\,\text{GHz} \cdot 0.16\,\text{cm}/30\,\text{GHz cm}\,\sqrt{2.32 - 1} \approx 1.0$. The crossing of the circle $R = 1$ with the tangent function delivers $k_{y_1}h \approx 0.88$ and $jk_{y_2}h \approx 0.46$. With $h = 1.6$ mm, we get for the wavenumbers $k_{y_1} = 0.55\,\text{mm}^{-1}$ and $k_{y_2} = 0.28\,\text{mm}^{-1}$. Then k_z is calculated with (3.175), $k_z = [(k_{y_1}^2 - \epsilon_r k_{y_2}^2)/(\epsilon_r - 1)]^{(1/2)} = 0.3\,\text{mm}^{-1}$. The wavelength of the surface wave is $\lambda_g = 2\pi/k_z = 2\pi/0.3\,\text{mm} = 19.89\,\text{mm}$. For comparison, the vacuum wavelength yields $\lambda_0 = c_0/f = 11.5\,\text{mm}$ and the wavelength in the dielectric medium $\lambda = c_0/\left(f\sqrt{\epsilon_r}\right) = 7.57\,\text{mm}$.

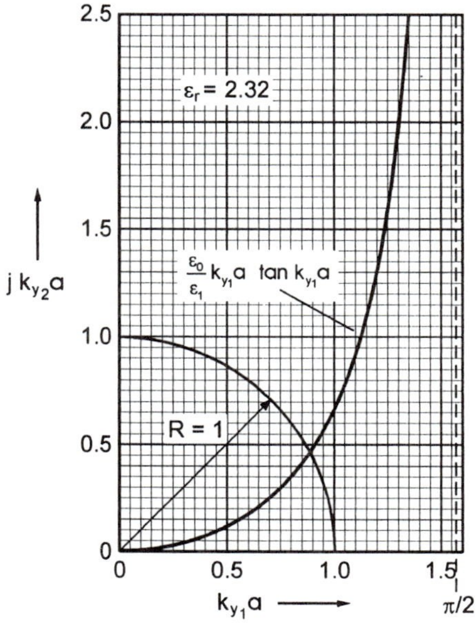

Figure 3.42 Graphical solution of Problem 3.5.

Understanding the physical mechanism of radiation in microstrip lines, in addition to the discussed surface waves, radiated waves and leaky waves must be taken into account [52]. Figure 3.43 illustrates the excitation of waves supposing a point source (Hertzian dipole). A source on the top surface of the strip radiates waves "upward," directly into open space (radiated waves). A source on the bottom surface of the strip will excite waves that travel between strip and ground plane in a zigzag path. At a conductor's edge, dependent on the incident angle of the waves, trapped surface waves may be excited and may propagate with small loss. They may produce spurious coupling between fairly distant circuit elements. When the incident angle (with respect to the normal of the dielectric–air interface) is small, then part of the excited wave will be reflected and another part transmitted so that the wave loses energy along its propagation path (leaky waves). Because of power loss, leaky waves only can exist locally.

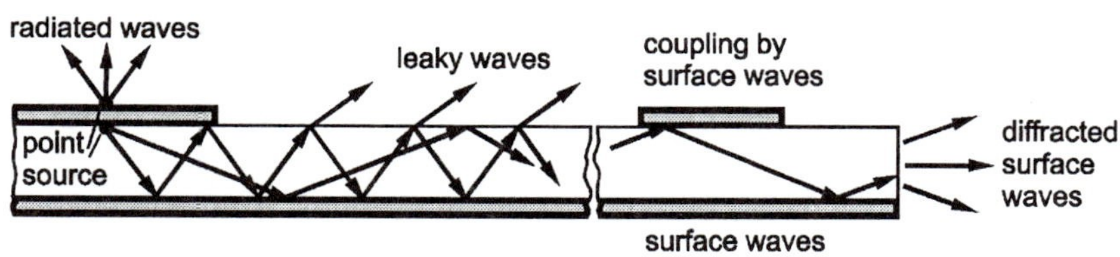

Figure 3.43 Graphical representation of surface waves, radiated waves, and leaky waves. (After [52].)

3.4.3 Shielded Microstrip Transmission Line Laterally Open

The structure under analysis is illustrated in Figure 3.44. It consists of two ground planes [one at the top ($z_2 = h_2$) and the other at the bottom ($z_1 = -h_1$)], and two dielectric layers with dielectric constants $\epsilon_1 = \epsilon_r \epsilon_0$ and $\epsilon_2 = \epsilon_0$, and an infinitely thin ($t = 0$) metallic strip placed at the interface of the two media. The ground planes and the dielectric materials are assumed to be lossless and infinite in length in the x and z directions. It is further noted that by letting $h_2 \to \infty$, this structure reduces to the open one. Due to boundary conditions, involved transverse inhomogeneity in the dielectric and the presence of the strip, no pure H (TE) or E (TM) mode may exist. The treatment of the given boundary problem will lead to the solution of coupled integral equations in the space domain. However, as will be shown, the analysis in the spectral domain will reveal essential advantages with respect to numerical calculation and efficiency for the reason that the integral equations are transformed to an algebraic set of equations.

The electromagnetic field in Figure 3.44 is bounded in the y direction by the ground and cap metal plate. The direction of propagation is the z direction. To the sides the transmission line is open and the field is infinitely expanded. From transmission line theory, it is known that an arbitrary field distribution within a finite space interval (field region limited by electric walls) can be represented by an infinite Fourier series expansion. Regarding the discussed laterally open (infinite space interval), the infinite field distribution is described by a Fourier integral. Because of given geometry, the potential functions can be written in terms of their Fourier transform $\tilde{\psi}(\alpha, y)$ with respect to x,

$$\psi(x,y,z) = \frac{1}{2\pi} \int_{-\infty}^{+\infty} \tilde{\psi}(\alpha, y, z)\, e^{-j\alpha x}\, d\alpha \tag{3.183}$$

where $\tilde{\psi}(\alpha, y, z)$ is the Fourier transform or spectral potential of $\psi(x,y,z)$. $\tilde{\psi}(\alpha, y, z)$ is given by the inverse Fourier integral

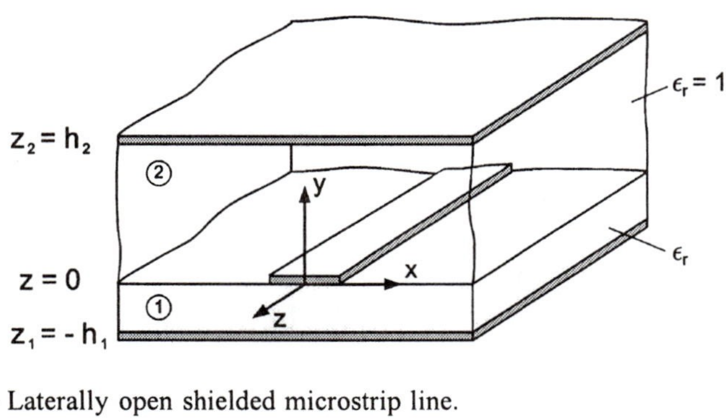

Figure 3.44 Laterally open shielded microstrip line.

$$\tilde{\psi}(\alpha,y,z) = \int_{-\infty}^{+\infty} \psi(x,y,z)\, e^{+j\alpha x}\, dx \tag{3.184}$$

The differential operators relate the electric and magnetic field strengths to the scalar potentials ψ^e [e.g., (D.171)]:

$$E_x^e(x) \propto \frac{\partial \psi^e}{\partial x} = \frac{\partial}{\partial x}\left(\frac{1}{2\pi}\int_{-\infty}^{+\infty} \tilde{\psi}^e(\alpha)\, e^{-j\alpha x}\, d\alpha\right) \tag{3.185}$$

If the above integral exists, then differentiation and integration can be interchanged,

$$\begin{aligned} E_x^e(x) &= \frac{1}{2\pi}\int_{-\infty}^{+\infty} \frac{\partial}{\partial x}\left(\tilde{\psi}^e(\alpha)\, e^{-j\alpha x}\right) d\alpha \\ &= \frac{1}{2\pi}\int_{-\infty}^{+\infty} \left(-j\alpha\,\tilde{\psi}^e(\alpha)\right) e^{-j\alpha x}\, d\alpha \\ &= \frac{1}{2\pi}\int_{-\infty}^{+\infty} \tilde{E}_x^e(\alpha)\, e^{-j\alpha x}\, d\alpha \end{aligned} \tag{3.186}$$

Thus, the field component $\tilde{E}(\alpha)$ in spectral (Fourier) domain is related to the spectral potential $\tilde{\psi}^e(\alpha)$ by the expression

$$\tilde{E}_x^e(\alpha) = -j\alpha\,\tilde{\psi}^e(\alpha) \tag{3.187}$$

Similarly, the x-component of the magnetic field $H_x^e(x)$ can be written as

$$H_x^e(x) = \frac{1}{2\pi}\int_{-\infty}^{+\infty} \tilde{H}_x^e(\alpha)\, e^{-j\alpha x}\, d\alpha \tag{3.188}$$

where $\tilde{H}_x^e(\alpha)$ is the Fourier transform of the magnetic field component H_x with respect to x, or shortly spectral magnetic field-strength. Equation (D.163) (Appendix D.2.2.2) reads for the three cartesian components in spectral domain (spectral wave equations):

$$\frac{\partial^2 \tilde{\psi}_i(\alpha,y,z)^{e,m}}{\partial x^2} + \alpha^2 \tilde{\psi}_i(\alpha,y,z)^{e,m} = 0 \tag{3.189}$$

$$\frac{\partial^2 \tilde{\psi}_i(\alpha,y,z)^{e,m}}{\partial y^2} + k_y^2 \tilde{\psi}_i(\alpha,y,z)^{e,m} = 0 \tag{3.190}$$

$$\frac{\partial^2 \tilde{\psi}_i(\alpha,y,z)^{e,m}}{\partial z^2} + k_z^2 \tilde{\psi}_i(\alpha,y,z)^{e,m} = 0 \tag{3.191}$$

The boundary conditions in the y direction read:

$$E_x^e, E_z^e \propto \frac{\partial \psi^e}{\partial y}, \quad H_x^m, H_z^m \propto \frac{\partial \psi^m}{\partial y} \tag{3.192}$$

The spectral scalar potentials defined as [53]

$$\tilde{\psi}_i^e(\alpha,y,z) = A_i^e(\alpha) \frac{\cos k_{y_i}(y - y_i)}{\cos(k_{y_i} y_i)} e^{-jk_z z} \tag{3.193}$$

$$\tilde{\psi}_i^m(\alpha,y,z) = A_i^m(\alpha) \frac{\sin k_{y_i}(y - y_i)}{\sin(k_{y_i} y_i)} e^{-jk_z z} \tag{3.194}$$

where $y_1 = -h_1$ and $y_2 = h_2$ already satisfy the boundary conditions at the bottom and top walls of the shielded microstrip line. The parameters $A_i^{e,m}(\alpha)$ $(i = 1,2)$ and k_z (which is implicit in k_{y_i}) are still unknown coefficients. The spectral field components can be derived from the spectral potential $\tilde{\psi}_i^{e,m}$ $(i = 1, 2)$. For brevity, the propagation term $e^{-jk_z z}$ is omitted in the following discussions. With (D.171) and (D.177), we get

$$\tilde{E}_{x_i}^e(\alpha,y,z) = \frac{1}{j\omega\epsilon_i} \frac{\partial^2 \tilde{\psi}_i^e}{\partial x \partial y} + \frac{\partial \tilde{\psi}_i^m}{\partial z}$$

$$= \frac{\alpha k_{yi}}{\omega \epsilon_i} \frac{\sin k_{yi}(y - y_i)}{\cos(k_{yi} y_i)} A_i^e(\alpha) - jk_z \frac{\sin k_{yi}(y - y_i)}{\sin(k_{yi} y_i)} A_i^m(\alpha) \tag{3.195}$$

$$\tilde{E}^e_{y_i}(\alpha,y,z) = \frac{1}{j\omega\epsilon_i}\left(\frac{\partial^2}{\partial y^2} + k^2\right)\tilde{\psi}_i^e$$

$$= \frac{k_i^2 - k_{yi}^2}{j\omega\epsilon_i}\frac{\cos k_{yi}(y - y_i)}{\cos(k_{yi}y_i)}A_i^e(\alpha) \tag{3.196}$$

$$\tilde{E}^e_{z_i}(\alpha,y,z) = \frac{1}{j\omega\epsilon_i}\frac{\partial^2\tilde{\psi}_i^e}{\partial y\partial z} - \frac{\partial\tilde{\psi}_i^m}{\partial x}$$

$$= \frac{k_z k_{yi}}{\omega\epsilon_i}\frac{\sin k_{yi}(y - y_i)}{\cos(k_{yi}y_i)}A_i^e(\alpha) + j\alpha\frac{\sin k_{yi}(y - y_i)}{\sin(k_{yi}y_i)}A_i^m(\alpha) \tag{3.197}$$

$$\tilde{H}^e_{x_i}(\alpha,y,z) = -\frac{\partial\tilde{\psi}_i^e}{\partial z} + \frac{1}{j\omega\mu_0}\frac{\partial^2\tilde{\psi}_i^m}{\partial x\partial y}$$

$$= jk_z\frac{\cos k_{yi}(y - y_i)}{\cos(k_{yi}y_i)}A_i^e(\alpha) - \frac{\alpha k_{y_i}}{\omega\mu_0}\frac{\cos k_{yi}(y - y_i)}{\sin(k_{yi}y_i)}A_i^m(\alpha) \tag{3.198}$$

$$\tilde{H}^e_{y_i}(\alpha,y,z) = \frac{1}{j\omega\mu_0}\left(\frac{\partial^2\tilde{\psi}_i^m}{\partial y^2} + k^2\tilde{\psi}_i^m\right)$$

$$= \frac{k_i^2 - k_{y_i}^2}{j\omega\mu_0}\frac{\sin k_{yi}(y - y_i)}{\sin(k_{yi}y_i)}A_i^m(\alpha) \tag{3.199}$$

$$\tilde{H}_{z_i}(\alpha,y,z) = \frac{\partial\tilde{\psi}_i^e}{\partial x} + \frac{1}{j\omega\mu_0}\frac{\partial^2\tilde{\psi}_i^m}{\partial y\partial z}$$

$$= -j\alpha\frac{\cos k_{yi}(y - y_i)}{\cos(k_{yi}y_i)}A_i^e(\alpha) - \frac{k_z k_{yi}}{\omega\mu_0}\frac{\cos k_{yi}(y - y_i)}{\sin(k_{yi}y_i)}A_i^m(\alpha) \tag{3.200}$$

At the air–dielectric interface the tangential field components [(3.195), (3.197), (3.198), and (3.200)] read for the dielectric medium ($y_1 = -h_1$):

$$\tilde{E}_{x_1}(\alpha,y,z) = \frac{\alpha k_{y1}}{\omega \epsilon_1} \tan(k_{y1} h_1) A_1^e(\alpha) + jk_z A_1^m(\alpha) \tag{3.201}$$

$$\tilde{E}_{z_1}(\alpha,y,z) = \frac{k_z k_{y1}}{\omega \epsilon_i} \tan(k_{y1} h_1) A_1^e(\alpha) - j\alpha A_1^m(\alpha) \tag{3.202}$$

$$\tilde{H}_{x_1}(\alpha,y,z) = jk_z \; A_1^e(\alpha) + \frac{\alpha k_{y1}}{\omega \mu_0} \frac{1}{\tan k_{y1} h_1} A_1^m(\alpha) \tag{3.203}$$

$$\tilde{H}_{z_1}(\alpha,y,z) = -j\alpha A_1^e(\alpha) + \frac{k_z k_{y1}}{\omega \mu_0} \frac{1}{\tan k_{y1} h_1} A_1^m(\alpha) \tag{3.204}$$

for the air medium ($y_2 = h_2$):

$$\tilde{E}_{x_2}(\alpha,y,z) = -\frac{\alpha k_{y2}}{\omega \epsilon_2} \tan(k_{y2} h_2) A_2^e(\alpha) + jk_z A_2^m(\alpha) \tag{3.205}$$

$$\tilde{E}_{z_2}(\alpha,y,z) = -\frac{k_z k_{y2}}{\omega \epsilon_2} \tan(k_{y2} h_2) A_2^e(\alpha) - j\alpha \tan(k_{y2} h_2) A_2^m(\alpha) \tag{3.206}$$

$$\tilde{H}_{x_2}(\alpha,y,z) = jk_z \; A_2^e(\alpha) - \frac{\alpha k_{y2}}{\omega \mu_0} \frac{1}{\tan(k_{y2} h_2)} A_2^m(\alpha) \tag{3.207}$$

$$\tilde{H}_{z_2}(\alpha,y,z) = -j\alpha A_2^e(\alpha) - \frac{k_z k_{y2}}{\omega \mu_0} \frac{1}{\tan(k_{y2} h_2)} A_2^m(\alpha) \tag{3.208}$$

The continuity of the electrical field at the interface (also along the strip) reads:

$$\frac{\alpha k_{y1}}{\omega \epsilon_1} \tan(k_{y1} h_1) A_1^e(\alpha) + jk_z A_1^m(\alpha) = -\frac{\alpha k_{y2}}{\omega \epsilon_2} \tan(k_{y2} h_2) A_2^e(\alpha) + jk_z A_2^m(\alpha) \tag{3.209}$$

$$\frac{k_z k_{y1}}{\omega \epsilon_1} \tan(k_{y1} h_1) A_1^e(\alpha) - j\alpha A_1^m(\alpha) = -\frac{k_z k_{y_2}}{\omega \epsilon_2} \tan(k_{y2} h_2) A_2^e(\alpha) - j\alpha \tan(k_{y2} h_2) A_2^m(\alpha) \tag{3.210}$$

With

$$T_{yi} = k_{yi} \tan(k_{yi} h_i) \tag{3.211}$$

(3.209) and (3.210) give

$$\frac{\alpha}{\omega \epsilon_1} T_{y1} A_1^e(\alpha) + jk_z A_1^m(\alpha) = -\frac{\alpha}{\omega \epsilon_2} T_{y2} A_2^e(\alpha) + jk_z A_2^m(\alpha) \tag{3.212}$$

$$\frac{k_z}{\omega \epsilon_1} T_{y1} A_1^e(\alpha) - j\alpha A_1^m(\alpha) = -\frac{k_z}{\omega \epsilon_2} T_{y2} A_2^e(\alpha) - j\alpha A_2^m(\alpha) \tag{3.213}$$

Thus, the potentials in region 2 can be related to the potentials in region 1:

$$A_2^m(\alpha) = A_1^m(\alpha) \tag{3.214}$$

$$A_2^e(\alpha) = -\frac{\epsilon_2}{\epsilon_1} \frac{T_{y1}}{T_{y2}} A_1^e(\alpha) \tag{3.215}$$

For reason of strip current, the tangential magnetic field is discontinuous in the region $-w/2 \le x \le w/2$. From Appendix D.1.1, we obtain in frequency domain

$$\tilde{J}_x = \tilde{H}_{z_2} - \tilde{H}_{z_1} \tag{3.216}$$

$$-\tilde{J}_z = \tilde{H}_{x_2} - \tilde{H}_{x_1} \tag{3.217}$$

Substituting the magnetic field components $\tilde{H}_{z_1}$ and $\tilde{H}_{z_2}$ of (3.204) and (3.208) in (3.216) and taking into account relations (3.214) and (3.215) result in

$$\tilde{J}_x = -j\alpha A_2^e(\alpha) - \frac{k_z k_{y2}}{\omega \mu_0} \frac{1}{\tan(k_{y2} h_2)} A_2^m(\alpha)$$

$$+j\alpha A_1^e(\alpha) - \frac{k_z k_{y1}}{\omega \mu_0} \frac{1}{\tan k_{y1} h_1} A_1^m(\alpha)$$

$$= j\alpha \left(1 + \frac{\epsilon_2}{\epsilon_1} \frac{T_{y1}}{T_{y2}}\right) A_1^e(\alpha) - \frac{k_z}{\omega \mu_0} \left(\frac{k_{y2}}{\tan(k_{y2} h_2)} + \frac{k_{y1}}{\tan(k_{y1} h_1)}\right) A_1^m(\alpha)$$

$$= j\alpha \frac{T_{y1}}{k_1^2}\left(\frac{k_2^2}{T_{y2}} + \frac{k_1^2}{T_{y1}}\right) A_1^e(\alpha) - \frac{k_z}{\omega\mu_0}\left(\frac{k_{y2}^{\ 2}}{T_{y2}} + \frac{k_{y1}^{\ 2}}{T_{y1}}\right) A_1^m(\alpha)$$

$$= j\alpha \frac{T_{y1}}{k_1^2} F_e A_1^e(\alpha) - \frac{k_z}{\omega\mu_0} F_m A_1^m(\alpha) \tag{3.218}$$

where

$$F_e = \frac{k_2^2}{T_{y2}} + \frac{k_1^2}{T_{y1}} \tag{3.219}$$

and

$$F_m = \frac{k_{y2}^{\ 2}}{T_{y2}} + \frac{k_{y1}^{\ 2}}{T_{y1}} \tag{3.220}$$

Similarly, we substitute the magnetic field components $\tilde{H}_{x_1}$ and $\tilde{H}_{x_2}$ of (3.203) and (3.207) in (3.217), and with relations (3.214) and (3.215) we obtain

$$-\tilde{J}_z(\alpha,y,z) = jk_z\ A_2^e(\alpha) - \frac{\alpha k_{y1}}{\omega\mu_0}\frac{1}{\tan(k_{y2}h_2)} A_2^m(\alpha)$$

$$- jk_z A_1^e(\alpha) - \frac{\alpha k_{y1}}{\omega\mu_0}\frac{1}{\tan(k_{y1}h_1)} A_1^m(\alpha)$$

$$= -jk_z\left(1 + \frac{\epsilon_2 T_{y1}}{\epsilon_1 T_{y2}}\right) A_1^e(\alpha) - \frac{\alpha}{\omega\mu_0}\left(\frac{k_{y2}}{\tan(k_{y2}h_2)} + \frac{k_{y1}}{\tan(k_{y1}h_1)}\right) A_1^m(\alpha)$$

$$= -jk_z \frac{T_{y1}}{k_1^2} F_e A_1^e(\alpha) - \frac{\alpha}{\omega\mu_0} F_m A_1^m(\alpha) \tag{3.221}$$

For simplification, we write (3.218) and (3.221) in an abbreviated form,

$$\tilde{J}_x = a A_1^e(\alpha) + b A_1^m(\alpha) \tag{3.222}$$

$$\tilde{J}_z(\alpha,y,z) = c A_1^e(\alpha) + d A_1^m(\alpha) \tag{3.223}$$

The system of equations can be solved for the parameters A_1^e and A_1^m,

$$A_1^e = \frac{d}{ad - bc}\tilde{J}_x - \frac{b}{ad - bc}\tilde{J}_z \tag{3.224}$$

$$A_1^m = \frac{c}{cb - ad}\tilde{J}_x - \frac{a}{cb - ad}\tilde{J}_z \tag{3.225}$$

Replacing the auxiliary coefficients a, b, c, and d by their respective expressions, the parameters A_1^e and A_1^m read as functions of the strip current components $\tilde{J}_x$ and $\tilde{J}_z$

$$A_1^e = -j \frac{k_1^2}{F_e T_{y1}} \frac{1}{\alpha^2 + k_z^2} (\alpha \tilde{J}_x + k_z \tilde{J}_z) \tag{3.226}$$

$$A_1^m = - \frac{\omega \mu_0}{F_m} \frac{1}{\alpha^2 + k_z^2} (k_z \tilde{J}_x - \alpha \tilde{J}_z) \tag{3.227}$$

Then, the field components $\tilde{E}_{x_1}$ and $\tilde{E}_{z_1}$ read from (3.201) and (3.202):

$$\begin{aligned} \tilde{E}_{x_1} &= \frac{\alpha T_{y1}}{j\omega\epsilon_1} \frac{k_1^2}{F_e T_{y1}} \frac{1}{\alpha^2 + k_z^2} (\alpha \tilde{J}_x + k_z \tilde{J}_z) - jk_z \frac{\omega \mu_0}{F_m} \frac{1}{\alpha^2 + k_z^2} (k_z \tilde{J}_x - \alpha \tilde{J}_z) \\ &= \frac{k_0^2}{j\omega\epsilon_0} \frac{1}{\alpha^2 + k_z^2} \left[\left(\frac{\alpha^2}{F_e} + \frac{k_z^2}{F_m} \right) \tilde{J}_x + \alpha k_y \left(\frac{1}{F_e} - \frac{1}{F_m} \right) \tilde{J}_z \right] \end{aligned} \tag{3.228}$$

with

$$\omega \mu_0 = \frac{k_0^2}{\omega \epsilon_0}$$

$$\frac{k_1^2}{\omega\epsilon_1} = \frac{k_0^2}{\omega \epsilon_0}$$

$$\begin{aligned} \tilde{E}_{z_1} &= \frac{k_z T_{y1}}{j\omega\epsilon_1} \frac{k_1^2}{F_e T_{y1}} \frac{1}{k_x^2 + k_z^2} (\alpha \tilde{J}_x + k_z \tilde{J}_z) + jk_x \frac{\omega \mu_0}{F_m} \frac{1}{k_x^2 + k_z^2} (k_z \tilde{J}_x - \alpha \tilde{J}_z) \\ &= \frac{k_0^2}{j\omega\epsilon_0} \frac{1}{\alpha^2 + k_z^2} \left[\alpha k_z \left(\frac{1}{F_e} - \frac{1}{F_m} \right) \tilde{J}_x + \left(\frac{\alpha^2}{F_m} + \frac{k_z^2}{F_e} \right) \tilde{J}_z \right] \\ &= \frac{k_0^2}{j\omega\epsilon_0} \frac{1}{\alpha^2 + k_z^2} \left[\left(\frac{\alpha^2}{F_e} + \frac{k_z^2}{F_m} \right) \tilde{J}_x + \alpha k_y \left(\frac{1}{F_e} - \frac{1}{F_m} \right) \tilde{J}_z \right] \end{aligned} \tag{3.229}$$

Equations (3.228) and (3.229) can be written in abbreviated form

$$\tilde{E}_{z_1} = G_{11}(\alpha,k_z)\tilde{J}_x + G_{12}(\alpha,k_z)\tilde{J}_z \tag{3.230}$$

$$\tilde{E}_{x_1} = G_{21}(\alpha,k_z)\tilde{J}_x + G_{22}(\alpha,k_z)\tilde{J}_z \tag{3.231}$$

$$G_{11}(\alpha,k_z) = G_{22}(\alpha,k_z) = \frac{k_0^2}{j\omega\epsilon_0}\frac{1}{\alpha^2 + k_y^2}\left[\alpha k_z\left(\frac{1}{F_e} - \frac{1}{F_m}\right)\right]$$

$$G_{12}(\alpha,k_z) = \frac{k_0^2}{j\omega\epsilon_0}\frac{1}{\alpha^2 + k_y^2}\left(\frac{\alpha^2}{F_m} + \frac{k_z^2}{F_e}\right)$$

$$G_{21}(\alpha,k_z) = \frac{k_0^2}{j\omega\epsilon_0}\frac{1}{\alpha^2 + k_y^2}\left(\frac{\alpha^2}{F_e} + \frac{k_z^2}{F_m}\right)$$

Matrix notation reads,

$$\begin{pmatrix}\tilde{E}_z \\ \tilde{E}_x\end{pmatrix} = \begin{pmatrix}G_{11} & G_{12} \\ G_{21} & G_{22}\end{pmatrix}\begin{pmatrix}\tilde{J}_x \\ \tilde{J}_z\end{pmatrix} \tag{3.232}$$

The two equations (3.230) and (3.231) contain four unknowns $\tilde{J}_x$, $\tilde{J}_z$, $\tilde{E}_x$, and $\tilde{E}_z$. The special properties of the functions, namely that the tangential electric field and the surface current should vanish in complementary parts of the interface, can be exploited to eliminate the two latter unknowns, $\tilde{E}_x$ and $\tilde{E}_z$. To repeat, the final aim is the determination of the propagating phase coefficient k_z from (3.232). Next we define the inner product [54]

$$\langle A(\alpha), B(\alpha)\rangle = \int_{-\infty}^{+\infty} A(\alpha)B(\alpha)^* d\alpha \tag{3.233}$$

and take the inner product of (3.230) and (3.231) with weighting functions $W_i(\alpha)$. If we choose

$$W_1(\alpha) = \tilde{J}_x(\alpha) \tag{3.234}$$

$$W_2(\alpha) = \tilde{J}_z(\alpha) \tag{3.235}$$

we obtain

$$\langle G_{11}\tilde{J}_x, \tilde{J}_x\rangle + \langle G_{12}\tilde{J}_z, \tilde{J}_x\rangle = \langle \tilde{E}_z, \tilde{J}_x\rangle \tag{3.236}$$

$$\langle G_{21}\tilde{J}_x, \tilde{J}_z\rangle + \langle G_{22}\tilde{J}_z, \tilde{J}_z\rangle = \langle \tilde{E}_x, \tilde{J}_z\rangle \tag{3.237}$$

It follows from Parseval's theorem [55] that the right-hand side of (3.230) and (3.231) is zero. For example, for $\langle \tilde{E}_z, \tilde{J}_x \rangle$ we obtain

$$\langle \tilde{E}_z, \tilde{J}_x \rangle = \int_{-\infty}^{+\infty} \tilde{E}_x(\alpha)\, \tilde{J}_x^*(\alpha)\, d\alpha = \int_{-\infty}^{+\infty} E_z(x)\, J_x(x)\, dx = 0 \tag{3.238}$$

because the tangential electric field vanishes on the strip, whereas the strip current is zero outside of the strip. Next, we expand the unknown surface current components in terms of known basis functions $\tilde{J}_{x_n}(\alpha)$ and $\tilde{J}_{z_n}(\alpha)$:

$$\tilde{J}_x(\alpha) = \sum_{n=1}^{\infty} a_n \tilde{J}_{x_n}(x) \tag{3.239}$$

$$\tilde{J}_z(x) = \sum_{n=1}^{\infty} b_n \tilde{J}_{z_n}(x) \tag{3.240}$$

For numerical calculation, the series is truncated after the nth term. The basis functions should be analytically Fourier transformable. As can be seen in Figure 3.45, the current is not constant across the strip width. In fact, its amplitude increases very rapidly at the edges of the strip. The longitudinal current terms are defined as

$$J_{z_n}(x) = \begin{cases} \dfrac{\cos\left((n-1)\pi \dfrac{x}{w/2}\right)}{\sqrt{1 - \left(\dfrac{x}{w/2}\right)^2}} & -w/2 < x < w/2 \\ 0 & |x| > w/2 \end{cases} \tag{3.241}$$

The amplitude of the transversal current varies sinusoidally in the x-direction. Thus, the transverse current term is described by

$$J_{x_n}(x) = \begin{cases} \dfrac{\sin\left(n\pi \dfrac{x}{w/2}\right)}{\sqrt{1 - \left(\dfrac{x}{w/2}\right)^2}} & -w/2 < x < w/2 \\ 0 & |x| > w/2 \end{cases} \tag{3.242}$$

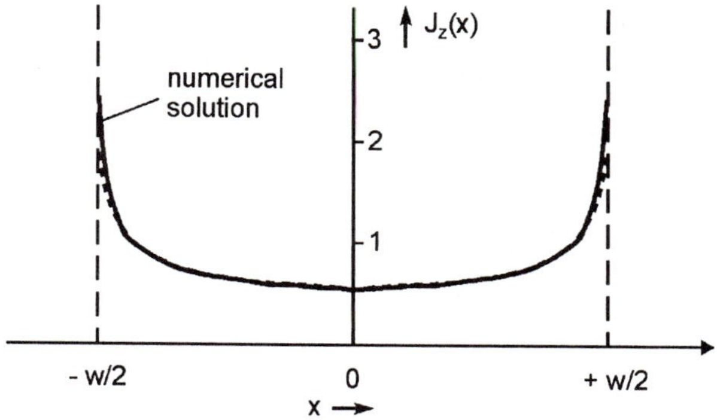

Figure 3.45 Longitudinal current distribution in the strip [solid line: numerical solution, dashed line: approximation of (3.241)]. (From: [27], © 1971 IEEE. Reprinted with permission.)

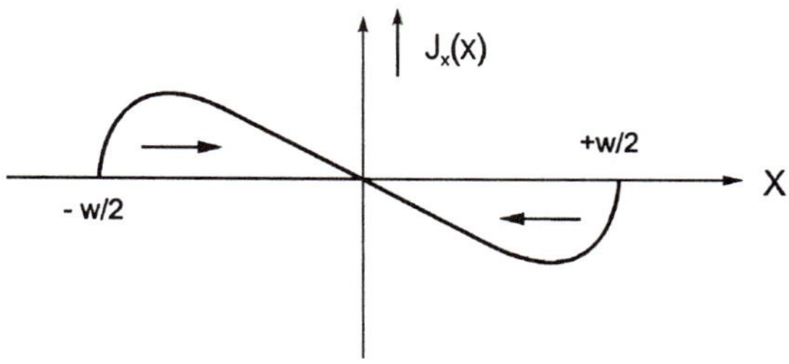

Figure 3.46 Transverse current distribution in the strip.

The transverse current of the fundamental mode shown in Figure 3.46 can be approximated as [27]

$$J_x(x) = J_{x0} \sin\frac{\pi x}{0.7w}, \quad |x| \leq 0.8\frac{w}{2}$$

$$J_x(x) = J_{x0} \cos\frac{\pi x}{0.2w}, \quad 0.8\frac{w}{2} \leq |x| \leq \frac{w}{2} \tag{3.243}$$

The amplitude of the transverse current component is proportional to the normalized strip width w/λ, which for most applications is very small. When this parameter is less than 0.1, the transverse current is at least an order of magnitude smaller than the longitudinal current.

After substituting the Fourier transforms of the transverse and longitudinal current terms, we take the inner products with the basis functions $\tilde{J}_{x_m}$ and $\tilde{J}_{z_m}$ for different values of m. Thus, for example, (3.236) reads:

$$\left\langle \sum_{n=1}^{\infty} G_{11}(\alpha,k_z)\, a_n \tilde{J}_{x_n}(\alpha),\ \tilde{J}_{x_m}(\alpha) \right\rangle + \left\langle \sum_{n=1}^{\infty} G_{12}(\alpha,k_z)\, b_n \tilde{J}_{z_n}(\alpha),\ \tilde{J}_{x_m}(\alpha) \right\rangle = 0 \tag{3.244}$$

with $m = 1, 2, \ldots, M$. After sorting and recalling (3.232) we obtain the following set of equations from (3.231) [56]:

$$\begin{aligned} \sum_{n=1}^{N} K_{mn}^{(11)} a_n + \sum_{n=1}^{N} K_{mn}^{(12)} b_n &= 0 \\ \sum_{n=1}^{N} K_{mn}^{(21)} a_n + \sum_{n=1}^{N} K_{mn}^{(22)} b_n &= 0 \end{aligned} \tag{3.245}$$

where, from the definition of the inner product,

$$\begin{aligned} K_{mn}^{(11)}(k_z) &= \int_{-\infty}^{+\infty} G_{11}(\alpha,k_z)\, \tilde{J}_{x_n} \tilde{J}_{x_m}^{*}\, d\alpha \\ K_{mn}^{(12)}(k_z) &= \int_{-\infty}^{+\infty} G_{12}(\alpha,k_z)\, \tilde{J}_{z_n} \tilde{J}_{x_m}^{*}\, d\alpha \\ K_{mn}^{(21)}(k_z) &= \int_{-\infty}^{+\infty} G_{21}(\alpha,k_z)\, \tilde{J}_{x_n} \tilde{J}_{z_m}^{*}\, d\alpha \\ K_{mn}^{(22)}(k_z) &= \int_{-\infty}^{+\infty} G_{22}(\alpha,k_z)\, \tilde{J}_{z_n} \tilde{J}_{z_m}^{*}\, d\alpha \end{aligned} \tag{3.246}$$

J_{x_n} and J_{y_n} are the basis functions of the surface current. J_{x_m} and J_{z_m} are the so-called test functions, which are identical with the basis functions (Galerkin method). The nontrivial solution of (3.244) is found by setting the determinant of the coefficient matrix equal to zero, and seeking the root of the resulting equation, which delivers the propagation coefficient value k_z at each frequency ω. With k_z known, the frequency-dependent effective dielectric constant can be determined:

$$\frac{k_z}{k_0} = \frac{2\pi}{\lambda}\frac{1}{k_0} = \frac{2\pi}{\lambda_0}\frac{\sqrt{\epsilon_{eff}(f)}}{k_0} = \sqrt{\epsilon_{eff}(f)} \tag{3.247}$$

Table 3.3
Comparison of the Discussed Quasi-Static Microstrip Analysis Methods

ϵ_r	*Height (mm)*	$Z_0(\varepsilon_r)$ (Ω) [49], [50]	$Z_0(\varepsilon_r)$ (Ω) [37]	$\Delta Z_0(\varepsilon_r)$ (%)	$Z_0(\varepsilon_r)$ (Ω) [38]	$\Delta Z_0(\varepsilon_r)$ (%)
2.32	0.794	25	24.891	–0.436	25.052	+0.209
		50	49.362	–1.276	50.200	+0.400
		75	75.321	+0.428	74.659	–0.455
9.8	0.635	25	24.704	–1.184	24.908	–0.364
		50	50.177	+0.354	49.498	–1.004
		75	75.236	+3.147	74.774	–0.30[illegible]
12.9	0.5	25	24.648	–1.408	24.870	–0.520
		50	50.166	+0.332	49.485	–1.030
		75	75.172	+0.229	74.981	–0.025

3.5 Comparison of Discussed Analysis Methods

Table 3.3 summarizes and compares the results obtained from the treated quasi-static microstrip analysis methods. The characteristic impedances of 25Ω, 50Ω, and 75Ω were calculated after Hammerstad with (3.120) and (3.121), and adopted by Kirschning applying full-wave analysis [50]. Wheeler's results have been derived from (3.68) and (3.70). Calculations after Schneider are based on (3.104), (3.105), and (3.109). A maximum relative error of 3.147% has been obtained using Wheeler's theory, whereas the results after Schneider show slightly better agreement with the full-wave analysis. In this case, a maximum relative error of 0-1.03% was calculated for the examples under consideration. Thus, it has been shown that the discrepancy of the results of various analysis methods are marginal, in particular regarding first step design. Final circuit design will be done with the aid of a computer.

References

[1] Gupta, K. C., R. Garg, and I. J. Bahl, *Microstrip Lines and Slotlines*, Dedham; MA: Artech House, Inc., 1979.

[2] Gupta, K. C., et al., *Microstrip Lines and Slotlines*, 2nd ed., Norwood, MA: Artech House, Inc., 1996.

[3] Assadourian, F., and E. Rimai, "Simplified Theory of Microstrip Transmission Systems," *Proc. I.R.E.*, December 1952, pp. 1651–1657.

[4] Wheeler, H. A., "Transmission-Line Properties of Parallel Wide Strips by a Conformal Mapping Approximation," *IEEE Trans.*, Vol. MTT-12, 1964, pp. 280–289.

[5] Wheeler, H. A., "Transmission-Line Properties of Parallel Strips Separated by a Dielectric Sheet," *IEEE Trans.*, Vol. MTT-13, 1965, pp. 172–185.

[6] Green, H. E., "The Numerical Solution of Some Important Transmission-Line Problems," *IEEE Trans.*, Vol. MTT-13, 1965, pp. 676–692.

[7] Stinehelfer, H. E., "An Accurate Calculation of Uniform Microstrip Transmission Lines," *IEEE Trans.*, Vol. MTT-16, 1968, pp. 439–444.

[8] Schneider, M. V., "Computation of Impedance and Attenuation of TEM-Lines by Finite Difference Methods," *IEEE Trans.*, Vol. MTT-13, 1965, pp. 793–800.

[9] Silvester, P., "TEM Wave Properties of Microstrip Transmission Lines," *Proc. IEE.*, Vol. 115, 1968, pp. 43–48.

[10] Yamashita, E., and R. Mittra, "Variational Method for the Analysis of Microstrip Lines," *IEEE Trans.*, Vol. MTT-16, 1968, pp. 251–256.

[11] Yamashita, E., "Variational Method for the Analysis of Microstrip-Like Transmission Lines," *IEEE Trans.*, Vol. MTT-16, 1968, pp. 529–535.

[12] Collin, R. E., *Field Theory of Guided Waves*, New York: McGraw-Hill Book Company, 1960.

[13] Kostriza, J. A., "Microstrip Components," *Proc. I.R.E.*, December 1952, pp. 1658–1663.

[14] Hewlett-Packard, *Stripline Component Measurements with the 8410A Network Analyzer*, Application Note 117-2, January 1971.

[15] Kompa, G., *Die frequenzabhängigen Übertragungseigenschaften von Streifenleitungs-Wellenwiderstandssprüngen, Filtern und Stichleitungen in Mikrostrip-Technik*, Doctoral Thesis, Aachen: RWTH, 1975.

[16] Kompa, G., "S-Matrix Computation of Microstrip Discontinuities with a Planar Waveguide Model," *Arch. Elektr. Übertr.*, Vol. 30, 1976, pp. 58–64.

[17] Hammerstad, E. O., and F. Bekkadal, *Microstrip Handbook ELAB Report*, STF 44 A74169, University of Trondheim, Norway, 1975.

[18] Menzel, W., *Die frequenzabhängigen Übertragungseigenschaften von unsymmetrischen Kreuz- und T-Verzweigungen, Y-Verzweigungen sowie 90°- und 120°-Winkeln in Mikrostreifenleitungstechnik*, Doctoral Thesis, University of Duisburg, 1977.

[19] Louisell, W. H., *Coupled Mode and Parametric Electronics*, New York: John Wiley & Sons, 1960.

[20] Watkins, D. A., *Topics in Electromagnetic Theory*, New York: John Wiley & Sons, 1958.

[21] Hartwig, C. P., D. Massé, and R. A. Pucel, "Frequency Dependent Behavior of Microstrip," *Proc. of the GMTT Microwave Symposium*, Detroit, 1968, pp. 110–116.

[22] Getsinger, W. J., "Microstrip Dispersion Model," *IEEE Trans.*, Vol. MTT-21, 1973, pp. 34–39.

[23] Wolff, I., G. Kompa, and R. Mehran, "Calculation Method for Microstrip Discontinuities and T Junctions," *Electron. Lett.*, Vol. 8, 1972, pp. 177–179.

[24] Wolff, I., G. Kompa, and R. Mehran, "Streifenleitungsdiskontinuitäten und Verzweigungen," *Nachrichtentechnische Zeitschrift*, Vol. 25, 1972, pp. 217–224.

[25] Kompa, G., and R. Mehran, "Planar Waveguide Model for Calculating Microstrip Components," *Electron. Lett.*, Vol. 11, September 1975, pp. 459–460.

[26] Hornsby, J. S., and A. Gopinath, "Numerical Analysis of a Dielectric Loaded Waveguide with a Microstrip Line—Finite Difference Methods," *IEEE Trans.*, Vol. MTT-17, 1969, pp. 684–690.

[27] Denlinger, E. J., "A Frequency Dependent Solution for Microstrip Transmission Lines," *IEEE Trans.*, Vol. MTT-19, 1971, pp. 30–39.

[28] Schmitt, H. J., and K. H. Sarges, "Wave Propagation in Microstrip," *Nachrichtentechnische Zeitschrift*, Vol. 24, 1971, pp. 260–264.

[29] Itoh, T., and R. Mittra, "Spectral-Domain Approach for Calculating the Dispersion Characteristics of Microstrip Lines," *IEEE Trans.*, Vol. MTT-21, 1973, pp. 496–499.

[30] Itoh, T., and R. Mittra, "A Technique for Computing Dispersion Characteristics of Shielded Microstrip Lines," *IEEE Trans.*, Vol. MTT-22, 1974, pp. 896–898.

[31] Caulton, M., J. J. Hughes, and H. Sobol, "Measurements on the Properties of Microstrip Transmission Lines for Microwave Integrated Circuits," *RCA Review*, Vol. 27, September 1966, pp. 377–391.

[32] Bronstein, L. N., and K. A. Semendjajew, *Taschenbuch der Mathematik*, 8. Auflage, Zürich und Frankfurt/M: Verlag Harri Deutsch, 1968.

[33] Oliner, A. A., "Equivalent Circuits for Discontinuities in Balanced Strip Transmission Line," *IEEE Trans.*, Vol. MTT-3, 1955, pp. 134–143.

[34] Altschuler, H. M., and A. A. Oliner, "Discontinuities in the Center Conductor of Symmetric Strip Transmission Line," *I.R.E. Trans.*, MTT, May 1960, 328–339.

[35] Wolff, I., *Einführung in die Mikrostrip-Leitungstechnik*, Aachen: Verlag H. Wolff, 2. Auflage, 1978.

[36] Pucel, R. A., D. J. Massé, and C. P. Hartwig, "Losses in Microstrips," *IEEE Trans.*, Vol. MTT-16, 1968, pp. 342–350.

[37] Wheeler, H. A., "Transmission-line Properties of a Strip on a Dielectric Sheet on a Plane," *IEEE Trans.*, Vol. MTT-25, 1977, pp. 631–647.

[38] Schneider, M. V., "Microstrip Lines for Microwave Integrated Circuits," *Bell System Technical Journal*, Vol. 48, 1969, pp. 1421–1444.

[39] Bellman, R., *A Brief Introduction to Theta Functions*, Athena Series—Selected Topics in Mathematics, New York: Holt, Rinehart and Winston, 1961.

[40] Oberhettinger, F., and W. Magnus, *Anwendung der Elliptischen Funktionen in Physik und Technik*, Springer-Verlag, 1949.

[41] Abramowitz, M., and I. A. Stegun, *Handbook of Mathematical Functions*, Applied Mathematics Series, No. 55, Tenth Printing, Washington, DC: National Bureau of Standards, 1972.

[42] Bowman, F., *Introduction to Elliptic Functions*, New York: Dover Publications, Inc., 1961.

[43] Tölke, F., Praktische Funktionenlehre. Zweiter Band: Theta-Funktionen und spezielle Weierstraßsche Funktionen. Springer-Verlag, 1966.

[44] Fettis, H. E., and J. C. Caslin, *Ten Place Tables of the Jacobian Elliptic Functions*, Aerospace Research Laboratories, Wright-Patterson Air Force Base, Ohio, Project 7071, ARL 65-180, Part I, September 1965.

[45] Hilberg, W., "Eine einfache und gute Näherung für den Wellenwiderstand paralleler Streifenleitungen," *Arch. Elektr. Übertr.*, Vol. 22, 1968, pp. 122–126.

[46] Schneider, M. V., B. Glance, and W. F. Bodtmann, "Microwave and Millimeter Wave Hybrid Integrated Circuits for Radio Systems," *Bell System Technical Journal*, Vol. 48, July–August 1969, pp. 1703–1726.

[47] Davis, W. A., *Microwave Semiconductor Circuit Design*, Van Nostrand Reinhold Company, 1984.

[48] Hammerstad, E., "Equations for Microstrip Circuit Design," *Proc. European Microwave Conf.*, 1975, pp. 268–272.

[49] Hammerstad, E., and O. Jensen, " Accurate Models for Microstrip Computer-Aided Design," *IEEE MTT-S Int. Microwave Symp. Digest*, 1980, pp. 407–409.

[50] Kirschning, M., *Entwicklung von Näherungsmodellen für den rechnergestützten Entwurf von hybriden und monolithischen Schaltungen und Mikrostreifenleitungstechnik*, Doctoral Thesis, University of Duisburg, 1984.

[51] Collin, R. E., *Foundations for Microwave Engineering*, 2nd ed., New York: McGraw-Hill, Inc., 1992.

[52] Gardiol, F. E., "Radiation from Microstrip Circuits: An Introduction," *Int. J. of Microwave and Millimeter-Wave Computer-Aided Engineering*, Vol. 1, 1991, pp. 225–235.

[53] Boukamp, J., *Spektralbereichsmethoden zur Berechnung von Strahlungseffekten in integrierten Mikrowellenschaltungen*, Doctoral Thesis, Aachen: RWTH, 1984.

[54] Knorr, J. B., and A. Tufekcioglu, "Spectral-Domain Calculation of Microstrip Characteristic Impedance," *IEEE Trans.*, Vol. MTT-23, 1975, pp. 725–728.

[55] Lüke, H. D., *Signalübertragung*, Springer-Verlag, Hochschultext, 1975.

[56] Mirshekar-Syahkal, D., *Spectral Domain Method for Microwave Integrated Circuits*, New York: John Wiley & Sons, 1990.

Chapter 4

Dispersion on Microstrip Line

The wave propagation of the quasi-TEM mode along the microstrip transmission line is mainly determined by the layered air–dielectric material system. As is known, lossless TEM transmission lines with homogeneous dielectric filling exhibit no dispersion effects. In this case, the phase velocity of the fundamental mode versus frequency is constant.

Regarding the microstrip transmission line with inhomogeneous dielectric filling in the transversal plane with respect to the direction of propagation, the electromagnetic field is partly located in air and in the dielectric material. At very low frequencies, near dc, the electromagnetic field can be assumed to be purely transversal. In this case, the phase velocity of the wave is related to the velocity of light (c_0) divided by the square root of the quasi-static effective dielectric constant [$\epsilon_{eff}(f=0)$], which is a function of the relative dielectric constant (ϵ_r) and thickness of the substrate material (h), and the dimensions of the upper strip of the microstrip line (w, t). Electrostatic field analysis can be applied to derive relations for the quasi-static effective dielectric constant [1, 2].

The power flow along the microstrip line occurs both in the air and dielectric regions. With increasing frequency, the portion of power in the dielectric is growing continuously. In the limited case of infinite frequency, the electromagnetic field can be supposed to be totally concentrated in the dielectric material. In this case, the phase velocity is calculated as $c_0/\sqrt{\epsilon_r}$.

Figure 4.1 illustrates the general frequency dependence of the phase velocity. We may introduce a transition frequency f_{tr} to distinguish between a quasi-static region at lower frequencies and a dispersive region above. Within the quasi-static region, the phase velocity shows only slight frequency dependence, so that the propagation properties of the quasi-TEM mode can be well approximated by the following relation:

$$v(f) \approx v(f=0) = \frac{c_0}{\sqrt{\epsilon_{eff}(f=0)}} \qquad (0 < f \leq f_{tr}) \tag{4.1}$$

The characteristic impedance of the line with substrate carrier in this quasi-static region, based on the characteristic impedance of an air-filled microstrip line [$Z_0(\epsilon_r=1)$], reads

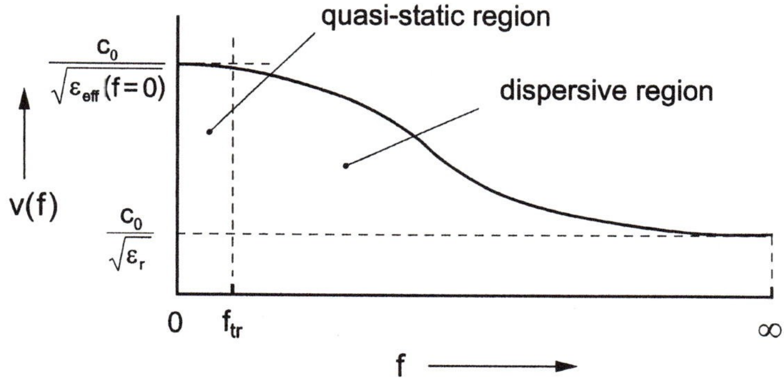

Figure 4.1 Frequency dependence of phase velocity.

$$Z_0(f, \epsilon_r) \approx Z_0(f = 0, \epsilon_r) = \frac{Z_0(\epsilon_r = 1)}{\sqrt{\epsilon_{eff}(f = 0)}} \tag{4.2}$$

The guide wavelength λ_g is approximated in the quasi-static region by

$$\lambda_g(f) \approx \frac{\lambda_0(f)}{\sqrt{\epsilon_{eff}(f = 0)}} = \frac{c_0}{f\sqrt{\epsilon_{eff}(f = 0)}} \tag{4.3}$$

The frequency f_{tr}, which characterizes the transition from the quasi-static to the dispersion region, depends on the value of the dielectric constant, thickness of the substrate, and the width of the microstrip under consideration. Regarding substrates commonly used for hybrid circuits like polyolefin ($\epsilon_r = 2.32$) and alumina ($\epsilon_r = 9.7$), the transition frequency for a characteristic impedance of a 50Ω line lies in the order of 1 GHz [3]. A more detailed relation comprising the characteristic impedance $Z_0(\epsilon_r)$, the substrate height h, and dielectric constant ϵ_r is given in [4],

$$f_{tr} = 0.95 \frac{\sqrt{Z_0(\epsilon_r)}}{h\sqrt{\epsilon_r - 1}} \tag{4.4}$$

where f_{tr} is in gigahertz and h is in millimeters. Using thin wafer material with a thickness of about 300 μm such as GaAs or HRS with a dielectric constant of 12.9 and 11.9, respectively, the transition frequency is much higher. This arises from the lower width-to-height ratio with higher value of ϵ_r, and decreasing width due to smaller substrate thickness. Using (4.4), the transition frequency for a 50Ω microstrip line on GaAs substrate material ($\epsilon_r = 12.9$, $h = 0.5$ mm) is calculated as 5.1 GHz, which, as we will see in Section 4.1, accurately characterizes the dispersion property of the respective material (Figure 4.16).

4.1 Frequency-Dependent Effective Dielectric Constant

The frequency dependence of the effective dielectric constant is directly related to the frequency-dependent phase velocity. As shown in Figure 4.2, the effective dielectric constant goes from a value $\epsilon_{eff}(f=0)$ at zero frequency to a value ϵ_r at infinite frequency, and the slope of $\epsilon_{eff}(f)$ with respect to frequency is zero at both extremes. The principal frequency dependence of $\epsilon_{eff}(f)$ can be characterized as follows:

$$\epsilon_{eff}(f)\big|_{f\to 0} = \epsilon_{eff}(f=0) \tag{4.5}$$

$$\epsilon_{eff}(f)\big|_{f\to\infty} = \epsilon_r \tag{4.6}$$

$$\frac{d\epsilon_{eff}(f)}{df} > 0 \qquad (f \neq 0, \infty) \tag{4.7}$$

$$\frac{d\epsilon_{eff}(f)}{df} = 0 \qquad (f = 0, f\to\infty) \tag{4.8}$$

Getsinger investigated a ridge waveguide with dielectric material beneath the ridge showing similar frequency-dependent dispersion properties [5]. He proposed the following analytical expression for the dispersion of a microstrip line:

$$\epsilon_{eff}(f) = \epsilon_r - \frac{\epsilon_r - \epsilon_{eff}(f=0)}{1 + P(f_n)} \tag{4.9}$$

where

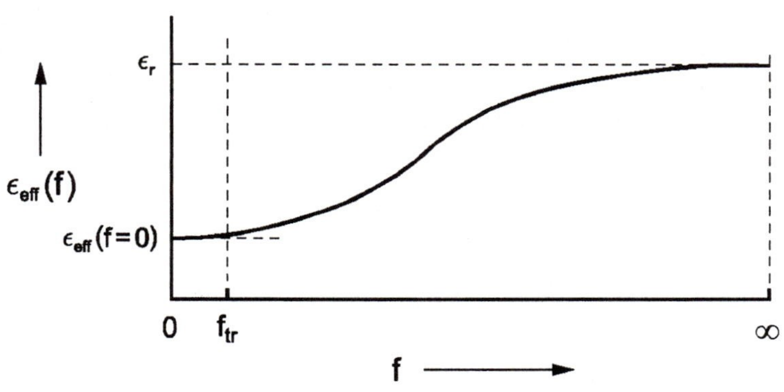

Figure 4.2 Principal frequency dependence of effective dielectric constant.

$$P(f_n) = G(f/f_g)^2 \quad \text{and} \quad f_g = \frac{Z_0(\epsilon_r, f=0)}{2\mu_0 h} = \frac{0.4\, Z_0/\Omega}{h/\text{mm}}\ \text{GHz} \tag{4.10}$$

The coefficient G is an empirical parameter and given as

$$G \approx 0.6 + 0.009\, Z_0(\epsilon_r)/\Omega \tag{4.11}$$

As stated in [6], the formula fits well with measurements of microstrip lines on alumina substrate. The frequency of the maximum slope between these two points is called the inflection point. The requirement that the second derivative of $\epsilon_{eff}(f)$ with respect to frequency must be equal to zero gives the inflection frequency f_i,

$$f_i = \frac{f_g}{\sqrt{3G}} \tag{4.12}$$

At the inflection frequency f_i, the value of the dielectric constant $\epsilon_{eff}(f_i)$ is calculated from (4.9) as

$$\epsilon_{eff}(f_i) = \frac{1}{4}\left[\epsilon_r + 3\,\epsilon_{eff}(f=0)\right] \tag{4.13}$$

The slope of $\epsilon_{eff}(f)$ at f_i is calculated as

$$\left.\frac{d\,\epsilon_{eff}(f_i)}{df}\right|_{f=f_i} = \frac{3}{8}\left[\epsilon_r - \epsilon_{eff}(f=0)\right] \tag{4.14}$$

Numerous other analytical formulas for the frequency-dependent effective dielectric constant have been published [6–11]. In 1984, Kirschning [7] proposed a dispersion expression that is still implemented in recent commercial microwave design software. The analytical function is in accordance with the dispersion equation (4.9) due to Getsinger, but with modification in the function $P(f_n)$. The frequency-dependent function $P(f_n)$ was determined by curvefitting with respect to the numerical results derived from the spectral domain method (see Section 3.4):

$$P(f_n) = P_3 P_4 \left[\left(0.1844 + P_5 P_6\right) f_n\right]^{1.5763} \tag{4.15}$$

where

$$f_n = (f/\text{GHz}) \cdot (h/\text{mm}), \qquad u = (w/h)$$

$$P_3 = 0.27488 + \left[0.6315 + 0.525/(1 + 0.0157 f_n)^{20}\right] u - 0.065683 \exp(-8.7513\, u)$$

$$P_4 = 0.33622 \left[1 - \exp\left(-0.03442\, \epsilon_r\right)\right]$$

$$P_5 = 0.0363 \exp(-4.6\, u) \cdot \left[1 - \exp\left(-(f_n / 38.7)^{4.97}\right)\right]$$

$$P_6 = 1 + 2.751 \left[1 - \exp\left(- (\epsilon_r / 15.916)^8 \right)\right]$$

With reference to the full-wave analysis, a maximum relative error of 0.6% in the range of $1.2 \le \epsilon_r \le 20$ for the dielectric constant, $0.1 \le w/h \le 100$ for the width-to-height ratio, and $0 \le h/\lambda_0 \le 0.13$ for the normalized wavelength (λ_0 = wavelength in vacuum) was found. Regarding Al_2O_3 substrate material with a thickness of $h = 0.635$ mm, (4.9) in connection with (4.15) can be applied for vacuum wavelengths $\lambda_0 \ge h/0.13 = 4.88$ mm. In this case, the maximum frequency is $f = c_0/\lambda_0 = 61.4$ GHz.

Figures 4.3 to 4.8 show the frequency-dependent effective dielectric constant versus the normalized frequency $f \cdot h$ in $\mathrm{GHz \cdot cm}$ for various w/h ratios of microstrip line on relevant substrate materials. Notice that the dielectric constant of alumina varies depending on the impurity in Al_2O_3 and the manufacturer. The specific resistivity ρ of HRS material depends on the doping level. HRS should have a minimum value of $1000\,\Omega\mathrm{cm}$ (see Chapter 9).

Knowing the frequency-dependent effective dielectric constant ($\epsilon_{eff}(f)$), the phase velocity of the quasi-TEM mode in the dispersion region can be written as

$$v(f) = \frac{c_0}{\sqrt{\epsilon_{eff}(f)}} \tag{4.16}$$

and the guide wavelength

$$\lambda_g(f) = \frac{\lambda_0(f)}{\sqrt{\epsilon_{eff}(f)}} \tag{4.17}$$

A special problem arises with the definition of a characteristic impedance in the dispersive frequency region. As the electromagnetic field of the fundamental microstrip mode evolves from a pure TEM field at dc into a hybrid mode with a preponderant longitudinal magnetic field component at higher frequency, a characteristic impedance based on the standard voltage-current definition cannot be uniquely calculated. This is because the contour integral of the magnetic field in the transverse plane is no longer path independent. Therefore, (4.2) cannot be simply extended to the dispersion region by substituting the quasi-static dielectric constant $\epsilon_{eff}(f=0)$ by a frequency-dependent one. Extension of (4.2) to a

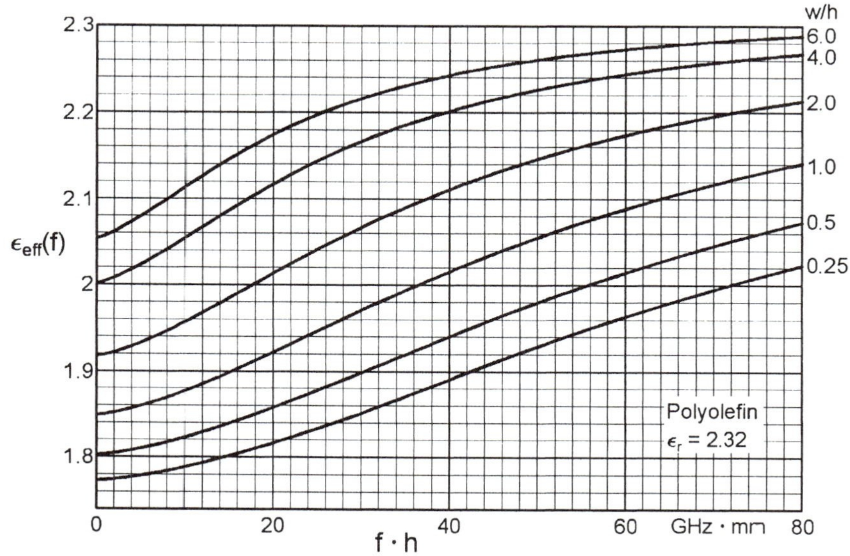

Figure 4.3 Frequency-dependent effective dielectric constant as a function of normalized frequency with *w/h* as parameter [substrate material: $\varepsilon_r = 2.32$ (Polyguide)].

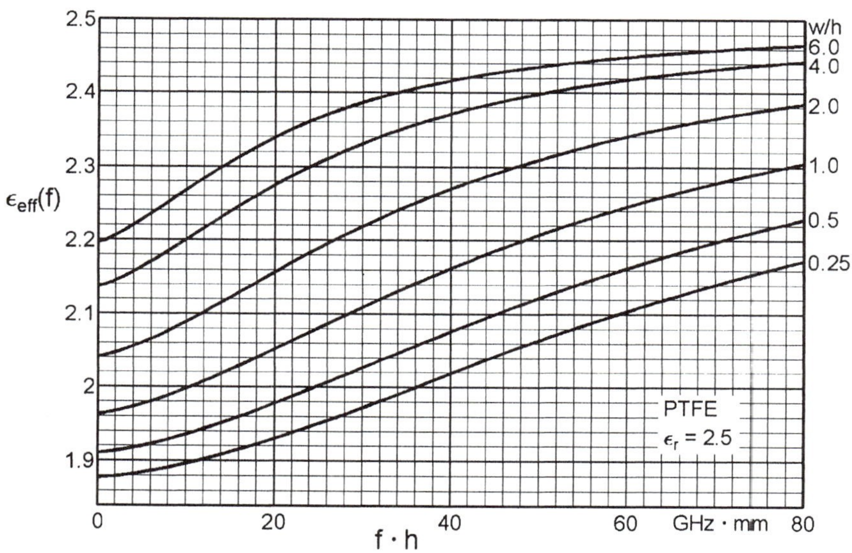

Figure 4.4 Frequency-dependent effective dielectric constant as a function of normalized frequency with *w/h* as parameter [substrate material: $\varepsilon_r = 2.5$ (PTFE)].

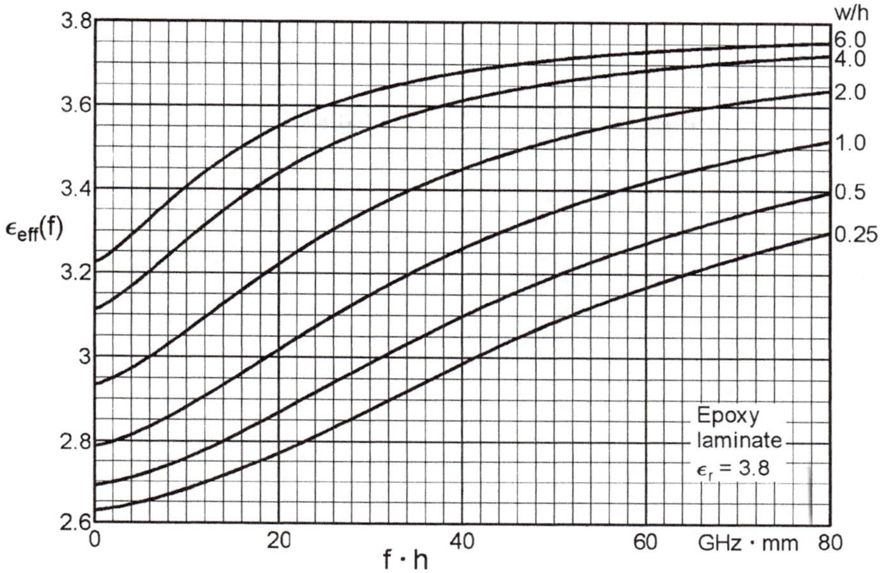

Figure 4.5 Frequency-dependent effective dielectric constant as a function of normalized frequency with *w/h* as parameter [substrate material: ε_r = 3.8 (Epoxy laminate)].

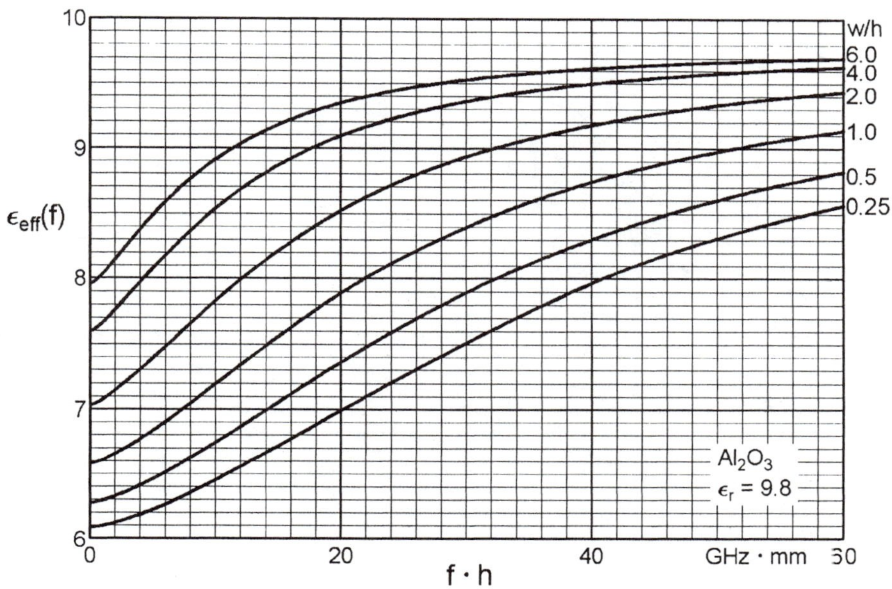

Figure 4.6 Frequency-dependent effective dielectric constant as a function of normalized frequency with *w/h* as parameter [substrate material: ε_r = 9.8 (Al_2O_3)].

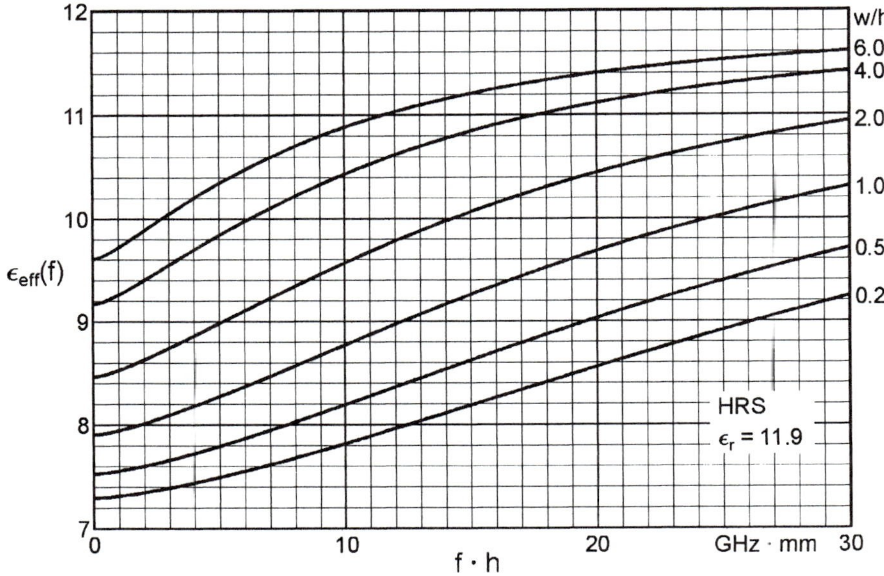

Figure 4.7 Frequency-dependent effective dielectric constant as a function of normalized frequency with *w/h* as parameter [substrate material: ε_r = 11.9 (HRS)].

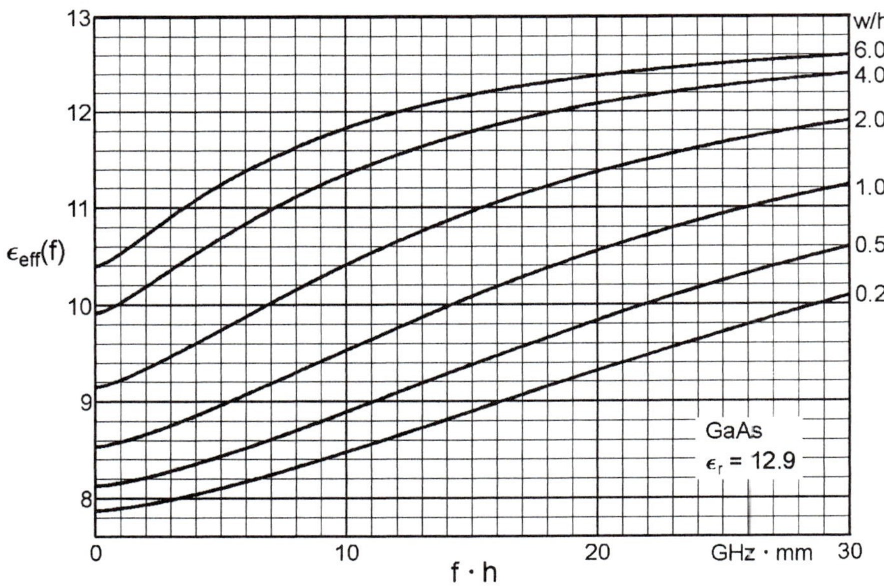

Figure 4.8 Frequency-dependent effective dielectric constant as a function of normalized frequency with *w/h* as parameter [substrate material: ε_r = 12.9 (GaAs)].

dispersive frequency region would mean that the characteristic impedance decreases with increasing frequency. However, as will be pointed out in the next section, this assumption would be inconsistent with simulated power transfer analysis of microstrip lines comprising lossless reciprocal two-ports. Therefore the following inequality holds:

$$Z_0(\epsilon_r, f) \neq \frac{Z_0(\epsilon_r = 1)}{\sqrt{\epsilon_{eff}(f)}} \qquad (f_{tr} < f < \infty) \tag{4.18}$$

Problem 4.1: Derive the inflection frequency f_i in (4.12) from (4.9).
Solution: In the inflection point, the second derivative of the frequency-dependent dielectric constant becomes zero. The first derivative reads

$$\frac{d\epsilon_{eff}(f)}{df} = -\left[\epsilon_r - \epsilon_{eff}(0)\right](-1)[1 + P(f)]^{-2} \cdot 2G\left(\frac{f}{f_g}\right) \cdot \frac{1}{f_g}$$

$$= \left[\epsilon_r - \epsilon_{eff}(0)\right]\frac{2G}{f_g^2}\,\frac{f}{[1 + P(f)]^2}$$

with the quasi-static dielectric constant $\epsilon_{eff}(0)$ in abbreviated form. From the second derivative being zero at the inflection point,

$$\frac{d^2\epsilon_{eff}(f)}{df^2} = \frac{2G[\epsilon_r - \epsilon_{eff}(0)]}{f_g^2}\,\frac{\left\{[1 + P(f)]^2 - f[2[1 + P(f)]]2G(f/f_g)\cdot 1/f_g\right\}}{[1 + P(f)]^4} = 0$$

for $f = f_i$, we obtain the condition

$$\left\{1 + G(f_i/f_g)^2\right\}^2 - 2f\left\{1 + G(f_i/f_g)^2\right\}2Gf_i/f_g^2 = 0$$

$$1 + G(f_i/f_g)^2 - 4G(f_i/f_g)^2 = 0$$

$$3G(f_i/f_g)^2 = 1$$

Thus,

$$f_i = \frac{f_g}{\sqrt{3G}}$$

4.2 Frequency-Dependent Characteristic Impedance

A lossless transmission line is characterized by its phase coefficient β and its characteristic impedance Z_0. Regarding a microstrip line, the phase coefficient reads

$$\beta(f) = \frac{2\pi}{\lambda_g(f)} = \frac{2\pi}{\lambda_0(f)/\sqrt{\epsilon_{eff}(f)}} = \frac{2\pi f}{c_0}\sqrt{\epsilon_{eff}(f)} \tag{4.19}$$

To approach a relation for a dispersive characteristic impedance of a microstrip line $[Z_0(f, \epsilon_r)]$, we first assume homogeneous dielectric filling. Furthermore, the dielectric and conductive material should be lossless, so that a pure TEM wave could propagate. The characteristic impedance of the line is related to the line voltage V and the line current I according to

$$\overline{Z_0} = \frac{V}{I} \tag{4.20}$$

which is also termed a voltage-current definition of the characteristic impedance. Line voltage V and line current I are calculated from the line integral

$$V = \int \vec{E} \cdot d\vec{s} \tag{4.21}$$

and the contour integral

$$I = \oint \vec{H} \cdot d\vec{s} \tag{4.22}$$

respectively. Since the electromagnetic field of a TEM wave is purely transversal, the integrals can be uniquely determined, independent of the chosen integration path. The path integral has to be taken between the electrodes, and the contour integral has to include the current of the upper conductor. The power (P) transported by the quasi-TEM wave in the direction of propagation is determined from the line voltage and line current,

$$P = \frac{1}{2} VI \tag{4.23}$$

Using (4.23) and substituting voltage or current in (4.20), the characteristic impedance also can be written as

$$Z_0^I = \frac{2P}{I^2} \tag{4.24}$$

which is the power-current definition of the characteristic impedance. If the current is substituted by the power, then we get the power-voltage definition,

$$Z_0^V = \frac{V^2}{2P} \tag{4.25}$$

It is seen that $\overline{Z_0}$ is the geometric mean of Z_0^I and Z_0^V,

$$\overline{Z_0} = \sqrt{Z_0^I \cdot Z_0^V} = \sqrt{\frac{2P}{I^2}\frac{V^2}{2P}} = \frac{V}{I} \tag{4.26}$$

The power P conveyed by the transmission line is obtained by integrating the z component of the complex Poynting vector $\vec{S} = \vec{E}_t \times \vec{H}_t^*$ over the cross section of the microstrip configuration:

$$P = \frac{1}{2} Re\left\{ \int (\vec{E}_t \times \vec{H}_t^*)\, d\vec{A} \right\} \tag{4.27}$$

where $d\vec{A}$ is the vector of an infinitesimal area of the cross section of the transmission line. In the Cartesian coordinate system, $d\vec{A}$ is written as $d\vec{A} = \vec{e}_z\, dx\, dy$, and it is oriented in the direction of propagation.

For a lossless TEM transmission line, (4.20), (4.24), and (4.25) provide same value for the characteristic impedance:

$$\overline{Z_0} = Z_0^I = Z_0^V = Z_0 \tag{4.28}$$

As we have already pointed out, the fundamental microstrip mode is not a pure TEM wave. This is illustrated in Figures 4.9 and 4.10, showing the magnetic and electric field of microstrip line of half-wavelength being operated in the dispersive frequency region. A maximum longitudinal surface current density occurs where the magnetic field is purely transversal (Figure 4.9). Conversely, a maximum transversal current density is found in the place where the magnetic field becomes longitudinal. Note that the amplitude of the transversal current is much lower than the longitudinal one, which is signified by shorter current density arrows. The electrical field of a dispersive microstrip line is shown in Figure 4.10. The field distributions show clearly that the integral values of (4.21) and (4.22) cannot be determined uniquely [12, 13]. Then, regarding the definition of a characteristic impedance, this cannot be exactly calculated for dispersive waveguiding structures. However, for the microstrip circuit design of matching networks, filters, and others, some value of characteristic impedance must be known, because the transfer characteristic is dependent on the impedance of the interconnecting transmission lines. For this design purpose, a characteristic impedance is modeled under the following specifications. The current I in the expressions for the characteristic

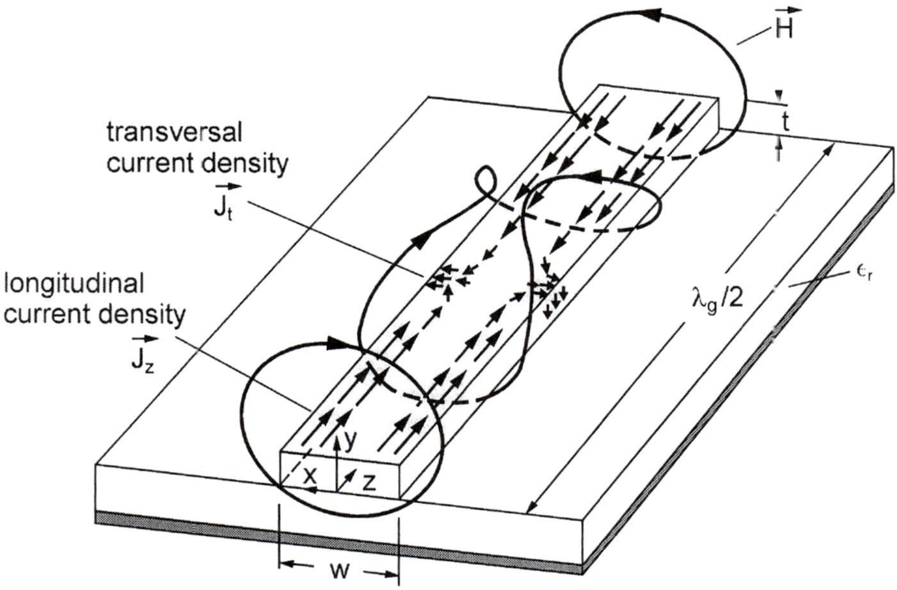

Figure 4.9 Magnetic field distribution and related surface current density (only upper strip surface and right strip edge are shown for clearness).

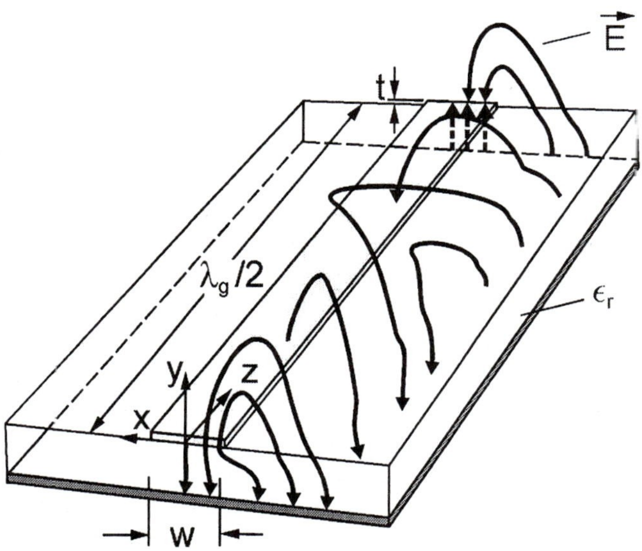

Figure 4.10 Electric field distribution of dispersive quasi-TEM mode (only one-half of the field is shown for clearness).

impedance is defined as the longitudinal current component of the total line current, and the voltage V is defined as the voltage on the microstrip line in the y–z plane of symmetry of the cross section at $x = 0$ (Figure 4.10). The current I is obtained simply by integrating the longitudinal surface current density $J_z(x)$ across the width of the strip:

$$I = \int_{-w/2}^{+w/2} J_z(x)\, dx \tag{4.29}$$

Based on the numerically calculated electromagnetic field distribution of a microstrip line using the spectral domain method (Section 3.4), the various definitions for the characteristic impedance [(4.20), (4.24), and (4.25)] were evaluated [14]. Figure 4.11 shows the frequency dependence of the characteristic impedance for different aspect ratio w/h on alumina. It can be seen that Z_0^I exhibits the smallest frequency dependence. $\overline{Z_0}$ is growing faster, and within a frequency range of 0 to 20 GHz, the characteristic impedance of a 50Ω line on Al_2O_3 ceramics increases about 10%. In the case of Z_0^V, the growth is largest and is approximately twice the value increase of $\overline{Z_0}$, which is the geometric mean of Z_0^V and Z_0^I. Thus, it has been shown that, no matter which definition is used, the characteristic impedance value increases with frequency, which confirms the inequality (4.18).

The decision for the acceptance of one of the defined relations for the characteristic impedance was made by the investigation of practical design tasks. The scattering coefficients of an asymmetric gap discontinuity and an abrupt impedance step were used for analysis. Figure 4.12 shows the calculated amplitude of the ratio of the forward and backward transmission coefficient of a capacitive microstrip gap, and shows an impedance step as well, using the discussed definitions for the characteristic impedance. Because the transmission line structure is supposed to be lossless and reciprocal, the condition $S_{12} = S_{21}$ must be fullfiled. Therefore, the magnitude of the calculated ratio of the transmission coefficients S_{12}/S_{21} should become one. The calculated curves in Figure 4.12 illustrate, taking into account the expected computation accuracy, only the definition of Z_0^I according to (4.24), fulfilling the reciprocity condition. Hence, the characteristic impedance is optimally defined in terms of the power P flowing along the longitudinal direction and the total longitudinal current I flowing on the strip. With the transported power known, the amplitude of the power wave is also fixed. The phase of the power wave is determined by V or I, whose phase is equal to the phase of the transversal electric and magnetic field, respectively.

The characteristic impedance is calculated after [7] on the basis of the power-current definition. Generally, the characteristic impedance (Z_0) can be determined from the expression

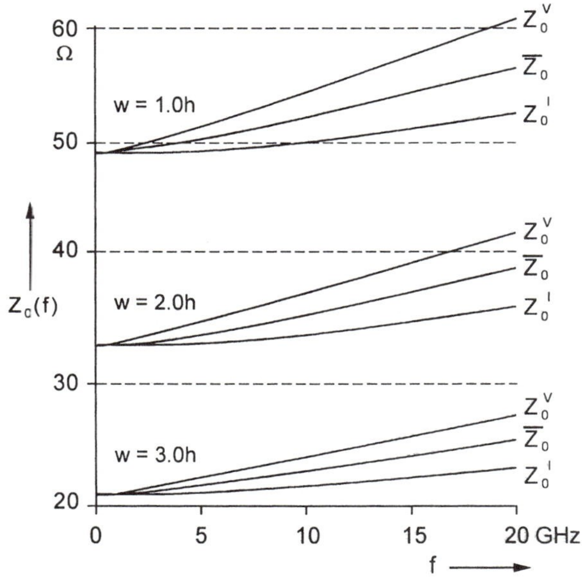

Figure 4.11 Characterisitic impedance as function of frequency and strip wicth w ($\varepsilon_r = 9.7$, $h = 0.635$ mm). (From: [14], © 1982 IEEE. Reprinted with permission.)

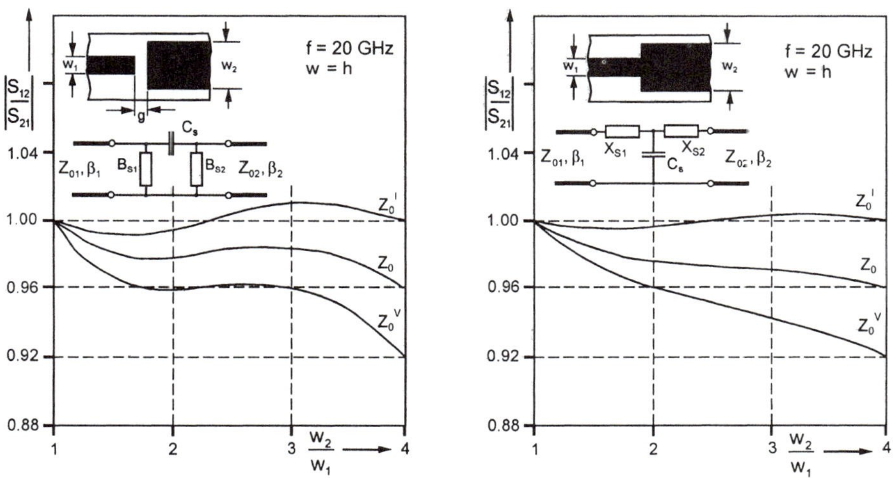

Figure 4.12 Numerically calculated amplitude of the transmission coefficient of an asymmetric capacitive gap (left) and symmetric impedance step on Al_2O_3 ceramics (right). (From: [14], © 1982 IEEE. Reprinted with permission.)

$$Z_0 = \frac{2P}{|I_z|^2} = \frac{Re\left\{\int (\vec{E}_t \times \vec{H}_t^*)\, \vec{e}_z\, dx\, dy\right\}}{\left|\int_{-w/2}^{+w/2} J_z(x)\, dx\right|^2} \tag{4.30}$$

Using an analytical approximation by curvefitting of numerically derived data, the following relation for the frequency-dependent characteristic impedance has been found:

$$Z_0(\epsilon_r, f) = Z_0(\epsilon_r, f=0) \cdot \left(\frac{R_{13}}{R_{14}}\right)^{R_{17}} \tag{4.31}$$

where

$$f_n = (f/\mathrm{GHz}) \cdot (h/\mathrm{mm}), \qquad u = (w/h)$$

$$R_1 = 0.03891\,\epsilon_r^{1.4}, \qquad R_2 = 0.267\,u^{7.0}, \qquad R_3 = 4.766\exp(-3.228\,u^{0.641})$$

$$R_4 = 0.016 + (0.0514\,\epsilon_r)^{4.524}, \qquad R_5 = (f_n/28.843)^{12.0}, \qquad R_6 = 22.20\,u^{1.92}$$

$$R_7 = 1.206 - 0.3144\exp(-R_1) \cdot \left[1 - \exp(-R_2)\right]$$

$$R_8 = 1 + 1.275\left\{1 - \exp\left[-0.004625\,R_3\,\epsilon_r^{1.674}\,(f_n/18.365)^{2.745}\right]\right\}$$

$$R_9 = 5.086\,R_4\,\frac{R_5}{0.3838 + 0.386\,R_4}\,\frac{\exp(-R_6)}{1 + 1.2992\,R_5}\,\frac{(\epsilon_r - 1)^6}{1 + 10\,(\epsilon_r - 1)^6}$$

$$R_{10} = 0.00044\,\epsilon_r^{2.136} + 0.0184, \qquad R_{11} = (f_n/19.47)^6/\left[1 + 0.0962\,(f_n/19.47)^6\right]$$

$$R_{12} = 1/(1 + 0.00245\,u^2), \qquad R_{13} = 0.9408\,\left[\epsilon_{eff}(f_n)\right]^{R_8} - 0.9603$$

$$R_{14} = (0.9408 - R_9)\,\left[\epsilon_{eff}(f=0)\right]^{R_8} - 0.9603, \qquad R_{15} = 0.707\,R_{10}\,(f_n/12.3)^{1.097}$$

$$R_{16} = 1 + 0.0503\,\epsilon_r^2\,R_{11}\left[1 - \exp(-u/15)^6\right],$$

$$R_{17} = R_7\left[1 - 1.1241\,\frac{R_{12}}{R_{16}}\,\exp\left(-0.026\,f_n^{1.15656} - R_{15}\right)\right]$$

After [7], a maximum relative error of 1% has been found in the range $2 \le \epsilon_r \le 18$, $0.1 \le w/h \le 10$, $0 \le h/\lambda_0 \le 0.1$, in comparison with field-theoretical results.

Figures 4.13 to 4.18 show the frequency-dependent characteristic impedance versus the normalized frequency $f \cdot h$ in GHz·cm for various w/h ratios of microstrip line on relevant substrate materials.

4.3 Measurement of Dispersion

The experimental determination of the frequency-dependent dielectric constant is useful for various reasons. The numerical calculation on the basis of full-wave analysis is time consuming and needs huge computer storage space. Calculation is often performed under the simplifying assumptions of zero thickness of metallization and lossless dielectric and conductive materials. Analytical expressions have often been found from numerically derived results applying curvefitting procedures. However, analytical functions are limited in their fitting capability, so that high fitting accuracy is often achieved only within a limited frequency range.

The following measurement techniques are generally not restricted to the considered frequency range and substrate materials. They are applicable for the experimental analysis of the dispersion properties of the quasi-TEM mode and higher-order modes, as well. Although single mode operation on microstrip lines is of the most practical importance, the lowest higher-order modes, HE_1 and HE_2, can severely influence the propagating properties of the fundamental quasi-TEM mode. This will become obvious in Section 6.2.2, which treats the discontinuity properties of an abrupt impedance step. It will be shown that the transferred power of the quasi-TEM mode decreases to zero (transmission zero) as the frequency approaches the cutoff frequency of a given higher-order mode. Regarding the symmetric impedance step, this will be identified as the first even higher-order mode (HE_2). In the asymmetric case, the first odd higher-order mode (HE_1) may contribute mainly to undesired parasitic effects. In the following, measurement techniques are discussed that are suitable to analyze the dispersion characteristics of the three lowest-order modes: quasi-TEM (HE_0), HE_1, and HE_2.

Figure 4.19 shows the approximate electric field distribution of the lowest order microstrip modes. The electric field of the quasi-TEM mode [Figure 4.19(a)] is symmetric with respect to the transverse plane of symmetry (dashed line), which characterizes an even mode. The electrical field lines are in parallel, and the magnetic field lines (not shown in the figure) are perpendicular to the symmetry plane. This behavior of the electric and magnetic fields signifies a magnetic wall (perfect open circuit). On a magnetic wall, the tangential magnetic field vanishes and the electrical field is purely tangential. Hence, the magnetic wall exhibits dual properties of an electric wall (short circuit).

Theoretically, the magnetic wall can be replaced by an ideal open circuit without changing the propagation properties of the wave under analysis. However, the model of a magnetic wall does not exist in reality, because of the fringing electric field that is always associated with a physical open circuit.

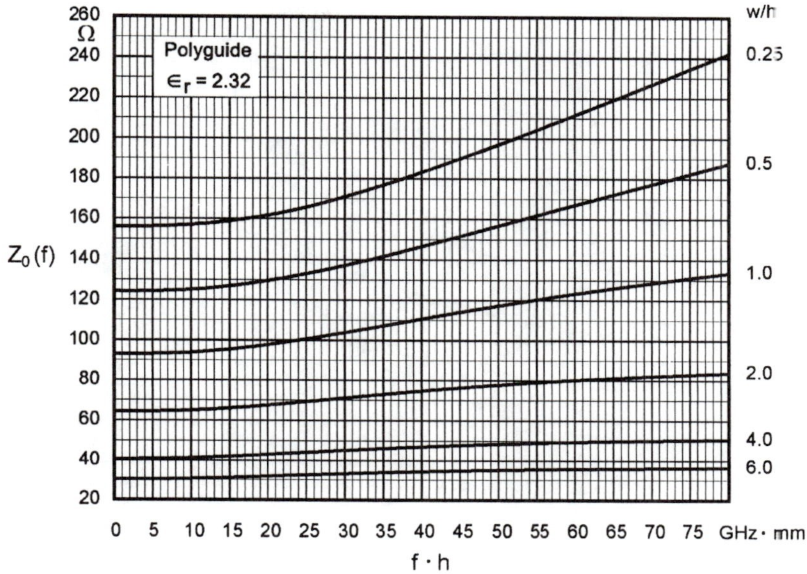

Figure 4.13 Frequency-dependent characteristic impedance as a function of normalized frequency with *w/h* as parameter [substrate material: $\varepsilon_r = 2.32$ (Polyguide)].

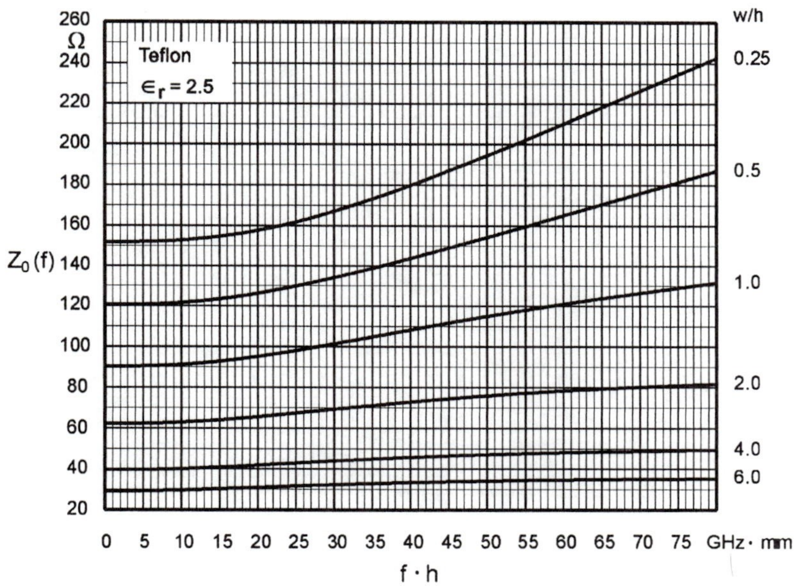

Figure 4.14 Frequency-dependent characteristic impedance as a function of normalized frequency with *w/h* as parameter [substrate materials: $\varepsilon_r = 2.5$ (PTFE)].

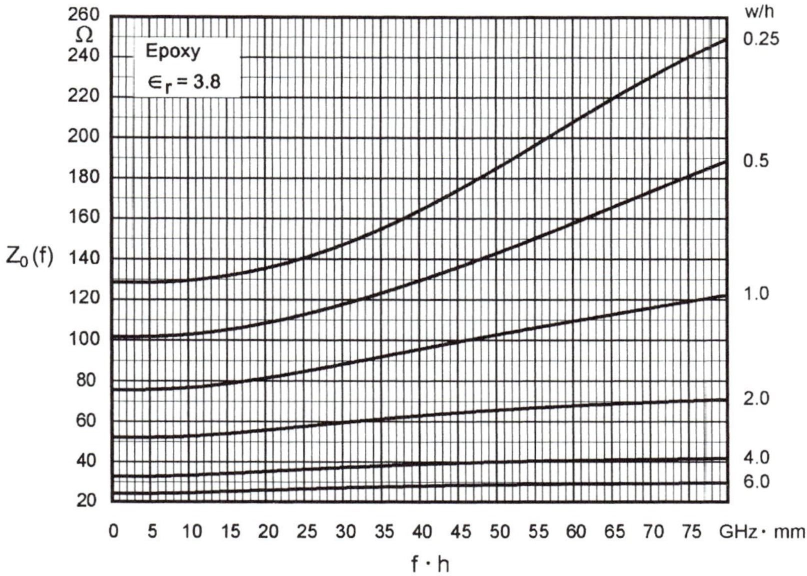

Figure 4.15 Frequency-dependent characteristic impedance as a function of normalized frequency with *w/h* as parameter [substrate material: $\varepsilon_r = 3.8$ (Epoxy laminate)].

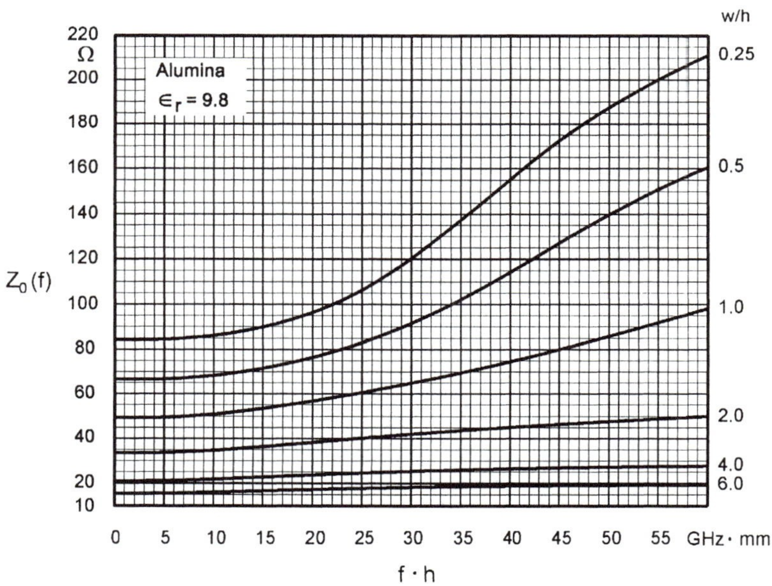

Figure 4.16 Frequency-dependent characteristic impedance as a function of normalized frequency with *w/h* as parameter [substrate material: $\varepsilon_r = 9.8$ (Al_2O_3)].

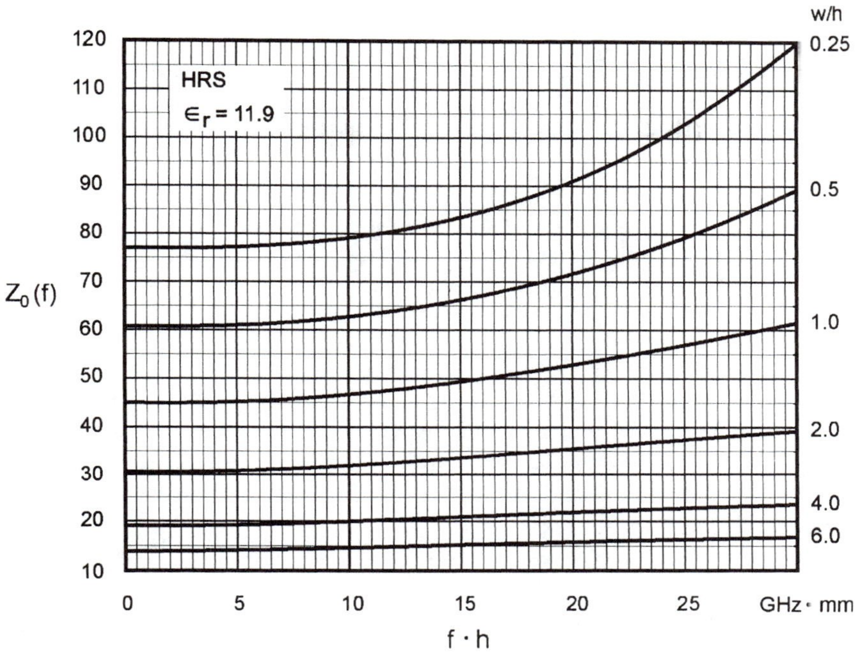

Figure 4.17 Frequency-dependent characteristic impedance as a function of normalized frequency with *w/h* as parameter [substrate material: ε_r = 11.9 (HRS)].

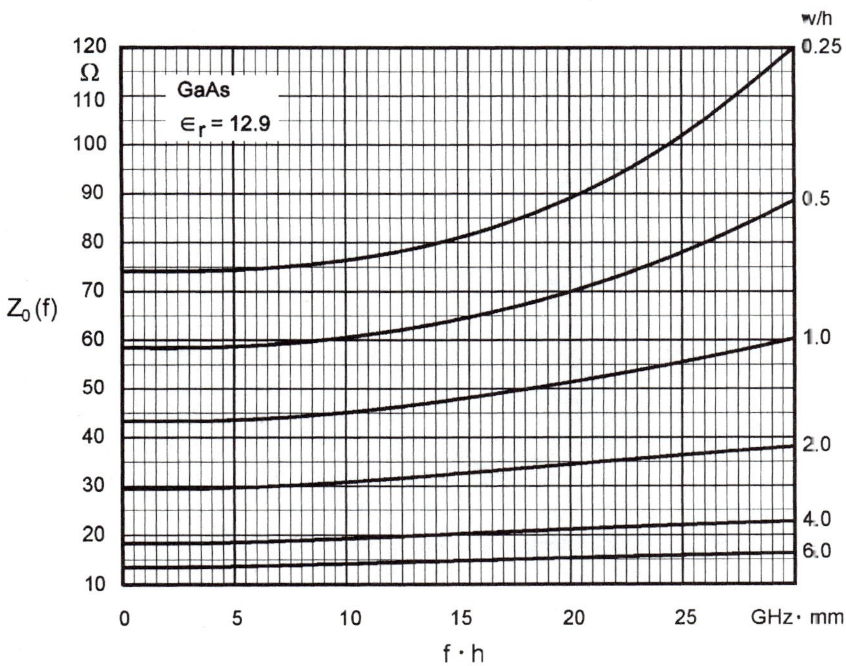

Figure 4.18 Frequency-dependent characteristic impedance as a function of normalized frequency with *w/h* as parameter [substrate material: ε_r = 12.9 (GaAs)].

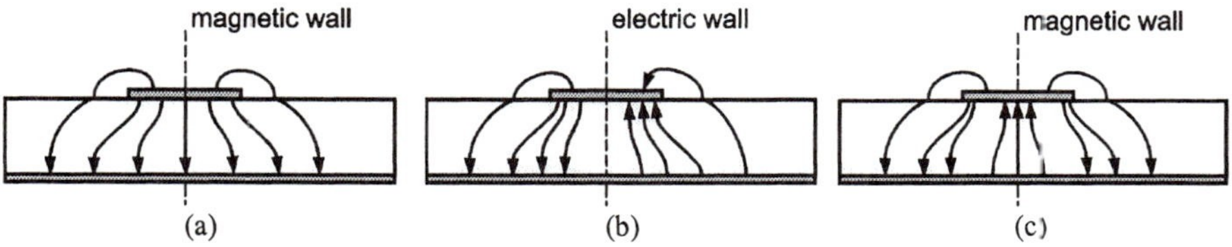

Figure 4.19 Approximate electric field distribution of the quasi-TEM and the first two higher-order modes on a microstrip line: (a) HE_0 mode or quasi-TEM mode, (b) HE_1 mode, and (c) HE_2 mode.

The electric field distribution of the HE_1 mode [Figure 4.19(b)] is asymmetric with respect to the plane of symmetry. This describes an odd mode. The variation of the transverse electric field is one-half wavelength. Since the electric field in the center is zero, the symmetry plane can be characterized as an electric wall. Electric walls can be replaced by short circuits without disturbing the propagation properties of the respective mode.

Figure 4.19(c) illustrates the second higher-order HE_2 mode with symmetric transverse electrical field distribution. The HE_2 mode is an even mode with a magnetic wall in the plane of symmetry. The distribution of the electrical field can be described as a full-wavelength transverse resonance.

In the lower frequency range, only the quasi-TEM mode can propagate. Above the cutoff frequency of the HE_1 and HE_2 modes, three waves can simultaneously occur, which makes the evaluation of measurement more challenging. However, since the microstrip is an open waveguiding structure, the electrical and magnetic fields of the different modes can be probed and identified from the top, close to the substrate surface. The first higher-order HE_1 mode can be separately analyzed by placing a metal plate as an electrical wall in the center of the microstrip line. This selective measure suppresses the even quasi-TEM and HE_2 modes (see Figure 4.33); the electromagnetic field distribution of the first higher-order HE_1 mode, however, will not be perturbed by this modification of line. In the following section, the most important mode characterization techniques, the sliding load and the resonance methods, are described.

4.3.1 Sliding Load Method

The sliding load technique is known as a measuring technique in microwaves to determine the eigenreflection coefficients S_{11} and S_{22} of a linear two-port [15] (see Appendix G). Supposing additionally that the two-port is reciprocal and lossless, the transmission coefficients are also known. This method is used to analyze the dispersive properties of a microstrip line. The test setup for the dispersion measurement of the quasi-TEM is shown in Figure 4.20. A microstrip surface absorber is used as a sliding load and moved along the microstrip line over several half-wavelengths. The measured input reflection coefficient Γ_1 describes

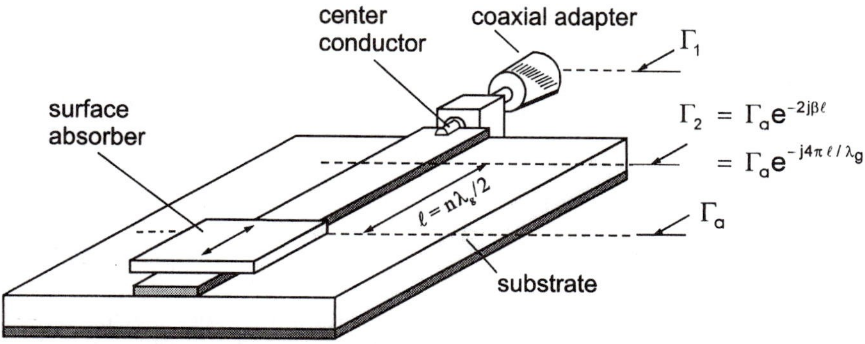

Figure 4.20 Measurement arrangement for the determination of the frequency-dependent effective dielectric constant of the quasi-TEM mode using the sliding load method. (From: [36], © 1975 Elsevier. Reprinted with permission.)

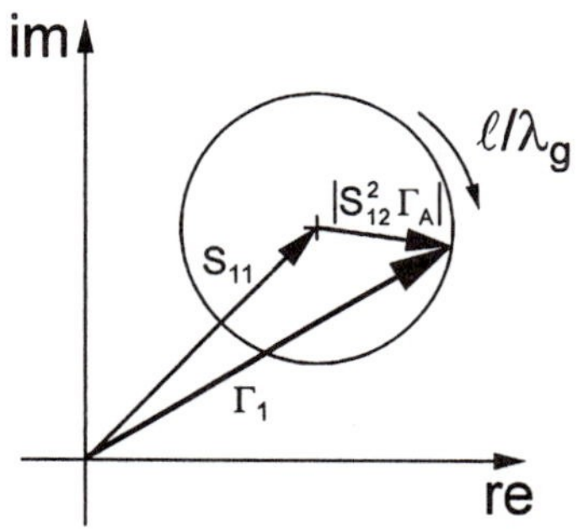

Figure 4.21 Reflection coefficient circle in the polar diagram when moving the surface absorber along the microstrip line.

a small circle with the radius $|S_{12}^2 \Gamma_a| \ll 1$ on the polar display of a vector network analyzer (Figure 4.21), where S_{12} is the transmission coefficient of the coaxial-to-microstrip adapter, and Γ_a is the absorber reflection coefficient.

Figure 4.22 shows different configurations of microstrip surface absorbers. The dielectric constant of the absorber material should be high to concentrate the electromagnetic field in the absorptive material. In Figure 4.22(a), the absorber is aligned with the microstrip line. The front of the absorber is symmetrically tapered to prevent strong reflection at the microstrip-absorber transition. Accurate guidance of the surface absorber is needed to attain reproducible circles for the input reflection coefficient. As described in [16], this is realized by plastic bars fixed on both sides of the microstrip line. The configuration in Figure 4.22(b) is less sensitive with respect to the orientation of the surface absorber, and therefore well suited for quick manual matching of a microstrip line.

Several materials were investigated for their absorbing properties on a microstrip line. Working on alumina ($\epsilon_r = 9.7$) and polyolefin substrate ($\epsilon_r = 2.32$),

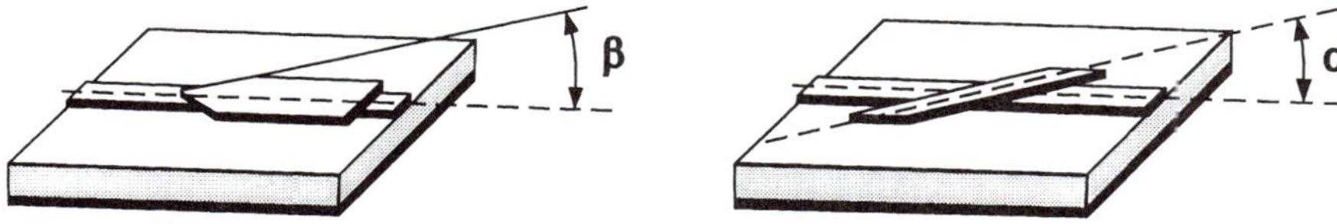

Figure 4.22 Different configurations of microstrip surface absorbers.

the following materials have proven to be most convenient in the investigated *X* and *K* bands.

Ferrite material is very hard and brittle. Using standard milling machines, the mechanical processing is extremely time consuming, because only a slow working feed will prevent any fracture of the machined workpiece. TT2-111 is a nickel-zinc ferrite material provided by Trans-Tech, Adamstown, MD [17]. In [18], the permittivity, permeability, and loss tangent of the TT2-111 material have been investigated experimentally with a focus on microwave absorption in the frequency range from 5.0 to 7.0 GHz. The dielectric constant ϵ_r varies weakly in the given frequency interval, from 12.83 to 12.92 ($12.5 \pm 10\%$ is specified in the manufacturer's data sheet). The dielectric loss tangent $\tan \delta_\epsilon$ is growing nearly continuously from 0.025 to 0.031 (≤ 0.001 is specified in the data sheet). The relative permeability μ_r changes slightly from 0.155 to 0.1361, and the magnetic loss tangent $\tan \delta_\mu$ decreases from 137.2 to 10.05 in the analyzed frequency range.

The ferrite material was shaped as shown in Figure 4.22. With an overall length of 9 cm, including a tapered front end with a tilt angle $\beta = 8°$, a width of 9 mm, and thickness of 2 mm, a low reflection coefficient of approximately 1% could be attained in the *X* band (50Ω line on alumina and Polyguide substrate, respectively). Regarding the simple construction and easy variation possibilities of the taper angle in Figure 4.22(b), practically no displacement of the reflection coefficient could be observed in the polar display when moving the absorber with a tilt angle of $\alpha = 10°$. Hence, the reflection coefficient is assumed less than or equal to 1% [19].

MF-124 is a lossy and magnetically loaded epoxide stock provided by Emerson and Cuming [20]. It can be readily machined with standard milling machines. The material data are as follows: The real part ϵ_r of the complex dielectric constant at 10 GHz is 23.6, and the dielectric loss tangent $\tan \delta_\epsilon$ is 0.03. The real part of the complex relative permeability μ_r is 1.5 at 10 GHz, and the magnetic loss tangent $\tan \delta_\mu$ is 1.4. MF-124 showed a similar performance as the ferrite material for the surface absorber configurations shown in Figure 4.22.

The microstrip sliding load configurations were tested on different substrate materials, such as polyolefin, alumina, and further plastic materials. The same surface absorber material has been used to calibrate a test-fixture up to 26.5 GHz [21] and 40 GHz, respectively [22]. The calibration process uses a very short surface absorber with a length of only 20 mm, without any tapered front end.

Figure 4.23 shows the block diagram of the complete measurement setup to investigate the dispersion of the quasi-TEM mode using the sliding load technique.

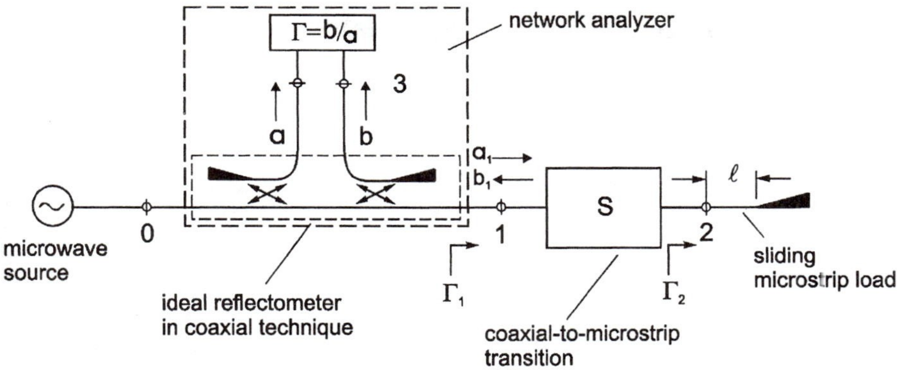

Figure 4.23 Sliding load measurement with a network analyzer.

The network analyzer is calibrated in the coaxial reference plane 1. The reference plane 2 is defined on the microstrip line sufficiently far away from the coaxial-to-microstrip transition to avoid any influence of evanescent higher-order microstrip modes. The load reflection coefficient Γ_2 (see Figure 4.20) reads

$$\Gamma_2 = \Gamma_a e^{-2j\beta l} = \Gamma_a e^{-j4\pi l/\lambda_g} \tag{4.32}$$

where Γ_a is the reflection coefficient of the surface absorber and l is the shifted length with respect to reference plane 2. A complete circle is observed after shifting the absorber over a length of $l = \lambda_g/2$. The input reflection coefficient Γ_1 is calculated from Γ_2 using the well-known two-port mapping relation

$$\Gamma_1 = S_{11} + \frac{S_{12}\,S_{21}\,\Gamma_2}{1 - S_{22}\,\Gamma_2} \tag{4.33}$$

The input reflection coefficient Γ_1 relates to the ratio of the reflected wave b_1 and the incident wave a_1,

$$\Gamma_1 = \frac{b_1}{a_1} \tag{4.34}$$

The network analyzer detects the complex power waves a and b in the detector reference plane 3 and determines the wave ratio $\Gamma = b/a$. It can be shown that a and b relate to the waves in the measurement reference plane 1 as

$$\begin{pmatrix} b \\ a \end{pmatrix} = \begin{pmatrix} A & B \\ C & D \end{pmatrix} \begin{pmatrix} a_1 \\ b_1 \end{pmatrix} \tag{4.35}$$

with $|A| \ll |B|$, $|D| \ll |C|$. Supposing a perfect network analyzer with $|A| = |D| = 0$, the detected waves (a,b) would correspond to the waves in the reference plane 1 (a_1, b_1). Otherwise, the transformation matrix in (4.35) must be determined by a calibration procedure.

The magnitude of the reflection coefficient of the sliding absorber is very small ($|\Gamma_2| = |\Gamma_a| \ll 1$). Commonly, the values of the eigenreflection coefficients of a coaxial-to-microstrip adapter are also very small ($|S_{mn}| \ll 1$, $m = n = 1, 2$). Thus, with $|S_{22}\Gamma_2| \ll 1$, and taking into account the reciprocity of the two-port ($S_{12} = S_{21}$), the input reflection coefficient can be approximated as

$$\Gamma_1 \approx S_{11} + S_{12}^2 \, \Gamma_a \, e^{-j4\pi l/\lambda_g} \tag{4.36}$$

The evaluation of the measurement is as follows. If the surface absorber is moved one-half wavelength ($\lambda_g/2$) along the microstrip, then the input reflection coefficient Γ_1 describes a complete circle in the polar diagram. Measuring the corresponding physical shift on the microstrip line, the effective dielectric constant easily can be determined. To increase the measurement accuracy, a multiple of half-wavelengths should be regarded. It reads

$$l = n \cdot \frac{\lambda_g}{2} = \frac{n}{2} \frac{\lambda_0}{\sqrt{\epsilon_{eff}(f)}} = \frac{n}{2} \frac{c_0}{f\sqrt{\epsilon_{eff}(f)}} \tag{4.37}$$

Solving (4.37) for the frequency-dependent effective dielectric constant yields

$$\epsilon_{eff}(f) = \left(\frac{n}{2l} \frac{c_0}{f} \right)^2 \tag{4.38}$$

The uncertainty of values of the effective dielectric constant extracted from measurement depends on the absorbing material and the width of the microstrip line under analysis. The experience has shown that the method is applicable to narrow microstrip lines. However, with increasing strip width, the absorption capability decreases, because the portion of the air fringing field with respect to the total field also decreases. Figure 4.24 shows an example of the dispersion measurement on polyolefin. It can be seen that the measurement uncertainty increases with increasing width of the analyzed microstrip line. Regarding the narrowest width $w = 4.37$ mm, which corresponds to a 50Ω impedance, a rather smooth curvature is obtained. A large amount of the electromagnetic field is located in the air region and penetrates the absorber material, which results in high loss of electromagnetic energy. With growing microstrip width, however, the electromagnetic field is mainly beneath the upper microstrip, so that absorption by the sliding load becomes worse, resulting in larger measurement uncertainties. The sliding load technique is only applicable in the single mode operation. Above the cutoff frequency of a higher-order mode, the measurement is no longer unique.

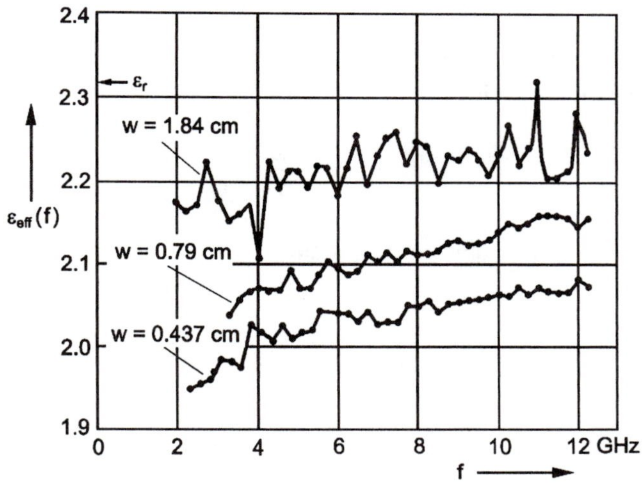

Figure 4.24 Dispersion measurement of microstrip lines on Polyguide substrate material ($\varepsilon_r = 2.32$, $h = 1.59$ mm) using the sliding load method.

4.3.2 Resonance Method

The frequency dependence of the effective dielectric constant easily can be accomplished by the use of a microstrip resonator in the form of a straight open- or short-ended microstrip line or a ring.

Open-Ended Transmission Line Resonator

Figure 4.25 shows the physical structure (top view) of a straight double-sided open-ended microstrip line of length l. Supposing perfect open ends with no electrical stray field, the configuration resonates when the length l is an integer number of half-wavelengths:

$$l = n\frac{\lambda_g}{2} \tag{4.39}$$

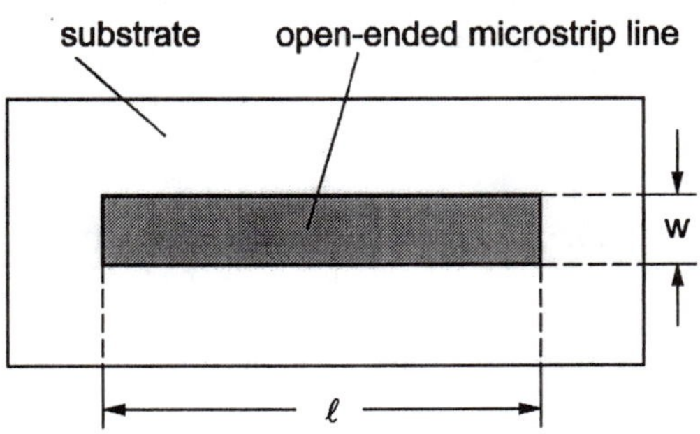

Figure 4.25 Physical structure of an open-ended transmission line resonator.

Since maximum electrical field occurs at the open ends, the resonant voltage wave along the line is described by a multiple of half-wavelengths. With the guide wavelength λ_g in (4.17), it follows from (4.39)

$$l = \frac{n}{2} \frac{c_0}{f_{0n}\sqrt{\epsilon_{eff}(f_{0n})}} \tag{4.40}$$

Solving for the effective dielectric constant at the measured resonant frequency f_{0n} yields

$$\epsilon_{eff}(f_{0n}) = \left(\frac{n\,c_0}{2\,l f_{on}}\right)^2 \tag{4.41}$$

The fringing field at the open end causes an effective extension Δl of the physical resonator length l. Thus, the electrical length that determines the resonant frequencies can be written as

$$l_{eff} = l + 2\Delta l \tag{4.42}$$

With (4.42), an improved relation for the effective dielectric constant is found,

$$\epsilon_{eff}(f_{0n}) = \left(\frac{n\,c_0}{2 l_{eff}\, f_{on}}\right)^2 \tag{4.43}$$

Generally, the end effect is frequency-dependent. It has been shown that Δl can be assumed to be frequency independent over a wide range of frequency [7].

In [23], it has been shown that the effective dielectric constant can be determined from resonant frequencies of open-ended resonant circuits without necessary knowledge of open-end effect. This can be done as follows.

Figure 4.26 shows two open-ended transmission line resonators with the lengths l_1 and l_2, where $l_2 \approx 2l_1$. Both test resonators are constructed on the same substrate. Since the resonance frequencies f_{01} and f_{02} are approximately equal, it can be assumed that the end effects are of the same order in both cases.

The line resonators fulfill the following resonance conditions,

$$l_1 + 2\Delta l = \lambda_1/2 = \frac{c_0}{2f_{01}\sqrt{\epsilon_{eff}(f_{01})}} \tag{4.44}$$

$$l_2 + 2\Delta l = \lambda_2 = \frac{c_0}{f_{02}\sqrt{\epsilon_{eff}(f_{02})}} \tag{4.45}$$

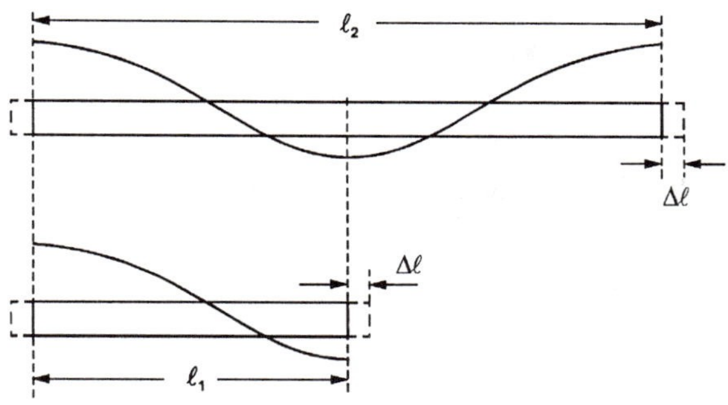

Figure 4.26 Voltage distribution at resonance frequency.

Δl can be eliminated by taking the difference of both equations. Thus, we obtain

$$l_2 - l_1 = \frac{c_0}{\sqrt{\epsilon_{eff}(f_0)}} \left(\frac{1}{f_{02}} - \frac{1}{2f_{01}} \right) \tag{4.46}$$

with

$$f_{01} \approx f_{02} \approx f_0 \tag{4.47}$$

Solving (4.46) for the effective dielectric constant results in

$$\sqrt{\epsilon_{eff}(f_0)} = \frac{c_0}{l_2 - l_1} \left(\frac{1}{f_{02}} - \frac{1}{2f_{01}} \right) \tag{4.48}$$

Wolff [24] treated the double-sided open-ended transmission line as a capacitively loaded resonator, and derived exact equations for the two resonator configurations. As can be seen from Figure 4.26, at λ-resonance, the voltage is a maximum and the current becomes zero in the plane of symmetry. In the case of the λ/2-resonance we have maximum current and zero voltage at the midpoint of the resonator.

Recall the relation for the transformation of a load Z_L over a transmission line length l [25],

$$Z(l) = \frac{Z_L + jZ_0 \tan\beta l}{1 + jZ_L/Z_0 \tan\beta l} \tag{4.49}$$

The input admittance of the λ-resonator at the midpoint of the transmission line is zero due to the electrical open. Thus, from (4.49) it follows with $l = l_2/2$

$$1 + jZ_L/Z_0 \tan\beta l_2/2 = 0 \tag{4.50}$$

where $\beta l_2/2 = \pi f\sqrt{\epsilon_{eff}}/c_0$. With $Y_L = 1/Z_L = j2\pi f_0 C_e$, where C_e is the open-end capacitance, the resonance condition (4.50) reads

$$2\pi f_0 C_e = -\frac{1}{Z_0}\tan\left(\pi f_0\sqrt{\epsilon_{eff}}\,\frac{l_2}{c_0}\right) \tag{4.51}$$

Regarding the $\lambda/2$-resonator, an electrical short is given in the plane of symmetry so that the input impedance becomes zero. In this case, from (4.49) with $l = l_1/2$ the resonance condition is obtained as

$$2\pi f_0 C_e = \frac{1}{Z_0}\cot\left(\pi f_0\sqrt{\epsilon_{eff}}\,\frac{l_1}{c_0}\right) \tag{4.52}$$

Similar to the approach after Deutsch and Jung [23], the end-capacitance C_e can be eliminated by taking the ratio of (4.51) and (4.52):

$$\frac{f_{0_2}}{f_{0_1}} + \frac{\cot\left(\pi f_{0_2}\sqrt{\epsilon_{eff_2}}\,l_2/c_0\right)}{\tan\left(\pi f_{0_1}\sqrt{\epsilon_{eff_1}}\,l_1/c_0\right)} = 0 \tag{4.53}$$

where $l_2 \approx 2l_1$ and $\epsilon_{eff}(f_2) \approx \epsilon_{eff}(f_1)$. Assuming a measurement error both in f_1 and f_2 of $4\times10^{-3}\%$, an error in $\epsilon_{eff}(f)$ of about 2% was estimated [24]. An additionalerror is contributed by the error in the determination of the resonator length. Assuming a minimum error of $4\times10^{-3}\%$ in measurement of resonator length, which may need optical measurement methods, an additional error in $\epsilon_{eff}(f)$ of about 1.5% is estimated. Thus, it is difficult to determine the effective dielectric constant with an error less than 3%. Using (4.51) or (4.52) for the determination of the end capacitance C_e, an error of about 25% is expected.

Short-Circuited Transmission Line Resonator

The fringing and radiation effects at the open end of a microstrip resonator (see Section 6.1) can be avoided, if the resonator is short-circuited at both ends. Short circuits generally can be realized in coaxial and waveguide technique with high quality and low expenditure, whereas the realization of a microstrip short must be performed very carefully. The experience shows that good technical microstrip shorts on Al_2O_3 substrate can be realized at least up to the X band [14]. The use of plastic substrates like polyolefin ($\epsilon_r = 2.32$) needs special precautions to realize a satisfactory electrical short. This can be done by providing the short-circuit plate

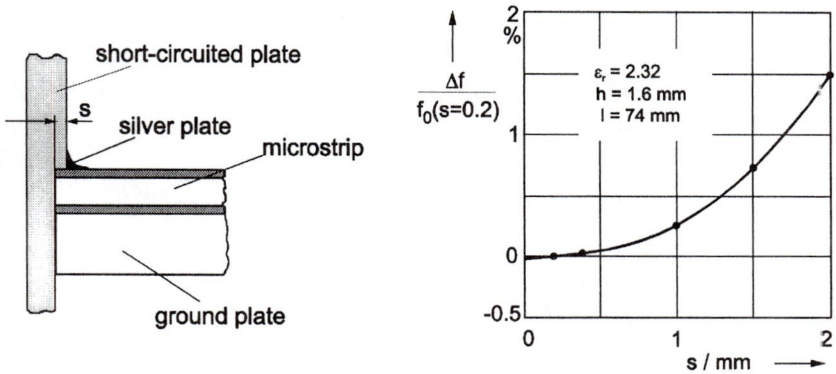

Figure 4.27 (a) Realized short-circuit on Polyguide substrate, (b) influence of ledge s on the measured resonance frequency.

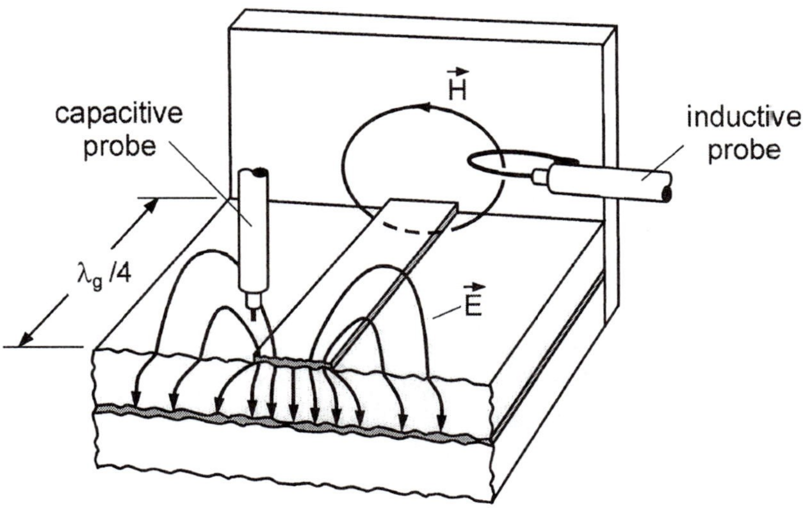

Figure 4.28 Electromagnetic resonant field of short-circuited microstrip resonator and coupling situation.

with a ledge of about 0.2 mm, which presses the soft substrate onto the fixture ground plate (Figure 4.27).

Coupling of the transmission line resonator is accomplished by miniaturized magnetic and electric field probes [19, 26–29]. Figure 4.28 illustrates the resonant field over a length of a quarter-wavelength. Near the short-circuited plates, the magnetic field becomes maximum, whereas no transversal electric field exists. This is not valid for the longitudinal electric field, as can be seen from Figure 4.10. The maximum electric field occurs at quarter-wavelength distance from the short-circuited plate where the magnetic field is zero. Note that the magnetic and electric resonant fields are not only shifted in space, but also in time, by 90°. When there is maximum electric field, the magnetic field is zero, and vice versa—when the magnetic field is maximum, there is no electric field.

For excitation a miniaturized inductive loop is placed at one of the short-circuit plates of the microstrip resonator as illustrated in Figure 4.28. The probe is simply made of a semi-rigid coaxial cable with the inner conductor uncovered at the open end, bent and soldered to the outer conductor to form a short-circuited loop. Optimally, excitation is given when the loop area is perpendicular to the direction of the magnetic field line.

Coupling out can be achieved capacitively at the sides of the microstrip line by using a miniaturized open-ended semirigid coaxial cable with the inner conductor extended to a small monopole. By this arrangement, it is possible to suppress the direct coupling of the probes to a high degree. The harmonic number n is found by moving the electric field probe along the microstrip line while observing the detected signal on the oscilloscope. A voltage peak indicates a maximum of electric field.

The wavelength at the measured frequency is simply twice the physical length of the microstrip resonator divided by the harmonic number n:

$$\lambda_{0n} = \frac{2l}{n} = \frac{c_0}{f_{0n}\sqrt{\epsilon_{eff}(f_{0n})}} \tag{4.54}$$

where

λ_{0n} = resonant wavelength
l = physical length of the resonator
n = harmonic number (number of half-wavelengths)

Then, from (4.54) the effective dielectric constant reads

$$\sqrt{\epsilon_{eff}(f_{0n})} = \frac{n\,c_0}{2lf_{0n}} \tag{4.55}$$

Ring Resonator

Ring resonators also are useful for measurement of dispersion and wavelength in microstrip transmission lines. They do not include any discontinuity effect of either open- or short-circuit terminations. As has been proposed by Troughton [30], the ring resonator should be sufficiently large, at least five wavelengths long at the frequency of interest, to avoid the effects of mutual inductance.

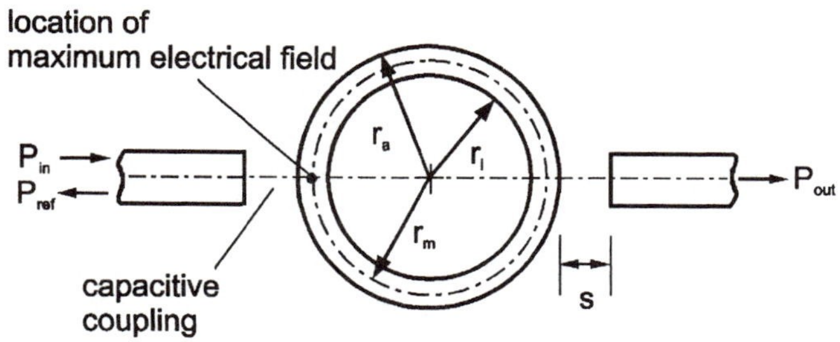

Figure 4.29 Top view of the loosely coupled ring resonator with transmission coupling lines.

The ring resonator is stimulated via a 50Ω input and output coupling line with a coupling gap s (Figure 4.29). Q-factor measurements (resonant frequency, bandwidth of resonant curve), in general, include the electrical properties of the resonating circuit, and the coupling network as well. To reduce the measurement to the Q factor of the resonator itself, the influence of the coupling network must be reduced to a minimum. This is achieved by decreasing the coupling gap continuously by starting, for example, with an initial gap of 2/100 mm (for Al_2O_3 substrate). Transmission Q-factor measurements are made using a sweep generator. The Q-factor of the resonator is then measured when no change of the resonant curve is observed (weak coupling condition). Under this condition, the coupling transmission lines are not seriously perturbing the field of the ring resonator, and the coupling has no severe influence on the resonance frequency.

Because of the weak coupling, the transmitted signal is attenuated severely, making measurements with a sweep generator very difficult because of the low power levels involved [31]. The frequency-dependent effective dielectric constant is determined as follows. The mean length of the ring is calculated from

$$2\pi r_m = 2\pi \frac{r_a + r_i}{2} \tag{4.56}$$

The formula works well with small widths of microstrips. However it becomes an approximation for larger line widths because of the increasing effect of curvature of the ring. The mean circumference of the resonator is an integer number of resonant wavelength,

$$n \cdot \lambda_{0n} = n \frac{c_0}{f_{0n}\sqrt{\epsilon_{eff}(f_{0n})}} = 2\pi r_m \tag{4.57}$$

Then the resonance frequency is obtained as

$$f_{0n} = \frac{n}{2\pi} \frac{c_0}{\sqrt{\epsilon_{eff}(f_{0n})}} \frac{1}{r_m} = \frac{n}{\pi} \frac{1}{r_a + r_i} \frac{c_0}{\sqrt{\epsilon_{eff}(f_{0n})}} \tag{4.58}$$

From (4.58), the effective dielectric constant reads

$$\sqrt{\epsilon_{eff}(f_{0n})} = \frac{n \cdot c_0}{\pi f_{0n} (r_a + r_i)} \tag{4.59}$$

The reliable determination of the frequency-dependent effective dielectric constant depends on the correctness of the defined mean-radius r_m. Wolff and Knoppik [32] investigated the influence of the curvature of the microstrip transmission line (and dispersion properties as well) on the resonance frequencies. Figure 4.30 shows the difference between the measured resonance frequencies of microstrip resonators on Polyguide substrate ($\epsilon_r = 2.32$, $h = 1.5$ mm), and the frequencies calculated from (4.58) having approximated the frequency-dependent effective dielectric constant by the quasi-static value after Wheeler. The parameter m in Table 4.1 indicates the number of wavelengths on the resonators at a resonance of 3 GHz. Because dispersion has not been taken into account, some deviation between calculated and measured results would have been expected. However, beyond this, the curves in Figure 4.30 also give information about the strong influence of the curvature of the microstrip line on the resonance frequencies. Curves e, f, g, and h indicate microstrip lines with same impedance of 20Ω. At 3 GHz, the mean length of the resonator (h) is shortest (m = 2), and (e) is longest (m = 5). It is clearly seen that

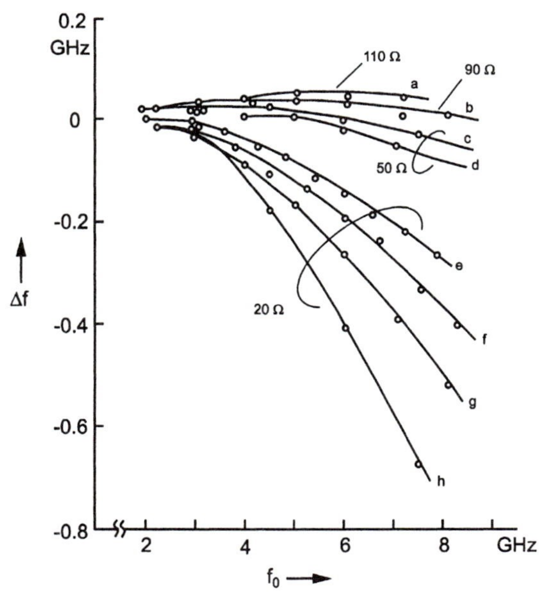

Figure 4.30 Difference of measured resonance frequencies and frequencies calculated from the mean-length approach. (From: [32], © 1971 IEE. Reprinted with permission.)

Table 4.1
Analyzed Characteristic Impedances and Electrical Lengths (Number of Wavelengths) of the Resonators

Curve	*Impedance* (Ω)	*m* (*Number of Wavelengths on the Ring*)
a	110	3
b	90	3
c	50	2
d	50	3
e	20	5
f	20	4
g	20	3
h	20	2

(*After*: [32].)

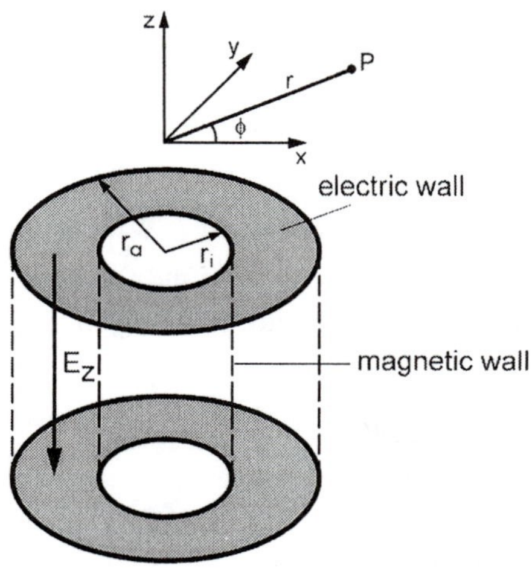

Figure 4.31 Magnetic wall model for the ring resonator after Wolff and Knoppik. (From: [32], © 1971 IEE. Reprinted with permission.)

the shortest resonator is mostly influenced by the curvature. A further consideration can be done analyzing same electrical lengths of resonator but having different characteristic impedance (microstrip width) values: curve (*a*), 110Ω; curve (*b*), 90Ω; curve(*d*), 50Ω; and curve (*g*), 20Ω. For same electrical length, the resonant frequency decreases with growing microstrip width. For relatively narrow microstrip lines as the 50Ω line [curves (*c*) and (*d*)], a remarkable deviation has occurred due to the curvature effect.

In the literature, theories have been developed to account for the effect of curvature on the resonant frequency providing more accurate dispersion measurement. Wolff and Knoppik [32] proposed a magnetic wall model for the ring resonator shown in Figure 4.31, which principally accounts for the curvature effect. The ring resonator can be considered as a cavity resonator with electric walls on the top and the bottom, and cylindrical magnetic walls at the inner ($r = r_i$) and outer radius ($r = r_a$). The resonant modes of this resonator model, which have a transversal field distribution similar to that of the quasi-TEM mode of the microstrip line, are the z independent modes. (The z axis is the cylinder axis of the resonator, as shown in Figure 4.31.) These modes have a pure longitudinal electric field, so that an E_z component is present. The magnetic field strength is perpendicular to the z axis. A solution of Maxwell equations for the problem considered is

$$E_z = \left[A J_m(kr) + B N_m(kr)\right] \cos(m\phi) \tag{4.60}$$

$$H_r = \frac{m}{j\omega\mu_0 r}\left[A J_m(kr) + B N_m(kr)\right] \sin(m\phi) \tag{4.61}$$

$$H_\phi = \frac{k}{j\omega\mu_0}\left[A J_m'(kr) + B N_m'(kr)\right] \cos(m\phi) \tag{4.62}$$

where J_m is the Bessel function of order m, and N_m is the Neumann function of order m. J_m' and N_m' are the derivatives of the functions of the argument (kr). The eigenvalue equation of the resonator resulting from the boundary conditions reads

$$\frac{J_m'(kr_a)}{J_m'(kr_i)} - \frac{N_m'(kr_a)}{N_m'(kr_i)} = 0, \qquad k = \omega\sqrt{\epsilon_0 \epsilon_r \mu_0} \tag{4.63}$$

where r_a is the outer radius and r_i the inner radius of the ring resonator. The solutions of (4.63) are resonant modes, which are commonly termed E_{mnq} modes ("E" emphasizes the existence of a longitudinal electrical field component) or TM_{mnq} modes ("TM" underlines the transverse character of the magnetic field). The first index m describes the field dependence on the ϕ variable of the polar coordinate system, the second index n describes the radial dependence, and the last index q describes the variation of the field in z direction. Using this magnetic wall model and measured resonance frequencies, frequency-dependent effective dielectric constant values have been determined.

A more accurate waveguide model for the ring resonator, which not only takes into account the curvature of the microstrip ring but also the influence of its microstrip width, was proposed in [33, 34]. Figure 4.32 shows the dispersive magnetic waveguide model for a ring resonator, which consists of two perfectly conducting plates having a frequency-dependent effective radial waveguide width

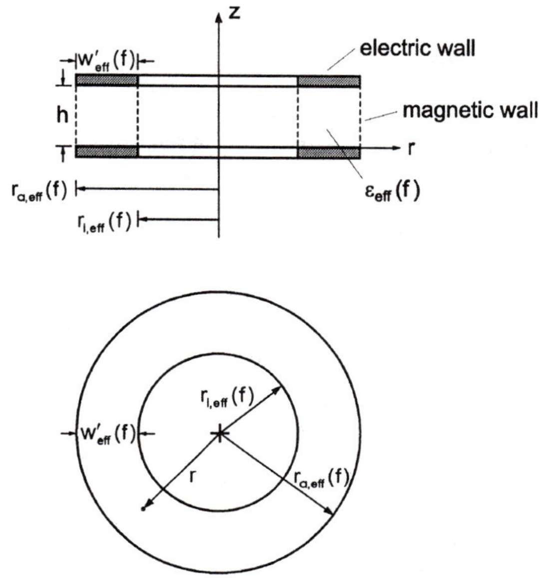

Figure 4.32 Dispersive magnetic wall model for ring resonator after Khilla [33].

$w'_{eff}(f)$, and perfect magnetic cylindrical sidewalls at $r = r_{i,eff}$ and $r = r_{a,eff}$. The plates are separated by the substrate thickness, h, and filled with the frequency-dependent effective dielectric constant $\epsilon_{eff}(f)$. $\epsilon_{eff}(f)$ and $Z_0(f)$ can be obtained from (4.9), (4.15), and (4.31). The frequency-dependent width $w_{eff}(f)$ of a straight microstrip line is calculated from

$$w_{eff}(f) = \frac{120\pi\, h}{Z_0(f)\sqrt{\epsilon_{eff}(f)}} \tag{4.64}$$

The effective radii of the modeled ring resonator are defined as (Figure 4.33)

$$r_{i,eff}(f) = r_i - \frac{\Delta w(f)}{2} \cdot \frac{r_i}{r_a} \tag{4.65}$$

$$r_{a,eff}(f) = r_a + \frac{\Delta w(f)}{2} \tag{4.66}$$

with

$$\Delta w(f) = w_{eff}(f) - w \tag{4.67}$$

The effective mean radius reads

$$r_{m,eff} = \frac{r_{a,eff} + r_{i,eff}}{2} \tag{4.68}$$

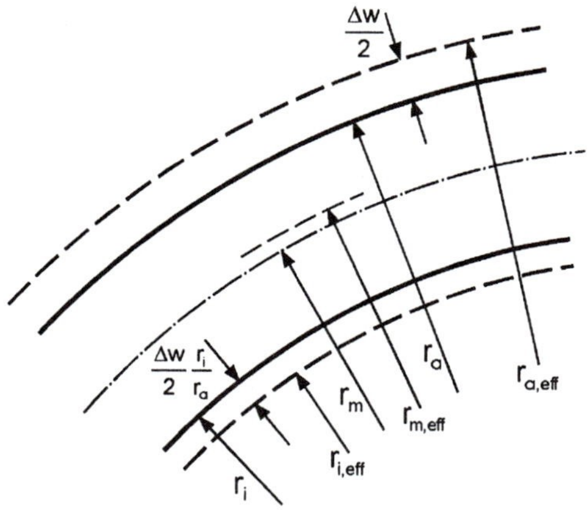

Figure 4.33 Frequency-dependent effective model parameters for curved microstrip lines.

and the effective width of the curved modeled microstrip line $w'_{eff}(f)$ is

$$w'_{eff}(f) = r_{a,eff}(f) - r_{i,eff}(f) = r_a - r_i + \frac{\Delta w(f)}{2}\left(1 + \frac{r_i}{r_a}\right) \tag{4.69}$$

From (4.69), it can be seen that with $r_i \to r_a$, the effective width of the curved microstrip line approaches the effective width of a straight microstrip line,

$$w'_{eff}\Big|_{r_i \to r_a} = r_a - r_i + \Delta w = w + \Delta w = w_{eff}(f) \tag{4.70}$$

The material wavenumber k_{eff} is defined as $k_{eff} = \omega/c_0\sqrt{\epsilon_{eff}}$. The proposed model includes the case of a disk resonator (see Section 6.7). With r_i approaching zero, the radius $r_{i,eff}$ also tends to zero which is the case of a disk resonator. Taking into account the frequency-dependent inner and outer radii, the field-theoretical analysis of the ring resonator can be performed according to [32].

The numerical solution of the characteristic equation for individual resonant modes is tedious and time consuming. Therefore, closed-form expressions for the lower-order resonant modes E_{mnq} with $m = 1,2,3$, $n = 1$, and $q = 0$ have been derived from the numerical solution. The index m indicates the number of full-wavelengths over the circumference of the ring resonator. The zerovalue of q signifies no variation of the field in z direction, and $n = 1$, $m \geq 1$ characterizes a transversal electrical field distribution, which is similar to the quasi-TEM mode on a straight mircrostrip line.

For E_{m10} modes the following relations were found:

Table 4.2
Constants of Equation (4.71)

m	$A1_m$	$A2_m$	$A3_m$	$B1_m$	$B2_m$	$B3_m$	$B4_m$
1	0.9206	0.0493	0.0794	–0.4129	–1.0773	5.9931	4.5168
2	1.5271	1.42 E–4	0.4729	6.3852	5.6221	–1.9139	3.8091
3	2.1005	4.42 E–6	0.8995	10.6240	9.6195	–8.3029	1.8957

$$k_{eff} r_{m,eff} = \begin{cases} A1_m + A3_m & x = 0 \\ A1_m + A2_m \dfrac{(\sin \pi x)^{B3_m}}{x^{B2_m}} \dfrac{(\cos \pi x/2)^{B3_m}}{(1-x)^{B4_m}} + A3_m(1-x) \quad \text{for} & 0 < x < 1 \\ A1_m & x = 1 \end{cases} \quad (4.71)$$

where $x = 0.5\, w'_{eff} / r_{m,eff}$. The constants $A1_m$, $A2_m$, $A3_m$, $B1_m$, $B2_m$, $B3_m$, and $B4_m$ are given in Table 4.2. Compared to the results from the numerical solution, the accuracy has been found to be better than 0.06%. This is confirmed by the measured and calculated results shown in Figure 4.34, for microstrip ring resonators fabricated on Polyguide substrate material ($\epsilon_r = 2.32, h = 1.56$ mm) with fixed mean radius ($R = 2.61$ cm) and various microstrip widths $0.1 \le w/(2R) \le 0.9$.

In Figure 4.35, the field distributions of the electromagnetic field of the lowest order resonant modes are illustrated. Regarding the E_{010} (TM_{010}) mode, the electromagnetic field is independent of the azimuthal angle with $m = 0$. The magnetic field lines describe complete circles, and vanish at the inner and outer magnetic side walls. The electrical field is maximum at the magnetic walls with opposite polarity, and vanishes in the middle of the curved microstrip line. The location of the zero depends on the geometry of the ring resonator. The E_{010} (TM_{010}) mode and similar higher-order modes with $n > 1$ normally are not used for practical application, because the resonance frequencies become very large for small strip widths [34].

The E_{m10} (TM_{m10}) modes ($m = 1, 2, \ldots$) exhibit the lowest resonance frequencies and, therefore, are most relevant for practical application. It can be seen from Figure 4.35 that the field pattern is very similar to the field distribution of a quasi-TEM mode on a straight microstrip transmission line; however, with the major difference that the curvature effect causes some compression of the field at the inner magnetic wall. The index m signifies the number of full wavelengths over the ring circumference, which corresponds to respective change of polarity of the electrical field.

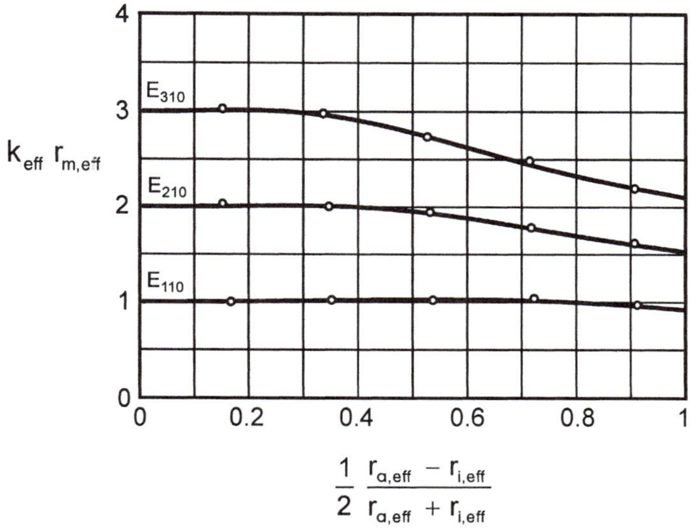

Figure 4.34 Experimental results compared with calculation of the E_{m10} (m = 1, 2, 3) modes (circles are measured data, solid line is calculated). (From [33], © 1981 European Microwave Association. Reprinted with permission.)

4.3.3 Dispersion Measurements of Higher-Order Modes

Resonance and sliding-load techniques also have been applied for dispersion analysis of higher-order modes. Regarding the measurement with the short-circuited microstrip line resonator, it is observed that radiation loss near the cutoff frequency is very high. In the case of the short-circuited microstrip resonator built on Polyguide substrate ($\epsilon_r = 2.32$), first resonance could be measured at a frequency where the harmonic number was from 3 to 5. Therefore, the resonance method has proved to be less useful for dispersion analysis near the cutoff frequencies of higher-order modes on low-dielectric substrates.

The sliding-load technique can be suitably used to investigate the dispersion of first higher-order mode. As in hollow waveguide, the cutoff frequency of a higher-order microstrip mode is characterized by a pure transverse resonance of the electromagnetic field (see Figure 5.3). The field does not change along the axis of the guide. When a short circuit is placed in the plane of symmetry of the microstrip line [Figure 4.19(b)], the quasi-TEM mode is suppressed. This new guide configuration has the first higher-order microstrip mode as its fundamental mode. A similar approach was reported by Oliner for a triplate system [35]. This is shown in Figure 4.36. It is clearly seen that the transition from evanescence to propagation is much smoother with the HE_1 microstrip mode than is observed on the modified (shielded) triplate transmission line, indicating increased electromagnetic radiation.

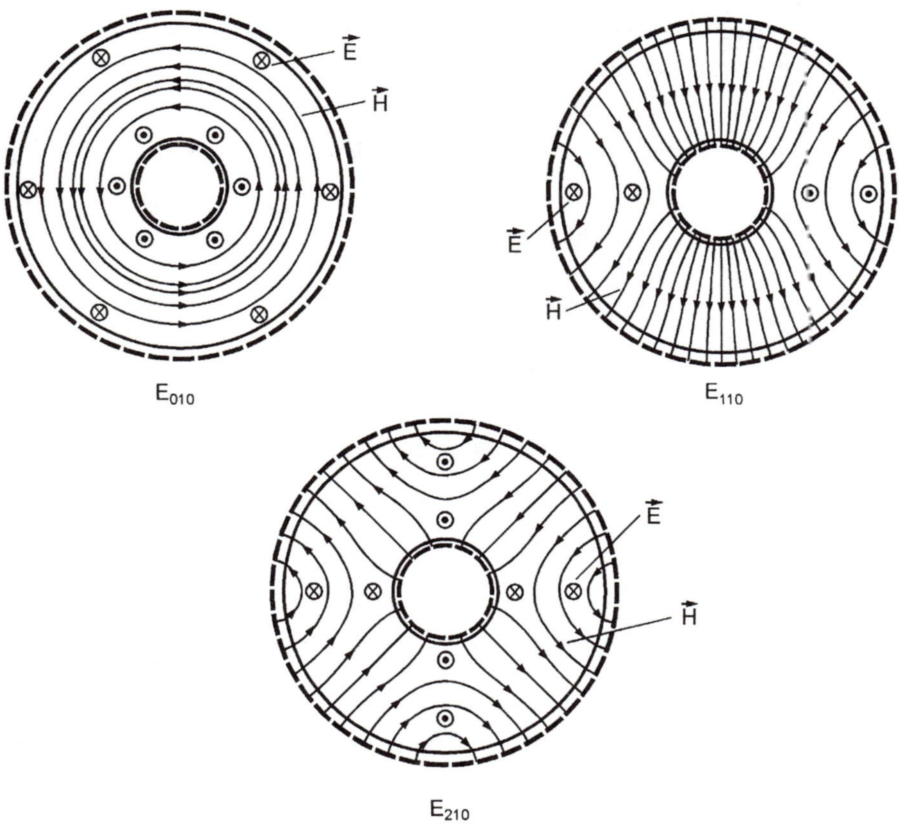

Figure 4.35 Field distribution of the electromagnetic field of the E_{010} mode and E_{m10} modes with $m = 1, 2$.

The procedure of measuring and evaluating the wavelength of the first higher-order mode using the sliding-load technique is similar to that of the procedure for the quasi-TEM mode. It should be noted that the circles of the input reflection coefficient (see Figure 4.21) degenerate to spiral loci due to increased radiation [36].

Figure 4.37 shows the measured wavelength λ_{HE_1} of the first higher-order microstrip mode on Polyguide substrate and alumina. At the cutoff frequency, the wavelength becomes infinite, and with growing frequency, the wavelength decreases. In the limit case $f \to \infty$, the wavelength would approach to $\lambda_{HE_1} \to \lambda = c_0 / f\sqrt{\epsilon_r}$. The measured dispersion curves show strong similarity with those known from standard rectangular waveguides. This observation suggests itself to model the microstrip line by an ideal parallel plate transmission line with magnetic sidewalls (see Chapter 5). Figure 4.38 shows the measured values of the

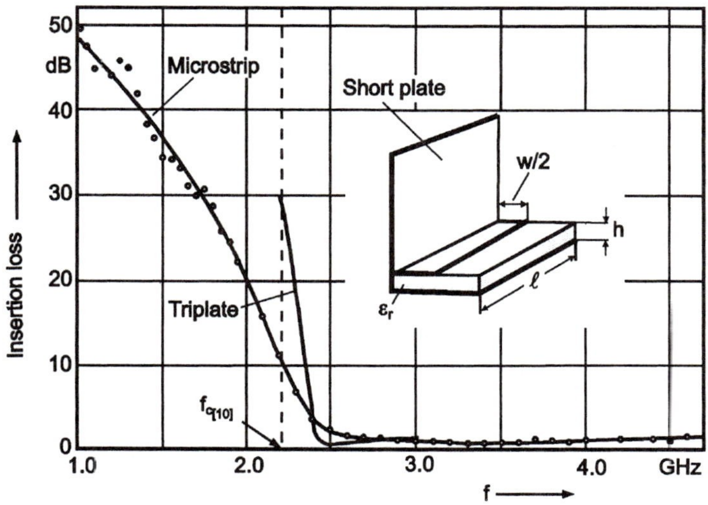

Figure 4.36 Measured insertion loss of the first higher-order microstrip mode in comparison with the triplate [35].

normalized phase coefficient $(\beta/k_0)^2$ versus frequency for the first three microstrip modes on Polyguide substrate and alumina. The solid curves show simulated results based on the waveguide model discussed in Chapter 5. It can be seen that the curves approach asymptotically to the dielectric constant ϵ_r with increasing frequency for any mode. $(\beta/k_0)^2$ is related to the wavelength,

$$(\beta/k_0)^2 = \left(\frac{2\pi}{\lambda_g}\frac{\lambda_0}{2\pi}\right)^2 = \left(\frac{\lambda_0}{\lambda_g}\right)^2 \tag{4.72}$$

The phase coefficient β is also referred to as longitudinal wavenumber k_z ($\beta = k_z$). Considering the first higher-order mode, it can be seen that utilizing the sliding-load technique, dispersion measurements are possible up to nearly the cutoff frequency. The spreading of the measured points increases with frequency, indicating the growing influence of the second odd higher-order mode at a frequency of approximately 6.0 GHz [see Figure 4.38(a)]. Regarding the measurement on alumina, because of small radiation losses in this case, the resonance method is also successful in measuring the dispersion near the cutoff frequencies.

Problem 4.2: Figure 4.39 shows a two-port circuit described by its scattering matrix S with $|S_{11}| = |S_{22}| \ll 1$. The output of the two-port (reference plane 2) is terminated by a sliding load, consisting of a variable length l and a fixed load with a reflection coefficient Γ_s. The input reflection coefficient Γ_1 is measured for five different lengths $l_1, l_2, ..., l_5$ (see following table). Determine graphically the

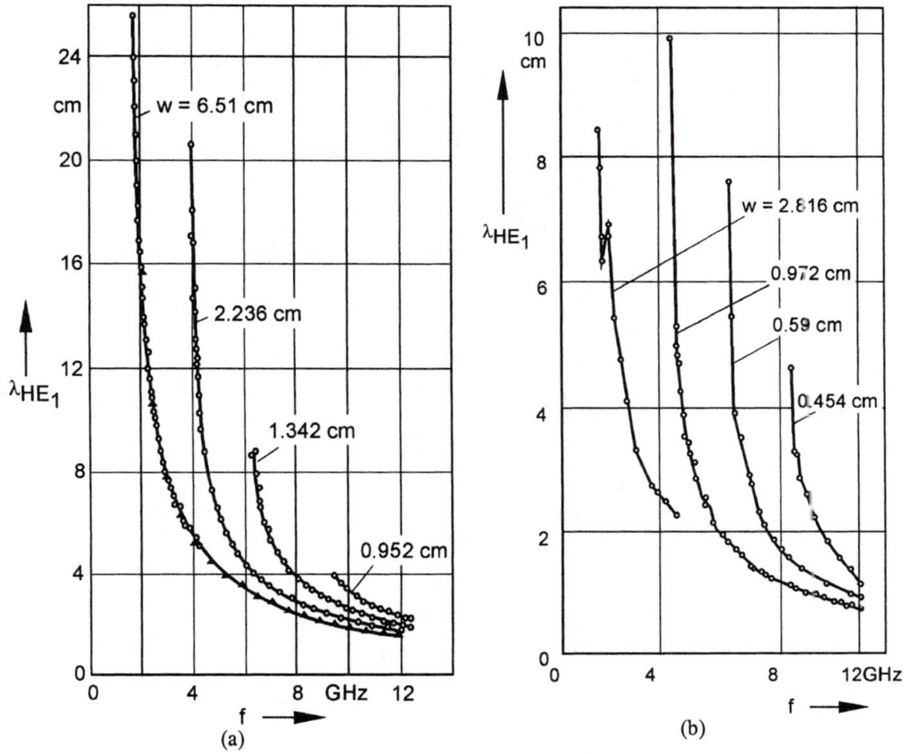

Figure 4.37 Measured wavelength λ_{HE1} of the first higher-order microstrip mode on (a) Polyguide substrate ($\varepsilon_r = 2.32$, $h = 0.162$ cm) and on (b) alumina ($\varepsilon_r = 9.7$, $h = 0.635$ mm) (triangles: resonator method, circles: sliding-load method). (From: [37], © 1976 Elsevier. Reprinted with permission.)

eigen-reflection coefficient S_{11} of the two-port, and estimate the magnitude of the reflection coefficient of the surface absorber (sliding load).

	l_1	l_2	l_3	l_4	l_5
Γ_1	$0.25\angle 95°$	$0.35\angle 82°$	$0.39\angle 50°$	$0.30\angle 30°$	$0.20\angle 24°$

Solution: The measured complex Γ_1 values are put into the enclosed polar diagram (Figure 4.40). Then, draw at least two center perpendiculars. The intersection is the center point of the circle through the measured Γ_1 points. The center of the circle equals S_{11}. From the graph, we read $S_{11} = 0.25\angle 60°$.

The radius of the circle is $|S_{12}|^2\,|\Gamma_s| = 0.15$ (see Figure 4.21). Then, $|\Gamma_s| = 0.15/|S_{12}|^2$. As the two-port is passive, it follows $|S_{12}| \le 1$, so that $|\Gamma_s| \ge 0.15$ is obtained.

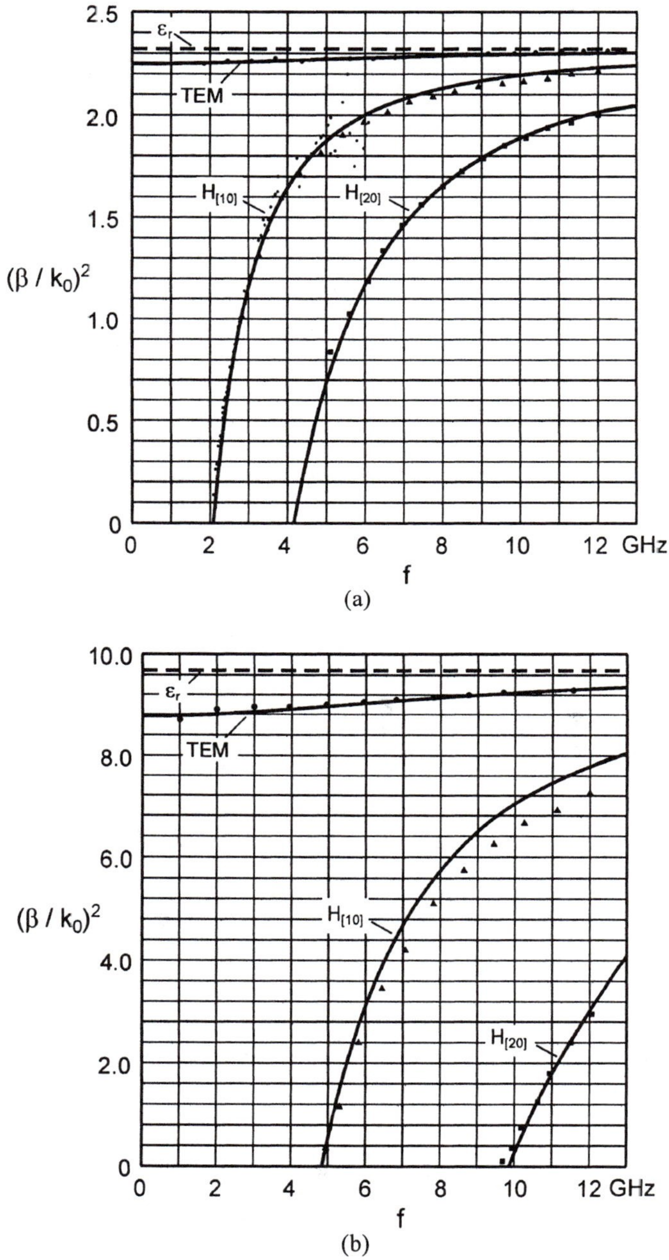

Figure 4.38 Measured normalized wavenumber of lowest-order microstrip modes as function of frequency on (a) Polyguide substrate ($\varepsilon_r = 2.32$, $w = 4.49$ cm, $h = 0.162$ cm), and on (b) alumina ($\varepsilon_r = 9.7$, $w = 9.15$ mm, $h = 0.635$ mm) (dots: sliding load method, triangles: resonance method).

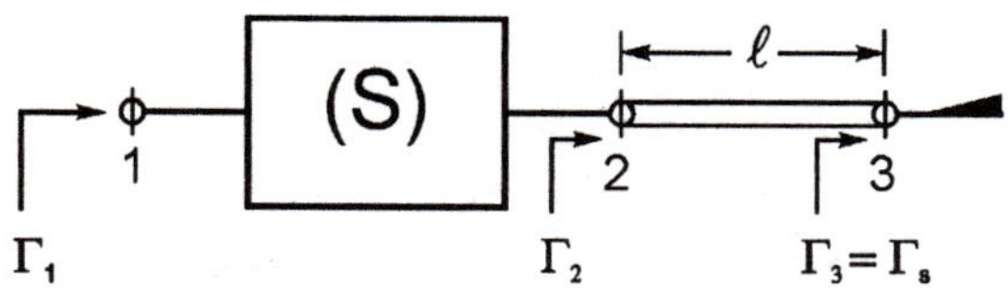

Figure 4.39 Circuit under analysis.

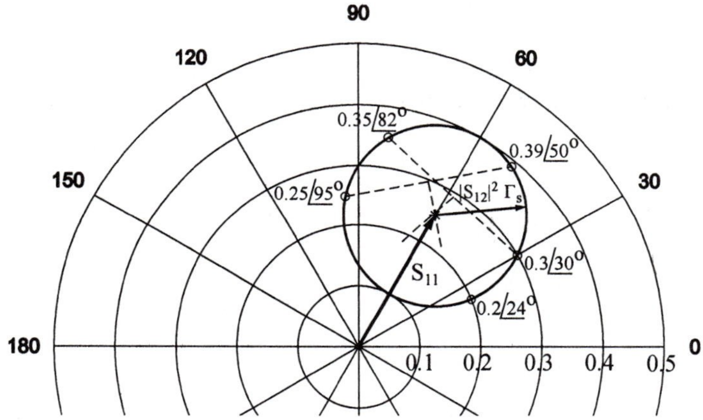

Figure 4.40 Solution of problem 4.2.

Problem 4.3: Determine the physical length l of a short-circuited microstrip transmission line resonator on Polyguide substrate ($\epsilon_r = 2.32$, $h = 1.6$ mm) shown in Figure 4.41, so that a λ-resonance occurs at a frequency of $f_0 = 4$ GHz. The microstrip width is chosen as $w = 3.2$ mm.

Solution: The normalized frequency $f \cdot h$ is calculated as 6.4 GHz·cm. The w/h ratio is given as 2.0. Then, from Figure 4.3, the effective dielectric constant for the considered substrate material is obtained as $\epsilon_{eff}(f_0) = 1.94$. The resonance condition at f_0 reads

$$l = \frac{n}{2} \frac{c_0}{f_0 \sqrt{\epsilon_{eff}(f_0)}} \tag{4.73}$$

Because of λ-resonance, the electrical length comprises two half-wavelengths, and therefore $n = 2$. Thus, we get from (4.73)

$$l = \frac{2}{2} \frac{30 \text{ GHz cm}}{4 \text{ GHz} \sqrt{1.94}} = 5.38 \text{ cm}$$

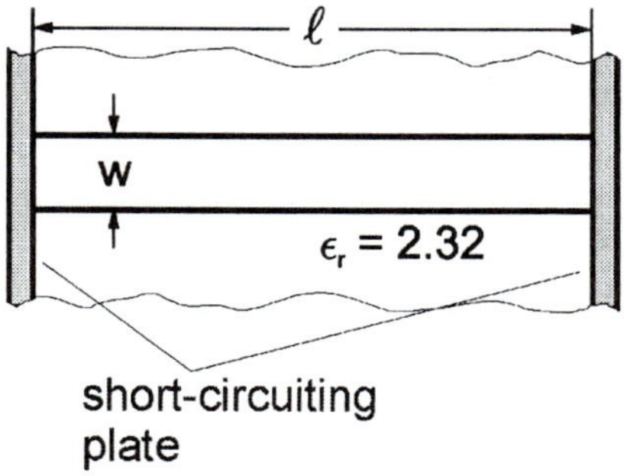

Figure 4.41 Measured short-circuited microstrip transmission line resonator.

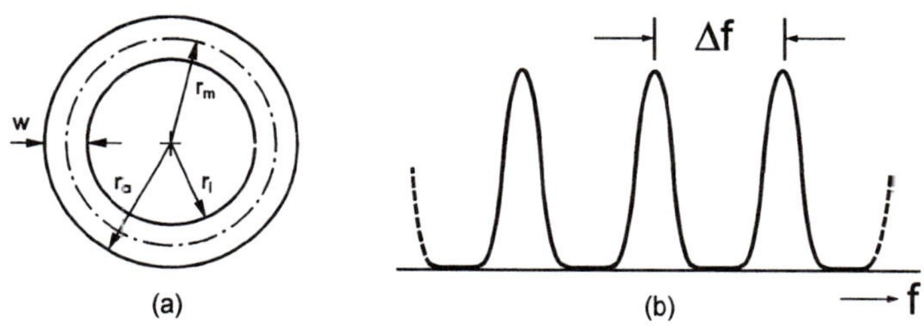

Figure 4.42 Ring resonator and measured resonant frequencies.

Problem 4.4: A ring resonator with the arithmetic mean radius r_m = 1.5 cm [Figure 4.42(a)] is stimulated by a microwave source. Increasing the frequency continuously, the resonance frequency pattern in Figure 4.42(b) is measured. The frequency difference of two adjacent resonances Δf is measured to be 1 GHz.

Determine the effective dielectric constant under the assumption that dispersion is neglected over the considered frequency interval Δf: $\epsilon_{eff}(f_{n_2}) \approx \epsilon_{eff}(f_{n_1}) \approx \epsilon_{eff}(f_n)$.

Solution: The distance between the resonant frequencies of two adjacent resonances read

$$f_{n_2} - f_{n_1} = \Delta f = \frac{(n+1)\,c_0}{\sqrt{\epsilon_{eff}(f_{n_2})}}\,\frac{1}{2\pi r_m} - \frac{n\,c_0}{\sqrt{\epsilon_{eff}(f_{n_1})}}\,\frac{1}{2\pi r_m} = \frac{c_0}{\sqrt{\epsilon_{eff}(f_n)}}\,\frac{1}{2\pi r_m} \tag{4.74}$$

where $\epsilon_{eff}(f_{n_2}) \approx \epsilon_{eff}(f_{n_1}) \approx \epsilon_{eff}(f_n)$. Then, $\epsilon_{eff}(f_n)$ is calculated from (4.74) as

$$\epsilon_{eff}(f_n) = \left(\frac{c_0}{2\pi\, r_m\, \Delta f} \right)^2 \tag{4.75}$$

With the given values, we obtain

$$\epsilon_{eff}(f_n) = \left(\frac{30\ \text{GHz cm}}{2\pi\ 1.5\ \text{cm}\ 1.0\ \text{GHz}} \right)^2 = 10.1$$

References

[1] Wheeler, H. A., "Transmission-line Properties of Parallel Wide Strips by a Conformal Mapping Approximation," *Trans. Inst. Elect. Electron. Engrs.* MTT-12, 1964, pp. 280–289.

[2] Wheeler, H. A., "Transmission-line Properties of Parallel Strips Separated by a Dielectric Sheet," *Trans. Inst. Elect. Electron. Engrs.* MTT-13, 1965, pp. 172–185.

[3] Rahmat-Samii, Y., T. Itoh, and R. Mittra, "A Spectral Domain Analysis for Solving Microstrip Discontinuity Problems," *IEEE* MTT-22, No. 4, 1974, pp. 372–378.

[4] Gupta, K. C., et al., *Microstrip Lines and Slotlines*, 2nd ed., Norwood, MA: Artech House, Inc., 1996.

[5] Getsinger, W. J., "Microstrip Dispersion Model," *IEEE Trans.*, Vol. MTT-21, 1973, pp. 34–39.

[6] Hammerstad, E. O., and F. Bekkadal, *Microstrip Handbook ELAB Report*, STF 44 A74169, University of Trondheim, Norway, 1975.

[7] Kirschning, M., *Entwicklung von Näherungsmodellen für den rechnergestützten Entwurf von hybriden und monolithischen Schaltungen und Mikrostreifenleitungstechnik*, Doctoral Thesis, University of Duisburg, 1984.

[8] Kirschning, M., and R. H. Jansen, "Accurate Model for Effective Dielectric Constant of Microstrip and Validity up to Millimeter-wave Frequencies," *Electron. Lett.*, Vol. 18, March 1982, pp. 272–273.

[9] Kobayashi, M., "A Dispersion Formula Satisfying Recent Requirements in Microstrip CAD," *IEEE Trans.*, Vol. MTT-36, 1988, pp. 1246–1250.

[10] Yamashita, E., K. Atsuki, and T. Ueda, "An Approximate Dispersion Formula of Microstrip Lines for Computer-Aided Design of Microwave Integrated Circuits," *IEEE Trans.*, Vol. MTT-29, 1979, pp. 1036–1038.

[11] Edwards, T., *Foundations for Microstrip Circuit Design*, 2nd ed., New York: John Wiley & Sons, 1992.

[12] Bianco, B., et al., "Frequency Dependence of Microstrip Parameters," *ALTA FREQUENZA*, Vol. XLIII, N.7, Lug. 1974, pp. 251E-413–254E-416.

[13] Bianco, B., M. Parodi, and S. Ridella, "On the Definitions of Characteristic Impedance of Uniform Microstrips," *ALTA FREQUENZA*, Vol. XLV, N.2, February 1976, pp. 23E-111–28E-116.

[14] Jansen, R. H., and N. H. L. Koster, "New Aspects Concerning the Definition of Microstrip Characteristic Impedance as a Function of Frequency," *IEEE* MTT-S Digest, Dallas, 1982, pp. 305–307.

[15] Brand, H., *Schaltungslehre linearer Mikrowellennetze*, Stuttgart, Germany: S. Hirzel Verlag, 1970, pp. 298–300.

[16] Kompa, G., M. Schlechtweg, and F. van Raay, "Einrichtung zur hochpräzisen meßtechnischen Bestimmung der Hochfrequenzeigenschaften von Schaltungseigenschaften in Streifenleitungstechnik, sowie Anordnungen zur Kalibrierung der Einrichtung," Patent Application P 3825081, 1988.

[17] Trans-Tech, *Trans-Tech Products*, http://www.trans-techin.com.

[18] Sun, D., and D. McGinnis, "Measurement and Simulation Results of Ti Coated Microwave Absorber," *Proc. 1999 Particle Accelerator Conference*, New York, 1999, pp. 854–856.

[19] Kompa, G., *Die frequenzabhängigen Übertragungseigenschaften von Streifenleitungs-Wellenwiderstandssprüngen, Filtern und Stichleitungen in Mikrostrip-Technik*, Doctoral Thesis, Aachen: RWTH, 1975.

[20] Emerson & Cuming, *Cuming Microwave Products*, ECCOSORB MF, May 24, 2004, http://www.eccosorb.com/.

[21] Kompa, G., M. Schlechtweg, and F. van Raay, "Precisely Calibrated Coaxial-to-Microstrip Transitions Yield Improved Performance in GaAs FET Characterization," *IEEE Trans. on Microwave Theory and Techniques*, MTT-38, 1990, pp. 62–68.

[22] Kompa, G., and M. Schlechtweg, "Generalized Modeling of GaAs MESFETs and MODFETs Based on Highly Accurate Broadband Measurements," *Proc. European Microwave Conf.*, 1989, pp. 179–186.

[23] Deutsch, J., and H. J. Jung, "Measurement of the Effective Dielectric Constant of Microstrip Lines in the Frequency Range from 2 GHz to 12 GHz," *Nachrichtentechnische Zeitschrift*, Vol. 12, 1970, pp. 620–624.

[24] Wolff, I., "Messung der effektiven Dielektrizitätszahlen des geraden und des ungeraden Wellentyps auf verkoppelten Mikrostrip-Leitungen," *Nachrichtentechnische Zeitschrift*, Vol. 27, 1974, pp. 30–34.

[25] Collin, R. E., *Foundations for Microwave Engineering*, 2nd ed., New York: McGraw-Hill, Inc., 1992.

[26] Osofsky, S., and S. E. Schwarz, "Design and Performance of a Non-contacting Probe for Measurements on High-frequency Planar Circuits," *IEEE Trans.*, Vol. MTT-40, 1992, pp. 1701–1708.

[27] Gao, Y. J., and I. Wolff, "A Miniature Magnetic Field Probe for Measuring Fields in Planar High-frequency Circuits," *1995 IEEE MTT-S Digest*, 1995, pp. 1159–1162.

[28] Gao, Y., and I. Wolff, "A Simple Electric Near Field Probe for Microwave Circuit Diagnostics," *1995 IEEE MTT-S Digest*, 1995, pp. 1159–1162.

[29] Hienonen, S., P. Vainikainen, and A. V. Räisänen, "Sensitivity Measurements of a Passive Intermodulation Near-field Scanner," *IEEE Antennas and Propagation Magazine*, Vol. 45, 2003, pp. 124–129.

[30] Troughton, P., "Measurement Techniques in Microstrip," *Electron. Lett.*, 1969, 5, pp. 25–26.

[31] Pucel, R., et al., "Losses in Microstrip," *IEEE Trans.*, Vol. MTT-16, June 1968, pp. 342–350.

[32] Wolff, I., and N. Knoppik, "Microstrip Ring Resonator and Dispersion Measurement on Microstrip Lines," *Electron. Lett.*, Vol. 7, No. 26, December 1971, pp. 779–780.

[33] Khilla, A. M., "Computer-Aided Design for Microstrip Ring Resonators," *Proc. 11th European Microwave Conference*, Amsterdam, 1981, pp. 677–681.

[34] Wolff, I., and V. K. Tripathi, "The Microstrip Open-Ring Resonator," *IEEE Trans.*, Vol. MTT-32, 1984, pp. 102–107.

[35] Oliner, A. A., "Theoretical Developments in Symmetrical Strip Transmission Line," *Symposium on Modern Advances in Microwave Techniques*, Brooklyn, N.Y., 1954, pp. 379–402.

[36] Kompa, G., "Dispersion Measurements of the First Two Higher-order Modes in Open Microstrip," *Arch. Elektr. Übertr.* 29, 1975, pp. 182–184.

[37] Kompa, G., "S-Matrix Computation of Microstrip Discontinuities with a Planar Waveguide Model," *Arch. Elektr. Übertr.*, Vol. 30, 1976, 58–64.

Chapter 5

Planar Waveguide Model

The planar waveguide model was devised in the 1970s to tackle the problem of the observed frequency dependence of discontinuities and junctions, such as microstrip impedance steps and T-junctions. At that time microstrip dispersion was under intensive investigation, and a rigorous field solution of the quasi-TEM mode was not known. This applies in particular to the higher-order modes that are the real cause for the frequency dependence of the scattering parameters of the devices under consideration. The introduction of a closed ideal parallel plate model for the open microstrip line with electrical walls at the top and bottom, and magnetic walls on both sides, cannot be truly justified from the theoretical point of view. The boundary conditions of both structures are totally different. However, in engineering work, any model may be acceptable under the condition that it delivers the expected results and passes the experimental test. Even for the strongly dispersive hollow waveguide, a characteristic impedance is known as an auxiliary design variable.

Meanwhile, the waveguide model has been proven to be a worthful analysis tool for a variety of microstrip discontinuites and junctions. The writer himself has treated thoroughly multistep discontinuities, such as quarter-wavelength transformers and tapered microstrip transitions, calculated with both the waveguide model and advanced 3D EM simulators. It seems that the waveguide model concept has still its advantage concerning speed of calculation.

It was shown in Chapter 1 that discontinuities and junctions are widely found in common microstrip circuits, such as amplifiers, oscillators, and multipliers. Therefore it is of urgent need to have accurate equivalent models for reliable computer aided design. At very low frequency, the dimensions of a given transmission line inhomogeneity are much smaller than the operating wavelength. In this quasi-static case, the equivalent models comprise frequency-independent reactive elements under the assumption that dissipation loss is negligible. However, with increasing frequency, the excitation of higher-order modes in the discontinuities will interfere with the propagation of the quasi-TEM mode. Commonly, the microstrip circuit operates in the single-mode region, so that higher-order modes only exist as evanescent modes, being localized near the respective discontinuity or junction. Electrical scattering fields at strip corners and edges or those associated with an abrupt change of strip width store electrical energy, which can be equivalently described by a lumped capacitance. Similarly, stored magnetic energy is equivalently described by a lumped inductance.

The distribution of the electromagnetic field of higher-order evanescent modes changes with frequency. This is also valid for the model elements of a given equivalent circuit. Equivalent circuits taking into account the frequency dependence of model elements are termed dynamic equivalent circuits.

Quasi-static equivalent circuit elements can be gained from stationary electric and magnetic field analysis [1–3]. However, derivation of broadband dynamic equivalent circuits of discontinuities and junctions needs a full-wave analysis on the basis of the complete Maxwell equations. As was discussed in Section 3.4.3, the solution of the microstrip boundary problem of a straight transmission line is already a difficult task for the fundamental quasi-TEM mode, leading to a set of coupled integral equations for the propagation coefficient. In [4], the analysis of microstrip modes also included higher-order modes derived from supposed respective current distributions on the strip.

An efficient numerical approach termed the orthogonal series expansion method [5–11] had been developed and successfully applied for the analysis of discontinuity and junction problems in hollow waveguide in the 1960s [5–10]. The method supposes a so-called complete set of eigenmodes for any wave guiding line connected to the discontinuity or junction under consideration. In hollow waveguides, these are identified with the waveguide modes and classified in *E* (or ***TM***) modes with purely transverse magnetic field, and *H* (or *TE*) modes with purely transverse electric field. The infinite number of modes represents a complete set of eigenfunctions, in the sense that any arbitrary transversal field distribution can be expanded into an infinite Fourier series comprising all eigenmodes with respectively tuned mode amplitudes. As is known, modes in any closed waveguide structure are orthogonal to each other [9], which simplifies the theoretical treatment of discontinuity problems, because integral expressions, including dot products of different vector mode functions vanish. The method of orthogonal series expansion concerning the calculation of the dynamic scattering coefficients of a microstrip impedance step will be discussed in more detail in Chapter 6.

Thus, in order to adopt the well-established orthogonal series expansion method to the analysis of microstrip discontinuity and junction problems, it had first to be confirmed that such a complete system of orthogonal modes really exists for the open microstrip line, and the microstrip waveguide model, respectively. It should be noted that the orthogonality property of modes is really not needed. Nevertheless, it strongly simplifies the application of the method under discussion.

A waveguide model with lateral magnetic walls had already been discussed earlier for the triplate [12–14]. Figure 5.1 shows the approximate model for the triplate as proposed in [12]. Supposing that only balanced discontinuity problems are of interest, only one-half of the transmission line needs to be considered. Therefore, the height of the model is one-half of the ground spacing b, and the effective width reads

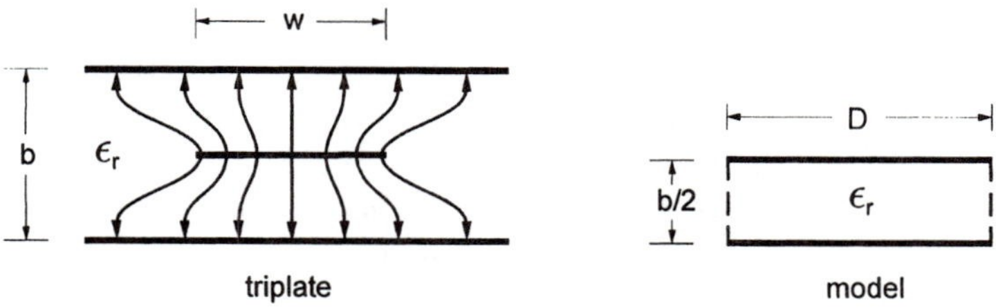

Figure 5.1 Waveguide model for the triplate [12].

$$D = b\,K(k)/K(k') \tag{5.1}$$

where

$$k = \tanh(\pi w/2b) \tag{5.2}$$

$K(k)$ is the complete elliptic integral described in Section 3.2. For wider strips ($w/b > 0.5$), the effective width D can be accurately approximated by

$$D = w + \frac{2b}{\pi}\ln 2 \tag{5.3}$$

Applying (5.1) to (5.3) to the microstrip system, we get

$$w_{eff} = 2h\,K(k)/K(k') \tag{5.4}$$

where

$$k = \tanh(\pi w/4h) \tag{5.5}$$

For $w/h > 1$, (5.4) can be approximated by

$$w_{eff} = w + \frac{4h}{\pi}\ln 2 = w + 2\,\Delta w \tag{5.6}$$

where

$$\Delta w = \frac{2h}{\pi}\ln 2 \approx 0.44\,h \tag{5.7}$$

Unlike the microstrip transmission line, the triplate with homogeneous dielectric material and shielded enclosure with top and bottom ground plates, propagates a pure TEM wave as a fundamental mode. This is quite different with the microstrip line comprised of mixed dielectric of air and substrate material, which causes dispersion, and is an open waveguiding structure with infinite extension of the electromagnetic field. Therefore, introducing a closed waveguide model for the microstrip line would rudely violate the actual boundary conditions. Nevertheless, engineers often have to follow the dictum "The end justifies the means" to find a feasible solution to a given sophisticated problem. Many attempts had been made to find a convenient microstrip line model that could deliver the observed frequency-dependent scattering coefficents of discontinuities and junctions, such as the step impedance and the T-junction. Figure 5.2 illustrates a collection of respective models proposed as equivalent waveguide counterparts for the open microstrip line. As can be seen, all models are closed waveguides with infinitely conductive top and ground plates, and magnetic walls at the lateral sides. The dielectric filling in models (b) to (e) is homogeneous, whereas model (f) reflects the inhomogeneous distribution of dielectric material as obtained by conformal mapping after Wheeler (Section 3.1) [15]. A first study concerning the fundamental dynamic electrical properties of an abrupt impedance step was accomplished in [16] on the basis of a magnetic wall model with physical strip width w and dielectric constant ϵ_r [Figure 5.1(b)]. In [17, 18], the effective width w_{eff} and effective dielectric constant ϵ_{eff} after Wheeler [19, 20] were used to reconcile in a better way simulated results with measurement at low frequencies. However, it was observed that Wheeler's effective parameters were not convenient to describe the electrical properties of discontinuities and junctions at higher frequencies, where the influence of higher-order modes becomes noticeable.

To obtain better agreement with the cutoff frequencies of higher-order modes, a model with dielectric constant ϵ_r and effective width $w_{eff} = w + 0.4h$ was proposed [21]. The respective effective width was used in [22] to estimate the cut-off frequency of the first higher-order (TE) mode with a transverse resonance of half-wavelength,

$$w_{eff} = w + 0.4h = \frac{1}{2}\frac{\lambda_{TE}}{\sqrt{\epsilon_r}} \tag{5.8}$$

Then from (5.8), the cutoff frequency for this mode reads

$$f_c = \frac{c_0}{\sqrt{\epsilon_r}}\frac{1}{w + 0.4h} \tag{5.9}$$

It should be noted that the effective width extension originating in the edge-fringing electrical field is only one-half the value proposed by Oliner [12]. As noted in [22], the effective value is only valid for sufficiently wide microstrip lines, and was taken from [23] concerning the radiation from parallel-plate waveguiding structure. From extensive experimental analysis of higher-order microstrip modes [24, 25], it was concluded that the electromagnetic properties of both the quasi-TEM mode and higher-order modes cannot be equivalently described with a unified closed waveguide model with sufficient accuracy over a broad frequency range. However, it was also observed that the interference of higher-order modes with the quasi-TEM mode is increasing as the operating frequency approaches the cutoff frequency of any higher-order mode. Thus, from experience, the requirements for

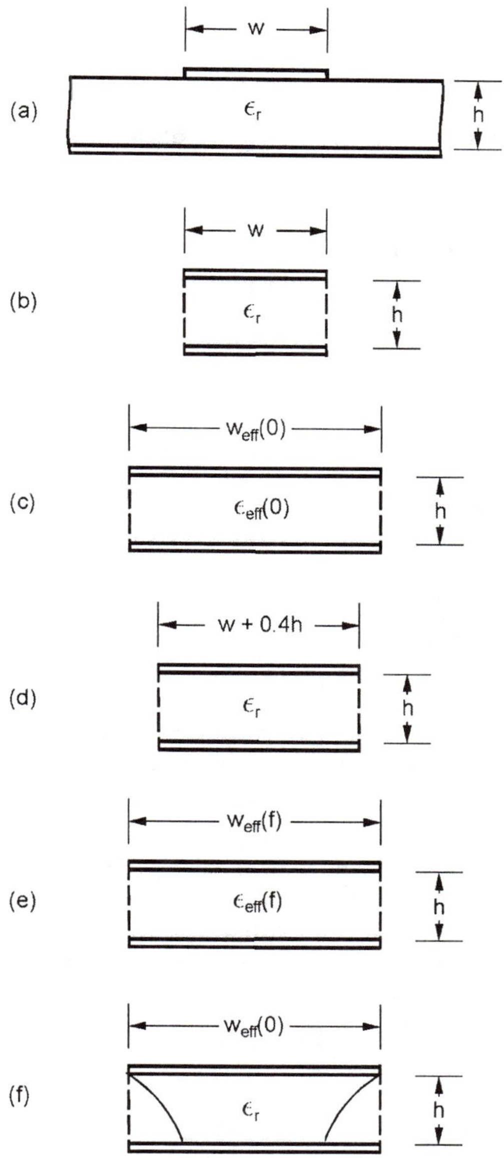

Figure 5.2 Various waveguide models proposed for the open microstrip line.

an acceptable waveguide model were reduced to the following points: (1) the model should be able to simulate the low-frequency (quasi-static) behavior of the quasi-TEM mode, and (2) it should be able to describe the cutoff frequencies of the higher-order modes with sufficient accuracy.

5.1 Definition of a Dynamic Waveguide Model

The definition of the frequency-dependent model parameters of the dynamic waveguide model was the result of several years, research of focusing on field analysis of microstrip discontinuities and junctions. Microwave components, such as impedance steps, offset-junctions, stubs, bends, T-junctions, and crossings are widely used in microstrip circuits. Therefore, knowledge of the frequency-dependent scattering coefficients and respective analytic expressions for such elements are of great practical importance for the circuit design.

The dynamic waveguide model was already developed in the 1970s and is still found in current commercial software packages, such as Advanced Design System (ADS) from Agilent Technologies [26] or Microwave Office supplied by Applied Wave Research, Inc. [27]. The definition of the model parameters was based on direct comparison of measured and simulated scattering coefficients of numerous microstrip discontinuities and junctions in an available frequency range of 1 GHz to 12.4 GHz, and on the experimental analysis of the lower-order modes in the microstrip line [25, 28, 29]. As discussed in Section 4.3, measurement methods are available to investigate the dispersion properties of the three lowest-order microstrip modes denoted as quasi-TEM (or HE_0), HE_1, and HE_2 mode [24].

Figure 4.23 shows typical results for the measured wavelength as a function of frequency for microstrip lines of different widths on Polyguide and alumina substrate. It can be seen that the curves are similar to those in rectangular hollow waveguides. A cutoff frequency is well recognized as the wavelength becomes infinite, and the wavelength approaches with increasing frequency the medium wavelength in the substrate. Since the height is much smaller than a quarter-wavelength ($h \ll 0.25\lambda_0/\sqrt{\epsilon_r}$), only higher-order modes with lateral transverse resonances are possible. As in hollow waveguide theory, the cutoff frequencies of the modes manifest in pure transverse resonances, being an integer number of one-half cutoff wavelengths. In propagation direction, the electromagnetic field of the respective mode is then constant at cutoff. Assuming a microstrip model with frequency-dependent effective width $w_{eff}(f)$ and frequency-dependent effective dielectric constant $\epsilon_{eff}(f)$, the cutoff frequency condition yields

$$w_{eff}(f_{c[m0]}) = m\frac{\lambda_{c[m0]}}{2} = \frac{m}{2}\frac{c_0}{f_{c[m0]}}\frac{1}{\sqrt{\epsilon_{eff}(f_{c[m0]})}} \tag{5.10}$$

where the subscript $[m0]$ denotes the respective higher-order TE (or H) mode with $m = 1, 2, \ldots$. The approximate transverse electrical field distribution of the lowest-order modes on the actual microstrip line and their corresponding waveguide modes are illustrated in Figure 5.3. Resolving for the cutoff frequency $f_{c[m0]}$ in (5.10) yields

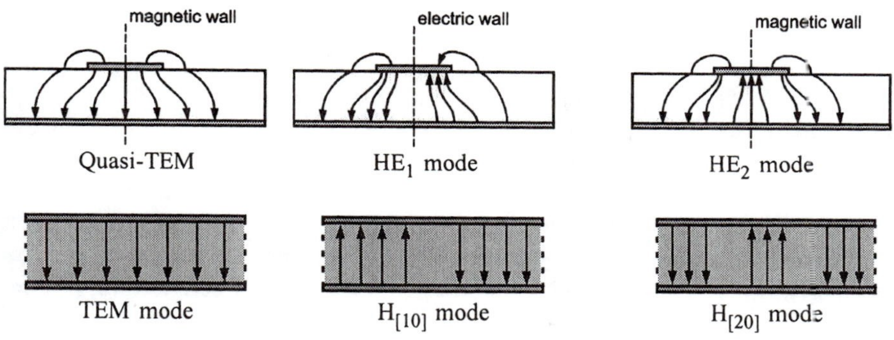

Figure 5.3 Approximate electrical field patterns of the lowest-order modes in the microstrip line and their corresponding modes in the planar waveguide model (quasi-TEM mode or HE_0 mode corresponds to the TEM mode, HE_1 mode to the $H_{[10]}$ mode, and HE_2 mode to the $H_{[20]}$ mode).

$$f_{c[m0]} = \frac{m \cdot c_0}{2\sqrt{\epsilon_{eff}(f_{c[m0]})}\ w_{eff}(f_{c[m0]})} \tag{5.11}$$

The wavenumber relation (D.88) reads with $k_y = 0$

$$k^2 = k_x^2 + k_z^2 \tag{5.12}$$

where

$$k_x = \frac{m\pi}{w_{eff}} \tag{5.13}$$

Resolved for k_z,

$$k_z = \frac{2\pi}{\lambda_g} = \sqrt{k^2 - k_x^2} \tag{5.14}$$

Then the guide wavelength λ_g reads

$$\lambda_g = \frac{2\pi}{k}\sqrt{\frac{1}{1 - (k_x/k)^2}} \tag{5.15}$$

With

$$\frac{k_x}{k} = \frac{f_c}{f} \tag{5.16}$$

we finally find

$$\lambda_{g[m0]} = \frac{\lambda_0}{\sqrt{\epsilon_{eff}(f)}\ \sqrt{1 - (f_{c[m0]}/f)^2}} \tag{5.17}$$

with $f_{c[m0]}$ given in (5.11). Equation (5.17) reveals that the wavelength is dependent on both the effective dielectric constant ϵ_{eff} and the effective width w_{eff}. Thus, a unique experimental determination of the model parameters is not possible. Therefore, different approaches had been discussed in [25] to define a dispersion model that is not only valid for the quasi-TEM mode, but also describes the dispersion properties of the higher-order microstrip modes with sufficient accuracy. Figures 5.4 and 5.5 show the fitting result under the assumption that both the effective width and effective dielectric constant are constant (i.e., not frequency dependent). It can be seen that the optimized values of the effective width (Figure 5.4) scatter around the effective width after Wheeler. Regarding the effective dielectric constant (Figure 5.5), only for very wide strips is there good agreement with Wheeler's results. The narrower the strip, the smaller the effective dielectric constant and the larger the deviation from Wheeler's results. It also can be seen that the effective dielectric constant values are dependent on the respective mode. Similar results had been obtained on Polyguide substrate [25].

Under assumption of a constant effective width $w_{eff}(0)$ after Wheeler, a frequency-dependent effective dielectric constant could be determined from the measured wavelength, according to (5.17). In this case, for very wide strips $w/h > 10$, quasi-TEM–like monotonously increasing effective dielectric constants could be evaluated for the different higher-order modes. However, regarding narrow strips, anomalous dispersion curves were evaluated, showing first a decrease of the effective dielectric constant with increasing frequency, before passing into a positive rise [25].

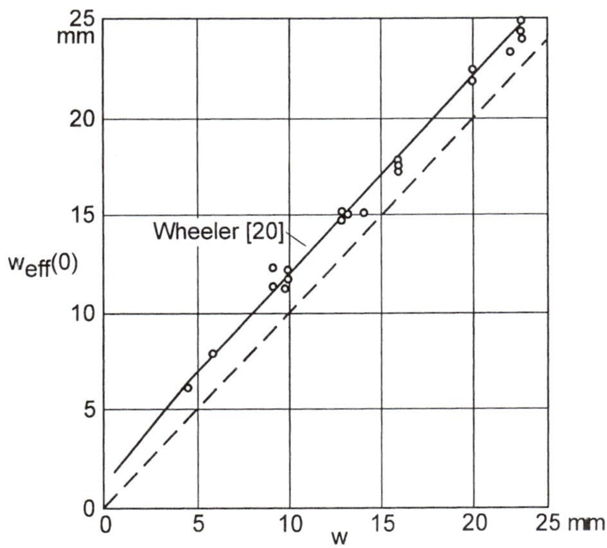

Figure 5.4 Extracted frequency independent effective width from wavelength measurements on alumina substrate ($\varepsilon_r = 9.7$, $h = 0.635$ mm).

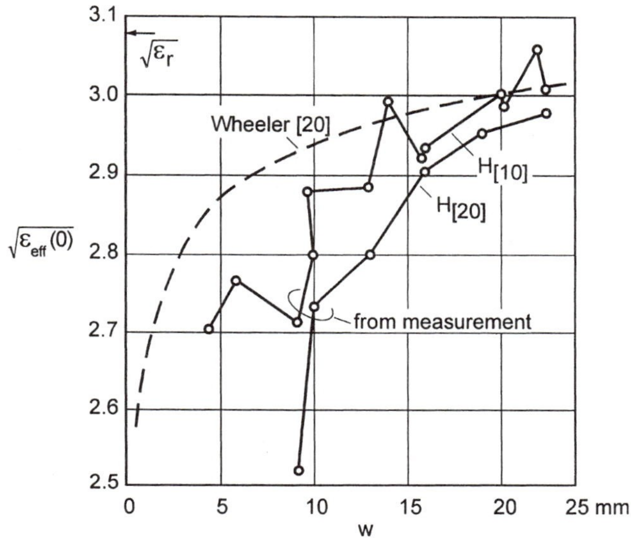

Figure 5.5 Extracted frequency independent effective dielectric constant from wavelength measurements on alumina substrate (ε_r = 9.7, h = 0.635 mm).

In a further simulation experiment, both the effective width and the effective dielectric constant were assumed to be frequency-dependent model parameters. It was supposed that the frequency-dependent effective dielectric constant for the quasi-TEM mode is also valid for the higher-order modes. As illustrated in Figure 5.6, the extracted values for the effective width, denoted by dots, show clear frequency dependence being largest near the cutoff frequency and decreasing with increasing frequency. The results confirm that the part of the electromagnetic field in air is shifted into the substrate region with growing frequency, and that the fringing edge field of the microstrip line decreases, so that the effective width approaches the physical value. In comparison with Figure 4.37, it can be noticed that the measured dispersion curves on alumina substrate are rather complete, in particular near the cutoff frequency, so that the evaluation of the measured data is very trustable. As can be seen, under assumption of same frequency-dependent effective dielectric constant for all modes, the effective widths of higher-order modes at cutoff have been extracted as being smaller than the effective width after Wheeler. The difference is rather large with very wide strips. An increase of the value would result in a decrease of the effective dielectric constant to fit the measured cutoff frequency. This result coincides entirely with the earlier analysis assuming constant effective model parameters. Considering Figure 5.6(b), which illustrates the evaluation of the measured frequency dependent wavelength on Polyguide substrate, the upper two widths can be interpretated in a similar way. The other curves with smaller widths (w < 18 mm), however, include evaluation

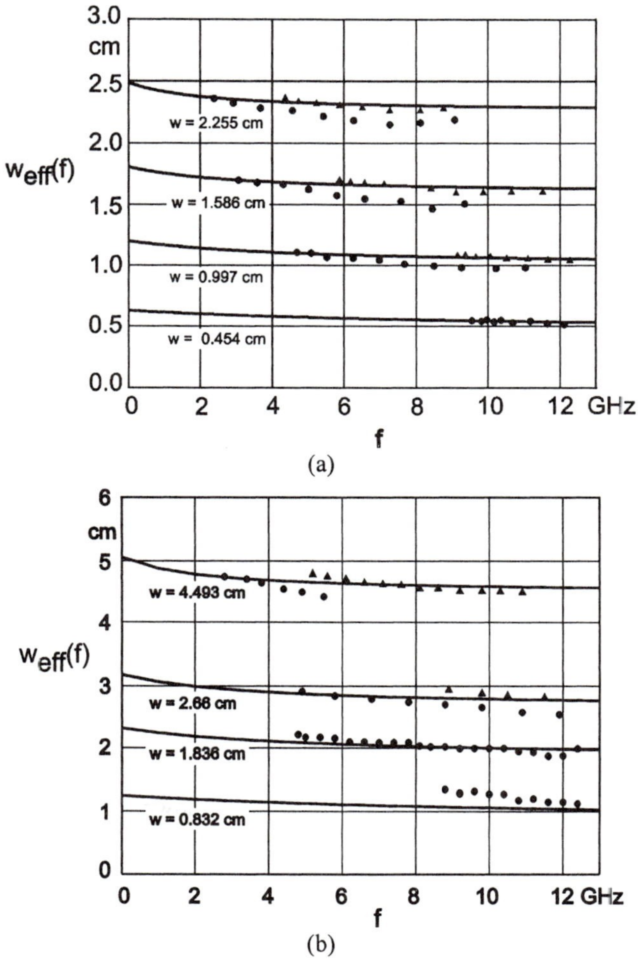

Figure 5.6 Computed and experimentally determined effective widths (a) on alumina substrate ($\varepsilon_r = 9.7$, $h = 0.635$ mm) and (b) Polyguide substrate ($\varepsilon_r = 2.32$, $h = 1.62$ mm) (model: solid lines; experiment: triangles for resonator and dots for sliding-load technique). (From [30], © 1975 IEE. Reprinted with permission.)

uncertainties because, as can be seen in Figure 4.37, the relevant measurement region near cutoff is partly missing, due to strong radiation effects on narrow microstrip lines.

Thus, it can be concluded that a unified microstrip waveguide model that satisfactorily describes both the propagation properties of the quasi-TEM mode and the higher-order modes as well does not exist. It should be noted that the wave propagation of higher-order modes in a microstrip circuit is undesired, so that an exact description of the higher-order mode properties above cutoff is not needed.

Nevertheless, the identification of the cutoff frequencies of the higher-order modes is of great importance because it defines the upper frequency limit for single-mode operation.

On the basis of numerous overall comparisons of measurement and simulation of a variety of discontinuities and junctions the following compromise solution for the frequency-dependent effective width was found [25, 28–30]:

$$w_{eff}(f) = w + \frac{w_{eff}(0) - w}{1 + f/f_g} \tag{5.18}$$

with

$$f_g = \frac{c_0}{2w\sqrt{\epsilon_r}} \tag{5.19}$$

where $w_{eff}(0)$ is the effective width calculated after Wheeler (3.44). f_g can be interpreted as the cutoff frequency of the first higher-order mode with model parameters w and ϵ_r at infinite frequency. The solid lines in Figure 5.6 have been calculated with (5.18). Regarding in particular the frequencies near cutoff of the first higher-order mode, it can be seen that the simulated effective widths agree with those derived from measurement. Note that this is not valid for narrower lines on Polyguide because of strong radiation [compare Figure 4.37(a)]. The simulated cutoff widths of the second higher-order mode, which exhibit nearly the same cutoff widths from measurement, are slightly smaller than the extracted values.

Since the frequency-dependent effective dielectric constant $\epsilon_{eff}(f)$ and frequency-dependent effective width $w_{eff}(f)$ are known, (5.11) can be applied to calculate the cutoff frequency of the first and second higher-order mode. This is plotted versus the microstrip width in Figure 5.7 for Polyguide and in Figure 5.8 for alumina substrate material. It can be seen that the calculated width-dependent cutoff frequency of the even higher-order mode $H_{[20]}$ is slightly above the measured data, which simply explains the different results derived from a spectral domain approach [31].

Knowing the model parameters, a frequency-dependent characteristic impedance $Z_0(f, \epsilon_r)$ can be calculated:

$$Z_0(f, \epsilon_r) = \sqrt{\frac{\mu_0}{\epsilon_0}} \frac{1}{\sqrt{\varepsilon_{eff}(f)}} \frac{h}{w_{eff}(f)} \tag{5.20}$$

Notice that the characteristic impedance of the waveguide model increases with growing frequency [32]. This is fully in agreement with the remarks in Section 3.4.

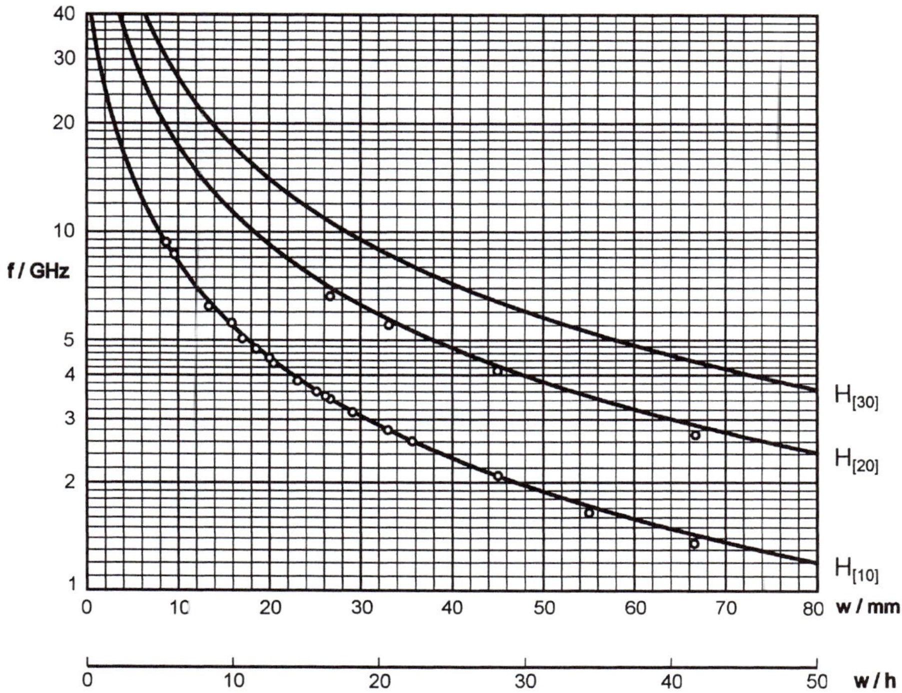

Figure 5.7 Measured and calculated cutoff frequencies of the first two higher-order microstrip modes on Polyguide substrate (ε_r = 2.32, h = 0.16 cm).

Owens [33] pointed out that the frequency-dependent characteristic impedance $Z_0(f)$ should be asymptotic to $Z_0(f=0)$ as the frequency approaches zero. Because of that it was proposed to replace f/f_g in (5.18) by $(f/f_g)^2$, thereby making $d[w_{eff}(f)]/df \to 0$ as $f \to 0$. As in this case, however, the frequency-dependent characteristic impedance would decrease with frequency. Owens recommended to replace the parameter f_g by f_p,

$$w_{eff}(f) = w + \frac{w_{eff}(0) - w}{1 + (f/f_p)^2} \tag{5.21}$$

where

$$f_p = \frac{c_0}{2w_{eff}(0)\sqrt{\epsilon_{eff}(0)}} \tag{5.22}$$

In [11], it is recommended to use (5.20) to derive the frequency-dependent effective width $w_{eff}(f)$ on the basis of known $Z_0(f, \epsilon_r)$ in (4.31) and $\epsilon_{eff}(f)$ in (4.9) with (4.15). This would indeed solve the nonzero slope problem at zero frequency and improve the agreement between measurement and simulation at lower

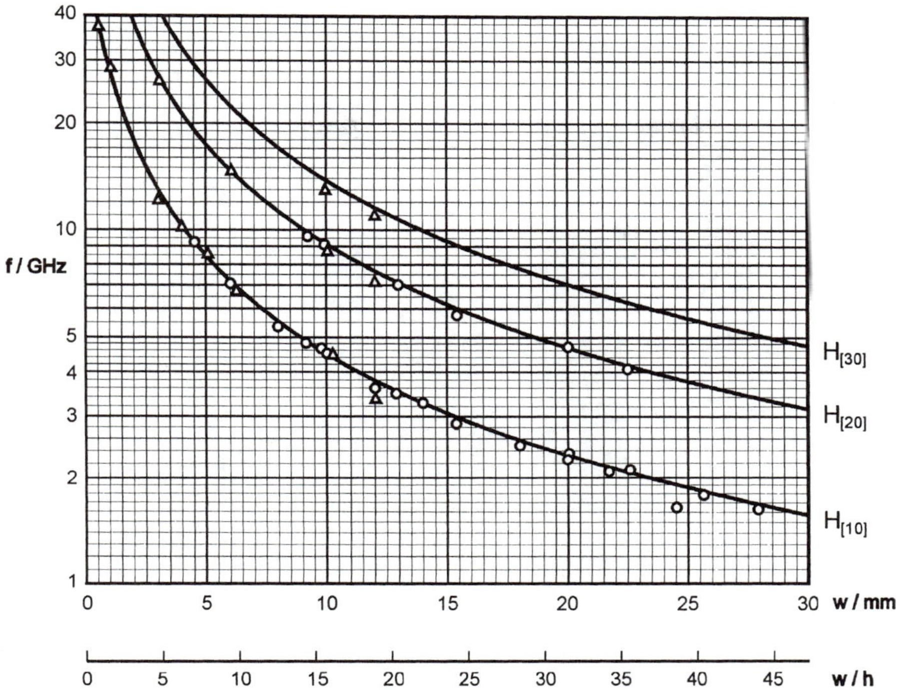

Figure 5.8 Measured and calculated cutoff frequencies of the first two higher-order microstrip modes on alumina substrate ($\varepsilon_r = 9.7$, $h = 0.635$ mm) (triangular dots after [15]).

frequencies. However, as will be discussed in Section 6.2.2, this is only valid in a frequency range in which the influence of higher-order modes is a second-order effect, which is supposed to be only one-third of the respective cutoff frequency.

In Figure 5.9, different defined frequency-dependent effective widths $w_{eff}(f)$ are plotted versus frequency for Polyguide substrate materials with the parameter w/h. Because single-mode operation is only relevant for practical application, the functions are limited by the cutoff frequency of the second higher-order mode $f_{c_{20}}$. It can be seen that Owens formula could describe the cutoff frequency of the first higher-order mode in the wider lines (w_1, w_2, w_3) as very good. This is also approximately valid for the second higher-order mode. In addition, it is seen that (5.20) with $Z_0(f,\epsilon_r)$ and $\epsilon_{eff}(f)$ after Kirschning [34] yields larger deviation of the simulated cutoff frequencies of the first two higher-order modes. As Kirschning's parameters are limited to about 18 GHz for claimed accuracy, cutoff frequency analysis cannot be performed, for example, for narrow lines w_5 and w_6. Owens' relation shows strong decrease of $w_{eff}(f)$ according to the smaller value f_p in contrast to the defined parameter f_g in (5.19). As an example, for the width w_4 from Figure 5.9, we read for the first cutoff frequency 12.5 GHz after Kirschning, 14.5 GHz using (5.18), and 16 GHz after Owens.

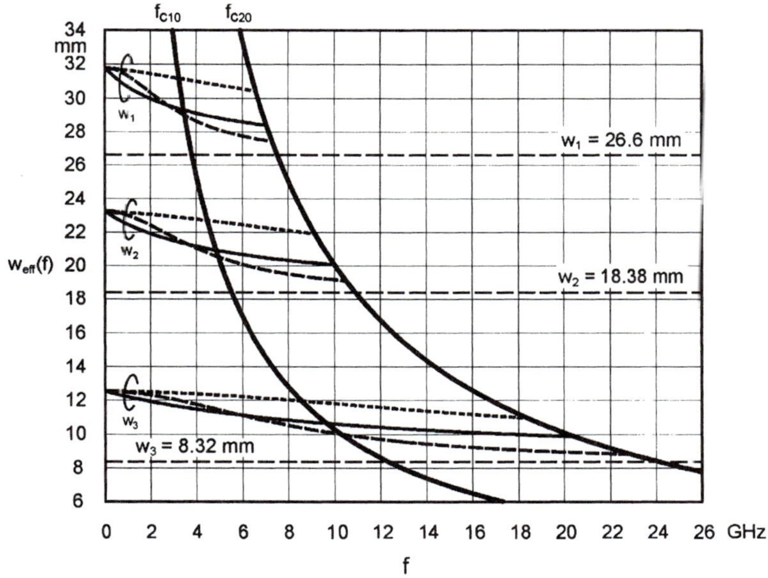

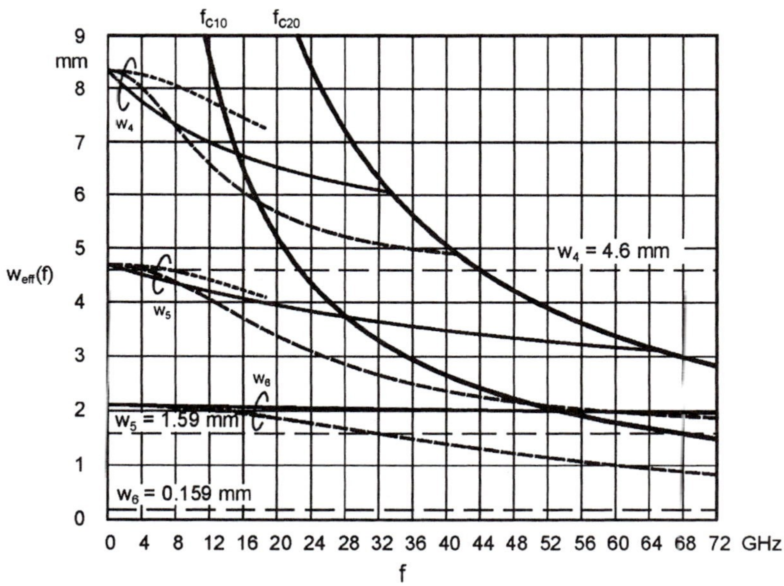

Figure 5.9 Frequency-dependent effective width of the waveguide model (solid line) for Polyguide substrate material ($\varepsilon_r = 2.32$, $h = 1.6$ mm) compared with Owens' result (dashed line), and the frequency-dependent effective width derived from $Z_0(\varepsilon_r, f)$ and $\varepsilon_{eff}(f)$ (dotted line).

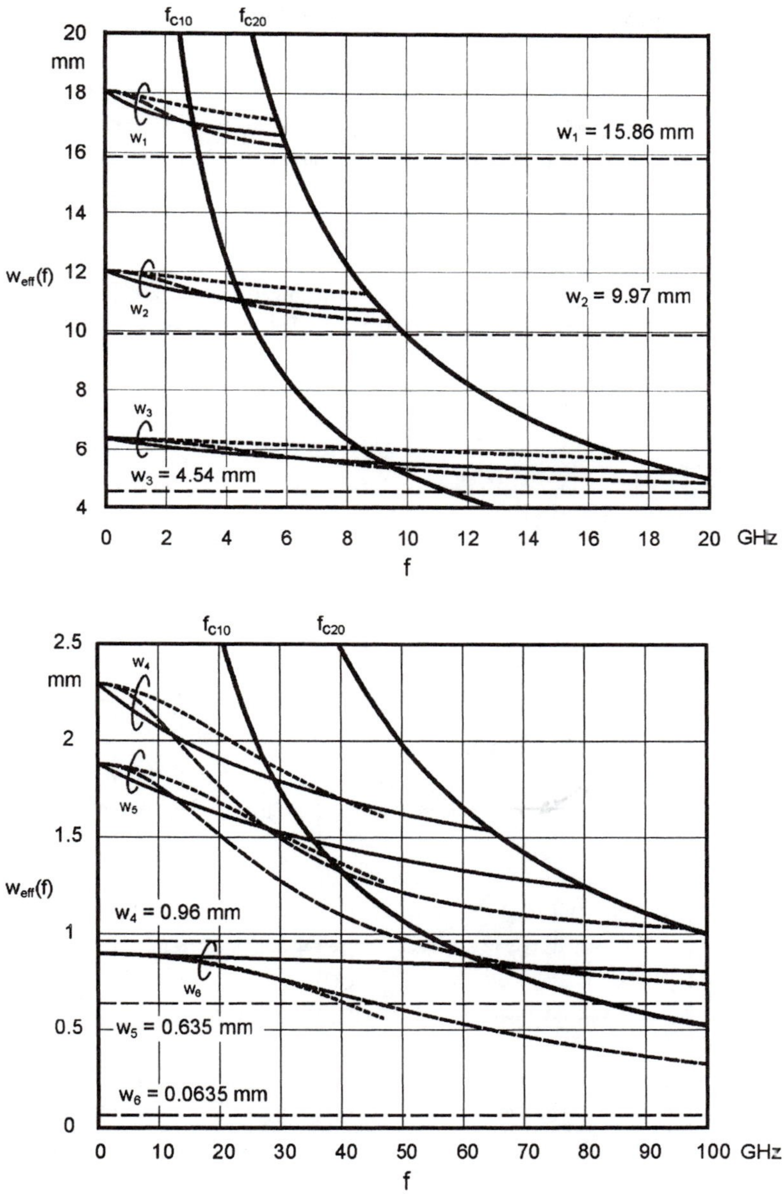

Figure 5.10 Frequency-dependent effective width of the waveguide model (solid line) for alumina substrate material ($\varepsilon_r = 9.7$, $h = 0.635$ mm) compared with Owens' result (dashed line), and the frequency-dependent effective width derived from $Z_0(\varepsilon_r, f)$ and $\varepsilon_{eff}(f)$ (dotted line).

Figure 5.10 shows a similar result on Al_2O_3 substrate material. It can be noticed that for wide microstrip lines, the cutoff frequency of the first higher-order mode is excellently described by (5.18) and (5.21). This is also approximately valid for $w_{eff}(f)$ of (5.20), using the dispersive microstrip line parameters after Kirschning. For the $H_{[20]}$ mode, the deviation of the cutoff frequencies calculated after Owens and Kirschning are not very large. In case of the smaller widths w_4 and w_5, the agreement of the $f_{c[10]}$ values calculated after Kirschning is rather good. In any case, for smaller widths, the deviation of discussed models varies significantly. As higher-order modes were investigated mainly at lower frequencies up to 12.4 GHz, further research would be useful at higher frequencies in the millimeter wave region.

References

[1] Gupta, C., and A. Gopinath, "Equivalent Circuit Capacitance of Microstrip Step Change in Width," *IEEE Trans.*, Vol. MTT-25, 1977, pp. 819–822.

[2] Gopinath, A., and C. Gupta, "Capacitance Parameters of Discontinuities in Microstriplines," *IEEE Trans.*, Vol. MTT-26, 1978, pp. 831–836.

[3] Gopinath, A., A. F. Thomson, and I. M. Stephenson, "Equivalent Circuit Parameters of Microstrip Step Change in Width and Cross Junctions," *IEEE Trans.*, Vol. MTT-24, 1976, pp. 142–144.

[4] van de Capelle, A. R., and P. J. Luypaert, "Fundamental- and Higher-Order Modes in Open Microstrip Lines," *Electron. Lett.*, Vol. 9, 1973, pp. 345–346.

[5] Piefke, G., and R. Strube, "Reflexion und Transmission bei Einfall einer H_{10}-Welle auf eine sprunghafte Änderung eines Rechteckhohlleiters in der E-Ebene," *Arch. Elektr. Übertr.*, Vol. 19, 1965, pp. 231–243.

[6] Knetsch, H. D., "Beitrag zur Theorie sprunghafter Querschnittsveränderungen von Hohlleitern," *Arch. Elektr. Übertr.*, Vol. 22, 1968, pp. 591–600.

[7] Knetsch, H. D., "T-Verbindung eines Rechteck- und eines Rundhohlleiters," *Arch. Elektr. Übertr.*, Vol. 24, 1970, pp. 1–9.

[8] Knetsch, H. D., "Wellentypwandler und Filteer mit Rechteck- und Rundhohlleitertechnik," *Nachrichtentechnische Zeitschrift*, Vol. 23, 1970, pp. 57–63.

[9] Bräckelmann, W., "Hohlleiterverbindungen für Rechteckhohlleiter," *Nachrichtentechnische Zeitschrift*, Vol. 23, 1970, pp. 2–7.

[10] Masterman, P. H., and P. J. B. Clarricoats, "Computer Field-Matching Solution of Waveguide Transverse Discontinuities," *Proc. IEE 118*, 1971, pp. 51–63.

[11] Itoh, T., *Numerical Techniques for Microwave and Millimeter-Wave Passive Structures*, New York: John Wiley & Sons, 1989.

[12] Oliner, A. A., "Theoretical Developments in Symmetrical Strip Transmission Line," *Symposium on Modern Advances in Microwave Techniques*, Brooklyn, N.Y., 1954, pp. 379–402.

[13] Oliner, A. A., "Equivalent Circuits for Discontinuities in Balanced Strip Transmission Line," *IEEE Trans.*, Vol. MTT-3, 1955, pp. 134–143.

[14] Altschuler, H. M., and A. A. Oliner, "Discontinuities in the Center Conductor of Symmetric Strip Transmission Line," *I.R.E. Trans.*, May 1960, pp. 328–339.

[15] Olbrich, G., *Näherungsweise Berechnung des Wellenfeldes und meßtechnische Ermittlung der Eigenschaften von Mikrowellen-Streifenleitungen*, Doctoral Thesis, Technical University of Munich, 1979.

[16] Kompa, G., *Microstrip Discontinuities* (in German), Diploma Thesis, Aachen: RWTH, 1970.

[17] Wolff, I., G. Kompa, and R. Mehran, "Calculation Method for Microstrip Discontinuities and T Junctions," *Electron. Lett.*, Vol. 8, 1972, pp. 177–179.

[18] Wolff, I., G. Kompa, and R. Mehran, "Streifenleitungsdiskontinuitäten und Verzweigungen," *Nachrichtentechnische Zeitschrift*, Vol. 25, 1972, pp. 217–224.

[19] Wheeler, H. A., "Transmission-Line Properties of Parallel Wide Strips by a Conformal Mapping Approximation," *IEEE Trans.*, Vol. MTT-12, 1964, pp. 280–289.

[20] Wheeler, H. A., "Transmission-Line Properties of Parallel Strips Separated by a Dielectric Sheet," *IEEE Trans.*, Vol. MTT-13, 1965, pp. 172–185.

[21] Kompa, G., "Excitation and Propagation of Higher-Order Modes in Microstrip Discontinuities," *Proc. European Microwave Conf.*, Paper B.7.6, 1973.

[22] Vendelin, G. D., "Limitations on Stripline Q," *Microwave J.*, Vol. 13, May 1970, pp. 63–69.

[23] Marcuvitz, N., *Waveguide Handbook*, New York: Dover Publications, Inc., 1951.

[24] Kompa, G., "Dispersion Measurements of the First Two Higher-Order Modes in Open Microstrip," *Arch. Elektr. Übertr.*, Vol. 29, 1975, pp. 182–184.

[25] Kompa, G., *Die frequenzabhängigen Übertragungseigenschaften von Streifenleitungs-Wellenwiderstandssprüngen, Filtern und Stichleitungen in Mikrostrip-Technik*, Doctoral Thesis, Aachen: RWTH, 1975.

[26] Agilent Technologies, *Advanced Design System (ADS)*, http://eesof.tm.agilent.com.

[27] Applied Wave Research (AWR), Inc., *Microwave Office*, http://www.microwaveoffice.com/products/.

[28] Mehran, R., *Die frequenzabhängigen Übertragungseigenschaften von Streifenleitungs-T-Verzweigungen und 90°-Winkeln*, Doctoral Thesis, Aachen: RWTH, 1974.

[29] Mehran, R., *Grundelemente des rechnergestützten Entwurfs von Mikrostreifenleitungs-Schaltungen*, Aachen: Verlag H. Wolff, 1983.

[30] Kompa, G., and R. Mehran, "Planar Waveguide Model for Calculating Microstrip Components," *Electron. Lett.*, Vol. 11, September 1975, pp. 459–460.

[31] Koster, N. H. L., and R. Jansen, "The Microstrip Step Discontinuity: A Revised Description," *IEEE Trans.*, Vol. MTT-34, No. 2, February 1986, 213–223.

[32] Schmitt, H. J., and K. H. Sarges, "Wave Propagation in Microstrip," *Nachrichtentechnische Zeitschrift*, Vol. 24, 1971, pp. 260–264.

[33] Owens, R. P., "Predicted Frequency Dependence of Microstrip Characteristic Impedance Using the Planar-Waveguide Model," *Electron. Lett.*, Vol. 12, May 27, 1976, pp. 269–270.

[34] Kirschning, M., *Entwicklung von Näherungsmodellen für den rechnergestützten Entwurf von hybriden und monolithischen Schaltungen und Mikrostreifenleitungstechnik*, Doctoral Thesis, University of Duisburg, 1984.

Chapter 6

Microstrip Discontinuities, Junctions, and Associated Circuit Elements

Discontinuities and junctions denote any change in the cross-section of a straight waveguiding structure. Waves are scattered, and transmitted power is reduced due to reflection loss at a given obstacle. Higher-order modes are excited as local (evanescent) waves, storing electromagnetic energy, which can be described by reactive lumped networks. The microstrip circuitry is an open configuration, so that radiation loss additionally occurs in discontinuities and junctions. At moderately low frequencies, radiation loss is commonly small and is therefore frequently negligible.

In other respects, discontinuities and junctions are the real elementary building blocks of any microstrip integrated circuit. They are found, for example, in microstrip hybrid couplers, filters, matching networks, and transformers. Similarly, the insertion of lumped hybrid elements, such as capacitive and inductive chips or transistors, and electrical contacting of devices with the microstrip line by using wire-bonds as well, signify a discontinuity that influences the wave propagation of the fundamental mode. The frequency-dependent quantification of discontinuities and junctions is of great practical importance for a reliable circuit design.

Frequency-independent lumped elements of equivalent circuits (capacitances, inductances, and resistances) are calculated under the assumption of static electric or stationary magnetic fields. The derived circuit models are often described by simple analytical functions that are very convenient for computer aided design. On the other hand, it must be noted that these quasi-static equivalent circuits are only applicable within a limited lower frequency region.

It is known that the theoretical treatment of waveguide discontinuities is rather difficult. An exact analysis of microstrip discontinuities is found only on the basis of a field-theoretical approach using Maxwell equations. Calculation methods have been described for less complex discontinuities, such as open, impedance step, and gap between microstrip lines [1–3].

In the case of the standard hollow waveguide with perfect electrically conducting walls, numerous discontinuity problems were solved using the well-known method of orthogonal series expansion (Section 6.2.2, Appendix H). Similarly, the microstrip waveguide model had been applied to many comparable discontinuity and junction boundary problems. An overview of the analyzed configurations is shown in Figure 6.1. If the frequency-dependent scattering parameters of a discontinuity or junction are known, then equivalent circuits with lumped elements can be derived.

The capacitive and inductive element values of the equivalent circuit of most discontinuities are generally very low (often less than 0.1 pF and 0.1 nH). Within the frequency range of a few gigahertz the parasitic effects may be negligible. However, at higher microwave and millimeterwave frequencies, the parasitics strongly influence the transfer characteristics and must be taken into account for the design. For example, an inductor of 0.1 nH yields a reactance value of about 0.628Ω at 1 GHz, and at 100 GHz the value takes 62.8Ω. Similarly, a capacitor of 0.1 pF yields a reactance value of 1.59kΩ at 1 GHz and 15.9Ω at 100 GHz.

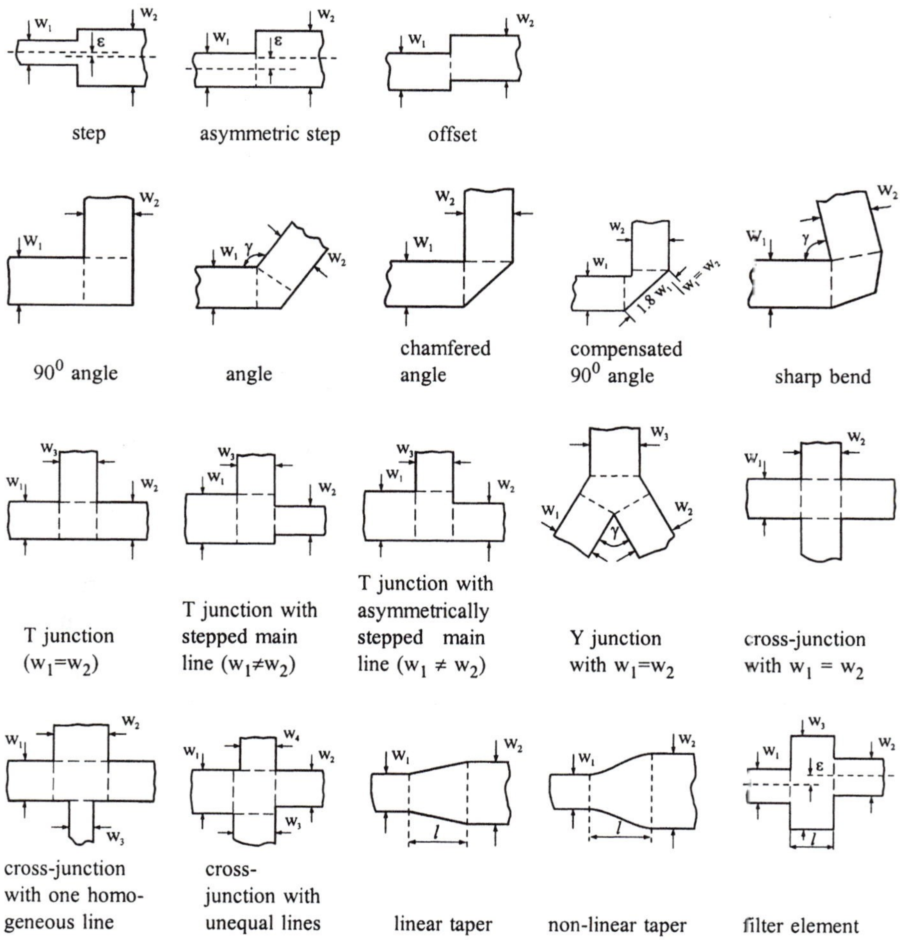

Figure 6.1 Microstrip discontinuities.

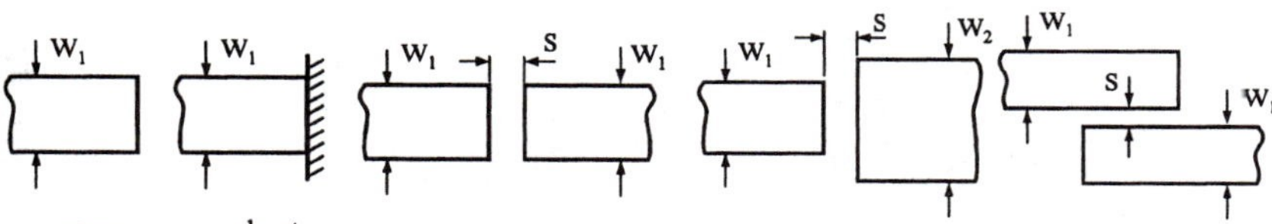

Figure 6.2 Other frequently used microstrip discontinuities.

In addition to the discontinuities illustrated in Figure 6.1, other frequently used discontinuities are shown in Figure 6.2.

6.1 Open End

The open-ended microstrip line, shown in Figure 6.3, is frequently used in microwave integrated circuits as a tuning stub, in multistub filters and transmission line resonators, or as radiating elements in microstrip antennas. The reference plane of the open is denoted by T and relates to the physical termination (Figure 6.4). Different physical effects occur at the open and must be considered in an equivalent circuit model. Similar to the fringing electrical field at the sides of the microstrip line, there is also an electrical stray field at the open termination, which can be equivalently described by an open end capacitance, as shown in Figure 6.5. This capacitance can be calculated on the basis of an electrostatic method [4]. Alternatively, the end capacitance also can be described by a short hypothetical line of length Δl with the same given microstrip line parameters as illustrated in Figure 6.6 (Section 6.1.1). Losses are caused by the excitation of substrate surface waves on the dielectrically loaded metal plate as discussed in Chapter 3, and in radiation fields. Both effects are equivalently described by a radiation resistance R_{rad} or the inverse of a radiation conductance G_{rad}, which will be discussed in the following section.

Comparing the electric field of the fundamental TM_0 surface mode (Section 3.4) with the scattering field of an open, strong similarity is recognized, which means that the surface wave might be easily launched if the operating frequency and given substrate parameters ϵ_r and h are appropriate to trap the wave.

The open end also can be characterized by its reflection coefficient Γ,

$$\Gamma = |\Gamma| e^{-j\phi_\Gamma} \tag{6.1}$$

Due to the radiation loss, the magnitude of the reflection coefficient is less than unity ($|\Gamma| < 1$). The phase term in (6.1) corresponds to the phase shift of the short line extension Δl.

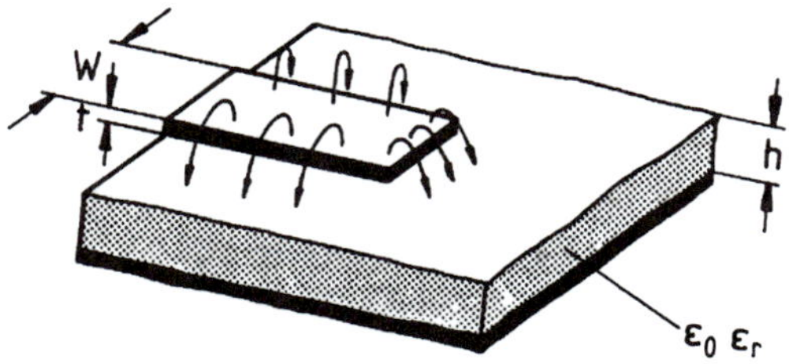

Figure 6.3 Open-ended microstrip line.

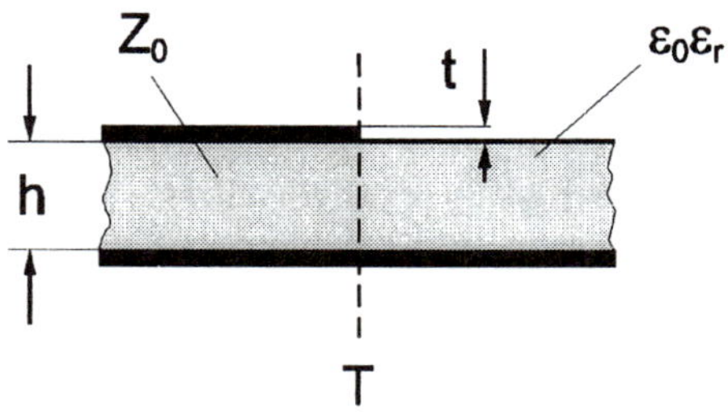

Figure 6.4 Definition of the reference plane.

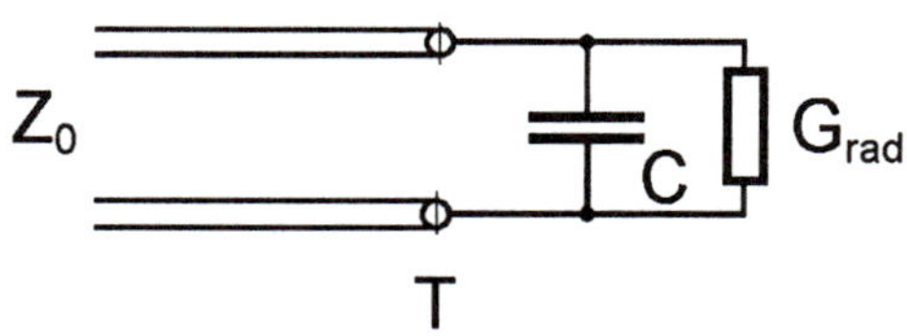

Figure 6.5 Modeling of the electrical stray fields at the physical open by capacitive loading.

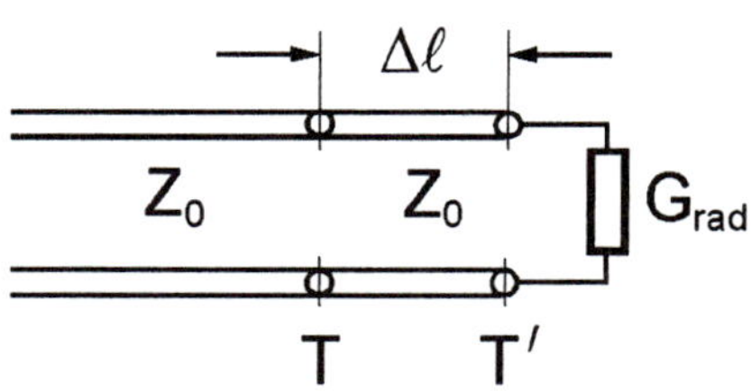

Figure 6.6 Hypothetical line extension of the physical line length.

6.1.1 Open End Capacitance and Excess Length

Altschuler and Oliner [5] had analyzed the open end of a triplate, resulting in an end-effect formula with frequency dependence and dependence of the dielectric constant of the substrate material. For a lower frequency range ($k\Delta l \ll 1$), the formula could be simplified. The end effect was dependent on the geometry, but was no longer related to frequency and dielectric material parameters. Taking $k\Delta l = 1$ as the limit of validity of the approximated formula, the following upper limit frequencies can be identified for Polyguide and alumina: (a) alumina ($\epsilon_r = 9.7$, $h = 0.635\,\text{mm}$): 16.8 GHz; (b) Polyguide ($\epsilon_r = 2.32$, $h = 1.58\,\text{mm}$): 43.2 GHz; and (c) Polyguide ($\epsilon_r = 2.32$, $h = 3.15\,\text{mm}$): 21.6 GHz.

Silvester and Benedek [4] analyzed the open-circuit capacitance on the basis of a purely electrostatic approach. The method included an integral equation with the excess-charge density at the open as integrand. Curvefitting procedure was applied to derive empirical expressions for six different values of dielectric material. Thus, further circumstantial interpolation was needed for general use.

Hammerstad and Bekkadal adopted the geometric expression after Altschuler, and added a factor describing the influence of the dielectric material on the end capacitance. Finally, curvefitting was performed with the results given in [4]. This led to the following simple open end relation [6]

$$\frac{\Delta l}{h} = 0.412 \, \frac{\epsilon_{eff} + 0.3}{\epsilon_{eff} - 0.258} \, \frac{w/h + 0.262}{w/h + 0.813} \tag{6.2}$$

It is claimed that a maximum relative error of 8.7% for $\epsilon_r = 1$ compared to the results given in [4]. In other cases, a maximum error of only 2.4% is obtained. For a very large dielectric constant ($\epsilon_r \gg 1$) and wide microstrip transmission line ($w/h \gg 1$), $\Delta l/h$ approaches the limit value $\Delta l/h = 0.412$. In the case of air ($\epsilon_r = 1$) and $w/h \gg 1$, $\Delta l/h$ becomes 0.72.

The excess length Δl can be equivalently described by a lumped capacitance C as follows. With

$$Z_0(\epsilon_r) = \frac{Z_0(\epsilon_r = 1)}{\sqrt{\epsilon_{eff}}} = \frac{120\,\pi}{\sqrt{\epsilon_{eff}}} \, \frac{h}{w_{eff}} \tag{6.3}$$

the fringing end capacitance C reads

$$C = \frac{\epsilon_0 \epsilon_{eff} \, \Delta l \, w_{eff}}{h} = \frac{\sqrt{\epsilon_{eff}} \, \Delta l}{Z_0(\epsilon_r) \, c_0} \tag{6.4}$$

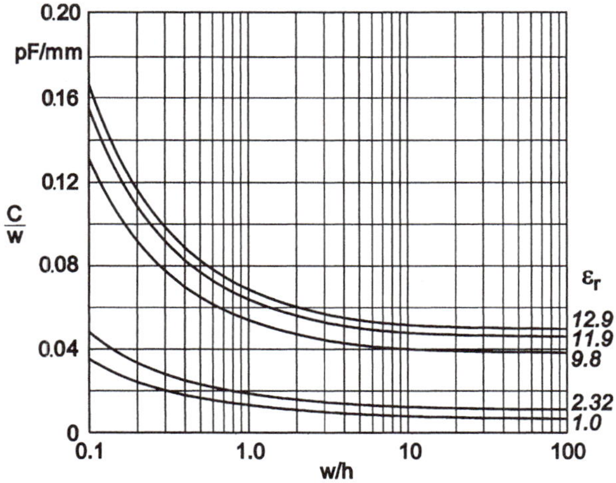

Figure 6.7 End effect capacitance as function of *w/h* ratio for different substrate materials.

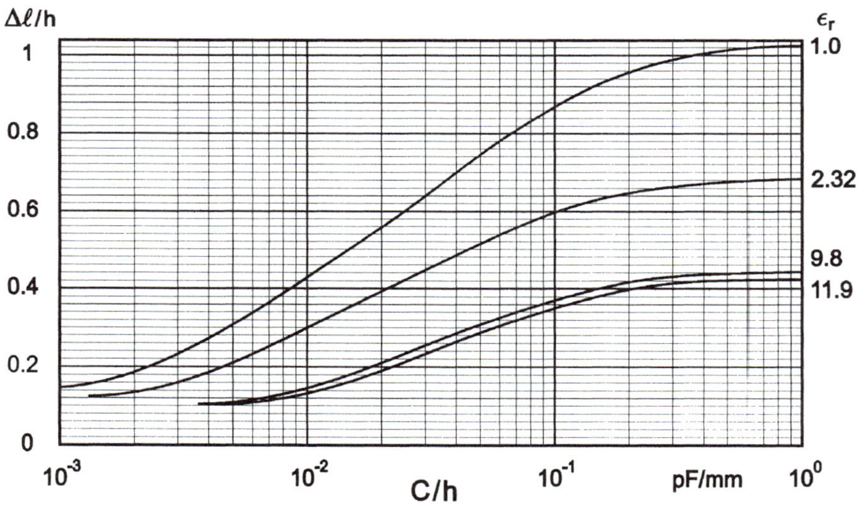

Figure 6.8 Open-ended extension length as a function of open end capacitance with ε_r as parameter.

In Figure 6.7, the end capacitance normalized to the strip width w is plotted versus the aspect ratio w/h with the parameter ϵ_r of relevant substrate materials. Combining (6.2) and (6.4), the extension length Δl can be related to the corresponding end capacitance C. This is plotted in Figure 6.8.

In [7], Kirschning has proposed an expression for the excess length $\Delta l/h$ derived from field-theoretical data applying a curvefitting procedure:

$$\Delta l/h = \frac{\xi_1 \cdot \xi_3 \cdot \xi_5}{\xi_4} \tag{6.5}$$

where

$$\xi_1 = 0.434907 \frac{\epsilon_{eff}^{0.81}(f=0) + 0.26}{\epsilon_{eff}^{0.81}(f=0) - 0.189} \cdot \frac{u^{0.8544} + 0.236}{u^{0.8544} + 0.87}, \qquad u = \frac{w}{h}$$

$$\xi_2 = 1 + \frac{u^{0.371}}{2.358\epsilon_r + 1}, \qquad \xi_3 = 1 + \frac{0.5274 \cdot \arctan\left(0.084 u^{1.9413/\xi_2}\right)}{\epsilon_{eff}^{0.9236}(f=0)}$$

$$\xi_4 = 1 + 0.0377 \arctan\left(0.067 u^{1.456}\right) \cdot \{6 - 5\exp[0.036(1 - \epsilon_r)]\}$$

$$\xi_5 = 1 - 0.218 \exp(-7.5u)$$

In Figure 6.9, the normalized open end line extension $\Delta l/h$ is plotted versus the w/h ratio in the range $0.01 \leq w/h \leq 1.0$, and $1 \leq w/h \leq 100$, respectively. Calculated results after Kirschning (6.2) are also shown for comparison.

As is stated in [7] the comparison with the results derived from a rigorous approach in the spectral domain yields a maximum relative error of 2.5% in the range $\epsilon_r \leq 50$ and $0.01 \leq w/h \leq 100$ at a frequency $f = 1\,\text{GHz}$. After [8], it can be assumed that the open end effect is nearly constant up to 20 GHz. This is confirmed in [9], analyzing the end-effect of an open microstrip transmission line on alumina substrate ($w/h = 1$, $\epsilon_r = 9.7$) on the basis of the spectral domain method. Only a slight increase of the end-effect length Δl was calculated, from 0.15 mm at 5 GHz to 0.21 mm at 20 GHz.

Wolff [10] investigated end capacitances of microstrip lines on Polyguide substrate at low frequencies up to approximately 6.5 GHz. Regarding an evaluation uncertainty of 25% applying (4.50) and (4.51), respectively, a frequency dependence of the end capacitances could not be observed (see Section 4.3.2). Olbrich [11] applied the measurement method after Wolff in a broad frequency range up to 40 GHz. He used alumina substrates ($h = 0.635\,\text{mm}$) with w/h values of approximately 1 to 10. The results are shown in Figure 6.10. For narrow strips, $w_1/h = 1$ and $w_2/h = 1.6$, no frequency dependence was observed. However, for $w_3/h = 4.72$ and $w_4/h = 9.45$, very strong decrease with frequency occurred. For w_4/h, the end capacitance changes about 50% over a frequency change of approximately 5 GHz, and for w_3/h, from 150 fF at 6.5 GHz to 15 fF at 38 GHz. The low-frequency values agree very well with the electrostatic values [4] shown by broken lines.

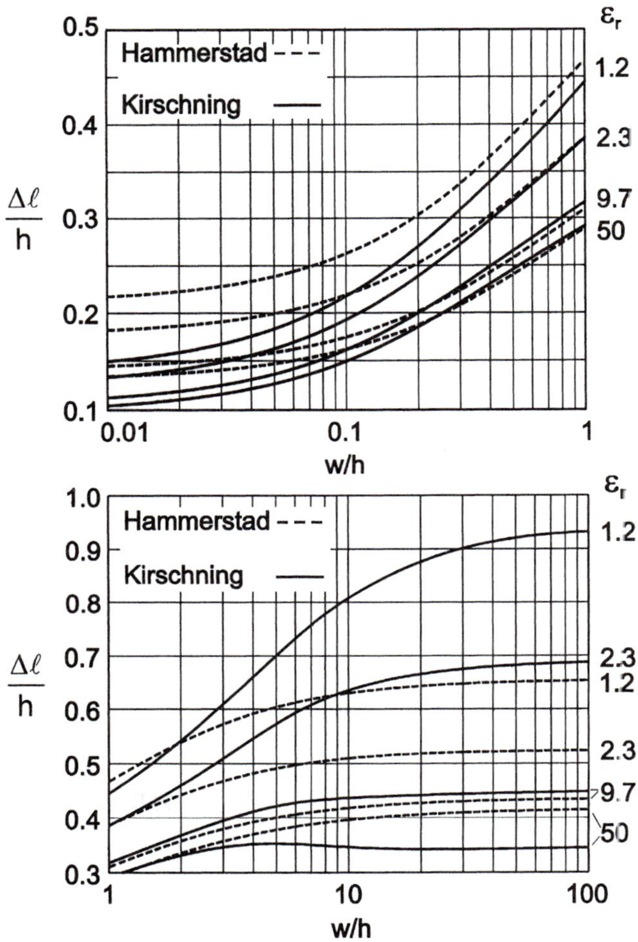

Figure 6.9 Open-end effect $\Delta l/h$ versus w/h after Kirschning (solid lines) in comparison with calculations after Hammerstad (dashed lines).

6.1.2 Radiation Resistance

High-efficient radiation from open-ended microstrip line is a key point for antenna application. However, in common microstrip circuit design, radiation is undesired, causes additional loss, and contributes to interference of separate parts. To derive a simple formula for the radiation resistance R_{rad}, and radiation conductance $G_{rad} = R_{rad}^{-1}$, respectively, we assume low radiated power, so that we can adopt the standing voltage and current distribution along the line in case of a perfect open (perturbation theory approach).

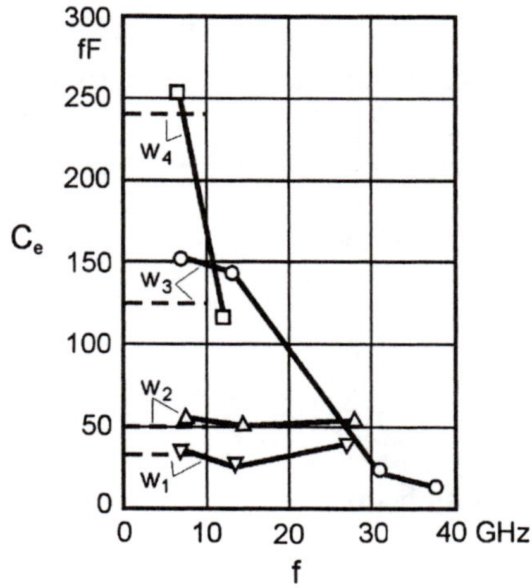

Figure 6.10 Measured frequency-dependent open-end capacitance of straight microstrip resonators on alumina substrate (ε_r = 9.7, h = 0.635 mm). (From: [11], by courtesy of Gerhard Olbrich.)

Figure 6.11 shows a lossless microstrip line ($\alpha = 0$) of characteristic impedance Z_0 and phase coefficient β. The line is terminated by a resistance $R_{rad} \gg Z_0$, which signifies the radiated power loss. At the input port of the line, an excitation source with open-circuit voltage V_0 and source resistance R_i is connected. The total line voltage V_2 at the output port is the superposition of the incident wave V_2^+ and reflected wave V_2^-. Because of the assumed perfect open, the load reflection factor Γ_L is unity, and therefore $V_2^+ = V_2^-$. Thus, we can write

$$V_2 = V_2^+ + V_2^- = 2\,V_2^+ \tag{6.6}$$

The incident power P_i is calculated from the forward traveling voltage wave,

$$P_i = \frac{\left(V_2^+\right)^2}{2\,Z_0} = \frac{\left(V_2\right)^2}{8\,Z_0} \tag{6.7}$$

The radiated power P_{rad} is the loss power in the radiation resistance, and reads with (6.7)

$$P_{rad} = \frac{1}{2}\,V_2^2\,G_{rad} = 4\,Z_0 G_{rad} P_i \tag{6.8}$$

Normalizing to the incident power P_i,

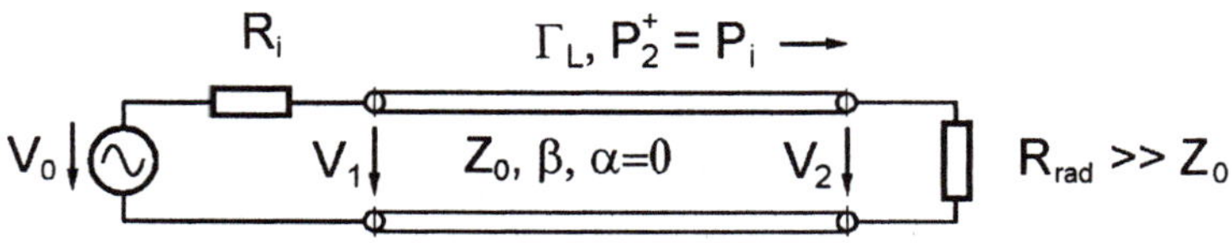

Figure 6.11 Derivation of the radiation resistance of an open-ended microstrip.

$$\frac{P_{rad}}{P_i} = 4\,Z_0 G_{rad} \tag{6.9}$$

the radiation conductance is obtained as

$$G_{rad} = \frac{1}{4Z_0}\,\frac{P_{rad}}{P_i} \tag{6.10}$$

With $P_i = P_r + P_{rad}$, where P_r is the power reflected at the load, and $P_r/P_i = |\Gamma_L|^2$, (6.10) reads

$$G_{rad} = \frac{1 - |\Gamma_L|^2}{4\,Z_0} \tag{6.11}$$

From (6.11), the load reflection coefficient Γ_L is obtained as

$$|\Gamma_L| = \sqrt{1 - 4\,Z_0\,G_{rad}} \tag{6.12}$$

Lewin [12] has considered radiation both from the current flowing in the microstrip and from an "effective polarization current" introduced to account for the presence of the substrate. Further approximations made in his approach are: (1) the strip is embedded in uniform dielectric constant ϵ_{eff}, (2) the TEM wave is propagating, and (3) the substrate thickness h is much less than the free-space wavelength ($h \ll \lambda_0$). The result given by Lewin, referred to as "radiated power," was

$$P' = 2\pi\eta_0\left(\frac{h}{\lambda_0}\right)^2 F(\epsilon_{eff}) = 240\pi^2\left(\frac{h}{\lambda_0}\right)^2 F(\epsilon_{eff}) \tag{6.13}$$

with $\eta_0 = 120\,\pi$, and where the radiation factor reads

$$F(\epsilon_{eff}) = \frac{\epsilon_{eff} + 1}{\epsilon_{eff}} - \frac{(\epsilon_{eff} - 1)^2}{2\,\epsilon_{eff}\sqrt{\epsilon_{eff}}}\ln\left[\frac{\sqrt{\epsilon_{eff}} + 1}{\sqrt{\epsilon_{eff}} - 1}\right] \tag{6.14}$$

As an example, given in [12], $P' = 1\Omega$ is calculated for $h = 1.58\,\text{mm}$, $f = 4\,\text{GHz}$, and $\epsilon_r = 2.25$, with $\lambda_0 = c_0/f = 7.5\,\text{cm}$ and $F_1(\epsilon_{eff}) = 1.07 \approx 1.0$. Regarding an open-ended microstrip line with a characteristic impedance of $50\,\Omega$, Lewin concluded that the radiated power was $P'/(Z_0 = 50\,\Omega) = 2\%$. Following this interpretation, the normalized radiated power can be written as

$$\frac{P_{rad}}{P_i} = \frac{P'}{Z_0} = 2\pi\,\frac{\eta_0}{Z_0}\left(\frac{h}{\lambda_0}\right)^2 F(\epsilon_{eff}) = \frac{240\pi^2}{Z_0}\left(\frac{h}{\lambda_0}\right)^2 F(\epsilon_{eff}) \tag{6.15}$$

Substituting P_{rad}/P_i in (6.9) yields

$$G_{rad}Z_0 = \frac{P'}{4Z_0} = \frac{60\,\pi^2}{Z_0}\left(\frac{h}{\lambda_0}\right)^2 F(\epsilon_{eff}) \tag{6.16}$$

Equation (6.16) is also found in [13].

Hammerstad [6], however, pointed out that the radiated power fraction should be $2P'/Z_0$ (i.e., twice that given by Lewin). This can be understood as follows. The forward current I_2^+ is related to the forward voltage V_2^+ as

$$V_2^+ = Z_0 I_2^+ \tag{6.17}$$

Then (6.7) reads

$$P_i = \frac{\left(V_2^+\right)^2}{2\,Z_0} = \frac{1}{2}\,Z_0\left(I_2^+\right)^2 \tag{6.18}$$

With

$$V_2 = 2\,V_2^+ = 2\,Z_0 I_2^+ \tag{6.19}$$

the radiated power in (6.8) reads

$$P_{rad} = 2\,Z_0^2\left|I_2^+\right|^2 G_{rad} \tag{6.20}$$

The radiated power corresponding to a forward current wave normalized to the incident current amplitude I_2^+ can be written as

$$\frac{P_{rad}}{\left|I_2^+\right|^2} = P' = 2\,Z_0^2\,G_{rad} \tag{6.21}$$

where $P_{rad}/(I_2^+)^2$ has the dimension in ohms and corresponds to P' in (6.13) after Lewin. Then the normalized radiation conductance reads

$$G_{rad}Z_0 = \frac{P'}{2Z_0} = \frac{120\pi^2}{Z_0}\left(\frac{h}{\lambda_0}\right)^2 F(\epsilon_{eff}) \tag{6.22}$$

which is in contrast to (6.15). With (6.10), the normalized radiated power is

$$\frac{P_{rad}}{P_i} = 4\pi\frac{\eta_0}{Z_0}\left(\frac{h}{\lambda_0}\right)^2 F(\epsilon_{eff}) \tag{6.23}$$

The same relation is also found in [14]. The logarithmic term of the radiation factor in (6.16) can be expanded in a series and approximated by [6]

$$F_1(\epsilon_{eff}) \approx \frac{8}{3\epsilon_{eff}}\left(1 - \frac{1}{5\epsilon_{eff}}\right) \approx \frac{8}{3\epsilon_{eff}} \tag{6.24}$$

Thus, (6.23) reads

$$\frac{P_{rad}}{P_i} = \frac{32\pi\,\eta_0}{3\epsilon_{eff}Z_0}\left(\frac{h}{\lambda_0}\right)^2 \tag{6.25}$$

Sobol [15] proposed a theory that should overcome the restriction of Lewin's approach to narrow microstrip lines. The following assumptions were made: (1) h/λ is much less than unity, (2) strip conductor is infinitely thin, (3) radiation flux is assumed to flow from the open end of the microstrip line directly into the substrate and then finally scatter into the air space above the substrate, (4) reflection at the air-dielectric boundary is neglected and the subsequent radiation into the air space is not considered, and (5) TEM-like electric field in the aperture is found by approximating the microstrip line by a simple planar waveguide model of width w. Effective dielectric constant after Wheeler has been used to take care of the fringe field of the microstrip line. Sobol regarded the open-ended microstrip line as a slot antenna, which radiates into space filled with a homogeneous dielectric constant ϵ_{eff}.

Sobol followed the concept of equivalent magnetic current source [16]. In this case, the homogeneous electrical field in the aperture of the terminated microstrip line can be assumed to be related to a magnetic surface current. The magnetic surface current replaces the actual sources to the left of the open end. The far-zone electromagnetic field can then be calculated. The radiated power is determined by Poynting's theorem. The respective radiation conductance then reads

$$G = \frac{\sqrt{\epsilon_{eff}}}{240\,\pi^2}\, I\left(\frac{w}{\lambda_0}\sqrt{\epsilon_{eff}}\right) \tag{6.26}$$

where

$$I = \int_0^{\pi} \frac{\sin^2\left(\dfrac{\pi\sqrt{\epsilon_{eff}}}{\lambda_0} w\cos\theta\right)\sin^2\theta}{\cos^2\theta}\, d\theta \tag{6.27}$$

For $\pi\sqrt{\epsilon_{eff}}\, w/\lambda_0 \ll 1$, (6.26) is approximated by

$$G \approx \frac{(\epsilon_{eff})^{3/2}}{180}\left(\frac{w}{\lambda_0}\right)^2 = \frac{(\epsilon_{eff})^{3/2}}{180}\left(\frac{w}{h}\right)^2\left(\frac{h}{\lambda_0}\right)^2 \tag{6.28}$$

The fraction of radiated power can be calculated as

$$\frac{P_{rad}}{P_i} = 1 - |\Gamma|^2 = 4\,Z_0(\epsilon_r)\,G_{rad} = \frac{(\epsilon_{eff})^{3/2}}{45}\,Z_0(\epsilon_r)\left(\frac{w}{h}\right)^2\left(\frac{h}{\lambda_0}\right)^2 \tag{6.29}$$

It is claimed that the relation is only valid for radiation loss less than 20%, and that the approximation is quite accurate for $\pi\sqrt{\epsilon_{eff}}\, w/\lambda_0 < 0.5$.

Evaluating the integral, Wood [14] proposed the following formula:

$$G = \frac{\sqrt{\epsilon_{eff}}}{240\,\pi^2}\, F_s\left(2\pi\frac{w}{\lambda_0}\sqrt{\epsilon_{eff}}\right) \tag{6.30}$$

where

$$F_s(x) = x\,Si(x) - 2\sin^2(x/2) - 1 + \sin(x)/x$$

with $x = 2\pi\, w/\lambda_0\sqrt{\epsilon_{eff}}$. Wood suggested replacing the physical width with the effective width (modified Sobol relation), and the agreement with measurement was found satisfying.

Easter [17] calculated the radiation loss of a half-wavelength open-circuit microstrip resonator. The assumptions made and analysis method applied were similar to that of Lewin. Confirmation of the results was carried out by a number of microstrip resonators using gold or copper conductors on 0.5-mm sapphire substrates ($\epsilon_r \approx 10$). As proposed earlier by Denlinger [18], the radiated power can be derived from Q factor measurements. However, as pointed out in [6, 17], relationships 3 and 4 in [18] appear to be in error, so that Denlinger's results cannot be used further. The Q factor is defined as [19]

$$Q = \frac{\omega W}{P} \tag{6.31}$$

and thus proportional to stored energy W and inversely proportional to the power loss P. P/f_0 is the energy that must be externally fed per cycle to the resonator to compensate for the loss. The total loss commonly comprises radiation, conductive, and dielectric loss. The resonator first may be measured in a shielding box resulting in a quality factor Q_s and power loss P_s. The measurement of the open configuration delivers a quality factor Q_o, power loss P_s, and radiated P_{rad}. Equation (6.31) reads for the open configuration

$$Q_o = 2\pi \frac{f_0 W}{P_s + P_{rad}} \tag{6.32}$$

and the inverse relation

$$\frac{1}{Q_0} = \frac{1}{2\pi}\left(\frac{P_s}{f_0 W} + \frac{P_{rad}}{f_0 W}\right) = \frac{1}{Q_s} + \frac{1}{Q_{rad}} \tag{6.33}$$

where $1/Q_{rad}$ can be written as

$$\frac{1}{Q_{rad}} = \frac{1}{2\pi}\frac{P_{rad}}{P_i} \tag{6.34}$$

Then (6.33) reads

$$\frac{P_{rad}}{P_{in}} = 2\pi\left(\frac{1}{Q_0} - \frac{1}{Q_s}\right) \tag{6.35}$$

Under assumption of high dielectric constant ($\epsilon_r \gg 1$), Easter [17] derived the following expression for the radiated power fraction

$$\frac{P_{rad}}{P_i} = \frac{64\eta\pi h^2}{3\epsilon_{eff} Z_0(\epsilon_r)\lambda_0^2} \qquad (\epsilon_r \geq 10) \tag{6.36}$$

Hammerstad [6] pointed out that the result after Easter is twice the value obtained from the corrected formula (6.25) after Lewin. Thus, he concluded that the radiated power from several discontinuites may be taken as the sum of the individual contributions even when the discontinuities are as close as one-half wavelength of the microstrip. Note that the open ends of the resonator in the air half-plane are much nearer. Supposing an effective dielectric constant $\epsilon_{eff} \approx 9$, then the overall length of the resonator in air medium would be $\lambda/2 = \lambda_0/(2\sqrt{\epsilon_{eff}}) = \lambda_0/6$ [18]. Hammerstad suggested taking the sum of the contributions to radiation loss given by Lewin and Sobol for commonly used substrates. The relation for P_{rad}/P_i reads [6]

$$\frac{P_{rad}}{P_i} = \left\{ \frac{32\pi\,\eta_0/\Omega}{3\,\epsilon_{eff}\,Z_0(\epsilon_r)/\Omega} + \frac{\epsilon_{eff}\sqrt{\epsilon_{eff}}\,Z_0(\epsilon_r)/\Omega}{45}\left(\frac{w}{h}\right)^2 \right\}\left(\frac{h}{\lambda_0}\right)^2 \tag{6.37}$$

where $\eta_0 = 120\pi\,\Omega$. As can be seen from (6.37), radiation loss depends on the dielectric constant, the substrate thickness, microstrip width, and frequency. The use of high-dielectric constant substrate materials reduces radiation loss, because most of the electromagnetic field is concentrated in the dielectric between the conductive strip and the ground plane.

The first term in (6.37) originates from the theory after Lewin [6, 12, 17] concerning the radiation of high-impedance microstrip transmission lines into space. For high-dielectric constant and low-impedance lines, a considerable amount of radiated power will be confined to the dielectric substrate as surface waves. This is not taken into account by Lewin's expression. Sobol [15] has analyzed the radiation of wide microstrip lines into the substrate, describing the open end as a slot antenna. The calculated fraction of radiated power is represented by the second term in (6.37). Figure 6.12 confirms that the first (Lewin) term dominates for high values of characteristic impedance and low values of ϵ_r, while, as shown in Figure 6.13, the second (Sobol) term predominates for low values of characteristic impedance and high values of ϵ_r. For intermediate impedance values, the two terms are of the same order of magnitude, and radiation loss may be overestimated [6].

In the following diagrams (Figure 6.14), the fractional radiated power P_{rad}/P_i is plotted versus frequency for various substrates and characteristic impedances. It should be noted that the plotted values only may be usable for weak and moderate radiation. This is in particular required with respect to the considered analysis for the determination of the radiation conductance. Roughly said, at values above 20% of calculated radiated power, (6.37) may become questionable.

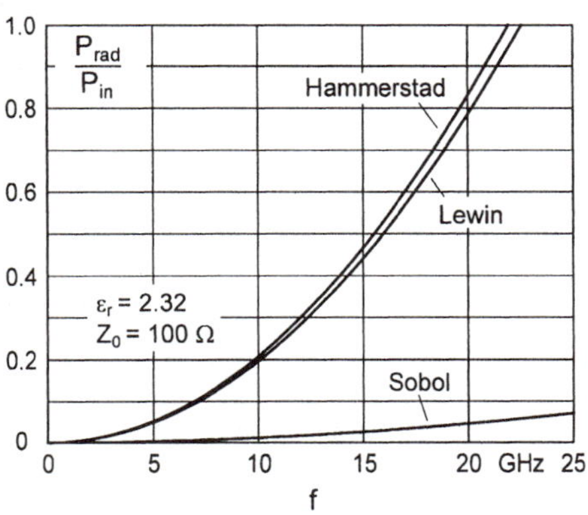

Figure 6.12 Fractional radiated power of microstrip line of high impedance and low dielectric constant.

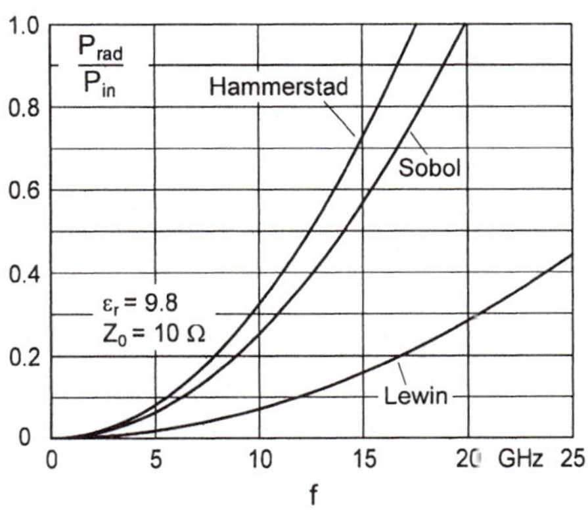

Figure 6.13 Fractional radiated power of microstrip line of low impedance and high dielectric constant.

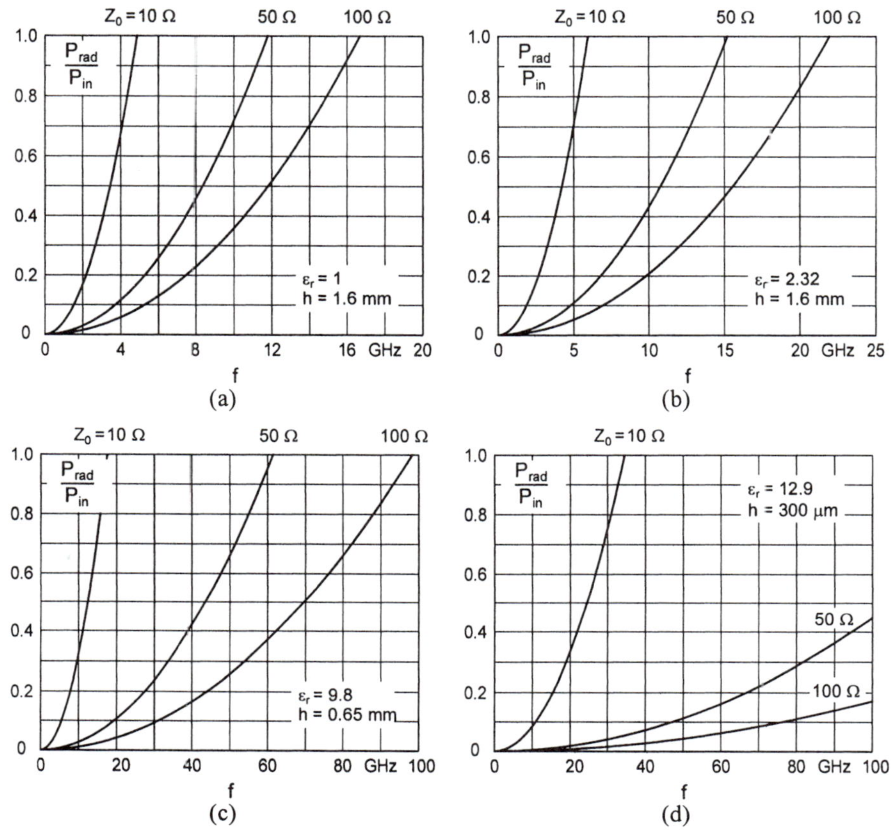

Figure 6.14 Fractional radiated power for an open-ended microstrip transmission line versus frequency with characteristic impedance as parameter [(a) air-filled line ($\varepsilon_r = 1$, $h = 1.6$ mm), (b) Polyguide substrate ($\varepsilon_r = 2.32$, $h = 1.6$ mm), (c) alumina substrate ($\varepsilon_r = 9.8$, $h = 0.65$ mm), and (d) GaAs substrate ($\varepsilon_r = 12.9$, $h = 300$ μm)].

In [20], the radiation of very wide open-ended microstrip lines, neglecting the fringing electrical field at both sides of the microstrip line, has been investigated. The microstrip line was approximately described by a magnetic wall model with physical width w. Thus, the discontinuity under analysis is similar to a section of an open-ended microstrip line of infinite width. An assimilable boundary problem had been already investigated by Angulo and Chang [21].

Figure 6.15 shows a parallel plate waveguide with terminated upper plate and infinite dielectric-loadedground plate. The given boundary problem was primarily considered to analyze the launching of a surface wave into a dielectric slab from a parallel plate waveguide. The slab thickness is denoted as h, and the spacing of the parallel plates as H, so that the air region is $H - h$. The results given in [21] can be applied to open-ended microstrip lines under the assumption that the air region becomes zero.

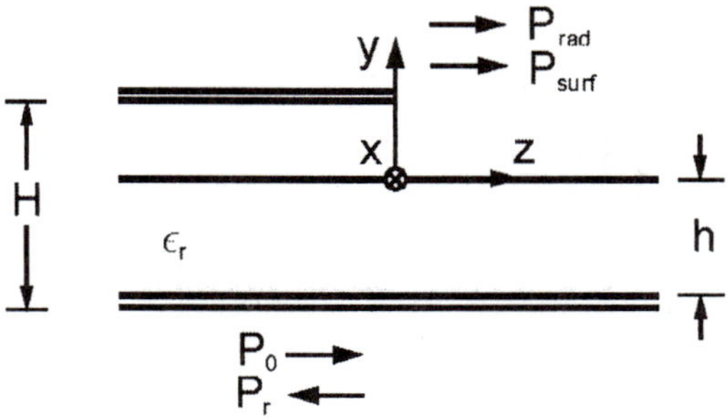

Figure 6.15 Grounded dielectric slab with terminated upper parallel plate used for open-ended microstrip line analysis.

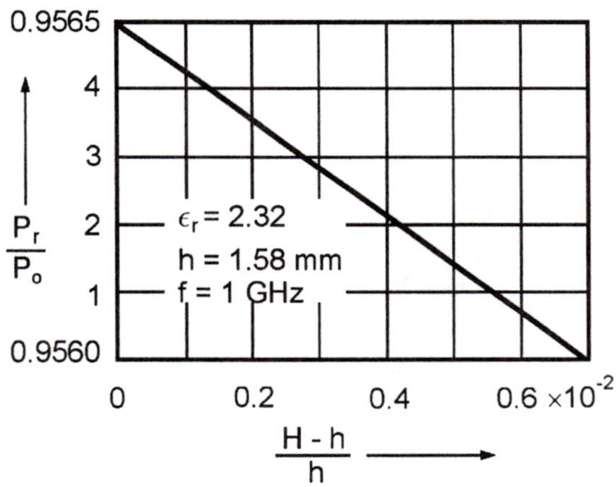

Figure 6.16 Reflected power at an open-ended dielectric-loaded parallel plate waveguide versus normalized air gap.

The boundary problem illustrated in Figure 6.15 was solved by the modal-analysis method in the Fourier domain. Having satisfied all boundary conditions at the discontinuity $z = 0$, a solution of the equation was found under the application of the Wiener-Hopf technique. Instead of inversion of the transforms, the integrals of the fields are evaluated for the far fields by the method of steepest descent for the limiting case of zero dissipation. The theory delivers an exact analytic solution [14, 21]. This is the reason why the results of the modified approach in [20] should provide a good check on the other analyses for large strip width [13, 14].

The reflected and radiated power, and the power in the surface wave per unit incident power, is found from Poynting's theorem. As a result, formulas for the reflected and transmitted power as a function of frequency and dielectric slab thickness were available. If, in the original structure in Figure 6.15, the height h of the dielectric slab is increased up to the height H of the parallel plate waveguide, then the geometry of an open-ended microstrip line as shown in Figure

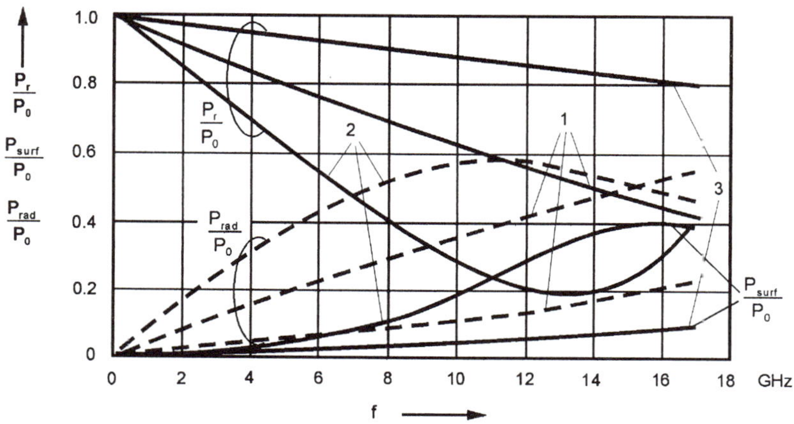

Figure 6.17 Simulated reflected, radiated, and surface wave power versus frequency of a wide open-ended microstrip line (see Table 6.1 for notation).

6.4 occurs. Figure 6.16 shows the influence of an air gap between the air-dielectric boundary and the upper plate of the parallel plate waveguide on the reflected power. As can be seen, the reflected normalized power P_r/P_0 varies only in the fourth significant digit if the relative gap $(H-h)/h$ is smaller than 1%. So the influence of a small gap is unimportant from the practical point of view. This result further justifies the following calculations, carried out with a relative gap of 0.2%.

Figure 6.17 shows the computed normalized reflected power P_r/P_0, radiated power P_{rad}/P_0, and the power radiated by surface waves P_{surf}/P_0 as a function of frequency up to 17 GHz at the open end of a wide microstrip line on Al_2O_3 ($\epsilon_r = 9.7$, $h = 0.635$ mm) and Polyguide substrate ($\epsilon_r = 2.32$), respectively. This is an extension to the results given in [20], with a limited upper frequency of 12.4 GHz. As can be seen, the reflected power decreases monotonically with increasing frequency. For the alumina substrate, the reflected power goes down to 87.5% at a frequency of 12.4 GHz. The results computed for Polyguide substrate with low relative permittivity shows rather high radiation loss in the X band (8 to 12.4 GHz). Considering a commonly used substrate with a height h equal to 1.58 mm, the reflected power decreases to approximately 50%. On the thicker substrate ($h = 3.16$ mm), the reflected power is less than 20%. The magnitude of the input reflection coefficient Γ, defined as

$$|\Gamma| = \sqrt{P_r/P_0} \tag{6.38}$$

can be expressed by the simple relation

$$|\Gamma| = 1 - f/k \tag{6.39}$$

Table 6.1
Definition of Radiation Parameter k

No	ε_r	*h/mm*	*k/GHz*
1	2.32	1.58	45
2	2.32	3.16	22.5
3	9.7	0.635	200

with f and k in gigahertz. The factor k is given in Table 6.1. It should be noted that (6.39) is only valid up to 12.4 GHz for the investigated substrates. As can be seen in Figure 6.18, the reflection coefficient for the thicker Polyguide substrate material increases again above 15 GHz. Wood derived a radiation conductance from the given reflection coefficient [14]:

$$\Gamma(f) = \frac{Z_{open} - Z_0}{Z_{open} + Z_0} = 1 - \frac{f}{k} \tag{6.40}$$

Solved for the open end impedance gives

$$Z_{open} = Z_0 \frac{2 - f/k}{f/k} = \frac{1}{G_{rad}} \tag{6.41}$$

Then the radiation conductance reads

$$G_{rad} = \frac{f/k}{Z_0 (2 - f/k)} \tag{6.42}$$

With the effective width

$$w_{eff} = \frac{120 \pi h}{Z_0/\Omega \sqrt{\epsilon_{eff}}}$$

we can define a radiation conductance per unit width [14]

$$G_{rad}/w_{eff} = \frac{5 \sqrt{\epsilon_{eff}}}{4\pi h \lambda_0 k \left(1 - \frac{f}{2k}\right)} \tag{6.43}$$

It should be noticed that (6.43) was found by curvefitting up to 12.4 GHz [20]. Therefore, extension up to 30 GHz [14] becomes somewhat questionable, because the monotonously decreasing reflection coefficient in the lower frequency range may increase again at higher frequencies, as is shown in Figure 6.18.

In Figure 6.19, measured and calculated radiation conductance of a high impedance open-ended microstrip line on Polyguide substrate ($w/h = 1.78$, $\epsilon_r = 2.32$) is plotted versus frequency. As can be seen, Sobol's relation does not really fit to measurement. Much better agreement is obtained by replacing the

Figure 6.18 Calculated magnitude of reflection coefficient of a wide open-ended microstrip line.

aperture width w by the effective width w_{eff}, as Wood has shown. Equation (6.43) does not predict the frequency dependence correctly, which confirms that this formula is not useful for narrow strip widths. It should be noted that with increasing width of the open-ended microstrip line, the excitation of higher-order modes becomes more probable.

The corrected Lewin relation gives good results for narrow microstrip lines. However, with increasing width and frequency, the disagreement of simulation and measurement increases due to the assumption of a line source in the analysis.

The total transmitted power consists of the power P_{surf}, which excites the TM_0 surface mode, and of the power P_{rad} radiated into the air space. Higher-order surface waves are not excited in the frequency range investigated. The cutoff frequency of the first higher-order TE_1 surface wave is 41.3 GHz for $\epsilon_r = 2.32$ and $h = 1.58$ mm. Figure 6.17 indicates that the excitation of the surface wave increases with the height of the grounded slab. It can be seen from the computed results that P_{surf}/P_0 is approximately 23% at 12.4 GHz for Polyguide substrate with a height $h = 3.16$ mm. On the thinner substrate ($h = 1.58$ mm) this value decreases to 4.5%. For Al_2O_3 ceramic, the power ratio under consideration becomes very small (0.85%), so that it was not plotted in Figure 6.17. It turns out that the total transmitted power is to a large extent concentrated in the continuous spectrum of the radiating field.

Figure 6.20 shows the power radiated from the edge of the upper metal plate of the wide open-ended microstrip line on Polyguide substrate material as a function of the observation angle (elevation angle) θ at a frequency of 1 GHz and 96 GHz, respectively. Very similar results were also found on alumina substrate. It can be seen that the radiation pattern at 1 GHz is symmetric with respect to the main beam axis $\theta = 0$. The half-power beamwidth is $90°$. With increasing frequency, the main beam direction inclines slightly to the horizontal and the beamwidth becomes much narrower. At a frequency of 96 GHz, an elevation angle $\theta = 25°$ has been obtained, and the beamwidth has reduced to $19.5°$. This corresponds with the observation

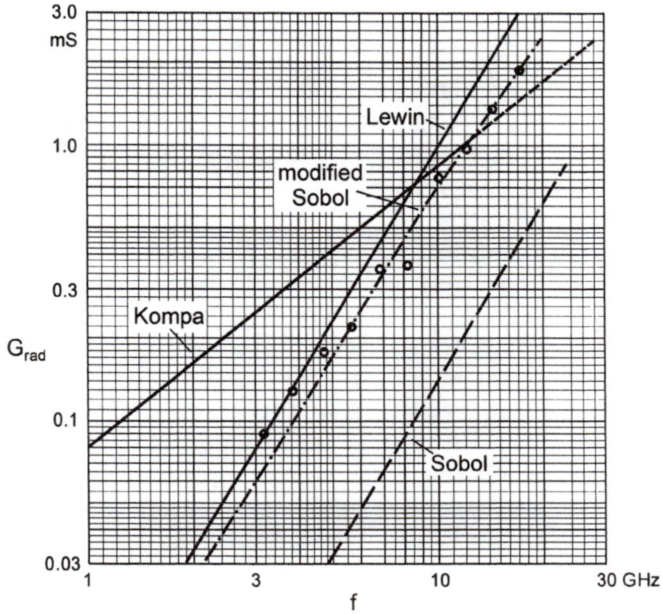

Figure 6.19 Radiation conductance of open-ended microstrip line after Wood (w = 2.82 mm, h = 1.58 mm, ε_r = 2.32). (From: [14], © 1978 IEE. Reprinted with permission.)

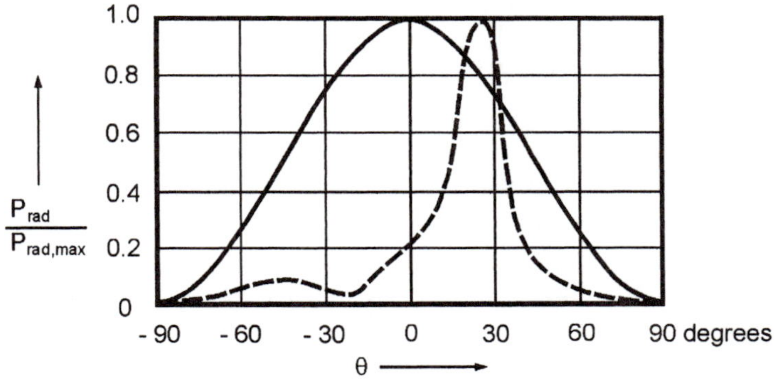

Figure 6.20 Simulated radiation pattern of power radiated per unit solid angle versus elevation angle θ for f = 1 GHz (solid line) and f = 96 GHz (dashed line) of a wide open-ended microstrip line.

that the measured reflection coefficient could be perturbed by a dielectric material (or hand) in near distance within an angle of approximately 30°, whereas movement of the same material over the substrate surface had only a minor effect.

With $P_{ref} = P_0 - (P_{surf} + P_{rad})$, (6.38) can also be written as

$$|\Gamma| = \sqrt{1 - \left(P_{rad} + P_{surf}\right)/P_0} \tag{6.44}$$

6.2 Step

Stepped microstrip components are often found in microstrip integrated circuits. They act as impedance transformers [e.g., in input and output matching networks of (high power) microwave transistor amplifiers], as compensating elements in microstrip power dividers, or as filter circuits comprised of lumped inductive and capacitive elements [22–25].

6.2.1 Quasi-Static Equivalent Circuit

An abrupt impedance step is established when two microstrip transmission lines with different characteristic impedances are cascaded. Figure 6.21 illustrates a symmetric structure of an impedance step. The capital letters T_1 and T_2 denote the reference planes on either side of the step. As Figure 6.22 shows, modeling of the step is usually in terms of an equivalent parallel capacitance C_p and a total series inductance L_s, both frequency independent. The shunted capacitance considers the stored excess charge and the electric fringing field at the front end of the partially open-ended wider microstrip transmission line of width w_1. The total series inductance accounts for current density crowding at the narrowing of the wider line to the smaller one.

Figure 6.23 shows a T-type network of an impedance step. The total inductance may be split into the partial inductances L_1 and L_2 (as a first-order approximation) as follows [26]

$$L_1 = \frac{L'_{w_1}}{L'_{w_1} + L'_{w_2}} L_s = \frac{1}{1 + w_{eff_1}/w_{eff_2}} L_s \tag{6.45}$$

$$L_2 = \frac{L'_{w_2}}{L'_{w_1} + L'_{w_2}} L_s = \frac{1}{1 + w_{eff_2}/w_{eff_1}} L_s \tag{6.46}$$

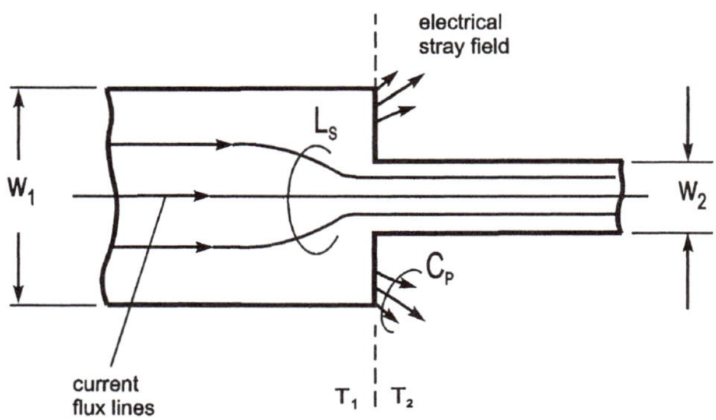

Figure 6.21 Symmetric impedance step.

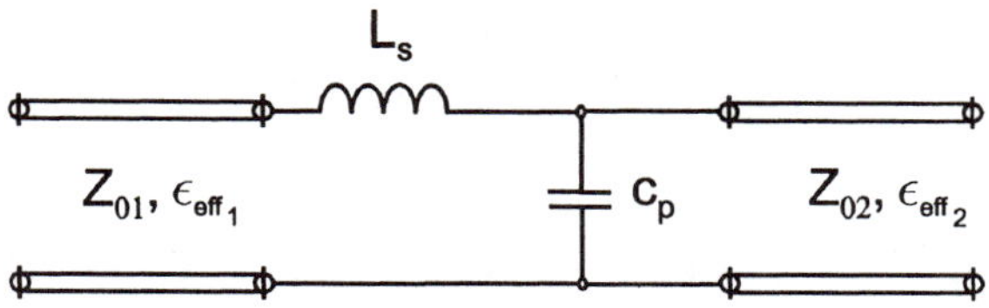

Figure 6.22 Equivalent circuit of the symmetric impedance step.

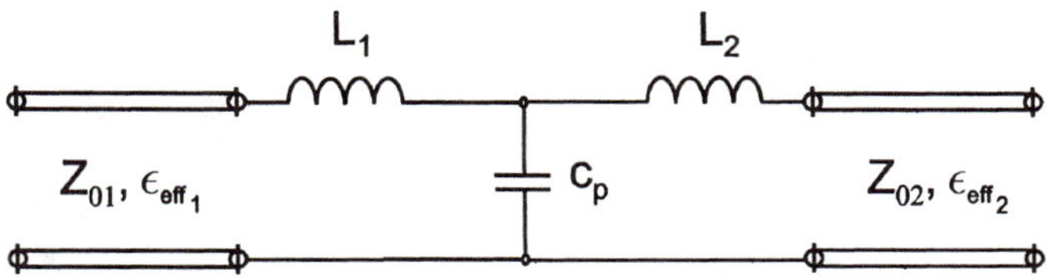

Figure 6.23 Equivalent T network of a symmetric impedance step.

with L'_{w_μ} $(\mu = 1, 2)$ the inductance per unit length of a microstrip line of width w_μ $(\mu = 1, 2)$. Note that the inductance per unit length L' is inversely proportional to the effective width w_{eff} (Section 3.1.5).

The inductance of an electrically short length of line (Δl) is calculated as $L = L'\Delta l$. With $\Delta l/h = L/(hL')$, it follows for the inductances L_1 and L_2, respectively:

$$\frac{\Delta l_\mu}{h} = \frac{L_\mu}{hL'_\mu} \qquad (\mu = 1,2) \tag{6.47}$$

From (6.47), it follows

$$\frac{\Delta l_1}{h} + \frac{\Delta l_2}{h} = \frac{L_1}{hL'_1} + \frac{L_2}{hL'_2}$$

$$= \frac{L'_1}{L'_1 + L'_2}\frac{L_s}{hL'_1} + \frac{L'_2}{L'_1 + L'_2}\frac{L_s}{hL'_2} = \frac{2L_s}{h\left(L'_1 + L'_2\right)} \tag{6.48}$$

As a first-order approximation, we may suppose [26]

$$\frac{\Delta l_1}{h} = \frac{\Delta l_2}{h} = \frac{L_s}{h\left(L'_1 + L'_2\right)} \tag{6.49}$$

Closed-form analytical expressions for the equivalent circuit elements were derived from numerical results by applying a curvefitting procedure [26]:

$$\frac{C_p}{h} = 10^{-3}\sqrt{\frac{w_1}{h}\frac{w_2}{h}}\left[\left(10.1\ \log\epsilon_r + 2.33\right)\frac{w_1/h}{w_2/h} - 12.6\log\epsilon_r - 3.17\right] \quad (6.50)$$

in the range $\epsilon_r \leq 10$ and $1.5 \leq w_1/w_2 \leq 3.5$ with C_p/h in picofarads per millimeter. The relative error is less than 10%. Regarding ceramic substrate material ($\epsilon_r = 9.6$), a more accurate relation has been found:

$$\frac{C_p}{h} = \sqrt{\frac{w_1}{h}\frac{w_2}{h}}\left[0.13\ \log\frac{w_1/h}{w_2/h} - 0.044\right] (\epsilon_r = 9.6) \quad (6.51)$$

in picofarads per millimeter in the range $3.5 \leq w_1/w_2 \leq 10$. In this case, the relative error is approximately less than or equal to 0.5%. Often the normalized capacitance $C_p/(C'_{w_1} h)$ is used in the literature, where $C'_{w_1} = \sqrt{\epsilon_{eff_1}} / \left[Z_{01}(\epsilon_r)\, c_0\right]$ is the capacitance per unit length of microstrip line. For the calculation of the capacitance per unit length, static values after Hammerstad were used.

The inductance is approximately calculated from

$$\frac{L_s}{h} = 0.405\left(\frac{w_1/h}{w_2/h} - 1.0\right) - 0.75\log\frac{w_1/h}{w_2/h} + 2\cdot10^{-3}\left(\frac{w_1/h}{w_2/h} - 1\right)^2 \quad (6.52)$$

with L_s in nanohenrys per centimeter. The relative error of this relation is less than 5% for $w_1/w_2 \leq 5$ und $w_2/h = 1.0$. On alumina substrate ($h = 0.637$ mm), inductance values in the range of 14 to 96 pH were measured for step discontinuities from 50Ω to 16Ω in the frequency range of 6.7 to 10.0 GHz [27].

Gupta et al. [28] also suggested more general expressions for C_p and L_s, reading

$$C_p = 1.37\frac{\sqrt{\epsilon_{eff}}}{Z_{01}}\left(1 - \frac{w_2}{w_1}\right)h\,\frac{\epsilon_{eff_1} + 0.3}{\epsilon_{eff_1} - 0.258}\,\frac{w_1/h + 0.264}{w_1/h + 0.8} \quad (\text{pF}) \quad (6.53)$$

$$L_s = 0.987\,h\left(1 - \frac{Z_{01}(\epsilon_r)}{Z_{02}(\epsilon_r)}\sqrt{\frac{\epsilon_{eff_1}}{\epsilon_{eff_2}}}\right)^2 \quad (\text{nH}) \quad (6.54)$$

with h in mm.

Figure 6.24 shows the shunt capacitance of a step discontinuity on Polyguide substrate ($\epsilon_r = 2.32$), normalized to the height h of the substrate, and the capacitance per unit length of the wider microstrip line C'_{w_1} as a function of the normalized smaller microstrip line width w_1/h with w_2/h as a parameter. The

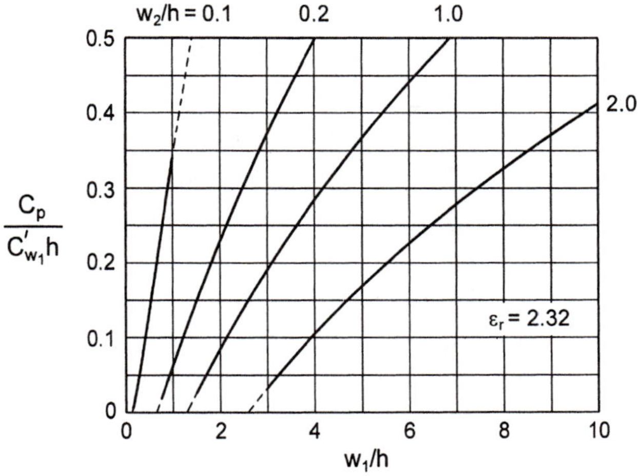

Figure 6.24 Parallel capacitance of a symmetric impedance step normalized to the substrate height and capacitance per unit length of the wider line as a function of the normalized wider line width w_1/h with the parameter w_2/h ($\varepsilon_r = 2.32$). Solid lines limit the valid parameter range.

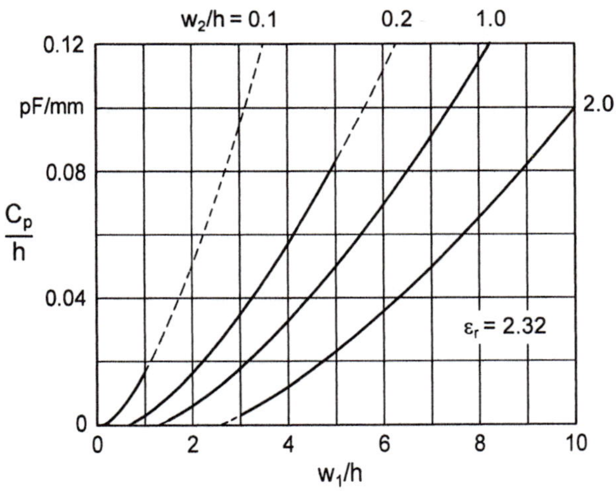

Figure 6.25 Parallel capacitance of a symmetric impedance step normalized to the substrate height as function of the normalized wider line width w_1/h with the parameter w_2/h ($\varepsilon_r = 2.32$). Solid lines limit the valid parameter range.

calculation was performed on the basis of (6.50). After introducing the capacitance per unit length described in Section 3.1.5, the value of C_p/h in dependence of the step parameter under study can directly be obtained from Figure 6.25. Figures 6.26

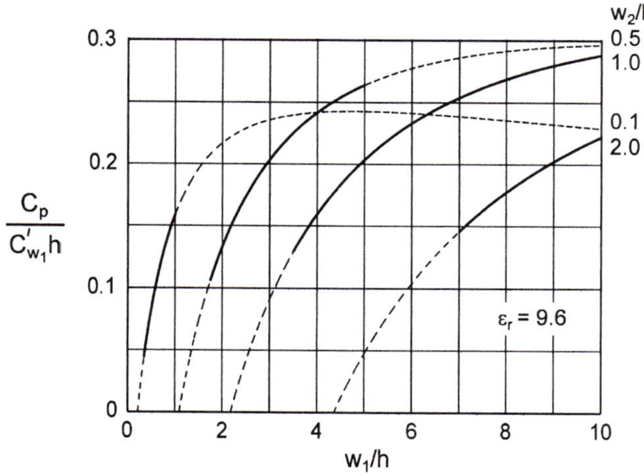

Figure 6.26 Parallel capacitance of a symmetric impedance step normalized to the substrate height and capacitance per unit length of the wider line as function of the normalized wider line width w_1/h with the parameter w_2/h ($\varepsilon_r = 9.6$). Solid lines limit the valid parameter range.

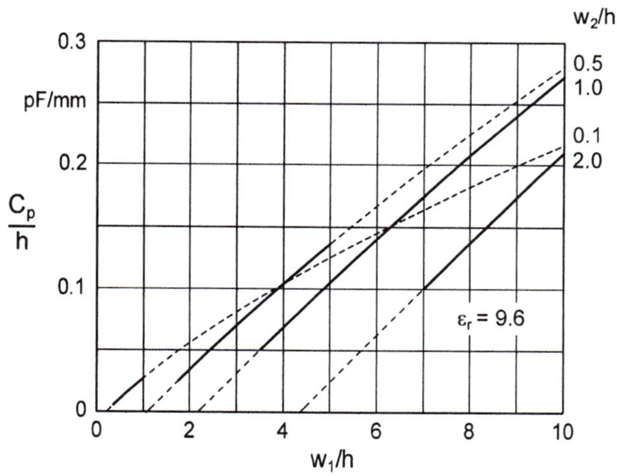

Figure 6.27 Parallel capacitance of a symmetric impedance step normalized to the substrate height as function of the normalized wider line width w_1/h with the parameter w_2/h (ε_r = 9.6). Solid lines limit the valid parameter range.

and 6.27 are similar plots for Al_2O_3 ceramic substrate ($\epsilon_r = 9.6$) using (6.51). Notice that relation (6.51) is only valid for values $w_1/w_2 > 3.5$. For decreasing ratios, it would give less accurate results. In particular, it is seen that the limit case

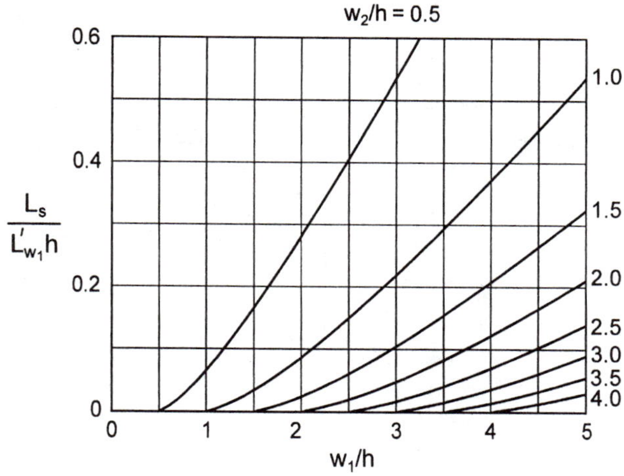

Figure 6.28 Series inductance of a symmetric impedance step normalized to the substrate height and inductance per unit length of the wider line as a function of the normalized wider line width w_1/h with the parameter w_2/h.

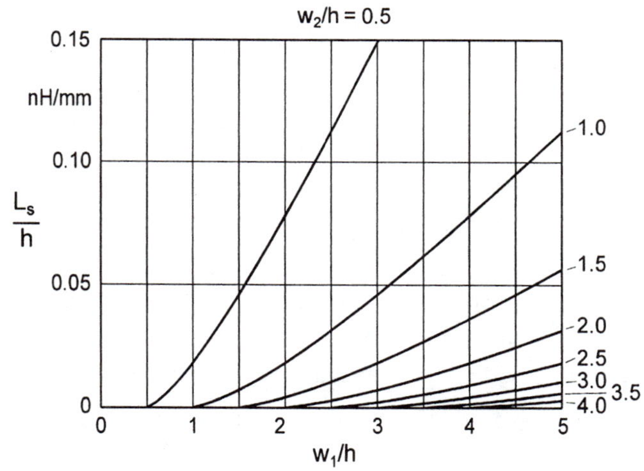

Figure 6.29 Series inductance of a symmetric impedance step normalized to the substrate height as a function of the normalized wider line width w_1/h with the parameter w_2/h.

$C_p \to \infty$ does not deliver $w_1 = w_2$. In Figure 6.28, the series inductance L_s normalized to the inductance per unit length L'_{w_1} of the wider line, and the height h is plotted versus w_1/h with parameter w_2/h. Figure 6.29 is similar. However, in this case the inductance is only normalized to the height h. Figure 6.30 shows the inductances L_1 and L_2 of the symmetric equivalent circuit as a function of w_1/h.

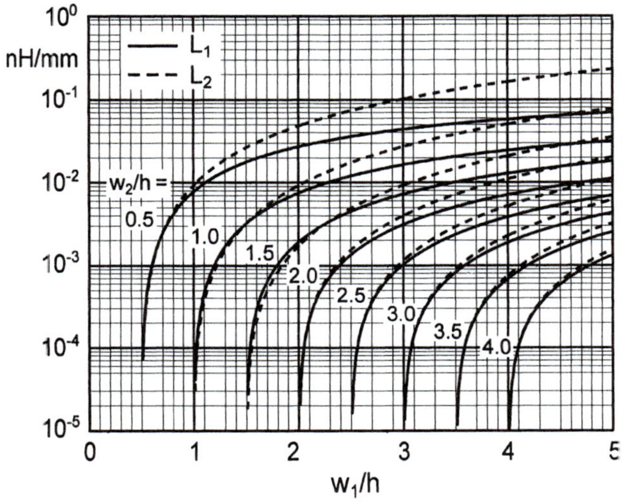

Figure 6.30 Partial inductances L_1 and L_2 of a symmetric impedance step normalized to the substrate height as a function of the normalized wider line width w_1/h with the parameter w_2/h.

Problem: Assume a symmetric impedance step on Polyguide substrate material ($\epsilon_r = 2.32$, $h = 1.56$ mm). The wider microstrip line has an impedance $Z_{01}(\epsilon_r)$ of 35Ω. The impedance of the narrower line $Z_{02}(\epsilon_r)$ is 65Ω. Determine the parallel capacitance C_p, the total series inductance L_s, and the partial inductances L_1 and L_2.

Solution: From Table 3.3, the widths of the step impedances are obtained as follows: (a) $Z_{01} = 35\Omega$: $w_1 = 7.694$ mm, $w_1/h = 4.93$; and (b) $Z_{02} = 65\Omega$: $w_2 = 3.070$ mm, $w_2/h = 1.96$. With the given values, the normalized parallel capacitance $C_p/(C'_{w_1}h)$ is read from Figure 6.24 as 0.1675. From Figure 3.25, the capacitance per unit length C'_{w_1} is obtained as 1.34 pF/cm. The capacitance then reads $C_p = 0.1675 \cdot C'_{w_1}h = 35$ fF. Regarding Figure 6.25, C_p/h is obtained as 0.0225 for the given parameters. Multiplication by the height of 1.56 mm delivers again a value of 35 fF.

For the step dimensions under consideration, from Figure 6.16, the normalized series inductance $L_s/(L'_{w_1}h)$ is read as 0.2. From Figure 3.24, the inductance per unit length L'_{w_1} is obtained as 1.65 nH/cm. Thus, the inductance turns out to be $L_s = 0.2 \cdot L'_{w_1}h = 0.2 \cdot 1.65\,\text{nH/cm} \cdot 0.156\,\text{cm} = 51\,\text{pH}$.

Regarding Figure 6.29, L_s/h is obtained as 0.3135 nH/cm for the given step structure. Multiplication by the height of 1.56 mm delivers a value of $L_s = 0.3135\,\text{nH} \cdot 0.156\,\text{mm} = 49\,\text{pH}$, which agrees well with the value calculated before.

6.2.2 Full-Wave Analysis Using the Planar Waveguide Model

Quasi-static equivalent circuits are only applicable at low frequencies, at which higher-order mode excitation at the discontinuity or junction under consideration is negligible. However, with increasing frequency, approaching the cutoff frequency of a first higher-order mode, the amplitudes of these parasitic modes are growing and can severely interfere the quasi-TEM mode response. In this chapter, we will analyze the dynamic properties of an impedance step based on the method of orthogonal series expansion using the planar waveguide model described in Chapter 5. The method is described in detail in [29, 30], but can be found meanwhile in various published books [1, 26, 28, 30]. Therefore, it is not intended to repeat any detail. However, the method itself should be presented as an example for solving a wide range of transverse waveguide discontinuity problems.

A propagating electromagnetic field is a vector field, comprising the electric field $\vec{E}(x,y,z)$ and magnetic field $\vec{H}(x,y,z)$. The field vectors can be decomposed in a longitudinal vector that is aligned with the propagation direction, and a transversal vector in a given cross-sectional plane of a waveguide,

$$\vec{E}(x,y,z) = \vec{E}_t(x,y,z) + \vec{E}_z(x,y,z) \tag{6.55}$$

$$\vec{H}(x,y,z) = \vec{H}_t(x,y,z) + \vec{H}_z(x,y,z) \tag{6.56}$$

where the index t denotes the transverse and z denotes the longitudinal field component. It is known that the longitudinal field components $\vec{E}_z(x,y,z)$ and $\vec{H}_z(x,y,z)$ can be calculated from the transverse electrical and magnetic fields, and vice versa [29]. The planar waveguide model, shown in Figure 6.31, is a closed waveguiding structure similar to the rectangular hollow waveguide. However, unlike the latter, the microstrip waveguide model is terminated on both sides by magnetic walls. The transverse electric and magnetic fields of mode ν can generally be expressed as [31]

$$\vec{E}_{t_\nu}(x,y,z) = \vec{t}_\nu(x,y)\, V_\nu(z) = \vec{t}_\nu \left(V_\nu^+ e^{-jk_{z\nu}z} + V_\nu^- e^{+jk_{z\nu}z} \right) \tag{6.57}$$

$$\begin{aligned} \vec{H}_{t_\nu}(x,y,z) &= \left[\vec{e}_z \times \vec{t}_\nu(x,y) \right] \frac{1}{Z_{F_\nu}} I_\nu(z) \\ &= \left[\vec{e}_z \times \vec{t}_\nu(x,y) \right] \frac{1}{Z_{F_\nu}} \left(V_\nu^+ e^{-jk_{z\nu}z} - V_\nu^- e^{+jk_{z\nu}z} \right) \end{aligned} \tag{6.58}$$

where $\vec{t}_\nu$ denotes the electrical vector mode function and V_ν^+ and V_ν^- the complex mode voltage of the incident, and backward traveling voltage wave, respectively (Figure 6.32). Z_{F_ν} is the wave impedance, and k_{z_ν} is the propagation coefficient of the νth mode. The power transported by a given mode reads

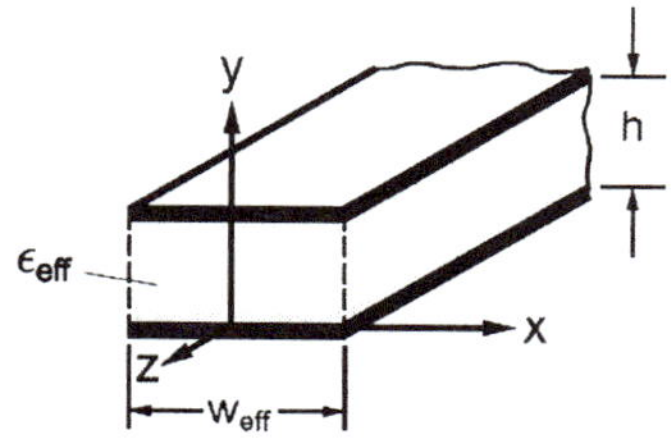

Figure 6.31 Microstrip waveguide model and definition of coordinate system.

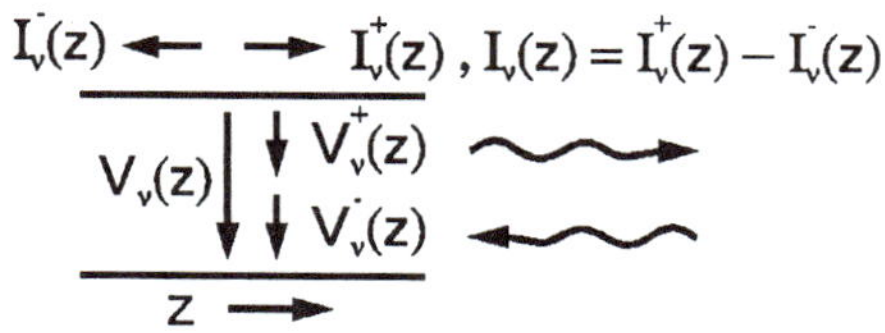

Figure 6.32 Definition of total current and voltage, and forward and backward current and voltage.

$$P_\nu^+ = \frac{1}{2} Re \int_A \left(\vec{E}_{t_\nu} \times \vec{H}_{t_\nu}^{+*} \right) \vec{e}_z \, dA \tag{6.59}$$

With $\vec{t}_\nu \times \left(\vec{e}_z \times \vec{t}_\nu \right) = \vec{e}_z \left(t_\nu \cdot t_\nu \right)$, it follows from (6.19)

$$P_\nu^+ = \frac{1}{2} \frac{\vec{V}_\nu^+ \vec{V}_\nu^{+*}}{Z_{F_\nu}} \int_A \vec{t}_\nu \cdot \vec{t}_\nu \, dA \tag{6.60}$$

$\vec{t}_\nu$ is specifically termed orthonormalized because the mode functions have orthogonality properties and are normalized to unity. This is expressed by the following relation:

$$\int_A \vec{t}_\nu \cdot \vec{t}_\nu \, dA = \delta_{\nu\mu} = \begin{cases} 1 & \text{for } \nu = \mu \\ 0 & \text{for } \nu \neq \mu \end{cases} \tag{6.61}$$

Thus, the active power transported by a wave simply reads

$$P_\nu^+ = \frac{1}{2} \frac{\vec{V}_\nu^+ \vec{V}_\nu^{+*}}{Z_{F_\nu}} \tag{6.62}$$

which is similar to the well-known power relation in classical transmission line theory. Thus, the amplitude V_ν of the νth mode can be termed the voltage of the mode. Similarly, the complex amplitude I_ν is called the current of the νth mode.

Regarding (6.58), the transverse magnetic field is perpendicular to the transversal electrical field, which is described by $(\vec{e}_z \times \vec{t}_\nu)$. The wave impedance is the ratio of the amplitude of the transversal electric field and the respective magnetic field, measured in ohms. Therefore, the wave impedance occurs in the denominator of (6.58). Often the minus sign in (6.58) causes misunderstanding. Both positive and negative signs are found in the literature. The positive sign is a pure mathematical approach, and expresses formally the total current being composed by a forward and backward current (superposition of incident and reflected current). When using the negative sign, the physical circumstances have already taken into account the fact that the partial currents are flowing in opposite direction.

In any closed waveguide, there exists an infinite number of modes (eigenmodes), which constitute a complete set of eigenfunctions. This means that a given arbitrary transversal electric and magnetic field near a discontinuity or junction can be described by a series expansion of the eigenmodes with suitably tuned mode amplitudes. In the planar waveguide model, the eigenmodes are TEM mode, **H** modes, and **E** modes [29, 32–34]. Because the height h of microstrip lines is very small with respect to the operating frequency, $E_{(mn)}$ modes, if at all excited, can be neglected. For that reason, $H_{[mn]}$ modes with $n \neq 0$ are also not regarded. Thus, modes taking into account in the step analysis are the quasi-TEM (TEM) as incident and excited modes, whereas HE_m ($H_{[m0]}$) modes ($m = 1,2,...$) exist only as stimulated modes. Because $E_{(mn)}$ modes do not appear in the following analysis, the indexed squared brackets will be omitted. The vector mode functions read for the TEM

$$\vec{t}_{00} = \frac{1}{\sqrt{w_{eff} h}} \vec{e}_y \tag{6.63}$$

and for the H_{m0} modes

$$\vec{t}_{m0} = \sqrt{\frac{2}{w_{eff} h}} \cos\left[\frac{m\pi}{w_{eff}}\left(x + \frac{w_{eff}}{2}\right)\right] \vec{e}_y \tag{6.64}$$

The mode functions of TEM and H_{m0} can be combined in the following relation

$$\vec{t}_{m0} = \sqrt{\frac{\xi_m}{w_{eff} h}} \cos\left[\frac{m\pi}{w_{eff}}\left(x + \frac{w_{eff}}{2}\right)\right] \vec{e}_y \tag{6.65}$$

where $m = 0, 1, 2, ...$ and

$$\xi_m = \begin{cases} 1 & \text{for } m = 0 \\ 2 & \text{for } m \neq 0 \end{cases}$$

It can be simply verified that the described mode functions fulfill the orthogonality condition in (6.61).

The scattering matrix of an n-port relates the outgoing power waves to the incident power waves. Figure 6.33 shows the top view of an offset step discontinuity under study (solid line) and its physical parameters. ε denotes the

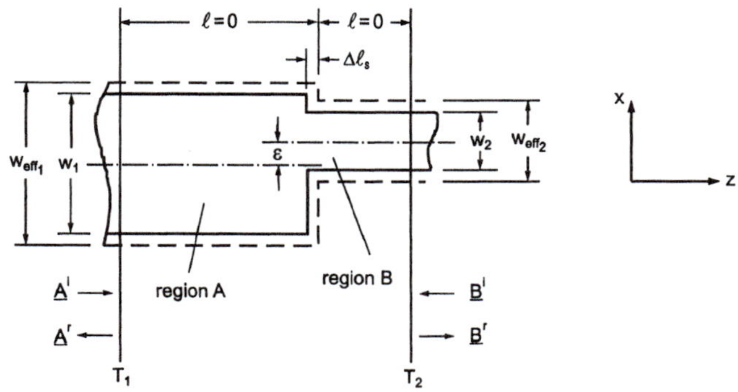

Figure 6.33 Impedance step discontinuity under study.

offset of the step transition. For analysis, the microstrip circuit is replaced by the planar waveguide model. The dashed lines indicate the magnetic wall contour of the microstrip waveguide model. The scattering electrical field in front of the step transition results in an equivalent shift Δl_s of the transition reference plane. Initial theoretical work mainly focused on the frequency-dependence of the amplitudes of the scattering parameters, thus neglecting this capacitive effect, which has indeed only an influence on the phase of the scattering parameters. $\boldsymbol{A}^r$ and $\boldsymbol{B}^r$ are the outgoing waves at port 1 (reference plane T_1) and port 2 (reference plane T_2), respectively. $\boldsymbol{A}^i$ and $\boldsymbol{B}^i$ are the stimulating waves at port 1 and port 2, respectively. $\boldsymbol{A}$ and $\boldsymbol{B}$ are column matrices which comprise the respective modes:

$$\boldsymbol{A}^i = \begin{pmatrix} a_{00}^i \\ a_{10}^i \\ \vdots \end{pmatrix}, \quad \boldsymbol{B}^i = \begin{pmatrix} b_{00}^i \\ b_{10}^i \\ \vdots \end{pmatrix}, \quad \boldsymbol{A}^r = \begin{pmatrix} a_{00}^r \\ a_{10}^r \\ \vdots \end{pmatrix}, \quad \boldsymbol{B}^r = \begin{pmatrix} b_{00}^r \\ b_{10}^r \\ \vdots \end{pmatrix} \tag{6.66}$$

The electric and magnetic field distribution on either side of the discontinuity can be written as follows:

Region A:

$$\vec{E}_t^A(z) = \sum_{m=0}^{\infty} V_m^A(z)\,\vec{t}_{m0} \tag{6.67}$$

$$\vec{H}_t^A(z) = \sum_{m=0}^{\infty} I_m^A(z)\left(\vec{e}_z \times \vec{t}_{m0}\right) \tag{6.68}$$

with

$$\sqrt{Y_{m0}^A}\,V_{m0}^A(z) = a_{m0}^A\,e^{-jk_{zm0}^A z} + b_{m0}^A\,e^{+jk_{zm0}^A z} \tag{6.69}$$

$$\sqrt{Z_{m0}^A}\,I_{m0}^A(z) = a_{m0}^A\,e^{-jk_{zm0}^A z} - b_{m0}^A\,e^{+jk_{zm0}^A z} \tag{6.70}$$

Region B:

$$\vec{E}_t^B(z) = \sum_{p=0}^{\infty} V_p^B(z)\, \vec{t}_{p0} \tag{6.71}$$

$$\vec{H}_t^B(z) = \sum_{p=0}^{\infty} I_m^B(z) \left(-\vec{e}_z \times \vec{t}_{p0}\right) \tag{6.72}$$

with

$$\sqrt{Y_{p0}^B}\, V_{p0}^B(z) = a_{p0}^B\, e^{-jk_{z_{p0}}^B z} + b_{p0}^B\, e^{+jk_{z_{p0}}^B z} \tag{6.73}$$

$$\sqrt{Z_{p0}^B}\, I_{p0}^B(z) = a_{p0}^B\, e^{-jk_{z_{p0}}^B z} - b_{p0}^B\, e^{+jk_{z_{p0}}^B z} \tag{6.74}$$

The negative sign in (6.72) results from the propagation direction of the wave in negative z direction. The step boundary conditions in the coupling aperture at $z = 0$ are:

(1)	$\vec{E}_t^A = \vec{E}_t^B$	or	$E_y^A = E_y^B$	in A_A
(2)	$\vec{H}_t^A = \vec{H}_t^B$	or	$H_x^A = H_x^B$	in A_A
(3)	$\vec{H}_t^B = 0$	or	$H_x^B = 0$	in $A_B - A_A$

A_A is the step coupling aperture (cross-section of waveguide A), and $A_B - A_A$ is the cross-section of waveguide B reduced by the coupling aperture. Multiplying (6.67) by the mode function $\vec{t}_{M0}$, and (6.72) by $(-\vec{e}_z \times \vec{t}_{P0})$, we obtain

$$V_{M0}^A = \int_{A_A} \vec{E}_t^A \cdot \vec{t}_{M0}\, dA \tag{6.75}$$

$$I_{P0}^B = \int_{A_B} \vec{H}_t^B \cdot \left(-\vec{e}_z \times \vec{t}_{P0}\right) dA \tag{6.76}$$

Introducing the boundary conditions results in

$$V_{M0}^A = \int_{A_A} \vec{E}_t^B \cdot \vec{t}_{M0}\, dA \tag{6.77}$$

$$I_{P0}^B = \int_{A_A} \vec{H}_t^A \cdot \left(-\vec{e}_z \times \vec{t}_{P0}\right) dA \tag{6.78}$$

Equations (6.77) and (6.78) describe two infinite systems of coupled equation for the amplitudes V_{M0}^A $(M=0,1,2,\dots,\infty)$, V_{P0}^B $(P=0,1,2,\dots,\infty)$, I_{M0}^A $(M=0,1,2,\dots,\infty)$, and I_{P0}^B $(P=0,1,2,\dots,\infty)$ of the electric and magnetic field on either side of the discontinuity. The coefficients of the equation system are integrals over the products of mode functions of different field regions; for example,

$$K_{m0,p0} = \int_A \vec{t}_{m0} \cdot \vec{t}_{p0}\, dA \tag{6.79}$$

With the defined coupling integrals, the set of equations can be written as

$$V_{M0}^A = \sum_{p=0} V_p^B K_{Mp} \tag{6.80}$$

$$I_{P0}^B = -\sum_{m=0} I_m^A K_{mP} \tag{6.81}$$

Introducing the power waves, we finally get

$$\sqrt{Z_M^A}\left(a_M^A + b_M^A\right) = \sum_{p=0}^{\infty} \sqrt{Z_p^B} K_{Mp}\left(a_p^B + b_p^B\right) \tag{6.82}$$

$$\sqrt{Y_B^B}\left(a_P^B - b_P^B\right) = -\sum_{m=0}^{\infty} \sqrt{Y_m^A} K_{mP}\left(a_m^A - b_m^A\right) \tag{6.83}$$

where $m, M, p, P = 0, 1, 2, \dots$. After rearranging, the reverse waves are related to the incident waves as follows

$$\begin{pmatrix} A^r \\ B^r \end{pmatrix} = \begin{pmatrix} S^{AA} & S^{AB} \\ S^{BA} & S^{BB} \end{pmatrix} \begin{pmatrix} A^i \\ B^i \end{pmatrix} \tag{6.84}$$

In more detail, (6.84) reads

$$\begin{pmatrix} a_{00}^r \\ a_{10}^r \\ \vdots \\ b_{00}^r \\ b_{10}^r \\ \vdots \end{pmatrix} = \begin{pmatrix} S_{00,00}^{AA} & S_{00,10}^{AA} & \cdots & S_{00,00}^{AB} & S_{00,10}^{AB} & \cdots \\ S_{10,00}^{AA} & S_{10,10}^{AA} & \cdots & S_{10,00}^{AB} & S_{10,10}^{AB} & \cdots \\ \vdots & \vdots & \vdots & \vdots & \vdots & \vdots \\ S_{00,00}^{BA} & S_{00,10}^{BA} & \cdots & S_{00,00}^{BB} & S_{00,10}^{BB} & \cdots \\ S_{10,00}^{BA} & S_{10,10}^{BA} & \cdots & S_{10,00}^{BB} & S_{10,10}^{BB} & \cdots \end{pmatrix} \begin{pmatrix} a_{00}^i \\ a_{10}^i \\ \vdots \\ b_{00}^i \\ b_{10}^i \\ \vdots \end{pmatrix} \tag{6.85}$$

The characterization of the step by the infinite scattering matrix is an extension of the commonly used scattering matrix used in the microwave network theory (Appendix G), but differs in that the dominant as well as higher order modes are included. The matrix, which is termed generalized scattering matrix, is, in general, of infinite order. The first upper index of the scattering elements denotes the port, and the first lower double-index denotes the mode of the outgoing wave. The second upper index denotes the port, and the second lower double-index denotes the mode of the incident wave. When all incident amplitudes with the exception of the quasi-TEM mode are chosen as zero, and unit amplitude is assumed for the quasi-TEM mode, then the amplitudes of the excited waves A^r and B^r refer to the scattering elements. Thus, for instance, $S^{AA}_{00,00}$ is the amplitude of the reflected quasi-TEM wave at port 1 when this is stimulated by the same mode with unit amplitude. $S^{AB}_{10,00}$ describes the excitation of the first higher-order mode (H_{10}) at port 1 when the quasi-TEM wave is incident at port 2.

The waveguide model neglects the radiation from the discontinuity and surface-wave excitation. Experiments have shown that these parasitic effects are less significant with impedance steps due to the reduced open end effect.

For numerical computation of the S matrix of the impedance step, only the quasi-TEM modes with normalized amplitude $a^A_{00} = 1$ or $a^B_{00} = 1$ are taken as incident (stimulated) waves. All other higher-order incident wave terms are set equal to zero. Figures 6.34 and 6.35 show the calculated magnitudes of the scattering coefficients for the quasi-TEM mode related to the physical step reference plane, as indicated in Figure 6.33. The small width w_2 was kept constant, whereas the wide strip varies. In the case of alumina substrate material, the excitation of higher-order modes is not relevant for impedance steps with moderate values ($w_1/w_2 \leq 4$) up to 20 GHz. This is not true for low-dielectric substrate material like Polyguide, as can be seen from Figure 6.35. For example, higher-order mode excitation already takes place for $w_1/w_2 \leq 4$ values, which are of importance in practical microstrip circuit design. It should be noticed that the excitation of higher-order modes becomes stronger with asymmetric impedance steps (Figure 6.1). In this case, the cutoff frequency of the first higher-order mode becomes one-half of the second one (compare Figures 5.7 and 5.8).

The parasitic effect of a step junction is similar to that of an open end. The effect of the fringing capacitance associated with the wider line of the step discontinuity is similar to an increase in the length of that line. The change in the electrical length can be estimated as a first-order approximation by using the equation (Figure 6.36) [6]

$$\Delta l_s = \Delta l(1 - w_2/w_1) \tag{6.86}$$

where w_2 is the width of narrowest line ($w_2 < w_1$). Δl represents the open end effect as described in Section 6.1.1. Substituting (6.2), proposed by Hammerstad, in (6.86) yields

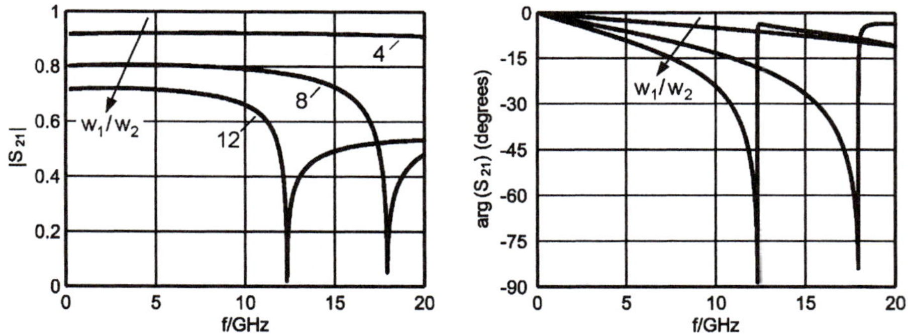

Figure 6.34 Magnitude and phase of transmission coefficient S_{21} of a symmetric step discontinuity on alumina substrate material for different values w_1/w_2 as parameter ($\varepsilon_r = 9.8$, $h = 0.635$ mm, $w_2 = 0.61$ mm).

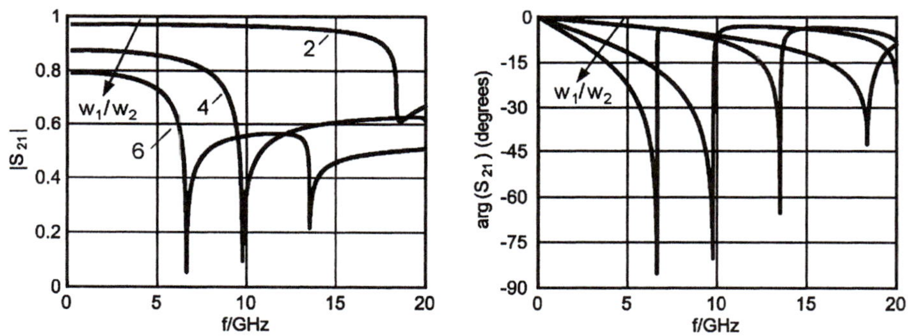

Figure 6.35 Magnitude and phase of transmission coefficient S_{21} of a symmetric step discontinuity on Polyguide substrate material for different values w_1/w_2 as parameter ($\varepsilon_r = 2.32$, $h = 1.58$ mm, $w_2 = 4.71$ mm).

$$\frac{\Delta l_s}{h} = (1 - w_2/w_1)\, 0.412\, \frac{\epsilon_{eff1} + 0.3}{\epsilon_{eff1} - 0.258}\, \frac{w_1/h + 0.262}{w_1/h + 0.813} \tag{6.87}$$

It is obvious that the shift of the step transition reference plane will always be less than that caused by an open end of a line with the wider microstrip width.

In [35], it is suggested to derive the step shift Δl_s from the edge-fringing electrical field. The total fringing capacitance per unit length is obtained from the total capacitance per unit length minus the capacitance per unit length, which corresponds to the strip region. Then the transition stray capacitance can be estimated as

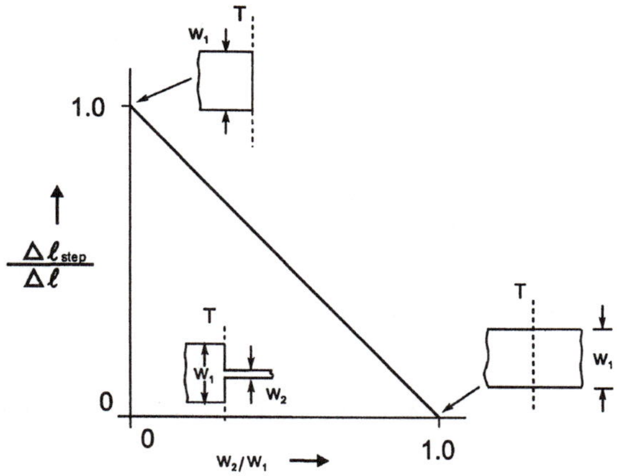

Figure 6.36 Definition of a step extension versus w_2/w_1 (6.86).

$$C_p = \frac{1}{2}\left(\frac{1}{Z_{01}(\epsilon_r)\, c_0/\sqrt{\epsilon_{eff}}} - \epsilon_0\epsilon_r\frac{w_1}{h}\right)(w_1 - w_2) \tag{6.88}$$

In [36], (6.88) was used to derive an equivalent step extension Δl_s. Equation (6.88) can be reformulated:

$$C_p = \frac{1}{2}\,\frac{\epsilon_0\epsilon_r\,(w_1 - w_2)}{h}\left(\frac{\epsilon_{eff_1}(f)}{\epsilon_r}\,w_{eff_1}(f) - w_1\right) \tag{6.89}$$

Supposing a magnetic wall model with physical widths w_1 and w_2 and respective material parameter ϵ_r, the stray capacitance can be equivalently described as

$$C_p = \frac{\epsilon_0\epsilon_r\,(w_1 - w_2)\,\Delta l_s}{h} \tag{6.90}$$

By comparison of (6.89) and (6.90), Δl_s is obtained as

$$\Delta l_s = \frac{1}{2}\left(\frac{\epsilon_{eff_1}(f)}{\epsilon_r}\,w_{eff_1}(f) - w_1\right) \tag{6.91}$$

Notice that the stray capacitance effect has turned out to be frequency-dependent. When $f \to \infty$, it follows that $\epsilon_{eff_1}(f) \to \epsilon_r$ and $w_{eff_1}(f) \to w_1$, and thus Δl_s, become zero at infinite frequency.

Rizzoli et al. describe in [37] a semiempirical formula for the microstrip impedance step derived from measurements on alumina ($\epsilon_r = 10$, $h = 0.25$ mm) and Teflon fiberglass ($\epsilon_r = 2.32$, $h = 0.5$ mm) using a resonance method [38]. Quasi-static characteristic impedance values used ranged from 20Ω to 100Ω. Scattering coefficients were measured from 1 up to 27 GHz. Measured data was compared with calculations based on the planar waveguide model, leaving Δl_s as a fitting parameter. Equation (6.86) was modified,

$$\Delta l_s = f(\epsilon_r)\, \Delta l\, (1 - w_2/w_1) \tag{6.92}$$

It was found that for best fit, $f(\epsilon_r = 2.32) = 1.5$ for the considered Teflon substrate material and $f(\epsilon_r = 10.0) = 1.0$ for alumina. These two-spot numerical values were used to approximate $f(\epsilon_r)$ by a function of the form

$$f(\epsilon_r) = 1.25 + 0.19\,\mathrm{atan}(6.16 - \epsilon_r) \tag{6.93}$$

For frequencies lower than approximately one-third of the cutoff frequency of the first higher-order mode (i.e., the H_{10} mode for the offset step discontinuity, and the H_{20} mode for the symmetric impedance step), the magnitudes of the scattering parameters can be calculated very simply [31], as illustrated in Figure 6.37. The eigenreflection coefficient S_{11} equals the input reflection coefficient Γ_{11}' when the step is matched at port 2-2′. Thus, the transmission line with characteristic impedance Z_1 is terminated by a load Z_2 at the output of the considered step discontinuity. Then, the magnitudes of the scattering coefficients are obtained as

$$|S_{11}| = |S_{22}| = \frac{Z_2/Z_1 - 1}{Z_2/Z_1 + 1} \tag{6.94}$$

$$|S_{12}| = \sqrt{1 - |S_{11}|^2} = |S_{21}| = \frac{2\,(Z_2/Z_1)^{1/2}}{Z_2/Z_1 + 1} \tag{6.95}$$

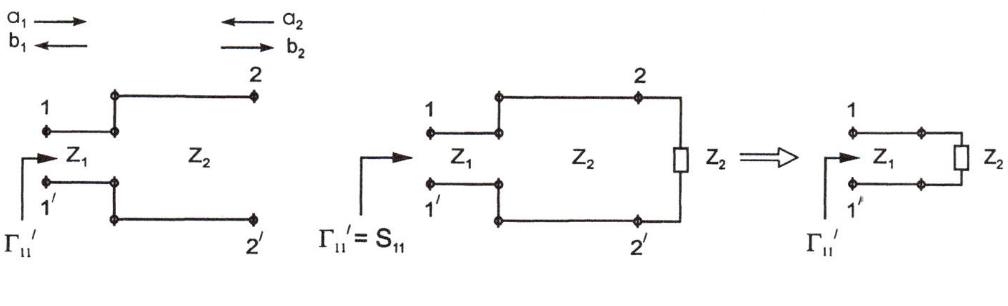

Figure 6.37 Calculation of the scattering coefficients of a lossless impedance step, neglecting parasitic step effects and loss.

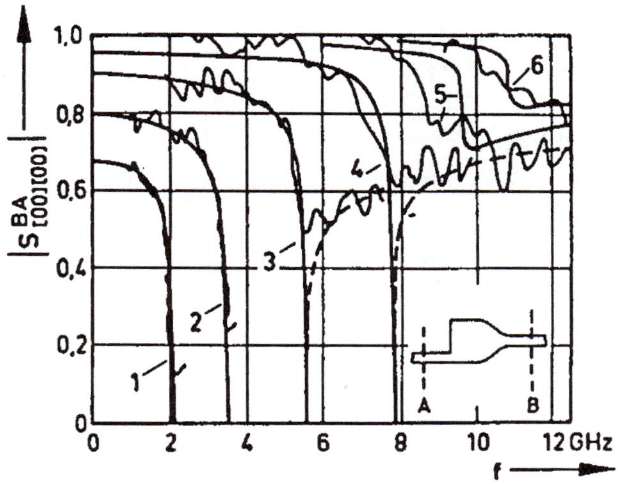

Figure 6.38 Measured and simulated magnitude of the transmission coefficient of the quasi-TEM mode of an asymmetric impedance step (ϵ_r = 2.32, h = 1.58 mm, w_1 = 4.61 mm) versus frequency [wider line widths w_2: (1) 47.57 mm, (2) 26.94 mm, (3) 15.66 mm, (4) 10.86 mm, (5) 0.832 cm, (6) 0.723 cm].

A simple prediction for the phases cannot be made. At zero frequency, the transmission phase becomes zero.

Figure 6.38 shows the measured and calculated transmission coefficient for an asymmetric impedance step on Polyguide substrate ($\epsilon_r = 2.32$, $h = 1.58$ mm) [32]. A vector network analyzer was used for measurement calibrated at its coaxial ports. Thus, measured data includes the electrical properties of the coaxial-to-microstrip transitions. The characterization of isolated step discontinuities in a 50Ω environment requires a broadband impedance transformer to match the low-impedance microstrip line ($Z_0 < 50\Omega$) to the 50Ω input port of the measuring equipment. A long microstrip taper with a length of $l = 15$ cm was inserted to avoid any severe matching problem. In an asymmetric step discontinuity, the H_{10} mode is stimulated as first higher-order mode. Transmission zeros occur at the cutoff frequency of the higher-order mode on the wider line of the regarded microstrip circuits. Above the cutoff frequencies, there exist two propagating waves on the wider line, the quasi-TEM and H_{10} mode.

Figure 6.39 shows the measured and calculated transmission coefficient for a symmetric impedance step. In this case, because of the symmetric (even) excitation of the wider line, the (odd) H_{10} mode cannot be stimulated by an incident quasi-TEM wave on either side. The first higher-order mode is now the even H_{20} mode with cosinusoidal transversal electrical field distribution. Strong higher-order mode excitation occurs for large w_2/w_1 ratios. With decreasing value of the width ratio, the measured transmission zero becomes less pronounced, intimating growing power transfer by the dominant mode.

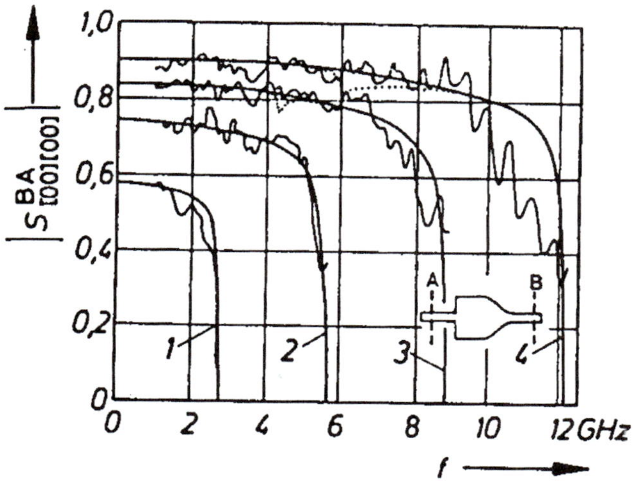

Figure 6.39 Measured and simulated amplitude of the transmission coefficient of the quasi-TEM mode of a symmetric impedance step ($\epsilon_r = 2.32$, $h = 1.58$ mm, $w_1 = 4.61$ mm) versus frequency [wider line widths w_2: (1) 7.339 cm, (2) 3.448 cm, (3) 2.192 cm, (4) 1.565 cm]. (From [62], © 1976 Elsevier. Reprinted with permission.)

As we have seen from the diagrams, the intensity of higher-order excitation depends strongly on the ratio of the widths of the step discontinuity. This is illustrated in Figure 6.40, showing an impedance step under different coupling settings. Figure 6.40(a) shows a symmetric step with a large w_2/w_1 ratio. We suppose an incident quasi-TEM wave on the narrow microstrip line. The approximated transverse electrical field distribution of the relevant modes is also drawn. The efficiency of excitation of modes depends on the field matching of the stimulating mode to the fields of eigenmodes in the connected wide line. In Figure 6.40(a), the excitation is symmetric to the wider line. Therefore, only symmetric modes can be stimulated. However, strong stimulation is only given if the aperture is much smaller than the broader width, because in the central region the stimulating field agrees well with the $H_{[20]}$ mode. In contrast, the asymmetric $H_{[10]}$ mode cannot be excited. Thus, the single-mode operation is limited by the occurrence of the second higher-order $H_{[20]}$ mode. In Figure 6.40(b), the same geometry is given; however, the aperture is located near the edge of the wider line. We would expect the same low-frequency behavior of the device as before. However, as all modes have maximum electrical field near the edge, the first excited higher-order mode will be the $H_{[10]}$ mode, followed by the $H_{[20]}$ mode at higher frequency. Thus, in this case the single-mode operation is limited by the cutoff frequency of the $H_{[10]}$ mode. Figure 6.40(c) shows an asymmetric transition with a ratio of the widths $w_1/w_2 > 0.5$. It is clear that the coupling of the quasi-TEM becomes much better. However, due to the transverse resonance properties

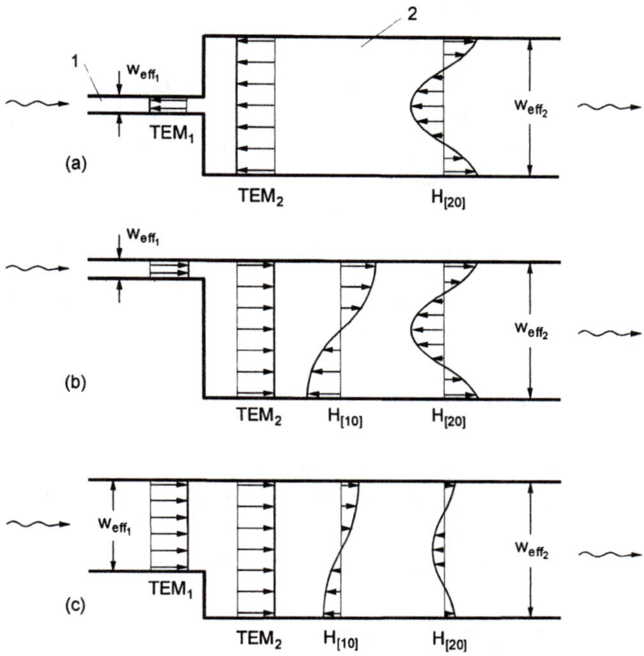

Figure 6.40 Coupling efficiency of higher-order modes: (a) narrow symmetric coupling aperture, (b) narrow asymmetric coupling aperture, and (c) wide asymmetric coupling aperture.

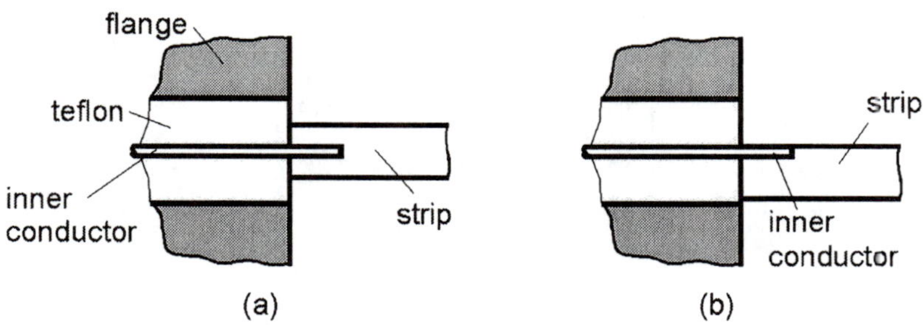

Figure 6.41 (a) Symmetric and (b) asymmetric implementation of different coaxial-to-microstrip transitions with different implication on higher-order mode excitation (top view).

of the higher-order modes and their respective sinusoidal transversal electric field distributions, the launching efficiency decreases with growing stimulating aperture.

Figure 6.41(a) shows a coaxial-to-microstrip transition. The contacting of the inner conductor of the coaxial line has been carried out symmetrically with respect to the upper conductor of the microstrip line. In this case, the first higher-order mode excited by the discontinuity would be the HE_2 mode or $H_{[20]}$ waveguide mode, respectively. The cutoff frequency may be far beyond the operating frequency range. A different situation is shown in Figure 6.41(b), where the inner conductor of the coaxial line is placed asymmetrically near the edge of the strip conductor of the microstrip line. In this case, the HE_1 mode or $H_{[10]}$ waveguide mode, respectively, would be stimulated as the first higher-order mode. Although the matching condition could have been attained at low frequencies, the first stimulated parasitic higher-order modes may severely influence the quasi-TEM response of the transition. This is in particular the case with relatively large widths of microstrip lines on plastic substrates.

6.2.3 Cascaded Step Discontinuities

Figure 6.42 shows two symmetric impedance steps interconnected by a homogeneous microstrip transmission line of length L. For analysis, the open microstrip structure is replaced by the planar waveguide model. The physical length L is extended by $2\Delta l_s$, taking into account the parasitic capacitive effect of either step transition.

When a quasi-TEM mode is incident from either side, the dominant mode as well as even higher-order modes are excited in the step discontinuites. The reference planes of the two-port therefore must be positioned sufficiently far away from the discontinuities, so that the amplitudes of the excited higher-order evanescent modes becomes negligible at the location of the input and output port.

Referring to the reference planes T_1 and T_2, the dominant mode two-port scattering matrix of the microstrip circuit is

$$\begin{pmatrix} a_{00}^r \\ c_{00}^r \end{pmatrix} = \begin{pmatrix} S_{00,00}^{AA} & S_{00,00}^{AC} \\ S_{00,00}^{CA} & S_{00,00}^{CC} \end{pmatrix} \begin{pmatrix} a_{00}^i \\ b_{00}^i \end{pmatrix} \tag{6.96}$$

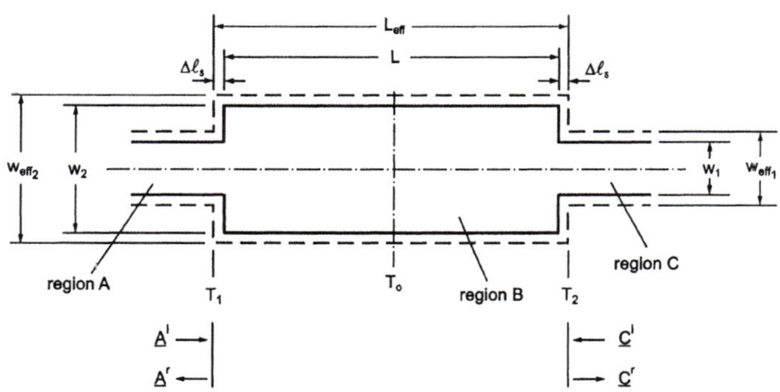

Figure 6.42 Top view of the cascaded step discontinuities under study.

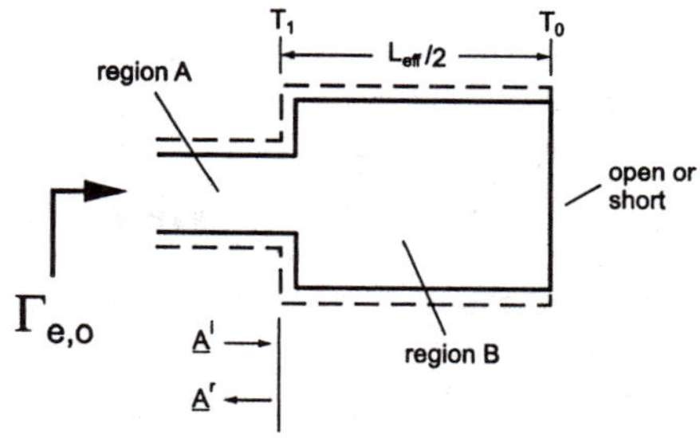

Figure 6.43 Analysis of cascaded steps of distance L reduced to the calculation of a step discontinuity terminated by an open-ended and short-circuited transmission line of length $L/2$, respectively.

The quantities a_{00}^r and c_{00}^r are related to the reflected quasi-TEM modes, and a_{00}^i and c_{00}^i to the incident modes. The column vectors A^i, A^r, C^i, and C^r in Figure 6.42 include higher-order evanescent modes in addition to the dominant mode. For efficient computation, the number of modes is limited to about 10.

For reason of geometric symmetry (plane T_0 in Figure 6.42), the analysis of the cascaded structure can be derived from the input reflection coefficient of a single step terminated by a line of length $L/2$. When a_{00}^r is in phase with c_{00}^r, the (even) excitation produces zero transversal magnetic field in the plane T_0, and thus, analysis can be restricted to a step discontinuity with the wider line terminated by an open-ended transmission line of length $L/2$. Γ_e is denoted as the input reflection coefficient of this one-port (see Figure 6.43). When the excitation in the ports are out of phase, the transversal electrical field vanishes, and the plane T_0 acts like a short (input reflection coefficient Γ_o). Then, the scattering coefficients of the cascaded steps (6.96) can be calculated from the reflection coefficients Γ_e and Γ_o, by superposition as follows:

$$
\begin{aligned}
S_{00,00}^{AA} &= S_{00,00}^{CC} = 0.5\left(\Gamma_e + \Gamma_o\right) \\
S_{00,00}^{AC} &= S_{00,00}^{CA} = 0.5\left(\Gamma_e - \Gamma_o\right)
\end{aligned}
\tag{6.97}
$$

Equation (6.97) includes reflection symmetry and reciprocity of the microstrip under study. Figure 6.44 shows the measured result of two cascaded symmetric impedance steps on Polyguide substrate material ($\epsilon_r = 2.32$, $h = 1.58$ mm) [32]. The geometric parameters are: $w_1 = 0.438$ cm, $w_2 = 2.57$ cm, $l = 9.0$ cm. It can be seen that in the dominant mode region ($f < f_{c[20]}$), the magnitude of the transmission coefficient $|S_{00,00}^{BA}|$ becomes unity when the condition $L = n\lambda/2$ ($n = 1,2,3,\ldots$) is fulfilled. Minimum values occur for $L = n\lambda/4$ ($n = 1,3,5,\ldots$). Because of increasing excitation of the first higher-order mode with frequency, the transmission coefficient decreases monotonously and becomes zero at the cutoff

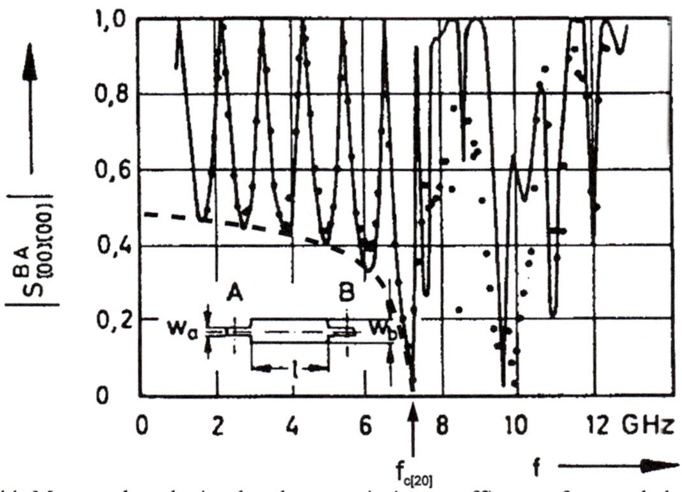

Figure 6.44 Measured and simulated transmission coefficent of cascaded impedance steps on Polyguide substrate (w_a = 4.38 mm, w_b = 25.7 mm, h = 1.55 mm, ε_r = 2.32, l = 90 mm) [32].

frequency of the higher-order mode. Above $f_{c[20]}$, both the quasi-TEM and H_{20} mode are propagating along the wider line. This multimode region commonly is not interesting for practical use.

6.2.4 Quarter-Wavelength Transformer

Transmission line impedance transformers are frequently found in microwave circuits. They are used to match unequal characteristic impedances. The narrow-band single-step impedance transformer is realized as a quarter-wavelength transformer, as illustrated in Figure 6.45 (top view of the microstrip circuit). In [39], calculation of the scattering matrix of a single-step transformer was done using the following methods:

1. Ideal transmission line theory [32];
2. Calculation of cascaded step discontinuities without taking into account any interaction of higher-order modes;
3. Full-wave analysis using the planar waveguide model.

The first method simply describes the step discontinuity by a transformer with constant transformer ratio. The second method includes step transition effects under the assumption that these are not interacting via higher-order modes, that is, the configuration can be considered as the cascade connection of two-step discontinuities. The third method represents a full-wave analysis for the microstrip circuit under study, so that, in particular, the interaction of the adjacent impedance

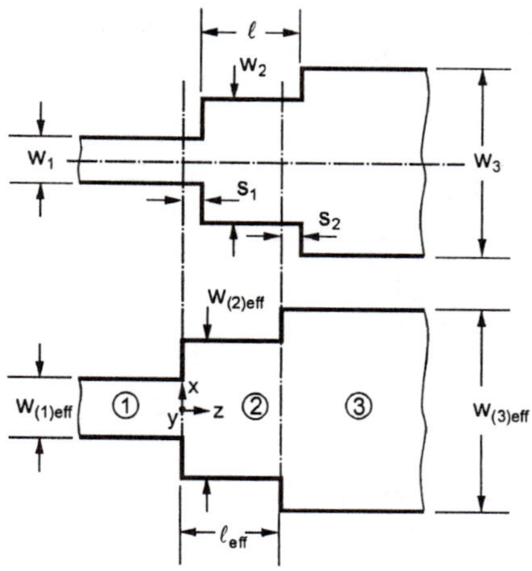

Figure 6.45 Quarter-wavelength transformer (above) and approximate model (below) with coordinate system.

steps via higher-order modes can be studied theoretically in dependence on frequency and geometry. Assuming ideal step discontinuities (the first method), zero reflection is obtained, as is well known, when the characteristic impedance Z_2 is equal to the geometric mean of the adjacent impedances Z_1 and Z_3, and the physical length l of the center line-section is nominally one-quarter wavelength long at a given frequency f_m,

$$Z_2 = \sqrt{Z_1 Z_3} \tag{6.98}$$

and

$$l = \frac{c_0}{4\sqrt{\epsilon_{eff_2}}\, f_m} \tag{6.99}$$

where f_m is the center frequency. The scattering coefficients can simply be derived from the wave transmission matrix T of the configuration, which is the product of the three matrices T_ν, with $\nu = 1, 2, 3$ describing the individual circuit sections. Figure 6.45, shows the approximate model for analysis. In the model, the effective dielectric constant and the effective widths are calculated as $\epsilon_{eff_\nu} = f(w_\nu, h, \epsilon_r)$ and $w_{eff_\nu} = f(w_\nu, h, \epsilon_r)$ $(\nu = 1, 2, 3)$. As can be seen from Figure 6.45 the electrical fringe field in the discontinuities moves the reference planes of the physical junctions. Equal amounts for the displacement at both steps are assumed, so that the effective length of the modeled section of line is equal to the physical length l. Neglecting the interaction of the adjacent steps, the characteristics can be simply evaluated from the known transmission scattering matrices of the homogeneous section of line and the step discontinuities.

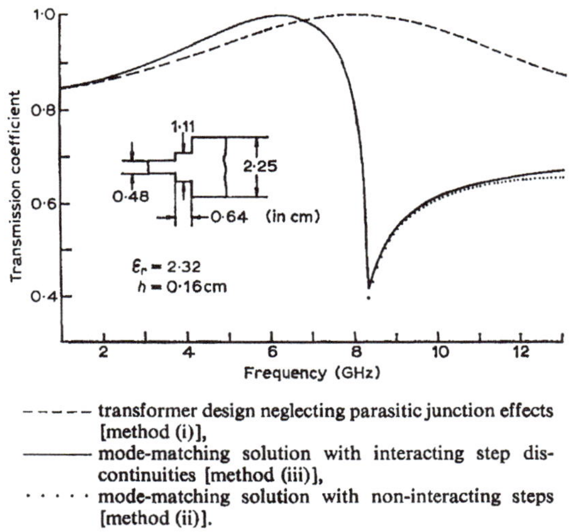

Figure 6.46 Simulated magnitude of the transmission coefficient of a quarter-wavelength transformer: (From: [39], © 1978 IEE. Reprinted with permission.)

Figure 6.46 shows an example for the calculated magnitude of the transmission coefficient of a quarter-wavelength transformer on Polyguide substrate ($\epsilon_r = 2.32$, h = 1.58 mm) [39]. The microstrip circuit, sketched in the diagram, has geometrical dimensions which may be used in practical microwave transistor amplifier design. The characteristic impedance of the output line is approximately 15Ω. The dashed line corresponds to simple transmission line theory (the first method), the center-frequency lying at 8 GHz. The solid line represents the mode-matching solution with interacting step discontinuities. In relation to the first method the computed transmission coefficient increases faster with frequency, indicating that the center-frequency is shifted to a lower frequency value. However, as can be seen from Figure 6.46, full matching is not obtained. The growing excitation of the first higher-order mode with frequency in the wider transmission line above approximately 6.5 GHz causes a sharp decrease of the dominant mode transmission coefficient. A minimum value of approximately 0.4 is found at 8.3 GHz. This frequency corresponds to the cutoff frequency of the first even higher-order mode in the wider line. As can be seen from the dotted line, which represents the analysis with noninteracting step discontinuities, the results are in good agreement with the third method. Thus, from the practical point of view, the steps may be considered to be decoupled, so that the calculation can simply be done by taking the product of transmission matrices of all line sections.

Figure 6.47 shows the microstrip circuit for measuring the scattering coefficients with a vector network analyzer. The wide transmission line is smoothly tapered on the right side to a 50Ω line over a length of about 10 cm to give a good matching to the output of 50Ω.

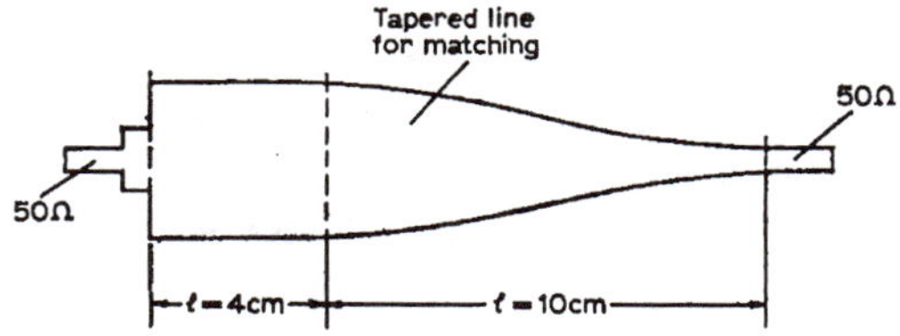

Figure 6.47 Measured microstrip quarter-wavelength transformer with tapered transition at the output.

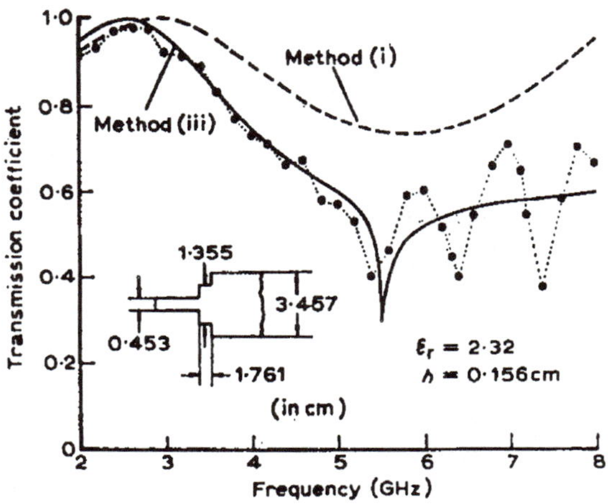

Figure 6.48 Simulated and measured magnitude of the transmission coefficient of a quarter-wavelength transformer with specified matching frequency of 3 GHz. Curve (i) is derived from ideal transmission line theory, and curve (iii) represents the full-wave analysis using the planar waveguide model. (From: [39], © 1978 IEEE. Reprinted with permission.)

Two microstrip circuits on Polyguide substrate ($\epsilon_r = 2.32$, $h = 1.58$ mm) have been designed to match a 50Ω microstrip line to a 10Ω transmission line at a frequency of 3 and 9 GHz, respectively. The geometric mean of the characteristic impedances yields a wave impedance of 22.4Ω for the quarter-wavelength section, the line length being 1.77 and 0.59 cm, respectively, including dispersion effects.

As can be seen from Figure 6.48, the measured center-frequency is obtained as 2.65 GHz. This is a deviation of approximately 12% from the value desired. However, full-wave analysis of the circuit shows good agreement with the measured data. Furthermore, a strong decrease of the transmission coefficient above the center-frequency is measured. This is due to the increasing higher-order mode excitation in the steps. A minimum of transmission is given at 5.5 GHz, which can be identified to be the cutoff frequency of the HE_2 mode in the widest line. Above

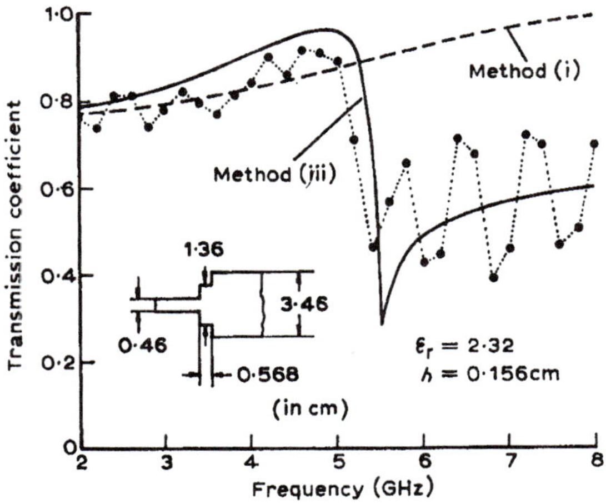

Figure 6.49 Simulated and measured magnitude of the transmission coefficient of a quarter-wavelength transformer with the center frequency above the cutoff frequency of the first even higher-order mode. (From: [39], © 1978 IEEE. Reprinted with permission.)

this frequency, the measured value oscillates, which is caused by HE_2 resonances between the largest step discontinuity and the inhomogeneous transformer (Figure 6.47).

Figure 6.49 shows the experimental and simulated result for the specified matching at 9 GHz. It can be seen that matching is not achieved. Due to the increasing excitation of the first higher-order mode with frequency, the dominant mode transmission coefficient strongly decreases approaching the cutoff frequency. The full-wave analysis shows good agreement with the experimental result. The results based on the second method above have not been plotted, because this is found to correspond closely to the solid line.

6.2.5 Stepped Microstrip Transformer

Figure 6.50 shows a straight-tapered stepped microstrip line. Because the change of widths of the step discontinuity is small, any interaction of the neighboring step discontinuities via higher-order modes can be neglected. The scattering coefficients of the dominant mode then can be simply derived from the overall wave transmission matrix T comprising the transmission matrices of the steps and the interconnecting lines. Figure 6.51 shows the simulated and measured transmission and reflection coefficients of an inhomogeneous transformer with a taperlength of 3 cm on Polyguide substrate (ϵ_r = 2.32, h = 1.58 cm). The smaller width was 0.46 cm and the wider 3.46 cm. The taper was divided into six equal sections of lines, as shown in Figure 6.50. The dashed lines reflect the simulation based on the ideal transmission line theory. The solid lines describe the simulated results using the

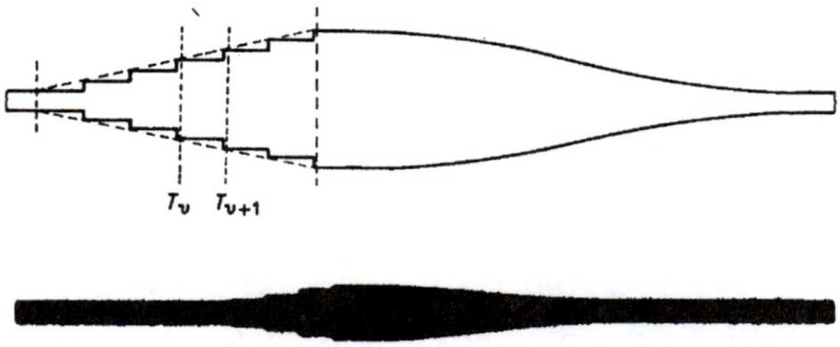

Figure 6.50 Stepped microstrip transformer comprising six abrupt impedance steps (above) and layout (below).

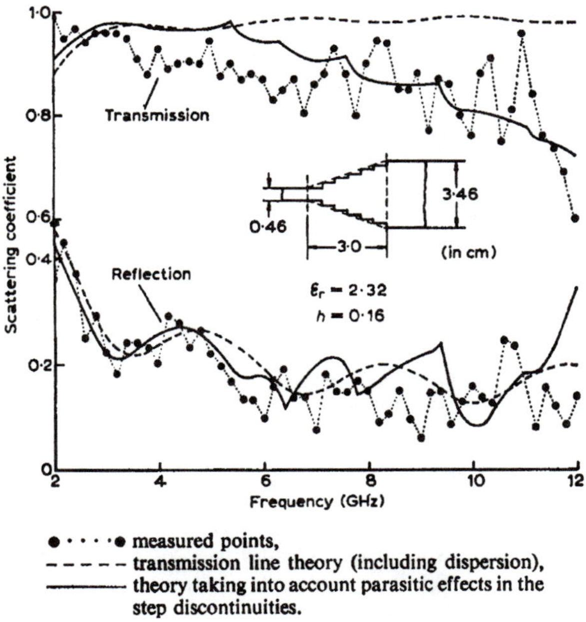

Figure 6.51 Computed and measured magnitude of the reflection and transmission coefficient of a stepped microstrip transformer on Polyguide. (From: [39], © 1978 IEE. Reprinted with permission.)

waveguide model. Higher-order mode interaction was not regarded. It can be seen that the latter approach closely agrees with the measured modulus of the transmission and reflection coefficient. The oscillations in the measured curves result from higher-order mode resonances above cutoff. It is also shown that the ideal transmission line theory is not suited to predict the frequency dependent electrical properties of the stepped transformer.

6.3 T-Junction

T-junctions are frequently found in microwave circuits—in stub filters, power dividers of antenna feeding networks, in stub matching networks, or branch-line couplers. In contrast to an ideal transmission line junction with no parasitic junction effects, the distributed microstrip T-junction includes parasitic capacitive, inductive, and radiation effects. Current flux lines on the strip and the ground plate and the electric and magnetic fields are disturbed in the vicinity of the junction.

Figure 6.52(a) shows a microstrip T-junction with different line widths in the main line. The symmetric structure shown in Figure 6.52(b) has the same characteristic impedance for the main through line and shunt branch arm. Figure 6.52(c) shows a T-junction with uniform main line and branch line with different characteristic impedance.

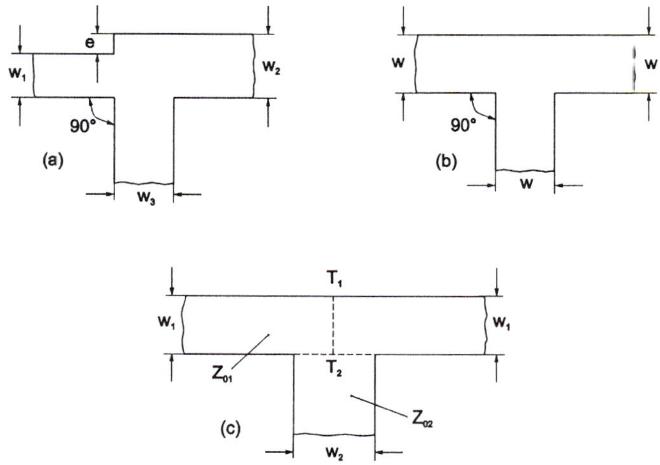

Figure 6.52 (a) T-junction with unequal connection lines, (b) symmetric T-junction with equal impedance of connection lines, and (c) T-junction with main line of width w_1 and shunted transmission line of width w_2.

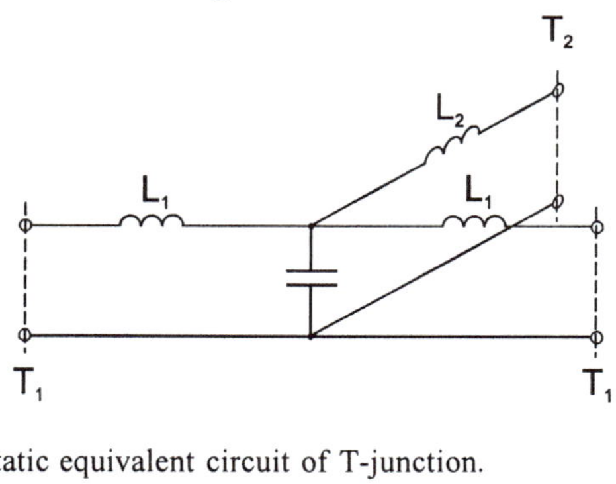

Figure 6.53 Quasi-static equivalent circuit of T-junction.

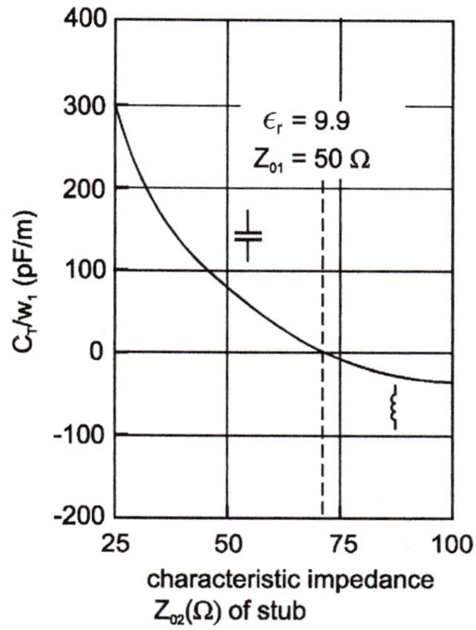

Figure 6.54 Capacitance of the equivalent circuit in Figure 6.53.

6.3.1 Quasi-Static Equivalent Circuit

Figure 6.52 shows different constructions of T-junctions often used in microstrip circuits. Figure 6.53 shows the quasi-static equivalent circuit of the T-junction with width w_1 for the main line and w_2 for the shunt branch. Formulas for the capacitance and for the inductances have been given by Garg and Bahl [40]. The junction capacitance theoretically calculated by Silvester and Benedek [41] is shown in Figure 6.54.

6.3.2 Dynamic Equivalent Circuit

Experimental work has been done by Hammerstad and Bekkadal on symmetric microstrip T-junctions on 1/8" Rexolite 2200 substrate [6]. Main-line impedance was 50Ω and stub impedance varied from 40Ω to 110Ω in 10Ω steps. They have published an equivalent circuit for the T-junction shown in Figure 6.55(a). It comprises a shunt admittance B_T and a transformer with transformation ratio n. The locations of the electrical reference planes are illustrated in Figure 6.55(b), where w_{eff_i} $(i = 1,2)$ is the electrostatic effective width of the planar waveguide model,

$$w_{eff_i} = \sqrt{\frac{\mu_0}{\epsilon_0}}\frac{1}{\sqrt{\epsilon_{eff_i}}}\frac{h}{Z_{0_i}} \qquad (i = 1,2)$$

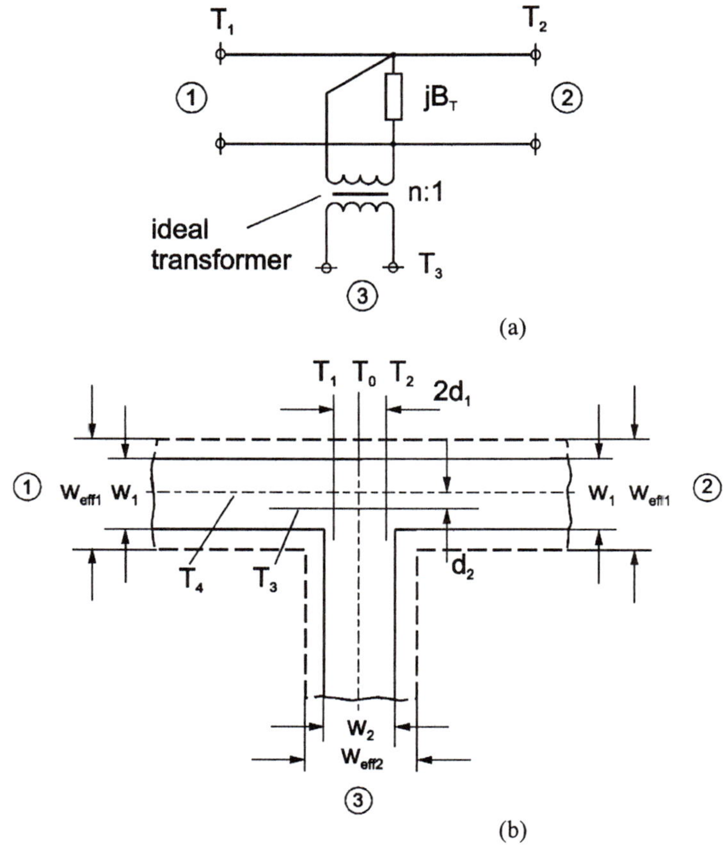

Figure 6.55 (a) Equivalent circuit of a symmetric T-junction and (b) definition of reference planes (After: [6]).

Semiempirical expressions have been given for the junction shunt admittance, the shifts in the locations of the electrical reference planes, as well as the transformation ratio n. The shunt admittance reads

$$B_T = \begin{cases} -\dfrac{w_{eff_1}}{\lambda_1 Z_{0_2}}\left(1 - 2\dfrac{w_{eff_1}}{\lambda_1}\right) & \text{for } Z_{0_1} \le 0.5\, Z_{0_2} \\[2ex] -\dfrac{w_{eff_1}}{\lambda_1 Z_{0_1}}\left(2 - 3\dfrac{Z_{0_1}}{Z_{0_2}}\right)\left(1 - 2\dfrac{w_{eff_1}}{\lambda_1}\right) & \text{for } Z_{0_1} \ge 0.5\, Z_{0_2} \end{cases} \tag{6.100a}$$

where

$$\lambda_1 = \lambda_0 / \sqrt{\epsilon_{eff_1}}$$

$$w_{eff_i} = \sqrt{\frac{\mu_0}{\epsilon_0}}\,\frac{h}{\sqrt{\epsilon_{eff_i}}\,Z_{0_i}(\epsilon_r)} = \sqrt{\frac{\mu_0}{\epsilon_0}}\,\frac{h}{Z_{0_i}(\epsilon_r = 0)} \qquad (i = 1,2)$$

The transformer turns ratio n is defined as

$$n = si(x) = \frac{\sin(x)}{x} \tag{6.100b}$$

with

$$x = \frac{\pi}{2}\,\frac{2w_{eff_1}}{\lambda_{\varepsilon_1}}\,\frac{Z_{0_1}}{Z_{0_2}}$$

and the placement of the reference planes read

$$d_1 = 0.05\, w_{eff_2}\,\frac{Z_{0_1}}{Z_{0_2}}\, n^2 \tag{6.100c}$$

$$d_2 = \frac{w_{eff_1}}{2} - 0.16\, w_{eff_1}\,\frac{Z_{0_1}}{Z_{0_2}}\left[1 + (2w_{eff_1}/\lambda_{\varepsilon_1})^2 - 2\ln\frac{Z_{0_1}}{Z_{0_2}}\right] \tag{5.100d}$$

In Figures 6.56 to 6.60 the T-junction model parameters are plotted versus the impedance ratio Z_{01}/Z_{02} for discrete values of $2w_{eff_1}/\lambda_1$ in the range from 0 to 0.9. Notice that the location of the reference planes T_1, T_2 (main-line reference plane displacements), and T_3, determined by the parameters d_1 and d_2, and the model parameters B_T and n are frequency dependent.

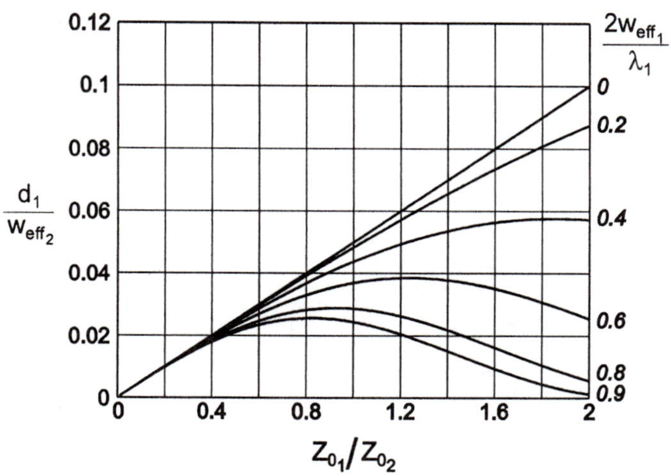

Figure 6.56 Location of reference plane T_1 and T_2 for a T-junction as a function of Z_{01}/Z_{02}.

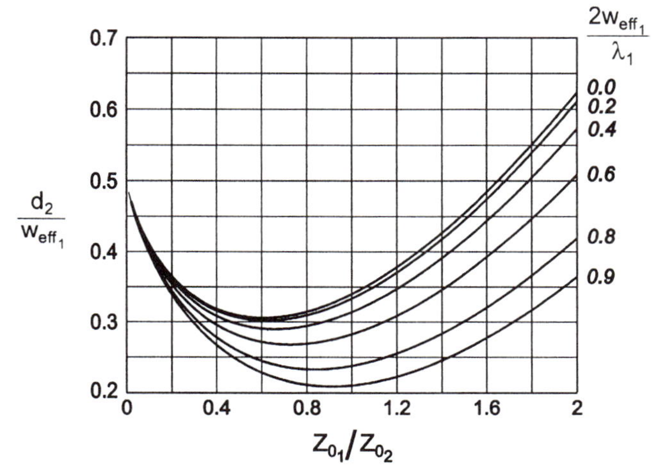

Figure 6.57 Location of reference plane T_3 (distance from symmetry axis) for a T-junction as a function of Z_{01}/Z_{02}.

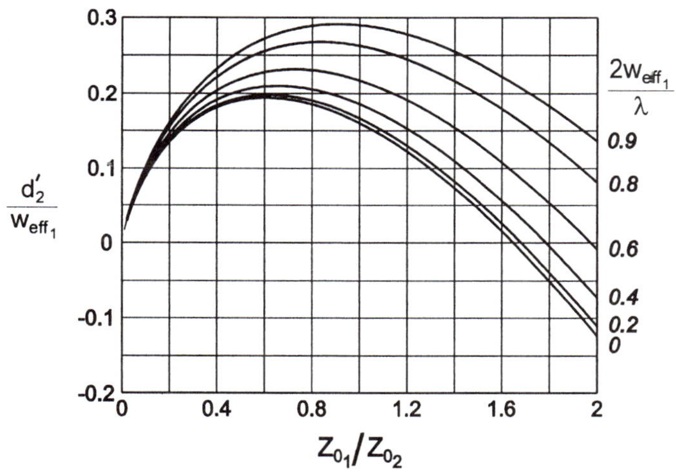

Figure 6.58 Location of reference plane T_3 (distance from the magnetic wall) for a T-junction as a function of Z_{01}/Z_{02}.

As has been pointed out by Hammerstad and Bekkadal, the proposed formulas yield good results if $Z_{01}/Z_{02} < 2$ (Z_{01} is large main transmission line). Otherwise, for increasing Z_{01}/Z_{02}, the values for d_2/w_{eff_1} become too large. Therefore, when $Z_{01}/Z_{02} > 2$, it is recommended to replace the Z_{01}/Z_{02} ratio by its inverse in the expression for d_2. The offset in the main line is usually very small, and the main effect is on the length of the stub.

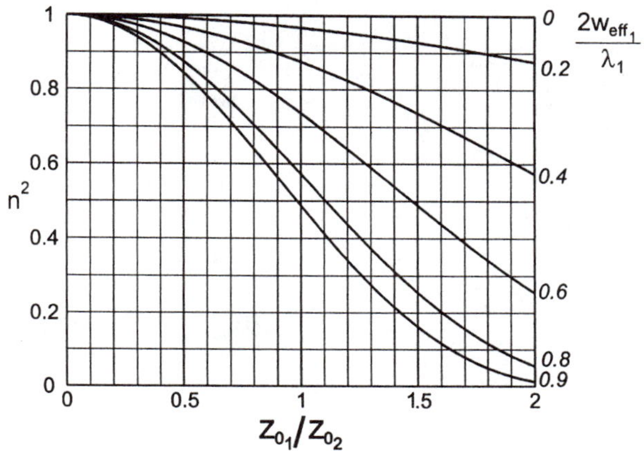

Figure 6.59 Transformation ratio for a T-junction as a function of Z_{01}/Z_{02}.

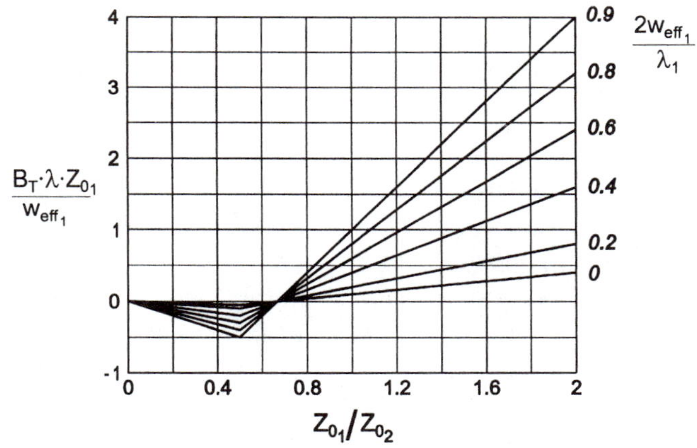

Figure 6.60 Normalized shunt susceptance B_T for a T-junction as a function of Z_{01}/Z_{02}.

6.3.3 Compensated T-Junction

The capacitance of a T-junction could be reduced by introducing a slit across the width of the main-through microstrip line [see Figure 6.62(d)], opposite the branch arm [42, 43]. Dydyk [44] has described a T-junction compensation technique illustrated in Figure 6.61. The reduction of the width of the main-straight line near the junction acts like a lumped inductance and therefore takes care of compensation of the parasitic shunt capacitance. The wider line length in the shunt branch acts as a quarter-wavelength transformer, and matches the branch line to the main line.

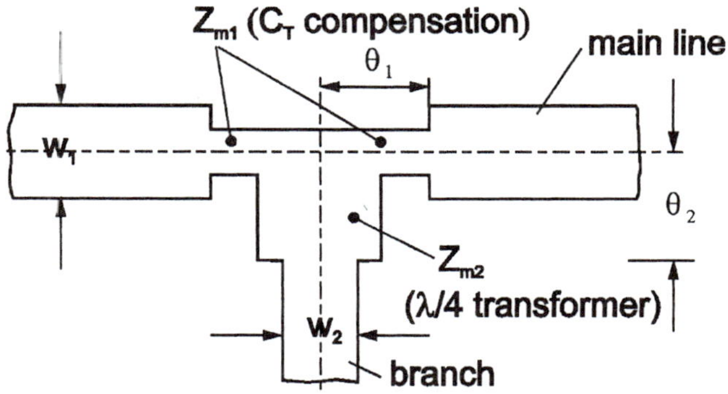

Figure 6.61 Compensated T-junction ($\theta_1 < \lambda/10$) [39].

The electrical lengths θ_1 and θ_2, and the characteristic impedances (Z_{m1}, Z_{m2}) of the modified T-junction, can be deduced from the reactances of the discontinuity [44]. The shift of the reference planes of the T-junction can be compensated by corresponding changes of the connecting lines.

6.3.4 Special Structures

In Figure 6.62, some special designs of microstrip junctions are presented [45]. Figure 6.62(a) describes a symmetric junction with connection lines of equal widths and equal inclination of 120°. The construction shown in Figure 6.62(b) is similar, however, with two different values of the characteristic impedance. Figure 6.62(c) presents a junction with arbitrary apex angle α.

A T-junction with reduced coupling aperture is shown in Figure 6.62(d). A similar construction has been used in [43] for the design of stub filter elements with a slit width $s = w_2$ and slit depth d as variable. It has been found that the filter response is no longer symmetric with respect to the stopband center frequency. The slit causes two effects. It reduces the coupling aperture of the resonating stub leading to a higher Q factor (smaller bandwidth) and thus a sharper transition from bandpass to bandstop, and vice versa. Moreover, the slit also supports the excitation of higher-order modes on the main line. Excitation conditions are similar to those occurring in offset junctions [46], where the first higher-order mode of a 50Ω line on Polyguide substrate can severely influence the overall response at X band frequencies (see Figure 5.8). The frequency-dependent transmission coefficient of the slit-designed stub looks like the characteristic of a lowpass filter.

Figure 6.62(e) shows a T-junction with rounded edge to minimize the excess fringing electrical field near the strip corner [47]. All shown junctions with the exception of that in Figure 6.62(e) could be analyzed using the planar waveguide model [30, 48].

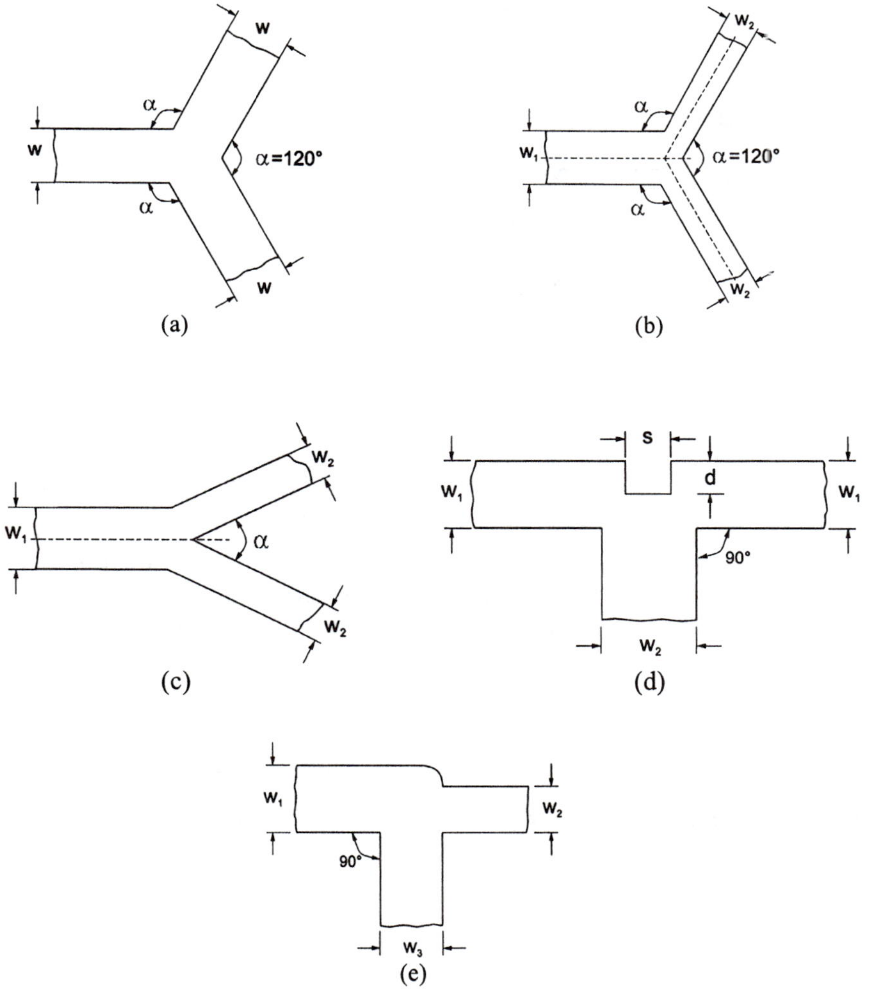

Figure 6.62 Special microstrip junctions. (a) 120° junction with microstrip transmission lines of equal characteristic impedance. (b) 120° junction with different microstrip transmission lines. (c) Junction with arbitrary angle α and different microstrip transmission lines. (d) T-junction with reduced coupling aperture. (e) T-junction with rounded edge to reduce parasitic field distortion.

6.4 Stub

When the branch arm of a T-junction is terminated by an open-ended microstrip transmission line, the composed circuit acts like a filter. As illustrated in Figure 6.63, the electrical length l_{leff} is calculated as

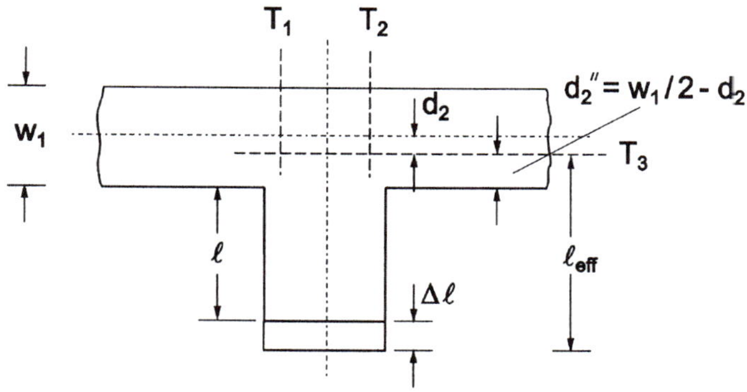

Figure 6.63 Reference planes and relevant geometric stub parameters.

$$l_{eff} = l + d_2'' + \Delta l \tag{6.101}$$

where l is the physical length of the shunt arm, d_2'' the location of the stub reference plane with respect to the strip edge of main line, and Δl the line extension of the open-ended stub. When the electrical length is

$$l_{eff} = (2n - 1)\frac{\lambda}{4} \tag{6.102}$$

where $n = 1, 2, 3, \ldots$, then the microstrip circuit behaves like a band-stop filter at the resonance wavelength λ. With the stub resonant frequency $f = c_0/\left(\lambda\sqrt{\epsilon_{eff}(f)}\right)$, the resonance condition (6.102) reads

$$f = \frac{(2n - 1)}{4}\frac{c_0}{\sqrt{\epsilon_{eff}(f)}\, l_{eff}} \tag{6.103}$$

At resonance frequency, no power is transmitted. All incident power on the main line is reflected. The circuit acts as a bandstop filter.

At resonance frequencies, where the effective stub length is a multiple of half-wavelengths,

$$l_{eff} = n\frac{\lambda}{2} \tag{6.104}$$

where $n = 0, 1, 2, \ldots$, all incident power is transmitted. The circuit acts as a bandpass filter. Figure 6.64 shows an example of the measured magnitude of

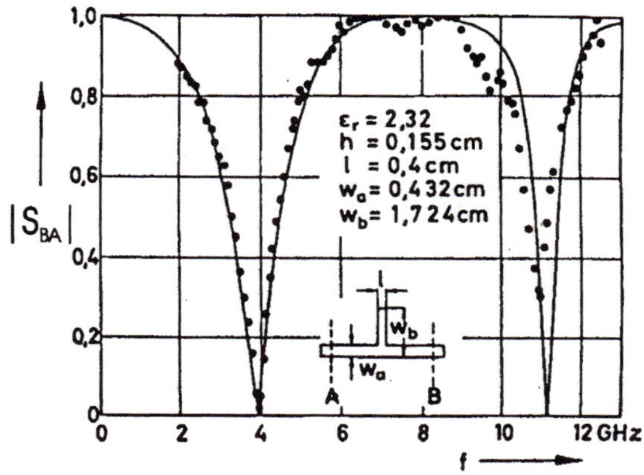

Figure 6.64 Measured and calculated (planar waveguide model) transmission coefficient of a stub on Polyguide substrate material (w_1 = 4.3 mm, w_2 = 4.0 mm, l = 12.9 mm).

transmission coefficient of a stub filter element on Polyguide substrate material ($\epsilon_r = 2.32$, $h = 1.6$ mm) [32]. A transmission zero is found at 4 GHz when the effective stub length is one quarter-wavelength. At the third quarter-wavelength resonance (approximately 11 GHz), no complete power transfer suppression is observed. This may be explained by the distributed nature of the T-junction. Because of a shorter wavelength with increasing frequency, obtaining an electrical short on the main line over its cross section becomes more difficult with ascending number of quarter-wavelengths.

Note that the range of usable characteristic stub impedance values is severely limited due to higher-order mode excitation. Therefore, operation should be far below the cutoff frequency of the first higher-order $H_{[10]}$ (HE_1) mode on the stub line (see Chapter 5). A very low impedance line is commonly avoided by shunting two stubs of double impedance value (open-ended cross-junction).

Problem: A quarter-wavelength stub filter element shown in Figure 6.65 shall be designed for a stopband frequency of 4 GHz. The substrate material used is Polyguide with a dielectric constant $\epsilon_r = 2.32$ and substrate height $h = 1.6$ mm. The characteristic impedance of the main line Z_{0_1} is assumed to be 50Ω, and that of the branch line $Z_{0_2} = 30\,\Omega$. Determine the physical stub length l.

Solution: The analysis of the problem under discussion is done under different assumptions, namely (1) design on a quasi-static approach, and (2) design taking into account dispersion effects. The final comparison with ADS simulation [49] will give us experience in the respective uncertainty of a design based on the data available in this book.

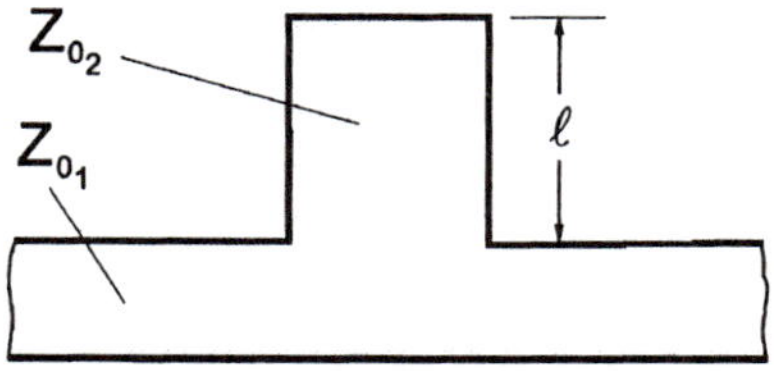

Figure 6.65 Design of single stub filter element.

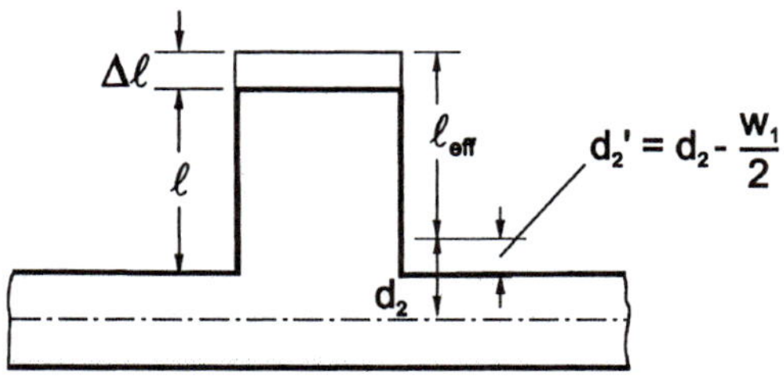

Figure 6.66 Relevant geometric parameters for microstrip stub design.

(1) *Quasi-static design*, using formula for characteristic impedance, effective dielectric constant, symmetric T-junction, and open-end effect after Hammerstad: From the table presented in Appendix F, we find for the given characteristic impedance values:

$Z_{0_1} = 50\,\Omega$: $w_1/h = 2.983\text{ mm},\ w_1 = 2.983h = 4.7728\text{ mm},\ \epsilon_{eff_1} = 1.968,$
$w_{eff_1} = 8.5994\text{ mm},\ \lambda_1 = 53.4267\text{ mm}$

$Z_{0_1} = 30\,\Omega$: $w_2/h = 6.044\text{ mm},\ w_2 = 6.044h = 9.6704\text{ mm},\ \epsilon_{eff_2} = 2.054$

where w_{eff_1} has been calculated according to (3.47), taking into account the effective dielectric constant,

$$w_{eff_1}(0) = \frac{120\pi}{\sqrt{\epsilon_{eff_1}(0)}} \frac{h}{Z_{0_1}(\epsilon_r)}$$

The wavelength λ_1 has been calculated due to

$$\lambda_1 = \frac{c_0}{f_0\sqrt{\epsilon_{eff_1}(0)}}$$

From the given resonance frequency of $f_0 = 4\text{ GHz}$ we calculate the effective stub length l_{eff},

$$l_{eff} = \frac{c_0}{4 f_0 \sqrt{\epsilon_{eff_2}(0)}} = 13.074 \text{ mm}$$

The open-end effect, calculated from (6.2) after Hammerstad, yields $\Delta l_1 = 0.4966 h = 0.7945$ mm. The reference plane for the stub is determined by the parameter d_2, defined in Figure 6.66 and Figure 6.55. Using the design formula (6.100) for the symmetric T-junction, d_2/w_{eff_1} is calculated as 0.473, and $d_2 = 0.483\, w_{eff_1} = 4.111$ mm. Then the parameter d_2' yields 1.725 mm. Thus, the physical length turns out to be

$$l_1/\text{mm} = l_{eff} + d_2' - \Delta l_1 = 13.074 + 1.725 - 0.794 = 14.005$$

Taking into account the open-end effect after Kirschning (see Figure 6.9), the open-end extension reads $\Delta l_2 = 0.6 h = 0.96$ mm. Thus, the physical length is changed to $l_2/\text{mm} = 13.839$.

(2) *Design including dispersion:* With the normalized frequency $f \cdot h$ = 4 GHz $\cdot$ 1.6 mm = 6.4 GHz mm, we read from Figure 4.3: $\epsilon_{eff_1}(4 \text{ GHz}) = 1.981$, $\epsilon_{eff_2}(4 \text{ GHz}) = 2.09$. The wavelength, effective stub length, and open-end extension are then calculated as $\lambda_1 = 53.251$ mm, $l_{eff} = 12.961$ mm, and $\Delta l_3 = 0.4943 h$ = 0.7908 mm, respectively. The dispersion of the characteristic impedances Z_{0_1} and Z_{0_2} is negligible, as can be seen in Figure 4.13. With $w_{eff_1} = 8.5711$ mm, the T-junction parameters read $d_2 = 0.47814 w_{eff_1} = 4.1085$ mm and $d_2' = 1.7221$ mm, respectively. Thus, the physical length taking into account dispersion effects reads

$$l_3/\text{mm} = l_{eff} + d_2' - \Delta l_3 = 13.074 + 1.774 - 0.794 = 14.054 \text{ mm}$$

Using open-end extension $\Delta l_4 = 0.96$ mm after Kirschning, results in a length $l_4 = 13.889$ mm.

For the derived stub lengths l_1, l_2, l_3, and l_4, schematic simulation was performed with ADS using the different models implemented for the T-junction, namely the MDS T-junction model MTEE and the Advanced Design System (ADS)/libra T-junction model. For comparison, 2.5D EM Momentum simulator was used to search for the optimum stub length resonating at the specified frequency of 4 GHz. Figure 6.67 shows the result. It can be seen that the MDS model implemented simulations reveal a relative frequency shift of approximately 0.1 GHz to lower frequencies. The relations used for the open-end effect yield a frequency shift of 0.05 GHz.

For completion, simulations were also performed with ADS Momentum. In Figure 6.67(c), it can be seen that the resonance is even more shifted to lower frequencies. Thus, optimizing the stub length for the specified center frequency of

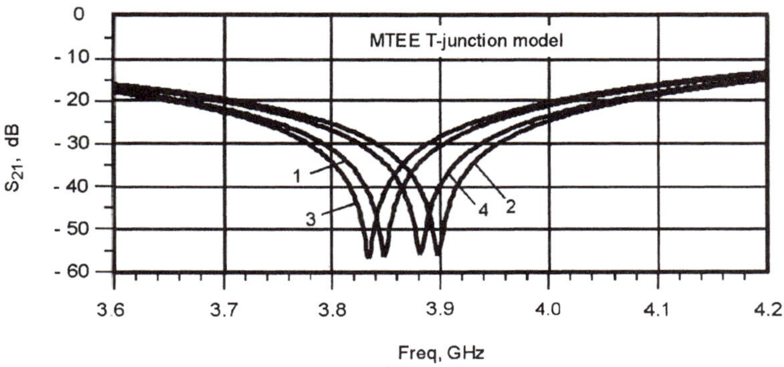

(a)

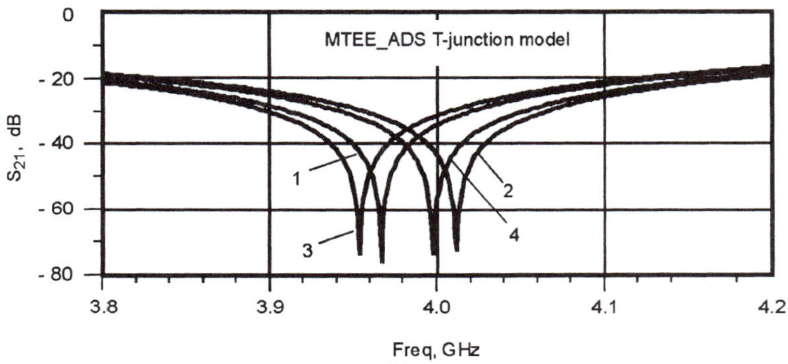

(b)

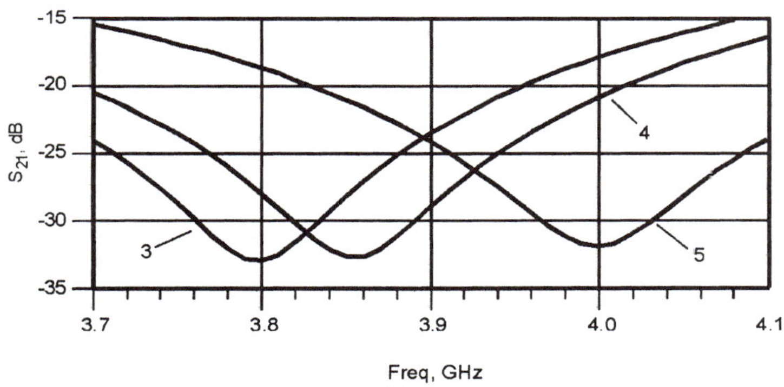

(c)

Figure 6.67 Schematic simulation with ADS [version 2003(C)] using (a) the MDS T-junction model MTEE, (b) ADS/libra T-junction model, and (c) 2.5D EM simulation using Momentum. Curve notation denotes as follows: (1) Purely quasi-static design (l_1 = 14.005 mm, $Z_0(\varepsilon_r)$, $\varepsilon_{eff}(f = 0)$, $\Delta l/h$ after Hammerstad); (2) Same as curve 1, but $\Delta l/h$ after Kirschning (l_2 = 13.839 mm); (3) Design including dispersion (l_3 = 14.054 mm, $Z_0(\varepsilon_r, f)$, $\varepsilon_{eff}(f)$, $\Delta l/h$ after Hammerstad); (4) Same as curve 3, but $\Delta l/h$ after Kirschning (l_4 = 13.889 mm); and (5) Optimization with Momentum (l_5 = 13.2 mm).

4 GHz yields a shortest stub length of l_5 = 13.2 mm, which differs strongly from the values discussed before. Thus, it is recommended to use EM simulators in addition to those in the schematic simulation, in particular with larger discontinuities and junctions, and when cross talk and coupling are expected. It is assumed that computing time for EM circuit analysis and optimization will be consistently improved, to solve design problems within an acceptable execution time.

Consulting the ADS manual, the definition of the MDS T-junction model is not fully clear. The equivalent circuit shown comprises reactances that indicate a quasi-static model [45], whereas the MTEE_ADS T-junction model is a dynamic model proposed by Hammerstad in 1982 [50]. This refers to the generalized asymmetric T-junction. However, in the special case of symmetric configuration as used in this example, it delivers approximately the same result as the model treated in Section 6.3.2. Frequency dispersion of the effective dielectric constant of the stub is calculated on data from [51–54]. Open-end effect is modeled after [7].

6.5 Bend

The manufacturing cost of a microstrip circuit is mainly determined by the material cost of the substrate. Therefore, it is important to consider efficient use of the substrate area in addition to the fulfillment of the electrical requirements. This makes it necessary that the circuit elements on a substrate often cannot be interconnected in a straight line. Circuit elements are commonly placed together with a high package density that requires the inclusion of bends in the interconnecting transmission lines.

6.5.1 Right-Angled Bend

A right-angled or 90° bend is shown in Figure 6.68. Figure 6.69 shows the quasi-static equivalent circuit. In the vicinity of the corner of the bend, the electromagnetic field and current flux lines are disturbed. Near the inner corner edge, current flux line density is increased. At the outer corner, excess charge is

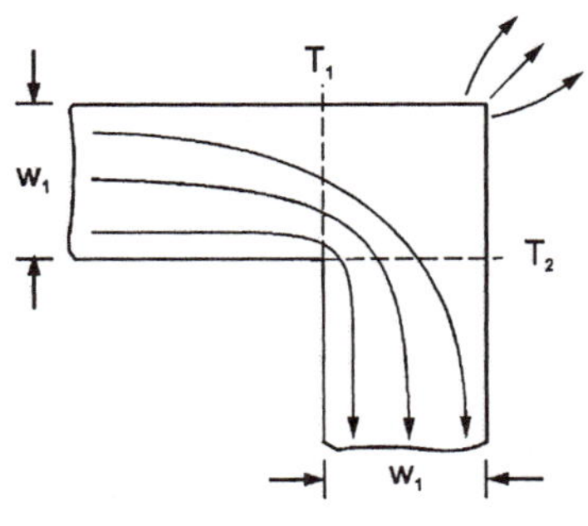

Figure 6.68 90° bend with equal microstrip lines.

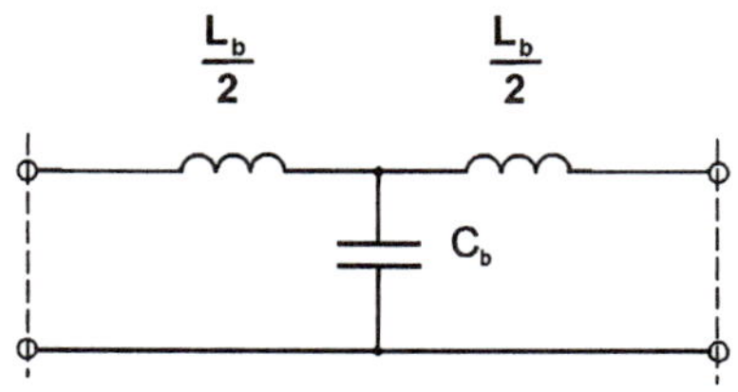

Figure 6.69 Quasi-static equivalent circuit of 90° bend with equal microstrip lines.

accumulated and additional electrical fringe field occurs. The series inductance L_b of the equivalent circuit represents the current flux distortion, and the parallel capacitance C_b represents the distortion of the electrical field. T_1 and T_2 denote the reference planes of the microstrip discontinuity.

Closed-form expressions for the right-angled bend capacitance and inductance have been published by Gupta et al. [26]:

$$\frac{C_b/w}{\mathrm{pF/m}} = \begin{cases} \dfrac{(14\epsilon_r + 12.5)\dfrac{w}{h} - (1.83\epsilon_r - 2.25)}{\sqrt{w/h}} + \dfrac{0.02\epsilon_r}{w/h} & \text{for} \quad w/h < 1 \\ (9.5\epsilon_r + 1.25)\dfrac{w}{h} + 5.2\,\epsilon_r + 7 & \text{for} \quad w/h \geq 1 \end{cases} \tag{6.105}$$

$$\frac{L_b/h}{\mathrm{nH/m}} = 100(4\sqrt{w/h} - 4.21) \tag{6.106}$$

Regarding the following geometry and material data, an accuracy of 5%, and 3% respectively, is quoted:

$$C_b: \quad 2.5 \leq \epsilon_r \leq 15, \qquad 0.1 \leq w/h \leq 5 \tag{6.107}$$

$$L_b: \quad 0.5 \leq w/h \leq 2 \tag{6.108}$$

Problem: Assume a right-angled bend with a microstrip line of 0.75 mm width, and fabricated on a 0.5-mm thick alumina substrate whose permittivity is 9.9. Calculate the values of the equivalent circuit elements.

Solution: With $w/h = \mathbf{0.75\,mm/0.5\,mm} = 1.5 > 1$, the lower equation of (6.105) must be applied. The capacitance C_b is $\mathbf{0.15\,pF}$. The bend inductance L_b is calculated from (6.106) and gives a value of $\mathbf{0.034\,nH}$. Although the equivalent circuit parameters may have small values, their reactances, even at 12 GHz, have a significant effect on the electrical properties of the complete microstrip circuit.

For instance, the inductance at 12 GHz:

$$\omega L = 2\pi \cdot 12 \cdot 10^9 s^{-1}\, 0.034 \cdot 10^{-9}\, \Omega \cdot s^{-1}$$
$$= 2.5\Omega$$

and for the capacitance:

$$\frac{1}{\omega C} = \frac{1}{2\pi \cdot 12 \cdot 10^9 s^{-1}\, 0.15 \cdot 10^{-12}\, \Omega^{-1} \cdot s}$$
$$= 88.4\Omega$$

It should be noted that both values are significant compared to the characteristic impedance of the microstrip line, having a value of some tens of ohms.

6.5.2 Compensated Bend

Bends should be optimally designed in such a way that an incident wave is fully transmitted. Therefore, for practical use in microstrip circuits, compensated bends are needed. Reduction of the parasitic effect of the capacitance can simply be achieved by mitering the bend.

Figure 6.70 shows a chamfered right-angled bend with the reference planes T_1 and T_2. The chamfer width is denoted as b and the chamfer length as w_c. b' is an auxiliary quantity. The equivalent quasi-static circuit is similar to that shown in Figure 6.69. Figure 6.71 illustrates the measured model parameter values of the equivalent electrical line-length l_c and shunt susceptance B, as a function of the chamfer fraction $1 - b/(\sqrt{2}w)$ [55, 56]. It can be seen that l_c increases with growing chamfering. Because of this, the degree of chamfer is generally restricted to around [42]

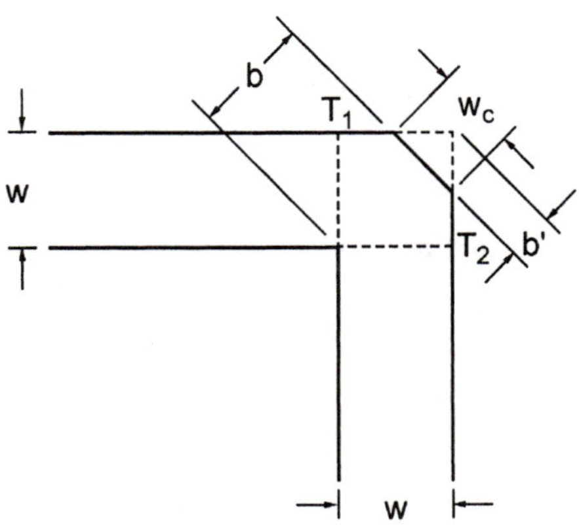

Figure 6.70 Geometric parameters of the right-angled bend.

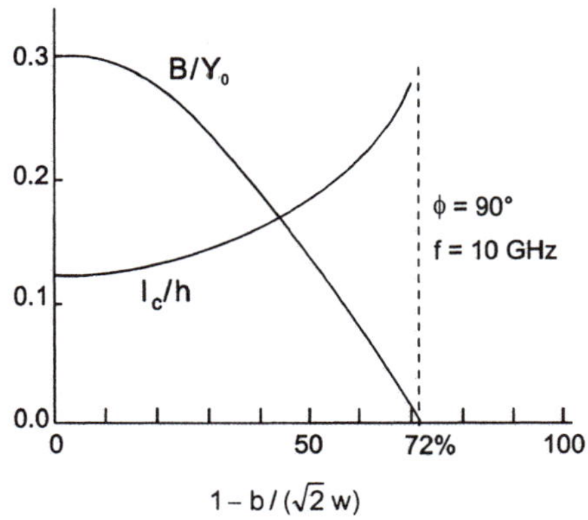

Figure 6.71 Equivalent circuit parameter variation as a function of amount of chamfer of a chamfered right-angled bend on Al_2O_3 substrate at a frequency of 10 GHz (Z_0 = 50Ω). (From: [55], © 1973 European Microwave Association. Reprinted with permission.)

$$1 - b/(\sqrt{2}w) \simeq 0.6 \tag{6.109}$$

so that $b \simeq 0.57\,w$ which corresponds to a chamfer length $w_c = 1.7\,w$. Although experimental study was performed on alumina substrate, comparable reduction of the capacitive effect with chamfering is also expected with other substrate materials [42].

Problem: Confirm that for a chamfer width $b/w \simeq 0.57$, the chamfer length is obtained as $w_c = 1.7\,w$.

Solution: From Figure 6.70, the following relations can be derived:

$$b' = w_c/2$$
$$b = \sqrt{2}\,w - b' = \sqrt{2}\,w - w_c/2$$
$$w_c = 2\,(\sqrt{2}\,w - b)$$

For $b = 0.57\,w$, w_c is calculated as $w_c = 2\,(\sqrt{2} - 0.57)\,w = 1.69\,w$.

In [57], a chamfer length of $w_c = 1.83$ was experimentally derived for low reflection chamfered bends on Al_2O_3 substrate. The voltage standing wave ratio (*VSWR*), also denoted as s, reads

$$\text{VSWR} = s = \frac{1 + |\Gamma|}{1 - |\Gamma|} \tag{6.110}$$

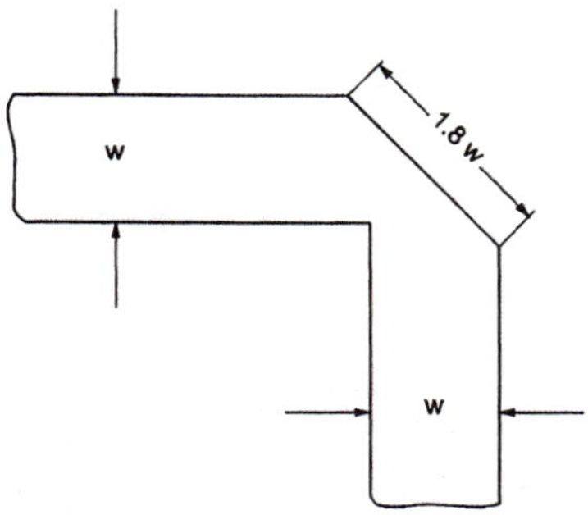

Figure 6.72 Compensated microstrip bend for practical use.

with the input reflection coefficient Γ. The magnitude of Γ as a function of s is then

$$|\Gamma| = \frac{s - 1}{s + 1} \tag{6.111}$$

In [57], a **VSWR** ≤ 1.11 was measured up to 12 GHz, which corresponds to a reflection of about 5%. Regarding various angles ($30°$, $60°$, $90°$, and $120°$) on rexolite with 50Ω connection lines, Hammerstad [6] showed that optimum value is about $w_c = 1.8w$. Thus, this value seems to be independent of the bend angle [26]. For practical design the layout illustrated in Figure 6.72 is recommended.

A more elaborate relation for the chamfer length w_c was published by Abrie [58]:

$$w_c/w = \sqrt{2}\left(1.04 + 1.3e^{-1.35\,w/h}\right) \tag{6.112}$$

This optimum miter can only be used when the microstrip line width is not too small (i.e., $w/h \geq 0.25$ and $\epsilon_r \leq 25$). The equivalent electrical length of the miter l_c can be estimated by using the equation [58]

$$l_c = w_c/\sqrt{2} \tag{6.113}$$

Figure 6.73 shows simulated results for chamfer length values w_c = 1.7, 1.8, and 1.97. The last value is calculated from (6.112) for the regarded alumina substrate ($\epsilon_r = 9.7$, $h = 0.635$ mm, $w = 0.611$ mm). The width has been chosen to deliver a characteristic value of 50Ω at a frequency of 500 MHz. Simulation was performed with ADS [Version 2003(C)]. It can be noticed that the miter length of 1.7 results in a negligible VSWR value up to 20 GHz.

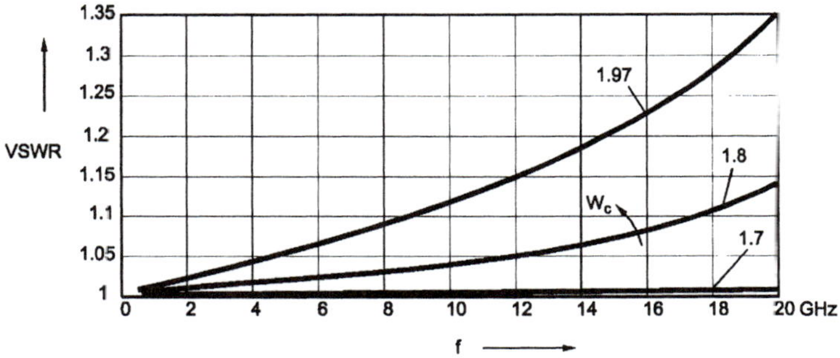

Figure 6.73 Simulated VSWR versus frequency of a microstrip bend with the chamfer length w_c as parameter.

6.5.3 Curved Microstrip Bend

Curving a microstrip transmission line as shown in Figure 6.74 is frequently a better option than mitering it [59]. When the curving radius is larger than twice the width of the line, the main parasitic effect is a change in the effective length. The effective length of the curve $(3 < R/w < 7)$ can be estimated by assuming the effective radius to be [56]

$$R_e = R_i + 0.3w \tag{6.114}$$

Curving a line also has the advantage that the direction of the line can be changed with an arbitrary angle. Theoretical work on the curved microstrip bend has been done by Weisshaar and Tripathi. using a planar waveguide model [60]. Figure 6.75(a) shows the circuit under investigation and Figure 6.75(b) shows the defined model for analysis. In the waveguide model, the effective mean radius of curvature R_e is related to the effective microstrip width w_e as

$$R_e = \frac{R}{2} + \frac{1}{2}\sqrt{R^2 + (w_e - w)\,w} \tag{6.115}$$

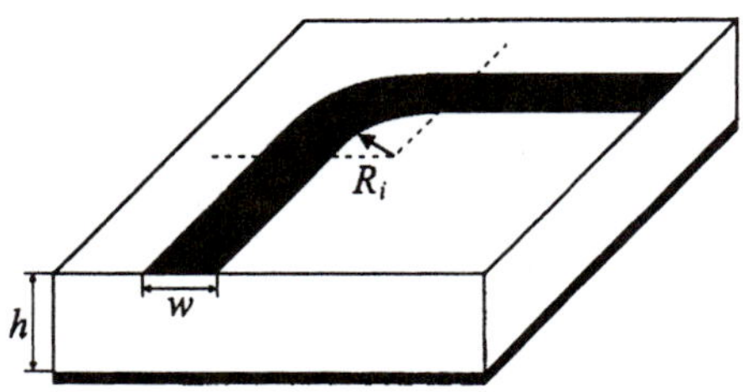

Figure 6.74 Curved microstrip bend.

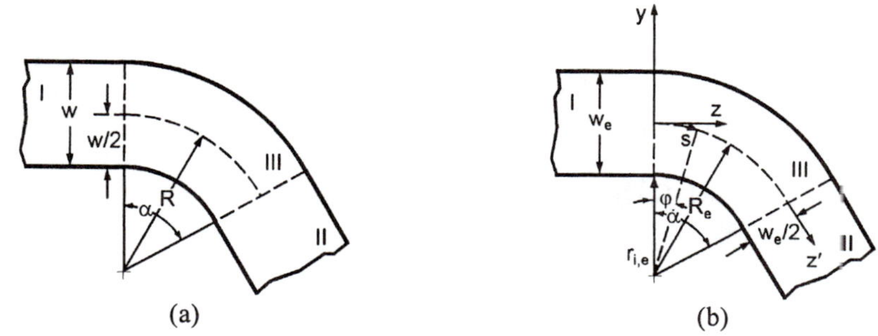

Figure 6.75 (a) Nomenclature of physical microstrip bend and (b) waveguide model for curved microstrip bend analysis. (From: [60], © 1990 IEEE. Reprinted with permission.)

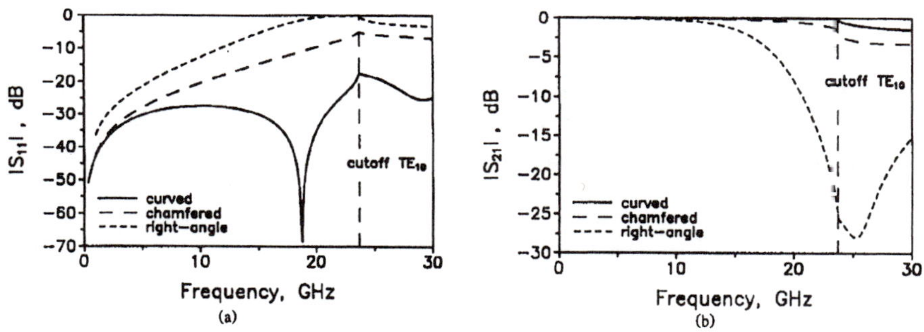

Figure 6.76 Magnitude of (a) the reflection coefficient and (b) the transmission coefficient as a function of frequency for a curved 90° microstrip bend with the line parameters w = 1.2 mm, h = 0.635 mm, ε_r = 9.8, R_i = 1.8 mm, $(R_a - w/2)/w$ = 2. (From: [60], © 1990 IEEE. Reprinted with permission.)

R_e is chosen such that there is always a positive effective inner radius regardless of actual inner and outer radii. The frequency-dependent behavior of the curved bend is determined using a second-order perturbation analysis of the waveguide model, together with a mode-matching technique that includes the higher-order modes [4].

Figure 6.76 shows the frequency-dependent properties of a curved microstrip bend on Al_2O_3 substrate material ($Z_0 = 35\Omega$). For comparison, the results for the uncompensated and chamfered right-angle are shown as well. It can be seen that the curved bend shows improved transmission properties. Referring to the solid line, at a frequency of approximately 23.5 GHz, a singularity is observed in the traces of the scattering coefficients, which originates in the cutoff frequency of the $H_{[10]}$ mode of the waveguide model used for analysis. A smooth round corner will produce no appreciable reflection. The radiation losses are of the order of 0.15 dB [60].

6.6 Quasi-Lumped Elements

Quasi-lumped capacitive and inductive components are commonly used up to approximately about 20 GHz in hybrid, and approximately 30 GHz in monolithic technology [61]. This is traced back to the growing influence of parasitics with frequency. The structures behave approximately as lumped elements as long as the maximum dimensions are small compared with the wavelength. As a rule of thumb, the dimensions should not exceed $\lambda/10$.

Quasi-lumped elements are often used in microstrip filters and bias networks. Unlike distributed components, quasi-lumped capacitive and inductive elements are less frequency-dependent, so that they are used for the design of broadband microstrip circuits. In the following section, we will see how to realize inductances and capacitances by short microstrip line lengths.

6.6.1 Series Inductance

Figure 6.77(a) shows the equivalent π-network of a transmission line of length l. γ is the propagation coefficient of the line. Assuming lossless transmission lines, the equivalent circuit becomes the pure reactive circuit shown in Figure 6.77(b). The reactances read

$$jX_1 = jX_3 = jZ_0 \cot(\beta l/2) = -j\frac{1}{\omega C} \tag{6.116}$$

$$jX_2 = jZ_0 \sin(\beta l) = j\omega L \tag{6.117}$$

where

$$Z_0 = Z_0(\epsilon_r) = Z_0(\epsilon_r = 1)/\sqrt{\epsilon_{eff}}$$

From (6.117), the inductance L for the line length l reads

$$L = Z_0/\omega \sin(\omega l/v_{ph}) \tag{6.118}$$

Or for a given L value, the necessary length l is

$$l = v_{ph}/\omega \arcsin(\omega L/Z_0) \tag{6.119}$$

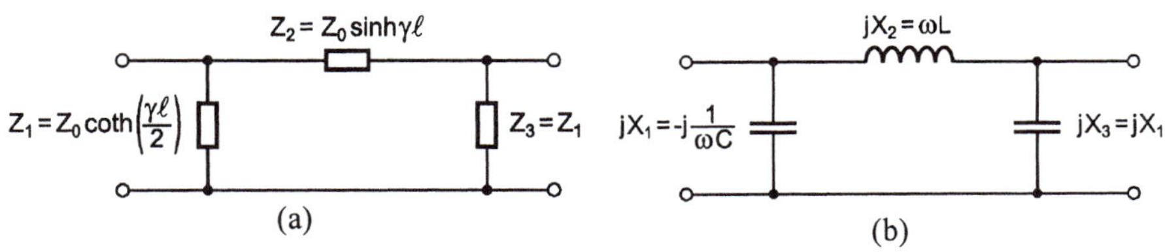

Figure 6.77 Equivalent π-network of (a) lossy and (b) lossless line of length l.

When the line length is very short,

$$\frac{2\pi l}{\lambda} \ll \frac{\pi}{4} \quad \text{or} \quad \frac{l}{\lambda} \ll \frac{1}{8} \tag{6.120}$$

then (6.118) and (6.119) read

$$L = Z_0 l / v_{ph} \tag{6.121}$$

$$l = \left(v_{ph}/Z_0\right) L \tag{6.122}$$

The shunt elements of the equivalent circuit in Figure 6.77(b) represent the capacitance per unit length multiplied by the line length l. From (6.116), we obtain

$$jB_c = j\frac{1}{Z_0}\tan(\beta l/2) \tag{6.123}$$

Thus,

$$B_c = \omega C_l = \frac{1}{Z_0}\tan(\beta l/2) \tag{6.124}$$

For short lengths,

$$C_l = \frac{l}{2\,Z_0\,v_{ph}} \tag{6.125}$$

The adjacent abrupt transitions to transmission lines of low impedance and wide microstrip lines represent significant step discontinuities. Expressions for the step capacitance were given in Section 6.2. Figure 6.78 shows the physical structure of a quasi-lumped inductive element with step discontinuities on both sides and the complete equivalent circuit comprising the π-network of the high-impedance line and the T-network of the impedance step. At sufficiently low frequencies, the step inductances L_s, which are usually much smaller than the series inductance L, can usually be neglected.

Alternatively, the quasi-lumped inductive structure in Figure 6.78 can be analyzed using the planar waveguide model [61]. As shown in Figure 6.79, the widths of the transmission lines are replaced by their corresponding effective widths. The capacitance of the step discontinuity can be equivalently described by a short section of transmission line of width w_1 and length Δ_s on both sides of the inductive line. Thus, the effective length l_{eff} of the high-impedance line is calculated as

$$l_{eff} = l - 2\Delta_s \tag{6.126}$$

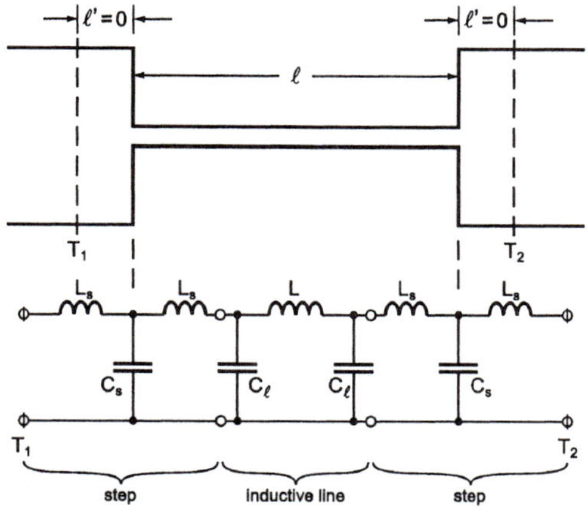

Figure 6.78 Complete equivalent network of a quasi-lumped inductance with adjacent step discontinuities.

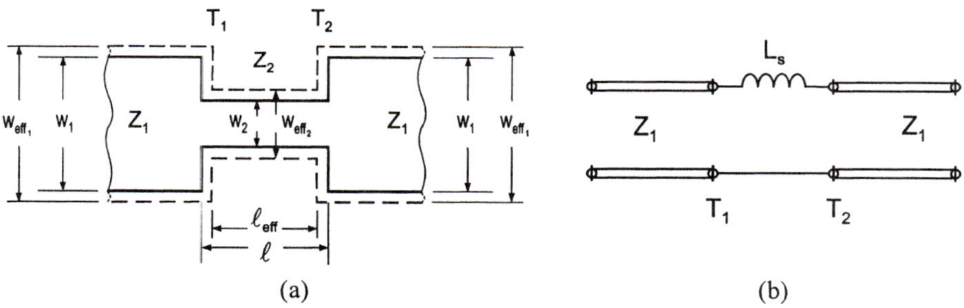

Figure 6.79 (a) Electrically short high impedance microstrip line (magnetic walls denoted by dotted lines), and (b) its equivalent circuit.

Both step inductance L_s and high impedance line capacitance C_l, as discussed above, are not taken into account by this approach. However, it is assumed that these are negligible with respect to the main inductance L and step capacitance C_s.

Under the assumption of very short line length l ($l \ll \lambda/4$) and high impedance ratio $Z_2/Z_1 \gg 1$ with

$$Z_{0n} = Z_{0n}(\epsilon_r) = Z_{0n}(\epsilon_r = 1)/\sqrt{\epsilon_{eff,n}}$$

where $n = 1, 2$, the inductance is calculated as

$$L = \frac{Z_2 \, l_{eff}}{c_0} \sqrt{\epsilon_{eff_2}} = \frac{Z_2(\epsilon_r = 1)}{c_0} \, l_{eff} \tag{6.127}$$

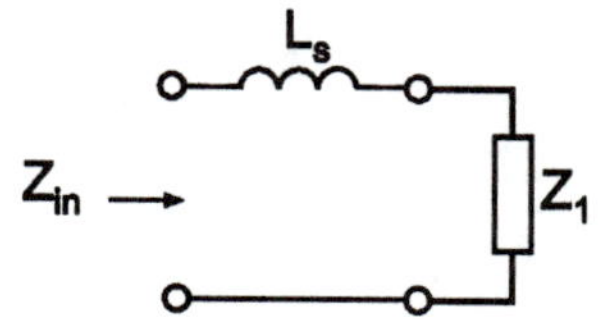

Figure 6.80 Lumped element equivalent circuit.

Figure 6.79(b) shows the equivalent circuit related to the reference planes T_1 and T_2 in Figure 6.79(a).

Problem 1: Derive (6.127) by calculating the input impedance at reference plane T_1 of a short inductive line of length l with adjacent low impedance microstrip lines.

Solution: Regarding the magnetic wall model in Figure 6.79(a), the input impedance at reference plane T_1 reads with (4.49)

$$Z_{in} = \frac{Z_1 + jZ_2 \tan \beta_2 l_{eff}}{1 + j\dfrac{Z_1}{Z_2} \tan \beta_2 l_{eff}} \tag{6.128}$$

With $Z_2 \gg Z_1$ and $\beta_2 l_{eff} \ll \pi/4$, the input impedance can be approximated as (Figure 6.80)

$$Z_{in} \simeq Z_1 + jZ_2 \beta_2 l_{eff} \tag{6.129}$$

Equation (6.129) describes the series circuit of characteristic impedance Z_1 and an inductance L. By comparison, we obtain

$$\omega L = Z_2 \beta_2 l_{eff} \tag{6.130}$$

Thus, the inductance L is

$$L = \frac{Z_2 l_{eff}}{c_0} \sqrt{\epsilon_{eff_2}} = \frac{Z_2(\epsilon_r = 1)}{c_0} l_{eff} \tag{6.131}$$

Problem 2: As is shown in Figure 6.81, a series inductance is realized by a bondwire with a diameter of 62.4 μm (note that this value was chosen to obtain a width ratio $w_2/w_1 = 0.1$, which is the minimum value accepted in the ADS program), and a length of 1 mm. The adjacent microstrip transmission lines have a characteristic impedance of $Z_{01} = 50\Omega$ and are fabricated on alumina substrate

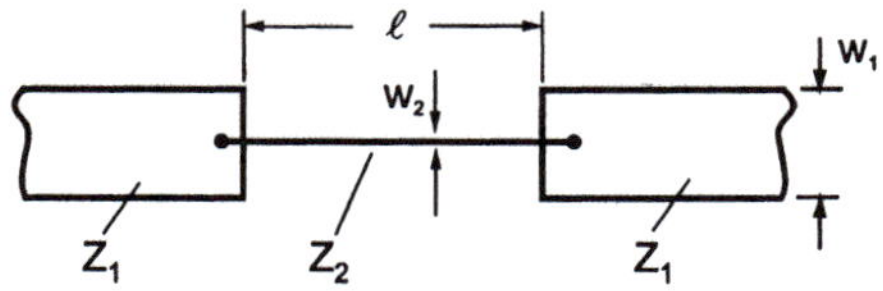

Figure 6.81 Series inductance realized by a short length of bondwire.

($\epsilon_r = 9.8$, $h = 0.635\,\text{mm}$). Determine the inductance value with and without fringing field effect in the step transition under the assumption that the wire is fixed closely on the substrate surface, and that dispersion is negligible.

Solution: The bondwire on the substrate can be approximately regarded as a microstrip of width $w_2 = 62.4\,\mu\text{m}$. The aspect ratio is $w_2/h = 0.098 \ll 1$. The effective dielectric constant yields, for instance, after Hammerstad with ($w/h \leq 1$) (3.115),

$$\varepsilon_{eff_2} = \frac{\varepsilon_r + 1}{2} + \frac{\varepsilon_r - 1}{2}\left\{\frac{1}{\sqrt{1 + 12h/w_2}} + 0.04\,(1 - w_2/h)^2\right\} = 5.58$$

The characteristic impedance is obtained, for instance, from (3.117) after Hammerstad ($w/h \leq 1$)

$$Z_{02} = \left(60\,\Omega/\sqrt{\epsilon_{eff_2}}\right)\ln\left(8h/w_2 + 0.25\,w_2/h\right) = 111.76\,\Omega$$

Then the inductance is calculated as

$$L = \frac{Z_{02}\,l_{eff}}{c_0}\sqrt{\epsilon_{eff_2}}$$

1. Suppose the fringing field is neglected in the step transition: $l_{eff} = l$.

$$L = \frac{Z_{02}\,l}{c_0}\sqrt{\epsilon_{eff_2}} = 0.88\,\text{nH}$$

Alternatively, we can apply the expression $L = Z_0(\epsilon_r = 1)\,l/c_0$. From the previous calculations, the characteristic impedance of the air-filled microstrip line reads

$$Z_{02}(\epsilon_r = 1) = Z_{02}(\epsilon_r)\sqrt{\epsilon_{eff_2}} = 263.97\,\Omega$$

Then we get

$$L = \frac{263.98\,\Omega \cdot 1\,\text{mm}}{29.98\,\text{GHz}\,\text{cm}} = 0.88\,\text{nH}$$

2. Assuming the capacitive step transition effect is taken into account, with $w_2/w_1 \ll 1$, the reference plane shift can be approximately determined from the open-ended 50Ω transmission line,

$$\frac{\Delta l_s}{h} \approx 0.412 \frac{\epsilon_{eff_2} + 0.3}{\epsilon_{eff_2} - 0.258} \frac{w_2/h + 0.262}{w_2/h + 0.813}$$

The diagram $Z_0 = Z_0(w/h)$ of Figure 3.17 delivers for a 50Ω line ($\epsilon_r = 9.8$, $h = 0.635$ mm) an aspect ratio $w_1/h = 0.98$. Then the open-end effect is calculated as $\Delta l_s/h = 0.2834$ and $\Delta l_s = 0.18$ mm, respectively. The effective length thus reads $l_{eff} = l - 2\Delta_s = 0.64$ mm, and the series inductance $L = l_{eff}/l \cdot L_l = 0.56$ nH.

Simulation of the scattering parameters of the configuration in Figure 6.81 with ADS and then searching for optimum value of inductance replacing the bondwire delivers the estimated value of exactly 0.56 nH.

6.6.2 Shunt Capacitance

Figure 6.82(a) shows the equivalent T-network of a transmission line of length l. Assuming loss-free transmission lines, the equivalent circuit becomes the pure reactive circuit shown in Figure 6.82(b). The reactances read

$$jY_1 = jY_2 = j\frac{1}{Z_0}\cot(\beta l/2) \tag{6.132}$$

$$jY_2 = j\frac{1}{Z_0}\sin(\beta l) = j\omega C \tag{6.133}$$

From (6.133), the capacitance C for the line length l reads

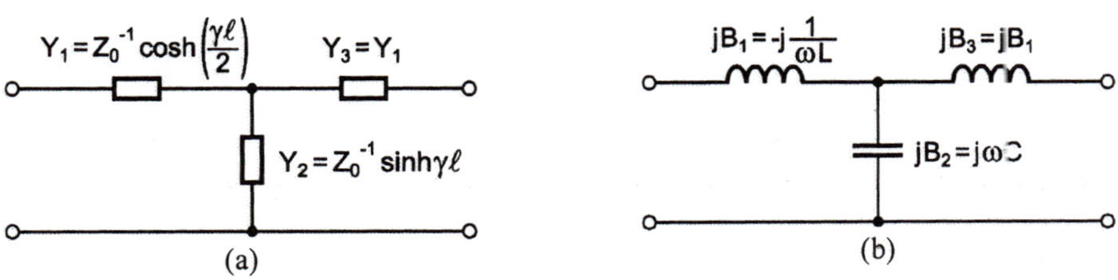

Figure 6.82 Equivalent T-network of (a) lossy and (b) lossless line of length l.

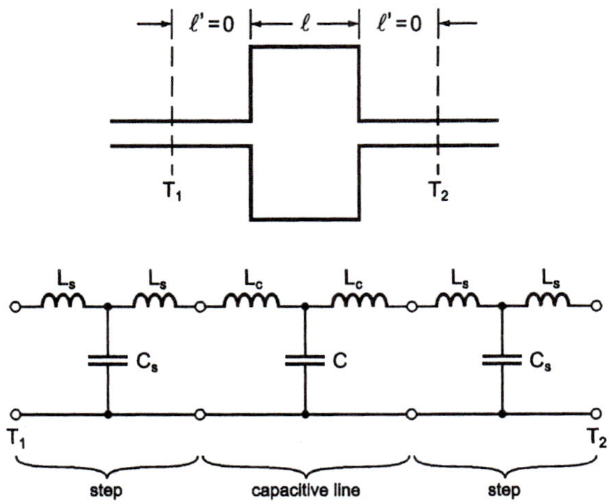

Figure 6.83 Complete equivalent network of a quasi-lumped inductance with adjacent step discontinuities.

$$C = \frac{1}{\omega Z_0} \sin(\omega l / v_{ph}) \tag{6.134}$$

For a given C value, the necessary length l is

$$l = v_{ph}/\omega \arcsin(\omega C Z_0) \tag{6.135}$$

When the line length is electrically very short,

$$\frac{2\pi l}{\lambda} \ll \frac{\pi}{4} \tag{6.136}$$

then (6.134) and (6.135) read

$$C = l/(Z_0 v_{ph}) \tag{6.137}$$

$$l = v_{ph} C Z_0 \tag{6.138}$$

The series elements of the equivalent circuit in Figure 6.82(a) represent the inductance per unit length multiplied by the line length l. From (6.132), we obtain

$$jX_c = j\frac{1}{Z_0} \tan(\beta l/2) \tag{6.139}$$

Thus,

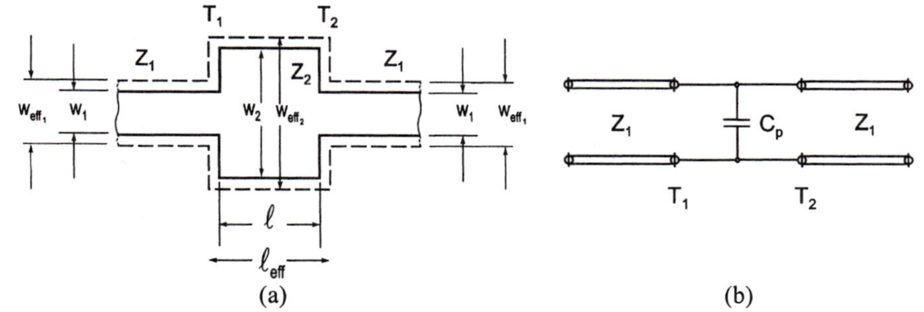

Figure 6.84 (a) Electrically short low impedance microstrip line (magnetic walls denoted by dotted lines), and (b) its equivalent circuit.

$$X_c = \omega L_c = Z_0 \tan(\beta l/2) \tag{6.140}$$

For electrically short lengths,

$$L_c = \frac{Z_0 l}{2 v_{ph}} \tag{6.141}$$

As was already discussed with the modeling of quasi-lumped inductances, the adjacent abrupt transitions to transmission lines of high impedance must be taken into account for the complete microstrip circuit modeling. Figure 6.83 shows the physical structure of a quasi-lumped capacitive element with step discontinuities on both sides, and the complete equivalent circuit comprising the T-network of the high-impedance line and the T-network of the impedance step. As is pointed out in [42], the inductance L_c can usually be neglected because of relatively small values of Z_0 and l in (6.142).

Alternatively, the quasi-lumped capacitive structure in Figure 6.83 can be analyzed using the planar waveguide model [62]. As shown in Figure 6.84(a), the width of the transmission lines are replaced by their corresponding effective widths. The capacitance of the step discontinuity can be equivalently described by a short section of transmission line of width w_2 and length Δ_s on both sides of the capacitive line. Thus, the effective length l_{eff} of the low impedance line is calculated as

$$l_{eff} = l + 2\Delta_s \tag{6.142}$$

Both step inductance L_s and low impedance line inductance L_c, as discussed above, are not taken into account by this approach. However, it is assumed that these are of negligible influence.

Under the assumption of very short line length l ($l \ll \lambda/4$) and high impedance ratio $Z_1/Z_2 \gg 1$, the capacitance is calculated as

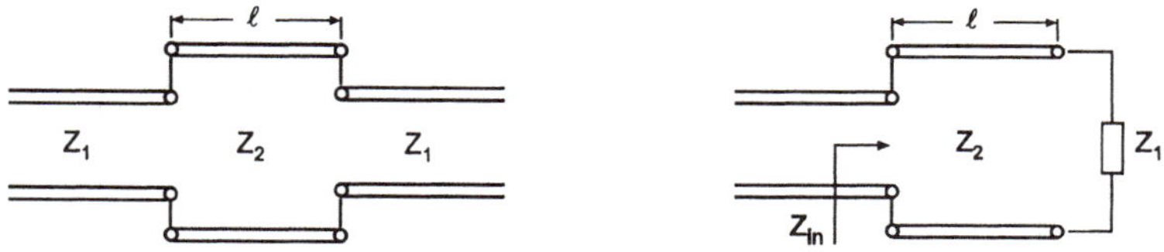

Figure 6.85 Two-wire transmission line representation of the quasi-lumped capacitance.

$$C = \frac{l_{eff}}{Z_2 c_0}\sqrt{\epsilon_{eff_2}} \tag{6.143}$$

Figure 6.84(b) shows the equivalent circuit related to the reference planes T_1 and T_2 in Figure 6.84(a).

Problem: Derive (6.143) by calculating the input impedance at reference plane T_1 of a short capacitive line of length l with adjacent high impedance microstrip lines.

Solution: Figure 6.85 illustrates the microstrip circuit under study using two-wire transmission line representation. The low impedance line is terminated by an infinite line with characteristic impedance Z_1; thus, we can terminate it by Z_1 at the reference plane T_2. Then, the input impedance at reference plane T_1 reads

$$Z_{in} = \frac{Z_1 + jZ_2 \tan\beta_2 l_{eff}}{1 + j\dfrac{Z_1}{Z_2}\tan\beta_2 l_{eff}} \tag{6.144}$$

With $Z_2 \ll Z_1$ and $\beta_2 l_{eff} \ll \pi/4$, the input impedance can be approximated as

$$Y_{in} \approx \frac{1}{Z_1}\left(1 + j\frac{Z_1}{Z_2}\beta_2 l_{eff}\right) \tag{6.145}$$

Equation (6.145) describes the shunt circuit of characteristic impedance Z_1 and capacitance C (Figure 6.86). By comparison we obtain

$$\omega C = \frac{1}{Z_2}\beta_2 l_{eff} \tag{6.146}$$

Thus, the capacitance is

$$C = \frac{l_{eff}}{Z_2 c_0}\sqrt{\epsilon_{eff_2}}$$

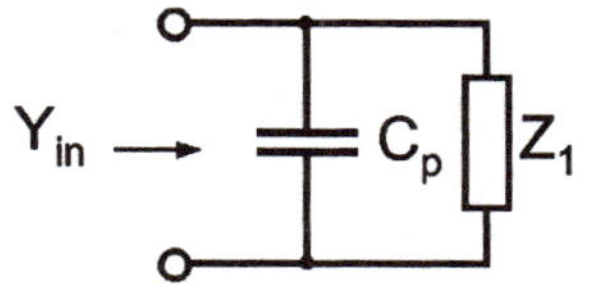

Figure 6.86 Input impedance equivalently described by lumped elements.

6.7 Circular Disk Resonator

Circular disk resonators are used, for example, as microstrip filters, circulators including ferrite material, and patch antennas. Resonating sectors of disks also offer an excellent RF ground for a limited bandwidth [63]. Sectoral disk configuration are frequently found in broadband bias networks (Section 6.8.4). Disk resonators are open-ended radial lines. Resonances can be understood as standing radial waves that are reflected at the open end of the radial line and in the cylindrical axis. Before we analyze the dynamic properties of a disk including the static case, we analyze the wave propagation in a radial transmission line and derive the input impedance for an open-circuited termination.

6.7.1 Radial Transmission Line

A radial transmission line, as illustrated in Figure 6.87, consists of two circular, parallel, conducting plates, separated by a dielectric substrate of height h. The line is stimulated by a source connected to the inner periphery at radius r_i. The line is terminated by a load impedance Z_L at the outer periphery at radius r_L following the notation in [64, 65]. We assume that the electromagnetic field has no variation along the circumference. Because the height is assumed to be very small ($h \ll \lambda$), waves with axial dependence are also excluded. This means also that all modes

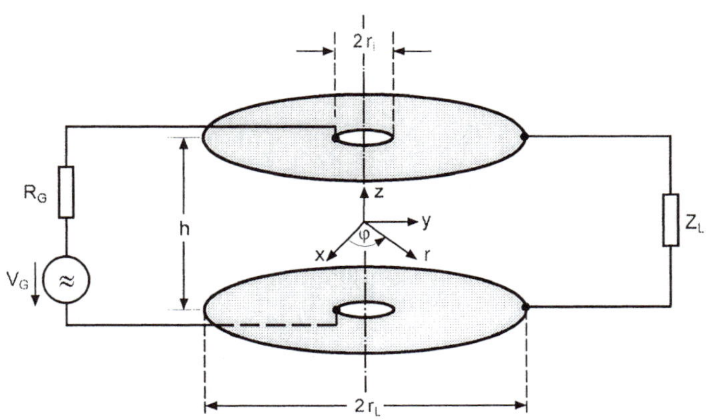

Figure 6.87 Open-circuited radial transmission line stimulated at the inner periphery.

with radial components vanish. Thus, we restrict our analysis to waves that possess only a longitudinal electrical field component E_z and a transversal magnetic field component H_φ.

Since we are interested in the input impedance of an open-circuited radial line, the electromagnetic field will be composed of a radial wave propagating outward direction and at the outer periphery a propagating wave reflected inward. As we suppose a finite inner radius $r_i > 0$, the radial standing field pattern is described by a linear combination of Bessel and Neumann functions,

$$H_m^{(1)}(kr) = J_m(kr) + jN_m(kr) \tag{6.147}$$

and

$$H_m^{(2)}(kr) = J_m(kr) - jN_m(kr) \tag{6.148}$$

where $H_m^{(1)}(kr)$ is the Hankel function of the first kind, $H_m^{(2)}(kr)$ is the Hankel function of the second kind, and k is the medium wavenumber. The Hankel functions can be approximated for large values of r as follows [64]:

$$H_m^{(1)}(kr) \approx \sqrt{\frac{2}{\pi kr}}\, e^{j\left(kr - \frac{\pi}{4} - \frac{m\pi}{2}\right)} \tag{6.149}$$

$$H_m^{(2)}(kr) \approx \sqrt{\frac{2}{\pi kr}}\, e^{-j\left(kr - \frac{\pi}{4} - \frac{m\pi}{2}\right)} \tag{6.150}$$

Thus, $H_m^{(1)}(kr)$ represents a wave traveling radially inward, while $H_m^{(2)}(kr)$ describes a wave traveling radially outward. Regarding the lowest order Hankel functions with $m = 0$, we find

$$H_0^{(1)}(kr) \approx \sqrt{\frac{2}{\pi kr}}\, e^{j\left(x - \frac{\pi}{4}\right)} \tag{6.151}$$

$$H_0^{(2)}(kr) \approx \sqrt{\frac{2}{\pi kr}}\, e^{-j\left(kr - \frac{\pi}{4}\right)} \tag{6.152}$$

In polar form, the complex Hankel functions can be written as [64]:

$$H_0^{(1)}(kr) = J_0(kr) + jN_0(kr) = G_0(kr)\, e^{j\theta(kr)} \tag{6.153}$$

$$H_0^{(2)}(kr) = J_0(kr) - jN_0(kr) = G_0(kr)\, e^{-j\theta(kr)} \tag{6.154}$$

with

$$G_0(kr) = \sqrt{J_0(kr)^2 + N_0(kr)^2}\,,\ \theta(kr) = \arctan\left[\frac{N_0(kr)}{J_0(kr)}\right] \tag{6.155}$$

Furthermore, we define,

$$j\,H_1^{(1)}(kr) = -N_1(kr) + j\,J_1(kr) = G_1(kr)\,e^{j\psi(kr)} \tag{6.156}$$

$$j\,H_1^{(2)}(kr) = -\left[-N_1(kr) - jJ_1(kr)\right] = -G_1(kr)\,e^{-j\psi(kr)} \tag{6.157}$$

with

$$G_1(kr) = \sqrt{J_1(kr)^2 + N_1(kr)^2}\,,\ \psi(kr) = \arctan\left[\frac{J_1(kr)}{-N_1(kr)}\right] \tag{6.158}$$

With the electrical field having only radial dependence, we can write

$$E_z(r) = A\,H_0^{(1)}(kr) + B\,H_0^{(2)}(kr) \tag{6.159}$$

where the second term describes a wave traveling outward, and the first term a wave propagating inward. The amplitude B depends on the excitation of the radial wave, and A, the magnitude of the reflected wave at the outer periphery, depends on the matching termination condition.

According to Section D.2.2.2, we find the magnetic field, deriving curl $\vec{A}^e$ for polar coordinates,

$$H_\varphi(r) = \frac{1}{j\omega\mu}\,\frac{dE_z(r)}{dr} \tag{6.160}$$

Applying the differentiation rule for cylindrical functions Z_n [66],

$$Z_{n+1} = -x^n\frac{d}{dx}\left[x^{-n}\,Z_n(x)\right] \tag{6.161}$$

which results in $H_1^{(\nu)}(x) = -d/dx\left[H_0^{(\nu)}(x)\right]$ $(\nu = 1,2)$ for the given Hankel functions. Then, (6.160) reads

$$H_\varphi(r) = -\frac{k}{j\omega\mu}\left[A\,H_1^{(1)}(kr) + B\,H_1^{(2)}(kr)\right] \tag{6.162}$$

From (6.153) to (6.159), $E_z(r)$ and $H_\varphi(r)$ can be rewritten as

$$E_z = G_0(kr)\left[A\,e^{j\theta(kr)} + B\,e^{-j\theta(kr)}\right] \tag{6.163}$$

$$H_\varphi = \sqrt{\frac{\epsilon_0 \epsilon_r}{\mu_0}}\, G_1(kr) \left[A e^{j\psi(kr)} - B e^{-j\psi(kr)} \right] \tag{6.164}$$

The wave impedance of a propagating wave Z_0 is commonly defined as the ratio of the transversal electric over the magnetic field component [32]. With (6.159), (6.162), and (6.153) and (6.156), respectively, we obtain

$$\frac{E_z^\pm}{H_\varphi^\pm} = -\frac{j\omega\mu}{k} \frac{H_0^{(1)}}{H_1^{(1)}} = \sqrt{\frac{\mu_0}{\epsilon_0 \epsilon_r}} \frac{G_0(kr)}{G_1(kr)} e^{j[\theta(kr) - \psi(kr)]} \tag{6.165}$$

The characteristic impedance of a radial transmission line is defined as the magnitude of (6.165) [64, 65]. Thus, with $G_0(kr)$ and $G_1(kr)$ of (6.155) and (6.158), we obtain

$$Z_0(kr) = \sqrt{\frac{\mu_0}{\epsilon_r \epsilon_0}} \sqrt{\frac{J_0^2(kr) + N_0^2(kr)}{J_1^2(kr) + N_1^2(kr)}} \tag{6.166}$$

With (6.166), the magnetic field H_φ can be rewritten as

$$H_\varphi = \frac{G_0(kr)}{Z_0(kr)} \left[A e^{j\psi(kr)} - B e^{-j\psi(kr)} \right] \tag{6.167}$$

Supposing higher positive voltage on the upper plate of the line, we may define a total line voltage according to

$$V = -E_z h \tag{6.168}$$

The total current, if outward current in the upper plate is considered as positive, is calculated as

$$I = 2\pi r H_\varphi \tag{6.169}$$

Thus, the input impedance of the radial transmission line $Z_i(r)$ at radius r reads

$$Z_i(r) = \frac{V(r)}{I(r)} = -\frac{h}{2\pi r} \frac{E_z(r)}{H_\varphi(r)} = -\frac{h}{2\pi r} Z_T(r) \tag{6.170}$$

where

$$Z_T(r) = \frac{E_z(r)}{H_\varphi(r)} \tag{6.171}$$

is called the input wave impedance at the radius r, and relates the total transversal electric field to the total transversal magnetic field, and should not be confused with the wave impedance in (6.166). Similarly, the total load wave impedance $Z_T(r = r_L) = Z_T(r_L)$ is defined as

$$Z_T(r_L) = \frac{E_z(r_L)}{H_\varphi(r_L)} \tag{6.172}$$

Substituting E_z and H_φ of (6.163) and (6.167) with $r = r_L$ in (6.172) results in

$$Z_T(r_L) = Z_0(r_L) \frac{A e^{j\theta(kr_L)} + B e^{-j\theta(kr_L)}}{A e^{j\psi(kr_L)} - B e^{-j\psi(kr_L)}} \tag{6.173}$$

The amplitudes A and B can be expressed by the electric and magnetic field components at the outer periphery determining an output wave impedance $Z_T(r_L)$. The input wave impedance $Z_T(r_i)$ at the inner periphery can then be written as

$$Z_T(r_i) = Z_0(kr_i) \frac{Z_T(r_L)\cos(\theta_i - \psi_L) + j Z_0(r_L)\sin(\theta_i - \theta_L)}{Z_0(r_L)\cos(\psi_i - \theta_L) + j Z_T(r_L)\sin(\psi_i - \psi_L)} \tag{6.174}$$

where

$$\theta_i = \theta(kr_i) = \arctan\left[\frac{N_0(kr_i)}{J_0(kr_i)}\right] \tag{6.175}$$

$$\theta_L = \theta(kr_L) = \arctan\left[\frac{N_0(kr_L)}{J_0(kr_L)}\right] \tag{6.176}$$

$$\psi_i = \psi(kr_i) = \arctan\left[\frac{J_1(kr_i)}{-N_1(kr_i)}\right] \tag{6.177}$$

$$\psi_L = \psi(kr_L) = \arctan\left[\frac{J_1(kr_L)}{-N_1(kr_L)}\right] \tag{6.178}$$

The input wave impedance $Z_T(r_i)$ reads, when terminated with an open $(Z_T(r_L) = \infty)$,

$$Z_T(r_i) = -j Z_0(r_i) \frac{\cos(\theta_i - \psi_L)}{\sin(\psi_i - \psi_L)} \tag{6.179}$$

Substituting (6.179) in (6.170) provides the input impedance of an open-circuited radial transmission line,

$$Z_i(r_i) = j\,\frac{h\,Z_0(r_i)}{2\pi r_i}\;\frac{\cos(\theta_i - \psi_L)}{\sin(\psi_i - \psi_L)} \tag{6.180}$$

Before we apply the results to sectoral open-ended radial transmission lines (radial stubs), we first analyze the circular disk capacitor in the following section.

6.7.2 Circular Disk Capacitor

The capacitance of an air-filled ($\epsilon_r = 1$) circular disk capacitor with radius R and height h is calculated as

$$C(R/h, \epsilon_r = 1) = \frac{\epsilon_0\,\pi\,R_{eff}^2(R/h)}{h} \tag{6.181}$$

where the effective radius R_{eff} is a function of the aspect ratio R/h, taking into account the edge fringing electrical field. In 1882, Kirchhoff [67] published formulas to calculate disk condensors. The edge scattering field was taken into account by introducing an effective radius R_{eff},

$$R_{eff} = R\left[1 + a\,(1.7726 - \ln a)\right]^{1/2} \tag{6.182}$$

with the geometry factor

$$a = \frac{2h}{\pi R} \tag{6.183}$$

Zinke [68] confirmed that the relation delivers satisfactory results under the condition $R > h$. This is always fulfilled with microstrip capacitors as well as resonators. In [69], the capacitance of a circular disk placed on top of a substrate is calculated by introducing an equivalent dielectric constant that describes the electrostatic properties of the mixed air-dielectric medium on the capacitance value. The total capacitance can be composed of a main capacitance beneath the disk metallization pattern and the fringing capacitance,

$$C(R/h, \epsilon_r) = C_m(R/h, \epsilon_r) + C_f(R/h, \epsilon_r) \tag{6.184}$$

The capacitance per unit length due to the fringing field at the edge of the disk can be approximately calculated using the conformal mapping concept for a transmission line [69, 70]. The capacitance per unit length of a microstrip line of width $2R$ reads

$$C' = \frac{1}{v\,Z_0(2R, h, \epsilon_r)} \tag{6.185}$$

where v is the phase velocity. With the main capacitance per unit length of the line calculated as

$$C'_{main} = \frac{\epsilon_0 \epsilon_r 2R}{h} \tag{6.186}$$

the fringing edge capacitance per unit length is obtained as the difference of the total capacitance per unit length and the main capacitance per unit length,

$$C'_f = C' - C'_m = \frac{1}{Z_0(2R, h, \epsilon_r)\, v} - \frac{\epsilon_0 \epsilon_r 2R}{h} \tag{6.187}$$

Multiplying (6.185) by one-half of circumference yields the edge fringing capacitance

$$C_f = C'_f R\pi \tag{6.188}$$

An equivalent dielectric constant can be derived from the ratios of the total capacitance of the disk with finite thickness t, and substrate carrier of height h to the capacitance of the air-filled structure,

$$\epsilon_{equ} = \frac{\epsilon_0 \epsilon_r \pi \dfrac{R^2}{h} + \left[\dfrac{1}{v Z_0(2R, h, t, \epsilon_r)} - \dfrac{2\epsilon_0 \epsilon_r R}{h}\right] \pi R}{\epsilon_0 \pi \dfrac{R^2}{h} + \left[\dfrac{1}{c_0 Z_0(2R, h, t, \epsilon_r = 1)} - \dfrac{2\epsilon_0 R}{h}\right] \pi R} \tag{6.189}$$

$Z_0(2R, h, t, \epsilon_r)$ is the characteristic impedance of a microstrip transmission line of width $w = 2R$ after Wheeler. The corresponding phase velocity then reads

$$v = c_0 \frac{Z_0(2R, h, t, \epsilon_r)}{Z_0(2R, h, t, \epsilon_r = 1)} \tag{6.190}$$

With (6.189) the capacitance of a circular disk capacitor with substrate carrier reads

$$C(R/h, t, \epsilon_r) = \frac{\epsilon_0\, \epsilon_{equ}(2R/h, t, \epsilon_r)\ \pi R^2_{eff}(R/h)}{h} \tag{6.191}$$

Measured capacitance of circular disk capacitors on Polyguide substrate ($\epsilon_r = 2.32$, $h = 1.5$ mm) shows good agreement with calculated results. As reported in [71], an error less than 3.6% was found for $10.0 \le R/h \le 26.5$.

6.7.3 Magnetic Wall Model After Watkins

A simple magnetic wall model was proposed by Watkins [72] to calculate the resonant frequencies of a circular disk resonator. As shown in Figure 6.88, for analysis the circular disk resonator is considered as a cavity resonator with electric walls represented by the upper conductor and bottom ground plate, and magnetic side walls at radius R. It is assumed that the electromagnetic fields are confined entirely inside the dielectric under the upper circular plate.

Typically, the height of microstrip substrate material is very small ($h \ll \lambda$). In this case, all field components of the resonance modes are independent on the z coordinate. The solution of the electric field for the internal region of the resonator can be derived from Maxwell's equations:

$$E_z = AJ_m(kr)\begin{Bmatrix} \cos m\varphi \\ \sin m\varphi \end{Bmatrix} \tag{6.192}$$

with $0 \le r \le R$ and the material wavenumber

$$k = \frac{\omega}{c_0}\sqrt{\epsilon_r} \tag{6.193}$$

The dependence of the electromagnetic resonant field on the angle φ is described by the harmonic functions $\cos m\varphi$ and $\sin m\varphi$, respectively. The general solution is a superposition of respective partial field solutions. The radial dependence is described by cylindrical functions. Both Bessel and Neumann functions describe radial standing wave pattern. However, because Neumann functions become infinite at the origin ($r = 0$), the longitudinal electrical field component is only described by Bessel functions. The amplitude A depends on the stimulation of the resonant field.

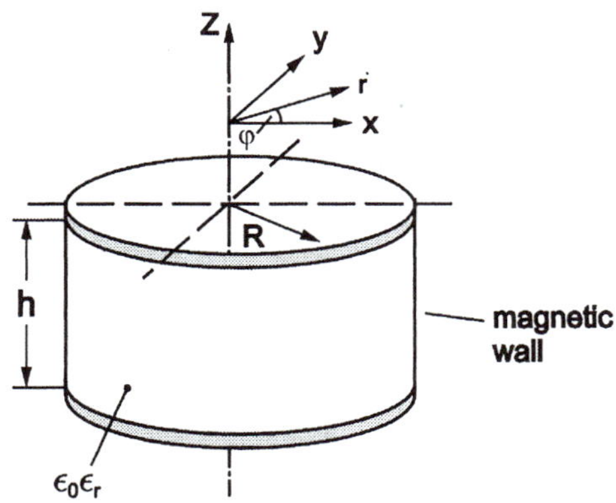

Figure 6.88 Magnetic wall model. (After: Watkins [72].)

$J_m(kr)$ is the Bessel function of the first kind and mth order. With the longitudinal electrical field component known, the transversal components (with respect to the z-axis) can be derived [Appendix D (D.100)]:

$$H_r = -\frac{1}{j\omega\mu r}\frac{\partial E_z}{\partial \varphi} = \frac{m}{j\omega\mu_0 r} A J_m(kr) \begin{Bmatrix} \sin m\varphi \\ -\cos m\varphi \end{Bmatrix} \tag{6.194}$$

$$H_\varphi = \frac{1}{j\omega\mu_0}\frac{\partial E_z}{\partial r} = \frac{k}{j\omega\mu_0} A J_m'(kr) \begin{Bmatrix} \cos m\varphi \\ \sin m\varphi \end{Bmatrix} \tag{6.195}$$

J_m' represents the derivative of J_m with respect to the argument. The boundary conditions require that the tangential magnetic field vanishes on the magnetic wall. Therefore, the boundary condition reads

$$H_\varphi(r = R) = 0 \tag{6.196}$$

for $0 \le \varphi < 2\pi$. From (6.195), it can be seen that the boundary condition only can be fulfilled if the derivative of the Bessel function J_m' vanishes at $r = R$,

$$J_m'(\eta_{mn}') = J_m'(k_{mn}R) = 0 \tag{6.197}$$

The boundary condition is fulfilled for discrete values of $k = k_{mn}$. With η_{mn}' representing the zeros of the derivative of the Bessel function $J_m(kr)$, k_{mn} reads

$$k_{mn} = \frac{\eta_{mn}'}{R} \tag{5.198}$$

η_{mn}' determines the order of the resonant frequencies. The prime signifies the argument of the differentiated Bessel function. The index m originates from the $\cos m\varphi$ function and denotes the number of full periods of field distribution along the disk periphery. m also signifies the Bessel function of mth order. The index n corresponds to the nth maximum of Bessel function J_m. The first four values of η_{mn}', in ascending order, are:

$$\eta_{11}' = 1.8412, \ \eta_{21}' = 3.0542, \ \eta_{01}' = 3.8318, \ \eta_{31}' = 4.2012 \tag{6.199}$$

Therefore, the resonant frequencies for the E_{mn0} modes can be written as

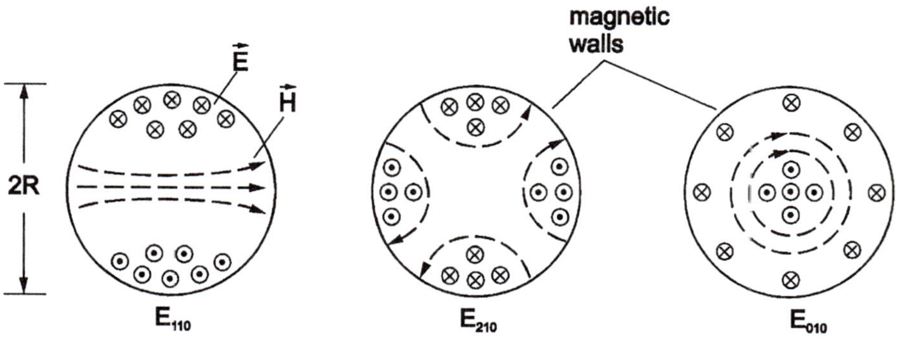

Figure 6.89 Field distribution of the resonance types E_{110}, E_{210}, and E_{010}.

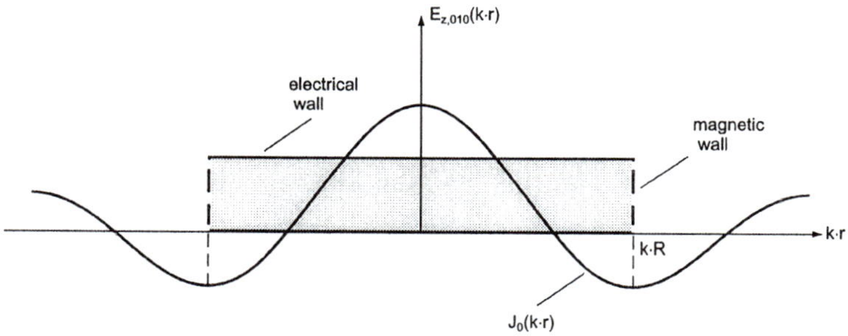

Figure 6.90 Field distribution of the longitudinal electrical field of the E_{010} resonant mode.

$$f_{mn0} = \frac{c_0}{2\pi\sqrt{\epsilon_r}} \frac{\eta'_{mn}}{R} \tag{6.200}$$

The four lowest order modes in ascending order are:

$$E_{110},\ E_{210},\ E_{010},\ E_{310}$$

The lowest order resonant mode is E_{110}, with a resonant frequency of

$$f_{E_{110}} = \frac{c_0}{2\pi\sqrt{\epsilon_r}} \frac{1.8412}{R} \tag{6.201}$$

Figure 6.89 shows the electromagnetic field distribution for different resonant modes. Figure 6.90 shows the field distribution of the longitudinal electrical field of the E_{010} resonant mode versus radius r.

The magnetic wall model after Watkins neglects any fringing fields. Fringing makes the disk appear electrically larger. Therefore, the physical radius must be corrected to an effective radius and the dielectric constant to an effective constant (Section 6.7.2).

6.7.4 Magnetic Wall Model After Wolff

Edge effects of the microstrip strip disk resonator can be approximately taken into account by replacing the relative permittivity ϵ_r by a dynamic permittivity ϵ_{dyn} and an effective radius R_{eff} (6.182). Figure 6.91 shows the magnetic wall model after Wolff. Figure 6.192 illustrates a transmission disk resonator configuration, and Figure 6.193 shows a quantitative figure of occurring resonant modes versus normalized frequency $f_n = f/[c_0/(2\pi R\sqrt{\epsilon_r})]$.

By comparison of the energy content of the electric field in a dielectric-filled resonator with an air-filled structure, a dynamic dielectric constant ϵ_{dyn} can be defined [70],

$$\epsilon_{dyn} = \frac{W_{el,\ mnq}(\epsilon_r)}{W_{el,\ mnq}(\epsilon_r = 1)} \tag{6.202}$$

With $h \ll \lambda$, the resonant modes are independent of the z variable ($q = 0$). The dynamic dielectric constant is a function of the geometry parameters R and h, the material parameter ϵ_r, and of the type of resonant mode, indexed by $mn0$. The stored electrical energy for a given mode is calculated from the voltage drop V between the top and bottom plates. For both the air-filled and dielectric-filled disk, it reads

$$W_{el,mnq}(\epsilon_r = 1) = \frac{1}{2} C_{dyn,mnq}(\epsilon_r = 1)\, V^2 \tag{6.203}$$

$$W_{el,mnq}(\epsilon_r) = \frac{1}{2} C_{dyn,mnq}(\epsilon_r)\, V^2 \tag{6.204}$$

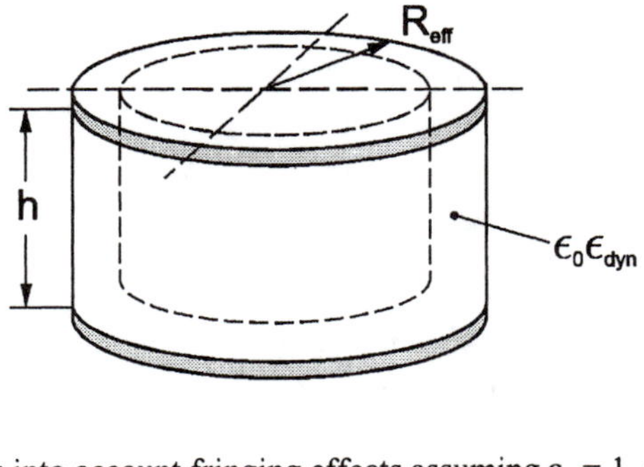

Figure 6.91 R_{eff} takes into account fringing effects assuming $\varepsilon_r = 1$, and ε_{dyn} considers the inhomogeneous field distribution. (After: Wolff [70].)

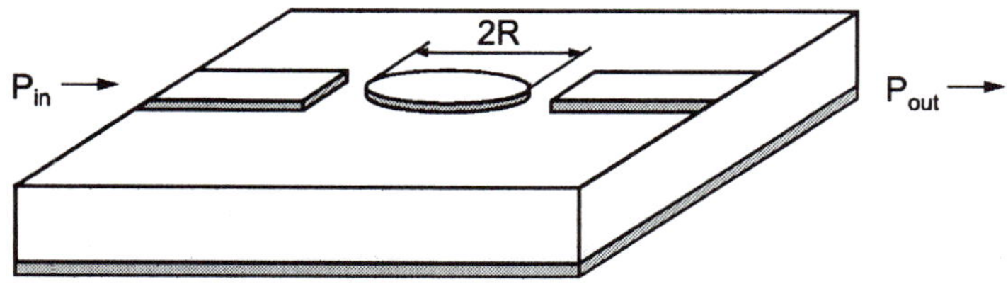

Figure 6.92 Transmission disk resonator configuration.

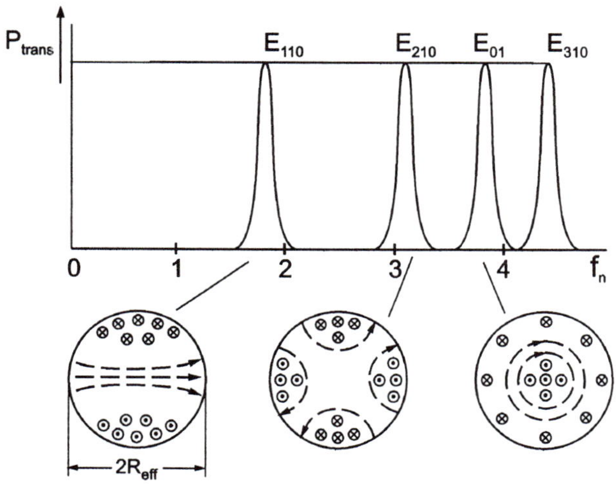

Figure 6.93 Lowest-order resonant modes.

Figure 6.94(a) illustrates the electrical field distribution of the second higher-order mode, and Figure 6.94(b) shows the corresponding dynamic waveguide model used to analyze the resonant frequency of the mode. The voltage V is defined at the disk edge,

$$V = E_z(r = R, \varphi = 0)\, h = A h J_m(kR) \tag{6.205}$$

Then (6.202) reads with (6.203) and (6.204)

$$\epsilon_{dyn} = \frac{C_{dyn,mn0}(\epsilon_r)}{C_{dyn,mn0}(\epsilon_r = 1)} \tag{6.206}$$

The total dynamic capacitance C_{dyn} is composed of the dynamic main capacitance $C_{o,dyn}$ and the dynamic edge capacitance $C_{e,dyn}$ (Figure 6.95),

$$C_{dyn,\,mn0} = C_{o,\,dyn,\,mn0} + C_{e,\,dyn,\,mn0} \tag{6.207}$$

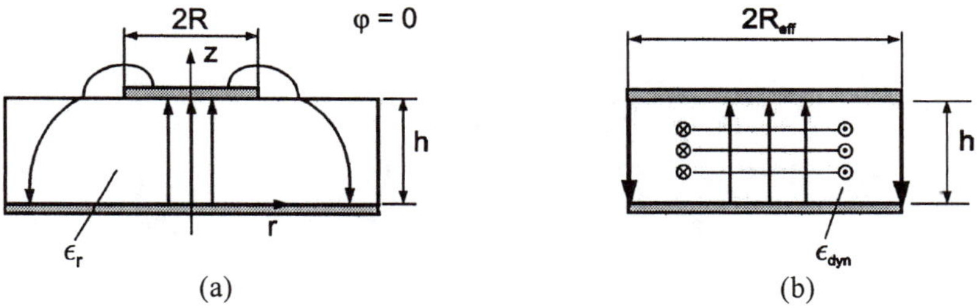

Figure 6.94 (a) Electric field distribution of second higher-order resonant mode, and (b) related E_{010} resonant mode in the planar waveguide model.

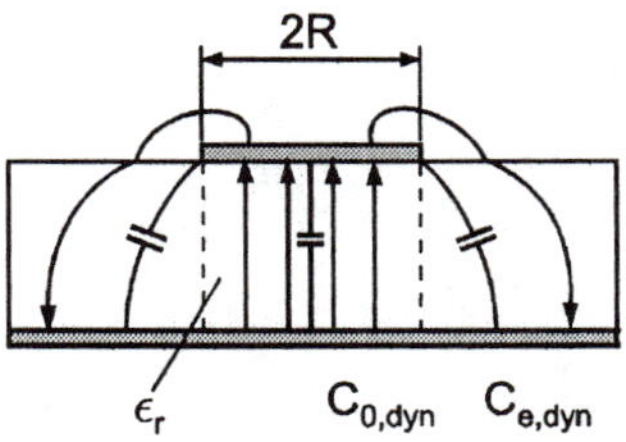

Figure 6.95 Partitioning of the total dynamic capacitance of second higher-order resonant mode.

The energy content of the main field region is calculated as ($q = 0$)

$$W_{el,mn0} = \frac{\epsilon_0 \epsilon_r}{\delta} A^2 h \pi \left\{ \frac{R^2}{2} \left[J_m^2(kR) - J_{m-1}(kR) J_{m+1}(kR) \right] \right\} \tag{6.208}$$

where

$$\delta = \begin{cases} 1 & for \quad m = 0 \\ 2 & for \quad m > 0 \end{cases}$$

Then the dynamic main capacitance $C_{o,dyn}$ is obtained as

$$C_{o,dyn} = \frac{W_{el}}{V^2/2} = \frac{\epsilon_0 \epsilon_r \pi R^2}{\delta h} \left\{ 1 - \frac{J_{m-1}(kR) J_{m+1}(kR)}{J_m^{\,2}(kR)} \right\} \tag{6.209}$$

For $m = 0$ (6.209) yields

$$C_{o,stat} = \frac{\epsilon_0 \epsilon_r \pi R^2}{h} = C_{o,dyn} \tag{6.210}$$

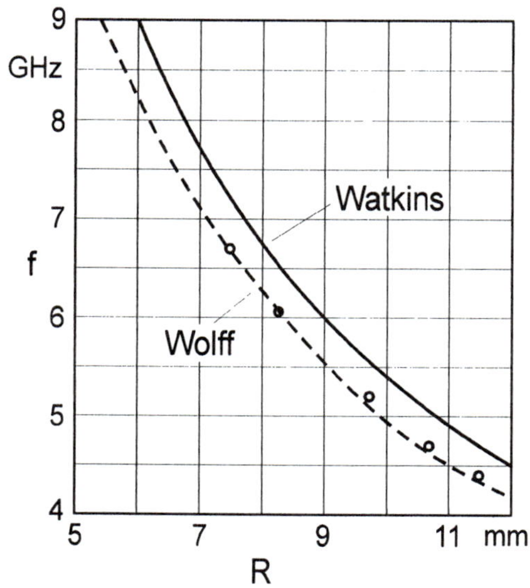

Figure 6.96 Calculated (dashed and solid line) and measured E_{110} resonant frequency of a circular disk resonator ($\varepsilon_r = 2.65$, $h = 0.159$ mm). (From: [71], © 1973 Elsevier. Reprinted with permission.)

For $m = 1, 2, 3$, we get

$$C_{o,dyn} = \begin{cases} 0.3525 \cdot C_{o,stat}, & m = 1 \\ 0.2856 \cdot C_{o,stat}, & m = 2 \\ 0.2450 \cdot C_{o,stat}, & m = 3 \end{cases} \qquad (6.211)$$

The dynamic fringing capacitance is treated as follows. The dynamic fringing capacitance is found by multiplying the static edge capacitance (6.188) with the φ-dependence of the considered mode, and averaging over the circumference,

$$C_{e,dyn,mn0} = \frac{1}{2\pi}\int_0^{2\pi} C_e \cos^2(m\varphi)\,d\varphi = \frac{1}{\delta}\,C_e \qquad (6.212)$$

With (6.219) and (6.212), the dynamic capacitance in (6.207) and thus the dynamic effective dielectric constant ϵ_{dyn} in (6.206) can be calculated for any E_{mn0} resonant mode. The resonant frequency is calculated from the eigenvalue-equation (6.197) substituting R by the effective radius R_{eff}. Then the eigenvalue η'_{mn} becomes

$$J'_m(k_{mn} R_{eff}) = J'_m(\eta'_{mn}) = 0 \qquad (6.213)$$

with

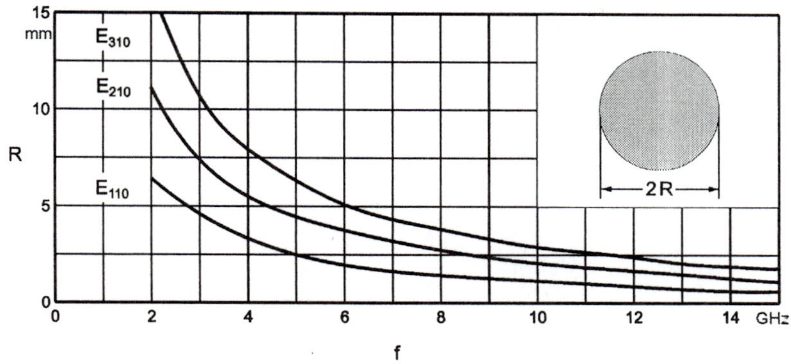

Figure 6.97 Lowest order resonant frequencies of a circular disk resonator on alumina substrate after Khilla ($\varepsilon_r = 9.8$, $h = 0.635$ mm). (From [63], courtesy of *Microwave Journal.* Reprinted with permission.)

$$\eta'_{mn} = k_{mn0} R_{eff} = \frac{2\pi f_{mn0} R_{eff} \sqrt{\epsilon_{dyn,\,mn0}}}{c_0} \tag{6.214}$$

From (6.214), the resonant frequencies read

$$f_{r,\,mn0} = \frac{c_0}{2\pi \sqrt{\epsilon_{dyn,\,mn0}}} \frac{\eta'_{mn}}{R_{eff}} \tag{6.215}$$

It is claimed that the difference between calculated and measured resonant frequencies is less than or equal to 1% for the following substrates investigated: Al_2O_3: $\epsilon_r = 10.4$, Polyguide: $\epsilon_r = 2.315$, Duroid: $\epsilon_r = 2.23$.

Figure 6.96 shows the measured and calculated resonant frequency of a disk as a function of the radius $R(\epsilon_r = 2.65,\ h = 1.59\text{ mm})$ [69]. Figure 6.97 shows design curves for a disk resonator on alumina substrate ($\epsilon_r = 9.8,\ h = 0.635\text{ mm}$) [69].

When a disk is cut through, an open circuit or magnetic wall is established. A disk resonator, if adequately partitioned, can resonate approximately at the same frequency as the full disk. This is illustrated in Figure 6.98. Regarding the E_{110} mode, a bisectional cut does not disturb the resonant field. The cut establishes an ideal open circuit on which the magnetic field is normal; no restriction is given for the electric field. It is clear that regarding the magnetic wall model after Watkins, the resonant frequency of the full and bisectional disk would be the same. However, because of different portions of edge fields, the resonant frequencies

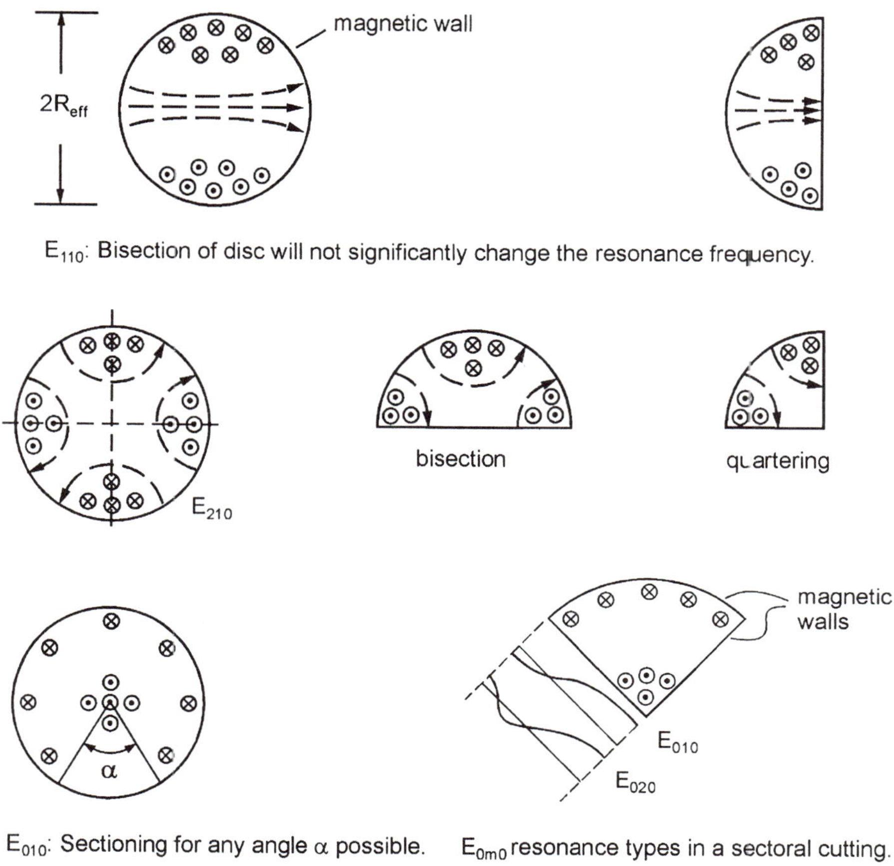

Figure 6.98 Field distribution of different resonant modes (solid contour lines represent magnetic walls).

would slightly differ. As illustrated for the E_{210} mode, bisectional and quartering of the disk is possible. Finally, for the E_{010} mode, arbitrary sectional angle α will produce identical (magnetic wall model after Watkins) or similar (taking into account edge fields) resonant frequencies.

6.8 Radial Line Stub

Radial stubs are typically found in bias networks of active microstrip circuits such as amplifiers, oscillators, and frequency multipliers. Unlike low impedance uniform microstrip stubs illustrated in Figure 6.99, radial stubs provide a well-defined point for radial wave excitation due to their narrow coupling aperture. As was already pointed out, the range of usable characteristic impedances with uniform line stubs

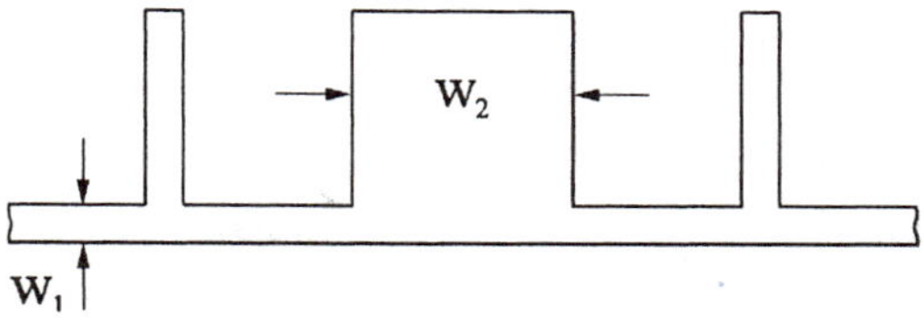

Figure 6.99 Filter with uniform transmission line stubs.

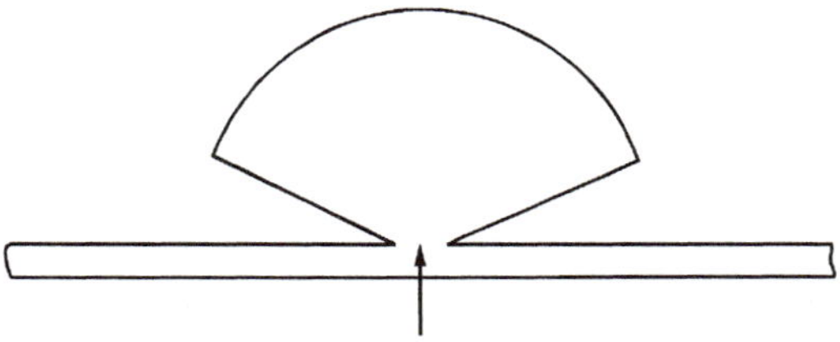

Figure 6.100 Narrow coupling aperture with radial stubs.

is severely limited. The limitation on low characteristic impedance becomes particularly troublesome at high operating frequencies when the stub width approaches one-half wavelength. Because the main and shunted line represent an asymmetric excitation problem, this would support the emergence of the first higher-order $HE_1(H_{[10]})$ mode in the stub. As was stated in Section 6.2.2, the operating frequency usually should not exceed one-third of the respective cutoff frequency $f_{c[10]}$ in order to avoid any severe intervention of higher-order modes. At higher frequencies, when the width of the stub becomes a substantial fraction of a wavelength of the stimulating quasi-TEM mode propagating along the main line, the electrical field in the coupling aperture of the stub becomes less uniform (TEM-like). This lowers the launching efficiency of the fundamental mode and allows for growing influence of the first higher-order parasitic mode on the circuit performance. Unlike with radial stubs the "point" of contact is poorly defined at higher frequencies (Figure 6.100). In many cases, this kind of problem can be avoided if radial line stubs are used as shunt elements. Radial stubs can have very small input radii that give a well defined "point" of contact, and can, at the same time, provide low input impedance over a substantial frequency range (Figure 6.100).

Figure 6.101 shows the simulated input impedance versus frequency both for a radial stub with a vertex angle $\alpha = 90°$ and uniform transmission line stubs with a characteristic impedance of $Z_0 = 35\,\Omega$ and $Z_0 = 50\,\Omega$ on Al_2O_3 substrate ($\epsilon_r = 9.6$, $h = 0.65$ mm, $t = 20\,\mu$m). The $\lambda/4$-resonance was specified as 10 GHz. Regarding the response of the transmission line stubs, it can be seen that the bandwidth can be enlarged by choosing a low stub-impedance. However,

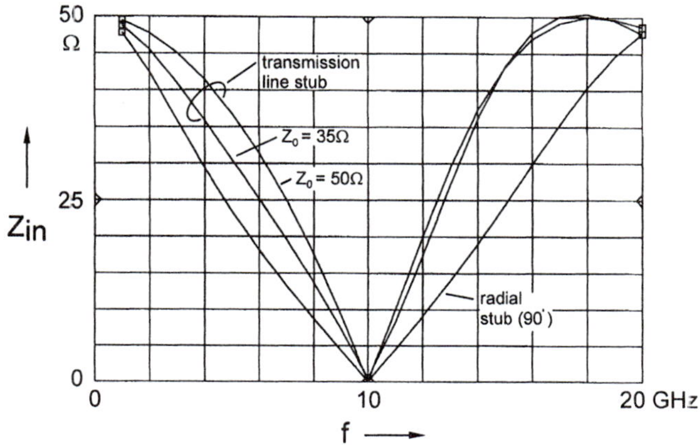

Figure 6.101 Comparison of the input impedance of uniform and radial stubs.

widening the microstrip line means lowering the cutoff frequency of the first higher-order HE_1 mode introducing increasing parasitic effects. Under worst conditions the cutoff frequency could be within the frequency range of consideration. The problem could be remedied by using two shunted stubs having a characteristic impedance that is twice the single stub. Regarding the simulated frequency-dependent input impedances of 50Ω, it can be clearly noticed that the radial stub exhibits a larger bandwidth.

The radial line stub is compact, usually having an outer radius smaller than the length of the $\lambda/4$-microstrip resonator for the same frequency and substrate [73].

6.8.1 Model After Vinding

Various analytical techniques have been employed in the literature to determine the resonant dimensions of the radial stub. Vinding [74] used the single-mode radial wave solution to calculate the input impedance of the stub. He assumed ideal magnetic wall boundaries at the periphery of the stub.

Recalling the analysis of the radial transmission line in Section 6.7.1, the line current was found by integration of the transversal magnetic field H_φ over the circumference (6.169). In the case of the sectoral radial line, this current has to be weighted by a factor $\alpha/360°$, where α is vertex angle in degrees,

$$I(\alpha) = I(\alpha = 360°)\,\frac{\alpha°}{360°} \tag{6.216}$$

Thus, with (6.180) the input impedance of an open-circuited microstrip radial stub with vertex angle α reads simply [74]

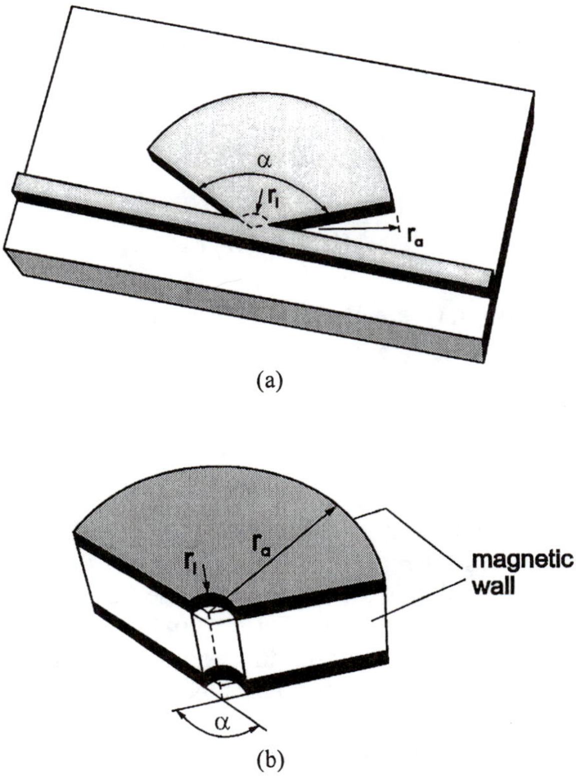

Figure 6.102 (a) Coupled radial stub and (b) magnetic wall model.

$$Z_i(r_i) = j\frac{h\,Z_0(r_i)}{2\pi r_i}\,\frac{\cos(\theta_i - \psi_L)}{\sin(\psi_i - \psi_L)}\,\frac{360}{\alpha} \tag{6.217}$$

The auxiliary variables and the characteristic impedance are defined in Section 6.7.1 [(6.175) to (6.178), and (6.166)]. It can be seen that the input impedance is directly proportional to the vertex angle α when all other dimensions are held constant.

The input impedance in (6.187) becomes zero for $(\theta_i - \psi_L) = n\pi/2$, $n = 1,2,\ldots$. For $n = 1$, the resonant condition reads

$$\theta_i - \psi_L = \arctan\left[\frac{N_0(kr_i)}{J_0(kr_i)}\right] - \arctan\left[\frac{J_1(kr_L)}{-N_1(kr_L)}\right] = \frac{\pi}{2} \tag{6.218}$$

With

$$\arctan\alpha - \arctan\beta = \arctan\frac{\alpha - \beta}{1 + \alpha\beta} \tag{6.219}$$

where $\alpha\beta > -1$ [66], we obtain from (6.218) a solution for the stub resonance condition

$$\frac{N_0(kr_i)}{J_0(kr_i)}\frac{J_1(kr_L)}{N_1(kr_L)} = 1 \tag{6.220}$$

where J_i and $N_i(i = 0, 1)$ are Bessel functions of the first and second kinds, respectively. r_i and r_L are the inner and outer radii.

$$k = \frac{2\pi f}{c_0}\sqrt{\epsilon_r} \tag{6.221}$$

is the radial wavenumber at the resonant frequency f in the substrate medium of relative dielectric constant ϵ_r, and c_0 is the velocity of light.

As would be expected from the negligence of the fringing electrical field at the sectoral side walls and the open-end effect, (6.220) has been found to be inaccurate for the determination of the resonant dimensions of the radial stub, leading in practice to stub resonance frequencies above the design value [73].

6.8.2 Model After Giannini

The actual radial stub, illustrated in Figure 6.103 (solid lines), is replaced by an effective shape. The effective angle is equal to the given vertex angle α. An effective radius and dynamic dielectric constant is used, as defined for the disk resonator in Section 6.7.3. From the geometry in Figure 6.103, the following relations for the effective inner and outer radius r_{ie} and r_{oe} can be derived [75]:

$$r_{ie} = \frac{w_e}{2\sin(\alpha/2)} \tag{6.222}$$

$$r_{oe} = r_0\left\{1 + \frac{2h}{\pi r_0}\left[\ln\left(\frac{\pi r_0}{2h}\right) + 1.7726\right]\right\}^{1/2} + \frac{w_e - w}{2}\begin{cases}\dfrac{1}{\sin(\alpha/2)} & \text{for } \alpha < \pi \\ 1 & \text{for } \pi < \alpha < 3\pi/2\end{cases} \tag{6.223}$$

The method of analysis is based on the electromagnetic field expansion in terms of resonant modes of the radial structure. It is assumed that the quasi-TEM mode in the main line excites only radial modes with only radial dependence. Then, the

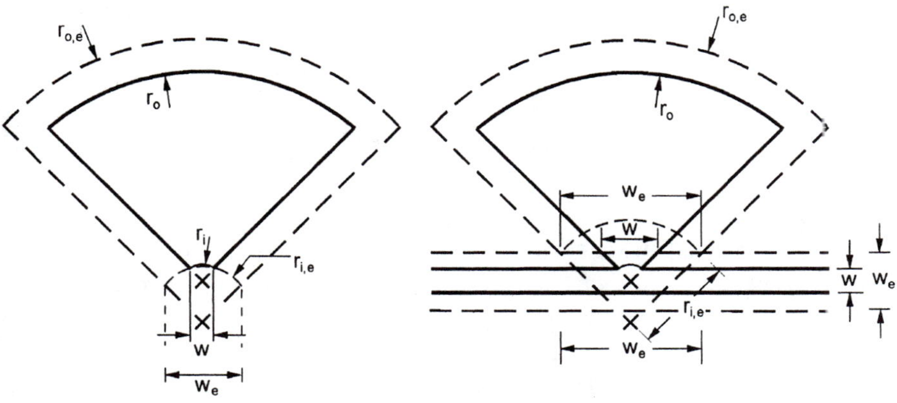

Figure 6.103 Magnetic wall model after Giannini. (From: [75], © 1984 IEEE. Reprinted with permission.)

input impedance Z_{in} normalized to the characteristic impedance of the main microstrip line can be expressed as

$$Z_{in} = -j \frac{k_g P_{oo}^2}{k^2 \epsilon_{do}} + jk_g \sum_{n=1}^{\infty} \frac{P_{on}^2}{k_{on}^2 - k^2 \epsilon_{dn}} \tag{6.224}$$

where

$$P_{on} = \sqrt{w_e}\left[A_{on} J_0(k_{on} r_{ie}) + B_{on} N_0(k_{on} r_{ie})\right]$$

$$A_{on} = \sqrt{\frac{2}{\alpha}} \left\{ r_{oe}^2 \left[J_0(k_{on} r_{oe}) + K_n N_0(k_{on} r_{oe})\right]^2 - r_{ie}^2 \left[J_0(k_{on} r_{ie}) + K_n N_0(k_{on} r_{ie})\right]^2 \right\}^{1/2}$$

$$B_{on} = K_n A_{on},\ K_n = -J_1(k_{on} r_{oe}) / N_1(k_{on} r_{oe}),\ k = \frac{\omega}{c_0},\ k_g = \frac{\omega}{c_0}\sqrt{\epsilon_{eff}}$$

P_{on} is the expansion coefficient (a measure for the coupling of the radial modes with the stimulating quasi-TEM mode), ϵ_{dn} is the dynamic effective dielectric constant after Wolff [70], and k_{on} is the eigenvalue of the radial modes.

Coimbra [76] proposed a radial line with input at the vertex of an isosceles triangle. For the computation of its input impedance, the element was modeled as a cascaded chain of microstrip transmission line segments of increasing width.

6.8.3 Model After Atwater

Atwater proposed to replace the physical radial stub structure by an effective contour and effective dielectric constant. This effective model is then similar to the approach after Coimbra [76], subdivided into line segments of increasing width, as shown in Figure 6.104.

The characteristic impedance Z_0 and dielectric coefficient ϵ_{eff} of the microstrip segments are calculated after Hammerstad. The outer radius R_1 is corrected prior to the calculation, by use of the open-end correction of a microstrip line with a width equal to the length of the outer arc of the stub at the radius R_1. The corresponding excess length ($\Delta l/h$ calculated after Hammerstad) is derived from the span of arc of circle. With the parameters in Figure 6.104, the effective arc $w(R_2)$ reads $(2\pi R_2/360°)\alpha$, with α in degrees, which can be interpreted as a microstrip width w. The number of subdivisions of the stub outline should be sufficiently high to approximate straight radial line contour.

In Figure 6.105, the stub outer radius R_2 is plotted as a function of the product $\sqrt{\epsilon_r}\, f_0$, where ϵ_r is the substrate relative dielectric constant and f_0 is the resonance frequency in gigahertz. The dashed-line curve shows the radii calculated by use of the magnetic wall after Vinding. The resonance frequency of a radial line stub of a given radius and substrate parameter varies only slowly with sectorial angle α, showing a broad minimum of the resonance frequency f_0 at $\alpha = 90°$. In typical cases, for given radii and substrate parameters, the resonance frequency of the stub varies less than 5% with an angle in a range of $60° \le \alpha \le 120°$. As can be seen from Figure 6.105, the dependence of the resonance frequency versus outer radius R_2 can be described by a relatively simple algebraic expression. Atwater has proposed the following closed-form relationship:

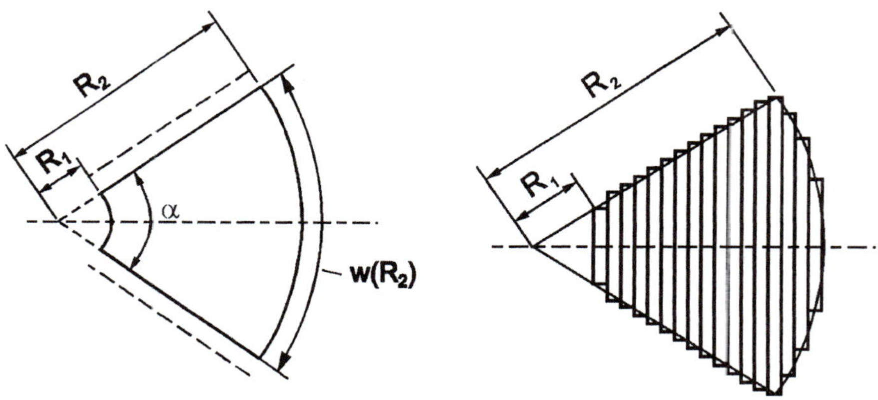

Figure 6.104 Dimensional parameters and segmentation of an open-ended radial line stub.

Table 6.2
Parameters of (6.225)

α	*A*	*B*	*C*	*D*
60°	–0.8232	0.0572	0.1169	–0.8082
90°	–0.8510	0.0614	0.0877	–0.8695

$$\log R_2 = A\log\left(\sqrt{\epsilon_r}f_0\right) + B\log(H) + C\log(R_1) + D \qquad (6.225)$$

The coefficients A, B, C, and D depend upon stub angle α. R_1, R_2, and H are the inner and outer radii and the substrate height in meters, and f_0 is the resonance frequency in gigahertz. The coefficients assume the values shown in Table 6.2. For substrate dielectric constants $2 \le \epsilon_r \le 15$ and for stub resonances in the frequency range $0.3 \le f_0/\mathrm{GHz} \le 30$, the accuracy of the model is claimed to be within 1% to 2% for $\alpha = 60°$ and 2% to 4% for $\alpha = 90°$.

Problem: A radial stub on 25-mil alumina substrate with height $H = 0.635$ mm and dielectric constant $\epsilon_r = 9.8$ shall be designed with a resonant frequency $f_0 = 10$ GHz [73].

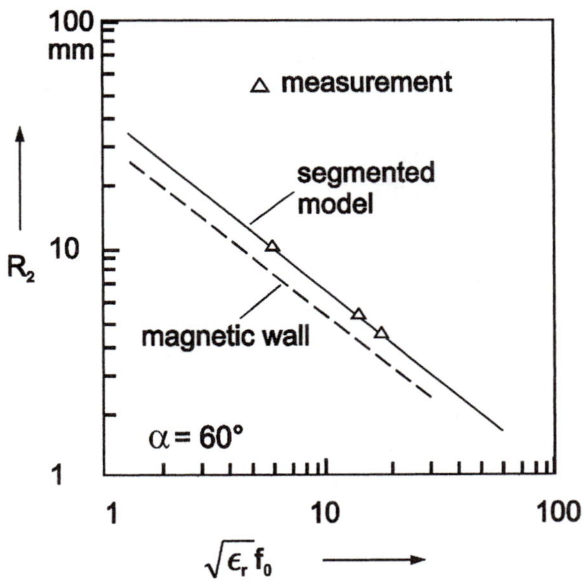

Figure 6.105 Resonance frequencies f_0 as function of the external effective radius R_2 ($\varepsilon_r = 9.8$, $h = 0.635$ mm, $R_1 = 50.8$ μm). (From: [73], courtesy of *Microwave Journal.*)

Solution: A 2-mil inner radius for the stub is selected, so that $R_1 = 50\,\mu\text{m}$. Then, substituting the parameters of Table 6.2 into the $\log R_2$ relation (6.225) for $\alpha = 60°$, we obtain

$$\begin{aligned}\log R_2 &= -0.8232\log\left(\sqrt{9.8}\cdot 10\right) + 0.0572\log\left(0.635\cdot 10^{-3}\right)\\ &\quad + 0.1169\log\left(50\cdot 10^{-6}\right) - 0.8082\\ &= -1.2311 - 0.18288 - 0.50279 - 0.8082\\ &= -2.7252\end{aligned}$$

Then the outer radius is calculated as $R_2 = 1.883\text{ mm}$.

In comparison with this value, the length of the $\lambda/4$ microstrip stub with a characteristic impedance of $50\,\Omega$ on this substrate, with dispersion and open-ended correction, is equal to 2.642 mm [73].

6.8.4 Bias Network

Bias networks are important devices in any active microstrip circuit. They should supply the active devices with the specified bias voltage and current. They can be realized in lumped form with lumped inductances and capacitances and as distributed networks. Figure 6.106 shows some examples used for microwave field-effect transistor (FET) circuits. Commonly, the gate electrode of the FET has negative potential with respect to ground, and the gate current is zero (class A amplifier operation). Unlike the gate the drain electrode is biased with a positive dc voltage, which causes a dc current.

The bias networks have to be implemented in the RF circuit without disturbing the RF signal paths. A typical bias network comprises a dc blocking capacitor, and an RF bias line in the form of a lowpass filter. Since the biasing must be achieved with minimum RF power loss to the bias line and dc source, the input impedance of the bias line should be an RF open circuit.

This can be realized, for example, by using a quarter-wave transformer that is virtually shorted by a resonating uniform transmission line stub. At the point of virtual ground, the dc power supply can be connected [Figure 6.106(a)]. When the design is performed carefully, the termination of the power supply will have no influence on the RF performance of the circuit under consideration. Generally the feeding point of the dc voltage should be a virtual short over a frequency range as broad as possible in order to achieve the required decoupling of power supply and the RF signal lines. Stubs of uniform microstrip transmission line with an electrical length of a quarter-wavelength, however, frequently do not exhibit sufficient bandwidth for the given application.

For active devices that must operate over a bandwidth greater than 100 MHz, the designed RF open circuit must also be broadband. In this case, as already discussed in this chapter, radial stubs are far more convenient for the design of broadband bias networks. Figure 6.106(b) shows a solution proposed in [65]. The radial stub has the shape of a half-moon radial line element. The broadband RF short circuit

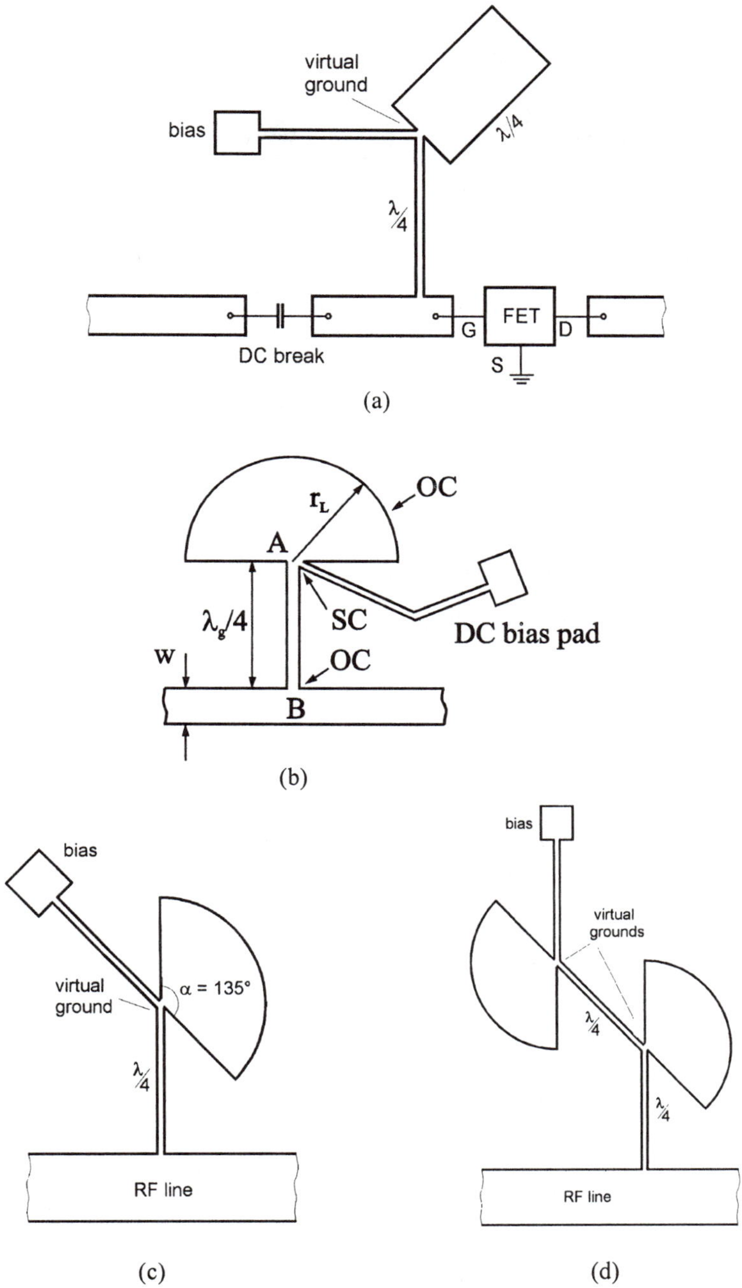

Figure 6.106 Various bias network configurations: (a) uniform stub, (b) design after Syrett, (c) radial stub with vertex angle of 135°, and (d) double radial stub.

at point A can be transformed into an RF open circuit at point B, with little sacrifice in bandwidth, by means of a quarter-wavelength microstrip line of high characteristic impedance. Again, the dc bias should be applied at point A through another line of high characteristic impedance, but with arbitrary length. Figure 1.106(c) shows a more appropriate solution, avoiding large vertex angle values α, because radial modes with an electromagnetic field depending on the azimuthal angle could be excited. Figure 1.106(d) shows a version with two cascaded radial stubs that provided appropriate power supply impedances over a 2-GHz bandwidth [77].

Problem: A capacitively coupled disk resonator, shown in Figure 6.107, is stimulated by a quasi-TEM wave at the input port (port 1). Input and output transmission lines are placed at 90° to each other. Determine the lowest two frequencies at which maximum power is transmitted from port 1 to port 2. For simplification, neglect any electrical stray field at the edge of the open resonator and use the geometric parameter R and material parameter ϵ_r for calculation.

Solution: Regarding the magnetic wall model after Watkins, the resonant frequencies read

$$f_{mn0} = \frac{c_0}{2\pi\sqrt{\epsilon_r}} \frac{\eta'_{mn}}{R}$$

The four lowest order modes in ascending order are: E_{110}, E_{210}, E_{010}, and E_{310}. Regarding the respective resonant field patterns in Figure 6.89, the following conclusions can be made (note that maximum electrical resonant field occurs near the stimulating port.):

E_{110}: Transmission of power is not possible with the given coupling configuration, because the electrical resonant field vanishes at an azimuthal angle of 90°.

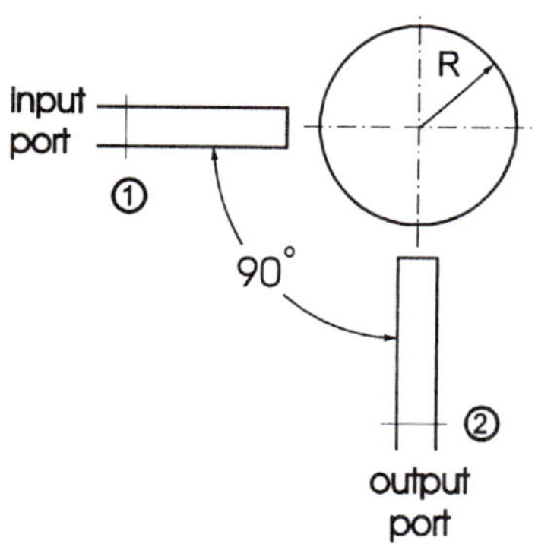

Figure 6.107 Disk resonator with coupled microstrip lines placed under 90° to each other.

Transmission would be possible only under 180° placement of the coupled lines.

E_{210}: Optimum transmission is possible. The circumferential electrical resonant field distribution comprises two wavelengths, which means that maximum power transfer occurs under 90°, as well as at integer multiples. The resonant frequency reads

$$f_{210} = \frac{c_0}{2\pi\sqrt{\epsilon_r}} \frac{\eta'_{21}}{R} = \frac{29.98 \text{ GHz cm } 3.054}{2\pi\sqrt{2.32}\ 2\text{cm}} = 4.78 \text{ GHz}$$

E_{010}: Power transfer occurs for any angle between the placed microstrip lines.

$$f_{010} = \frac{c_0}{2\pi\sqrt{\epsilon_r}} \frac{\eta'_{01}}{\eta'_{21}} = 4.78 \text{ GHz} \frac{3.831}{3.054} = 6.0 \text{ GHz}$$

6.9 Series Gap

The series gap interrupts the conductive strip of the microstrip line. The dc power cannot be transmitted. RF power transfer is accomplished by electrical field coupling. In Section 4.3.2, the gap was used to excite microstrip resonators. Figure 6.108 shows the top view of the frequently used symmetric gap structure with identical line width on both sides, and its equivalent circuit having purely capacitive equivalent circuit parameters. This is not given for the general case of an asymmetric gap of different line widths. Since it looks like a capacitively coupled abrupt impedance step, inductive equivalent circuit parameters have to be included [78].

The gap can be described as two coupled open-ended microstrip lines. Thus, the capacitance C refers to the open-end capacitance, and the series gap capacitance C_g describes the electrical coupling. Up to approximately 20 GHz, the frequency dependence of the equivalent capacitances is negligible [38, 78]. Figure 6.109 shows the equivalent capacitances of a symmetric gap on alumina substrate ($\epsilon_r = 9.6$) after Hoffmann [45] using a numerical static method [79, 80]. More accurate formulas for the capacitances have been published by Kirschning et al. [7, 45, 81]:

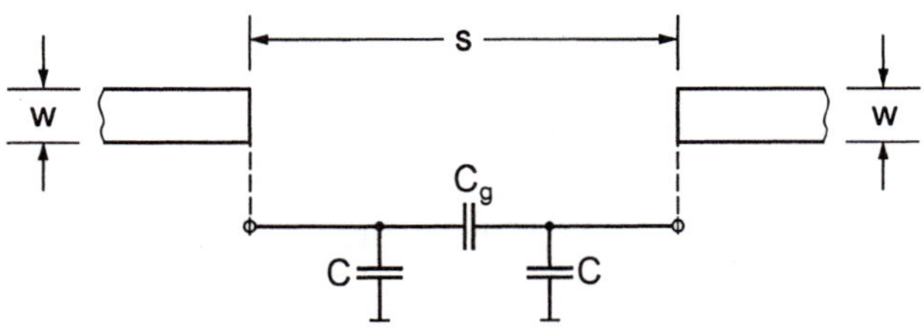

Figure 6.108 Symmetric gap structure and related equivalent circuit.

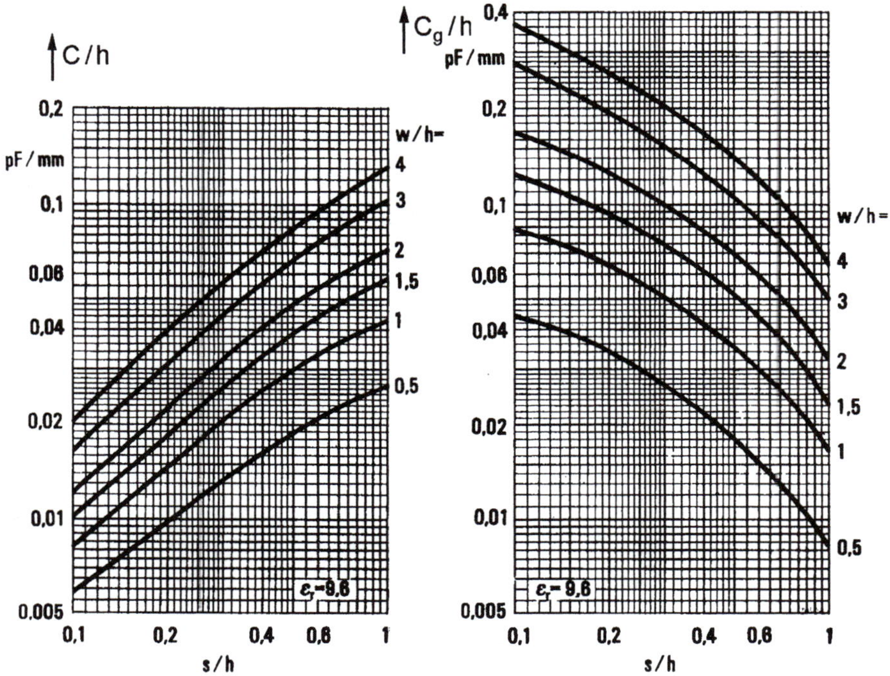

Figure 6.109 Equivalent circuit parameters of a symmetric gap. (From: [45], © 1983 Springer-Verlag. Reprinted with permission.).

$$C_g = 0.5\, h\, Q_1 \exp(-1.86\, s/h) \left\{ 1 + 4.19 \left[1 - \exp\left(-0.785 \sqrt{h/w_2}\; w_1/w_2\right) \right] \right\} \tag{6.226}$$

$$C_1 = C_{e1}(Q_2 + Q_4)/(1 + Q_2)$$

$$C_2 = C_{e2}(Q_2 + Q_3)/(1 + Q_2)$$

where

$$Q_1 = 0.04598\left[0.03 + (w_2/h)^{Q_5}\right](0.272 + 0.07\epsilon_r)$$

$$Q_2 = 0.107\left(\frac{w_2}{h} + 9\right)\left(\frac{s}{h}\right)^{3.23} + 2.09\left(\frac{s}{h}\right)^{1.05} \frac{1.5 + (0.3\, w_2/h)}{1 + (0.6 w_2/h)}$$

$$Q_3 = \exp\left[-0.5978\,(w_1/w_2)^{1.35}\right] - 0.55$$

$$Q_4 = \exp\left[-0.5978\,(w_2/w_1)^{1.35}\right] - 0.55$$

$$Q_5 = 1.23/\left\{1 + 0.12\left[(w_1/w_2) - 1\right]^{0.9}\right\}$$

C_{e1} and C_{e2} denote the respective capacitance of the open-end microstrip lines. Equation (6.226) was found by curvefitting to numerical data of full-wave analysis at a frequency of 4 GHz. The maximum error of 0.1 mS is obtained within the following parameter value ranges: $6 \le \epsilon_r \le 13$, $0.1 \le w_1/h \le 3$, $0.1 \le w_2/h \le 3$, $1 \le w_2/w_1 \le 3$, $0.2 \le s/h \le \infty$.

6.10 Cross Junction

The cross junction and its equivalent circuit are illustrated in Figure 6.110. It is commonly used in matching and filter networks. As discussed in Sections 6.8 and 8.1.3 low impedance stubs suffer from the excitation of the first higher-order mode $H_{[10]}$ (HE_1) on the stub. Although operating below cutoff the parasitic local wave can severely influence the quasi-TEM response. This can be avoided by replacing a low impedance stub by a cross junction with open-ended branches having twice the impedance value of the single stub, thus approximately doubling the cutoff frequency.

The following equations have been proposed by Garg and Bahl [40] for the equivalent circuit parameters referred to alumina substrate ($\epsilon_r = 9.9$).

$$\frac{C_+}{w_1} = \left\{\log\frac{w_1}{h}\left[86.6\frac{w_2}{h} - 30.9\left(\frac{w_2}{h}\right)^{1/2} + 367\right]\left(\frac{w_2}{h}\right)^3 + 74\frac{w_2}{h} + 130\right\}\left(\frac{w_1}{h}\right)^{-1/3}$$
$$- 240 + \frac{2}{w_2/h} - 1.5\frac{w_1}{h}\left(1 - \frac{w_2}{h}\right) \qquad (6.227)$$

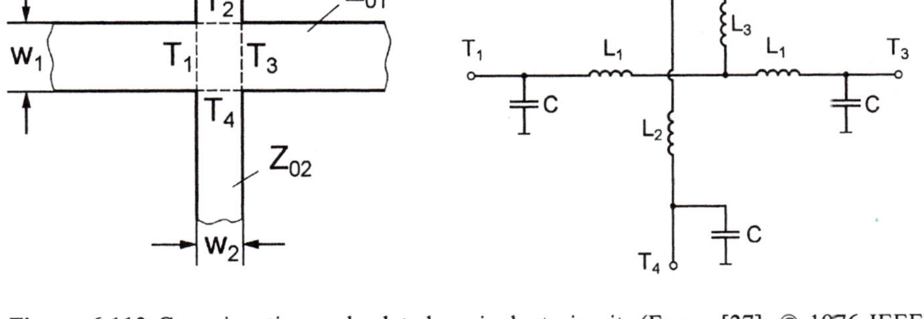

Figure 6.110 Cross junction and related equivalent circuit. (From: [27], © 1976 IEEE. Reprinted with permission.)

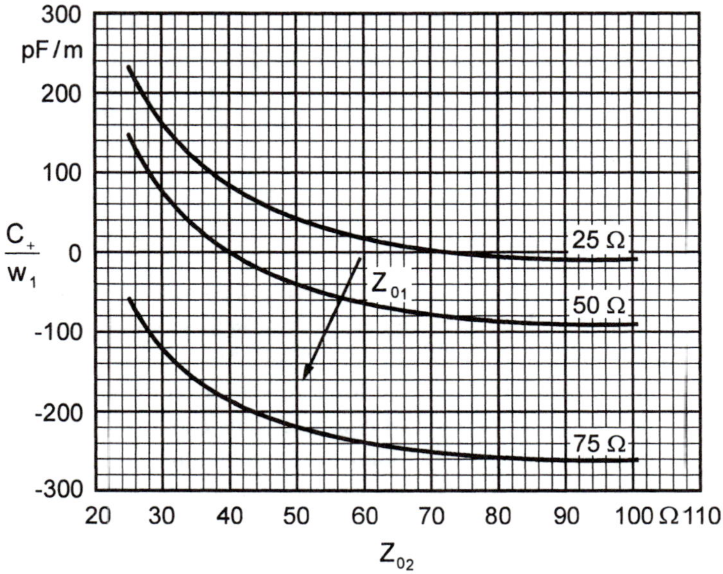

Figure 6.111 Calculated microstrip cross junction capacitance C_+ (ε_r = 9.9).

$$\frac{L_1}{h} = \left\{ \frac{w_1}{h} \left[165.6 \frac{w_2}{h} + 31.2 \sqrt{\frac{w_2}{h}} - 11.8 \left(\frac{w_2}{h} \right)^2 \right] - 32 \frac{w_2}{h} + 3 \right\} \left(\frac{w_1}{h} \right)^{-3/2} \tag{6.228}$$

$$-\frac{L_3}{h} = 337.5 + \left(1 + \frac{7}{w_1/h} \right) \frac{1}{w_2/h} - 5 \frac{w_2}{h} \cos\left[\frac{\pi}{2} \left(1.5 - \frac{w_1}{h} \right) \right] \tag{6.229}$$

where C_+ relates to C by the relation $C = C_+/4$. Accuracy is supposed to be within 5% in the following parameter range: C_+: $0.3 \le w_1/h \le 3.0$ and $0.1 \le w_2/h \le 3.0$; L_ν ($\nu = 1, 2, 3$): $0.5 \le (w_1/h, w_2/h) \le 2.0$. Normalized L_2 can be calculated from (6.228) by interchanging w_1 and w_2. From Figures 6.111 to 6.113, the equivalent circuit parameters can be read for given microstrip widths w_1 and w_2.

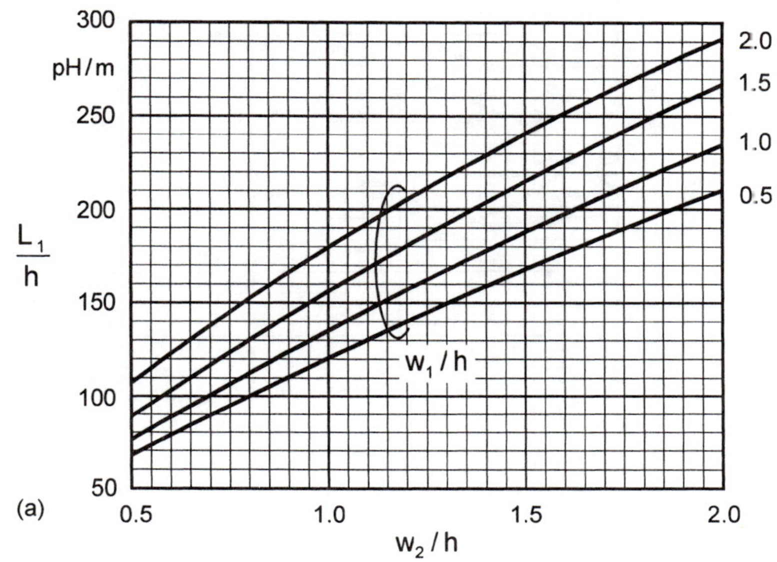

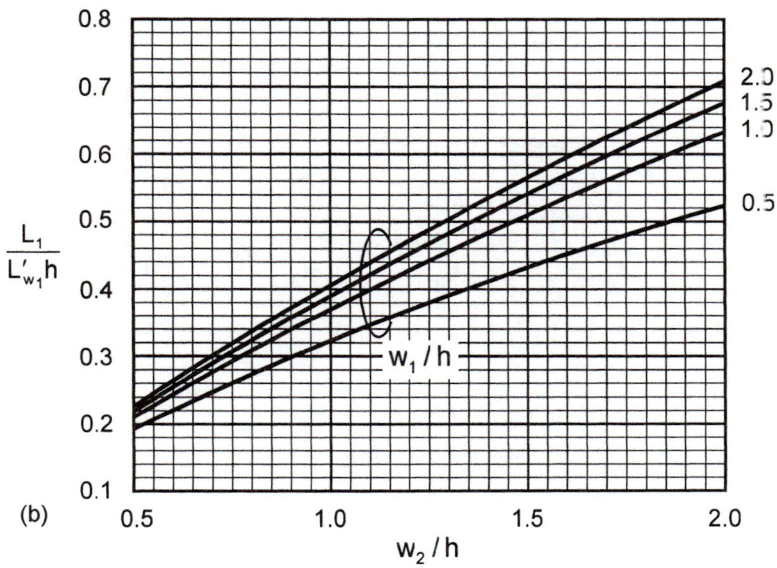

Figure 6.112 Normalized inductance L_1 of cross junction equivalent circuit ($\varepsilon_r = 9.9$).

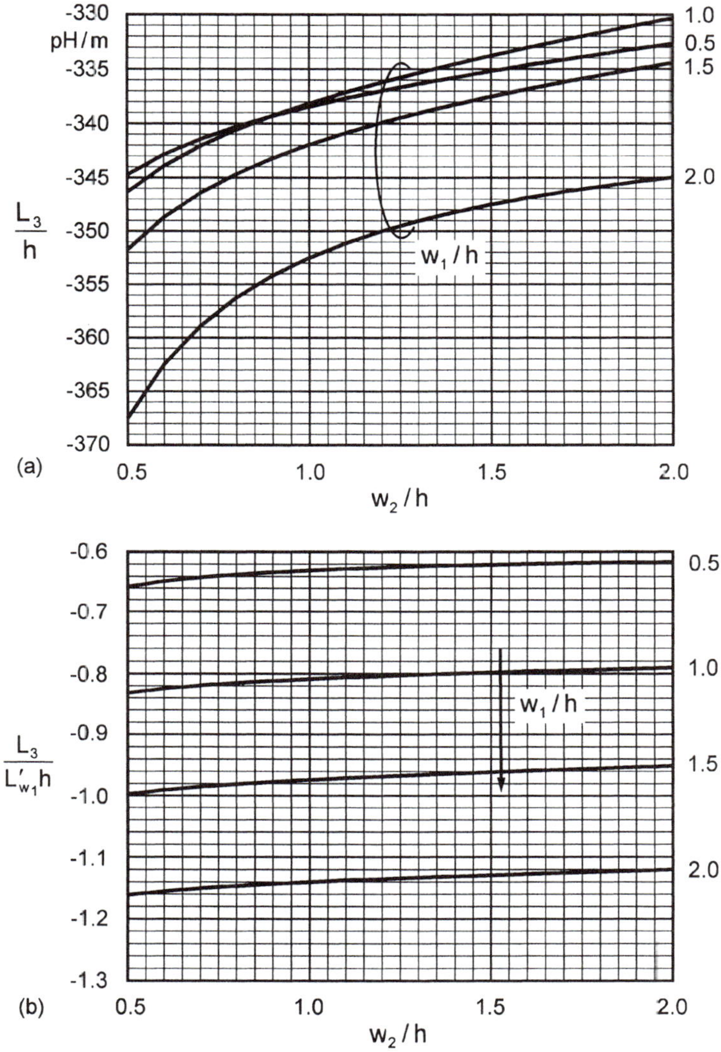

Figure 6.113 Normalized inductance L_3 of cross junction equivalent circuit ($\varepsilon_r = 9.9$).

References

[1] Itoh, T., *Numerical Techniques for Microwave and Millimeter-Wave Passive Structures*, New York: John Wiley & Sons, 1989.

[2] Schmidt, L.-P., *Zur feldtheoretischen Berechnung von transversalen Diskontinuitäten in Mikrostrip-Leitungen*, Doctoral Thesis, Aachen: RWTH, 1979.

[3] Koster, N. H. L., *Zur Charakterisierung der frequenzabhängigen Eigenschaften von Diskontinuitäten in planaren Wellenleitern*, Doctoral Thesis, University of Duisburg, 1984.

[4] Silvester, P., and P. Benedek, "Equivalent Capacitances of Microstrip Open Circuits," *IEEE Trans.*, Vol. MTT-20, 1972, pp. 511–516.

[5] Altschuler, H. M., and A. A. Oliner, "Discontinuities in the Center Conductor of Symmetric Strip Transmission Line," *IEEE Trans.*, Vol. MTT-8, 1960, pp. 328–339.

[6] Hammerstad, E. O., and F. Bekkadal, *Microstrip Handbook ELAB Report*, STF 44 A74169, University of Trondheim, Norway, 1975.

[7] Kirschning, M., *Entwicklung von Näherungsmodellen für den rechnergestützten Entwurf von hybriden und monolithischen Schaltungen in Mikrostreifenleitungstechnik*, Doctoral Thesis, University of Duisburg, 1984.

[8] Edwards, T., *Foundations for Microstrip Circuit Design*, 2nd ed., New York: John Wiley & Sons, 1992.

[9] Boukamp, J., and R. H. Jansen, "The High-Frequency Behaviour of Microstrip Open Ends in Microwave Integrated Circuits Including Energy Leakage," *Proc. European Microwave Conf.*, 1984, pp. 142–147.

[10] Wolff, I., "Messung der effektiven Dielektrizitätszahlen des geraden und des ungeraden Wellentyps auf verkoppelten Mikrostrip-Leitungen," *Nachrichtentechnische Zeitschrift*, Vol. 27, 1974, pp. 30–34.

[11] Olbrich, G., *Näherungsweise Berechnung des Wellenfeldes und meßtechnische Ermittlung der Eigenschaften von Mikrowellen-Streifenleitungen*, Doctoral Thesis, Technical University of Munich, 1979.

[12] Lewin, L., "Radiation from Discontinuities in Strip-Line," *The Institution of Electrical Engineers*, Paper No. 358 E, February 1960, pp. 163–170.

[13] James, J. R., and A. Henderson, "High-Frequency Behaviour of Microstrip Open-Circuit Terminations," *IEE J. Microwaves, Optics and Acoustics*, Vol. 3, September 1979, pp. 205–218.

[14] Wood, C., P. S. Hall, and J. R. James, "Radiation Conductance of Open-Circuit Low Dielectric Constant Microstrip," *Electron. Lett.*, Vol. 14, February 16, 1978, pp. 121–123.

[15] Sobol, H., "Radiation Conductance of Open-Circuit Microstrip," *IEEE Trans.*, Vol. MTT-19, 1971, pp. 885–887.

[16] Ramo, S., J. R. Whinnery, and T. van Duzer, *Fields and Waves in Communication Electronics*, 2nd ed., New York: John Wiley & Sons, 1984.

[17] Easter, B., and R. J. Roberts, "Radiation from Half-Wavelength Open-Circuit Microstrip Resonators," *Electron. Lett.*, Vol. 6, September 1970, pp. 573–574.

[18] Denlinger, E. J., "Radiation from Microstrip Resonator," *IEEE Trans.*, Vol. MTT-17, 1969, pp. 235–236.

[19] Knoppik, N., *Resonanzfrequenz und Gütefaktor von Resonatoren in unsymmetrischer Streifenleitungstechnik (Mikrostrip-Technik)*, Doctoral Thesis, Aachen: RWTH, 1975.

[20] Kompa, G., "Approximate Calculation of Radiation from Open-Ended Wide Microstrip Lines," *Electronics Lett.*, Vol. 12, April 1976, pp. 222–224.

[21] Angulo, C. M., and W. S. C. Chang, "The Launching of Surface Waves by a Parallel Plate Waveguide," *IRE Trans.*, Vol. AP-7, 1959, pp. 359–368.

[22] Neidert, R. E., and G. T. O'Reilly, "Very Large Impedance Steps in Microstrip," *IEEE Trans.*, Vol. MTT-22, 1974, pp. 808–810.

[23] Arntz, B., "Analyzing Abrupt Microstrip Transitions," *rfdesign*, March 2002, pp. 22–32 (www.rfdesign.com).

[24] Shirasaki, H., "Optimum Design by Waveguide Model and Mode-Matching Technique of Microstrip Line Taper Shapes for Satellite Broadcast Planar Antenna," *1999 International Symposium Digest*, IEEE Antennas and Propagation Society, 1999, pp. 238–241.

[25] Shirasaki, H., "Approximate Step Analysis of Linear and Raised Cosine Tapered Microstrip Lines by Using a Waveguide Model," *Electronics and Communications in Japan*, Part 2, Vol. 80, No. 12, 1997, pp. 57–63.

[26] Gupta, K. C., R. Garg, and I. J. Bahl, *Microstrip Lines and Slotlines*, Dedham, MA: Artech House, Inc., 1979.

[27] Gopinath, A., A. F. Thomson, and I. M. Stephenson, "Equivalent Circuit Parameters of Microstrip Step Change in Width and Cross Junctions," *IEEE Trans.*, Vol. MTT-24, 1976, pp. 142–144.

[28] Gupta, K. C., et al., *Microstrip Lines and Slotlines*, Norwood, MA: Artech House, Inc., 1996.

[29] Kompa, G, *Microstrip Discontinuites*, Diploma Thesis, Aachen: RWTH, 1970 (in German).

[30] Mehran, R., *Grundelemente des rechnergestützten Entwurfs von Mikrostreifenleitungs-Schaltungen*, Verlag H. Wolff, Aachen, 1983.

[31] Brand, H., *Schaltungslehre linearer Mikrowellennetze*, Stuttgart, Germany, S. Hirzel Verlag, 1970.

[32] Kompa, G., *Die frequenzabhängigen Übertragungseigenschaften von Streifenleitungs-Wellenwiderstandssprüngen, Filtern und Stichleitungen in Mikrostrip-Technik*, Doctoral Thesis, Aachen: RWTH, 1975.

[33] Wolff, I., G. Kompa, and R. Mehran, "Calculation Method for Microstrip Discontinuities and T Junctions," *Electron. Lett.*, Vol. 8, 1972, pp. 177–179.

[34] Wolff, I., G. Kompa, and R. Mehran, "Streifenleitungsdiskontinuitäten und Verzweigungen," *Nachrichtentechnische Zeitschrift*, Vol. 25, 1972, pp. 217–224.

[35] Storck, H., *Streifenleitungen auf Halbleitermaterialien*, Doctoral Thesis, Aachen: RWTH, 1971.

[36] Kompa, G., and R. Mehran, "Microstrip Filter Analysis Using a Microstrip Waveguide Model," *The Radio and Electronic Engineer*, Vol. 50, No. 1/2, January/February 1980, pp. 54–58.

[37] Rizzoli, V., F. Mastri, and D. Masotti, "Modeling the Microstrip Impedance Step from Direct Measurements in the 1–27-GHz Band," *Microwave and Optical Technology Letters*, Vol. 4, No. 13, December 1991, pp. 562–566.

[38] Rizzoli, V., and A. Lipparini, "A Resonance Method for the Broad-Band Characterization of General Two-Port Microstrip Discontinuities," *IEEE Trans.*, Vol. MTT-29, 1981, pp. 655–660.

[39] Kompa, G., "Design of Stepped Microstrip Components," *The Radio and Electronic Engineer*, Vol. 48, No. 1/2, January/February 1978, pp. 53–63.

[40] Garg, R., and I. J. Bahl, "Microstrip Discontinuities," *Int. J. Electronics*, 45, No. 1, 1978, pp. 81–87.

[41] Silvester, P., and P. Benedek, "Microstrip Discontinuity Capacitances for Right-Angle Bends, T-Junctions and Crossings," *IEEE Trans.*, Vol. MTT-21, 1973, pp. 341–346.

[42] Edwards, T., *Foundations for Microstrip Circuit Design*, 2nd ed., New York: John Wiley & Sons, 1992.

[43] Kompa, G., "Reduced Coupling Aperture of Microstrip Stubs Provides New Aspects in Stub-Filter Design," *Proc. 6th European Microwave Conf.*, Rome, 1976, pp. 39–43.

[44] Dydyk, M., "Master the T-Junction and Sharpen your MIC Design," *Microwaves*, Vol. 16, 1977, pp. 184–186.

[45] Hoffmann, R. K., *Integrierte Mikrowellenschaltungen*, Berlin: Springer-Verlag, Berlin, 1983.

[46] Kompa, G.; "Frequency-Dependent Behaviour of Microstrip Offset Junctions," *Electron. Lett.*, Vol. 11, October 1975, pp. 537–538.

[47] Malherbe, J. A. G., and A. F. Steyn, "The Compensation of Step Discontinuities in TEM-Mode Transmission Lines," *IEEE Trans.*, Vol. MTT-26, 1978, pp. 883–885.

[48] Menzel, W., *Die frequenzabhängigen Übertragungseigenschaften von unsymmetrischen Kreuz- und T-Verzweigungen, Y-Verzweigungen sowie 90°- und 120°-Winkeln in Mikrostreifen-leitungstechnik,* University of Duisburg, 1977.

[49] Agilent Technologies, *Advanced Design System* (ADS), Version 2003(C).

[50] Hammerstad, E., "Computer-Aided Design of Microstrip Couplers with Accurate Discontinuity Models," *IEEE MTT-S Int. Microwave Symp. Digest*, 1981, pp. 54–56.

[51] Kirschning, M., and J. Jansen, "Accurate Model for Effective Dielectric Constant of Microstrip with Validity up to Millimetre-Wave Frequencies," *Electron. Lett.*, Vol. 18, March 18, 1982, pp. 272–273.

[52] Kobayashi, M., "Frequency Dependent Characteristics of Microstrips on Anisotropic Substrates," *IEEE Trans.*, Vol. MTT-30, 1982, pp. 2054–2057.

[53] Kobayashi, M., "A Dispersion Formula Satisfying Recent Requirements in Microstrip CAD," *IEEE Trans.*, Vol. MTT-36, 1988, pp. 1246–1250.

[54] Yamashita, E., K. Atsuki, and T. Hirahata, "Microstrip Dispersion in a Wide-Frequency Range," *IEEE Trans.*, Vol. MTT-29, 1981, pp. 610–611.

[55] Easter, B. et al., "Resonant Techniques for the Accurate Measurement of Microstrip Properties and Equivalent Circuits," *Proc. 1973 European Microwave Conf.*, 1973, Paper B. 7.5.

[56] Easter, B. et al., "Theoretical and Experimental Methods for Evaluating Discontinuities in Microstrip," *Radio and Electronic Engineer*, 48, No. 1/2, January/February 1978, pp. 73–84.

[57] Groll H., and W. Weidmann, "Measurement of Equivalent Circuit Elements of Microstrip Discontinuities by Resonant Method," *Nachrichtentechnische Zeitschrift*, Vol. 28, 1975, pp. 74–77.

[58] Abrie, P. L. D., *Design of RF and Microwave Amplifiers and Oscillators*, Dedham, MA: Artech House, Inc., 1977.

[59] Arditi, M, "Experimental Determination of the Properties of Microstrip Components," *Electrical Communication*, 1953, pp. 283–293.

[60] Weisshaar, A., and V. K. Tripathi, "Perturbation Analysis and Modeling of Curved Microstrip Bends," *IEEE Trans.*, Vol. MTT-38, 1990, pp. 1449–1454.

[61] Ishii, T. K., *Handbook of Microwave Technology*, Vol. I, Components and Devices, Academic Press, 1995.

[62] Kompa, G., "S-Matrix Computation of Microstrip Discontinuities with a Planar Waveguide Model," *Arch. Elektr. Übertr.*, Vol. 30, 1976, pp. 58–64.

[63] Khilla, K. M., "Ring and Disk Resonator CAD Model," *Microwave J.*, November 1984, pp. 91–105.

[64] Ramo, S., J. R. Whinnery, and T. van Duzer, *Fields and Waves in Communication Electronics*, 2nd ed., New York: John Wiley & Sons, 1984.

[65] Syrett, B. A., "A Broad-Band Element for Microstrip Bias or Tuning Circuits," *IEEE Trans.*, Vol. MTT-28, 1980, pp. 925–927.

[66] Bronstein, L. N., and K. A. Semendjajew, *Taschenbuch der Mathematik*, 24. Auflage, Zürich und Frankfurt/M: Verlag Harri Deutsch, 1989.

[67] Kirchhoff, G., *Gesammelte Abhandlungen*, Leipzig: Johann Ambrosius Barth, 1882, pp. 101–113.

[68] Zinke, O., and H. Seither, *Widerstände, Kondensatoren, Spulen und ihre Werkstoffe*, 2. Auflage, Berlin, Heidelberg, New York: Springer-Verlag, 1982.

[69] Knoppik, N., *Resonanzfrequenz und Gütefaktor von Resonatoren in unsymmetrischer Streifenleitungstechnik (Mikrostrip-Technik)*, Doctoral Thesis, Aachen: RWTH, 1975.

[70] Wolff, I., and N. Knoppik, "Mikrostrip-Scheibenresonatoren," *Arch. Elektr. Übertr.*, Vol. 28, 1974, pp. 101–108.

[71] Wolff, I., "Statische Kapazitäten von rechteckigen und kreisförmigen Mikrostrip-Scheibenresonatoren," *Arch. Elektr. Übertr.*, Vol. 27, 1973, pp. 44–47.

[72] Watkins, J., "Circular Resonant Structures in Microstrip," *Electron. Lett.*, Vol. 5, 1969, pp. 524–525.

[73] Atwater, H. A., "The Design of the Radial Line Stub: A Useful Microstrip Circuit Element," *Microwave Journal*, 1985, pp. 149–156.

[74] Vinding, J. P., "Radial Line Stubs as Elements in Strip Line Circuits," *NEREM RECORD*, 1967, pp. 108–109.

[75] Giannini, F., R. Sorrentino, and J. Vrba, "Planar Circuit Analysis of Microstrip Radial Stub," *IEEE Trans.*, Vol. MTT-32, 1984, pp. 1652–1655.

[76] De Lima Coimbra, "A New Kind of Radial Stub and Some Applications," *Proc. European Microwave Conf.*, 1984, pp. 516–521.

[77] Jaques, R., "A Highly Efficient 24 GHz GaAs MESFET Oscillator," *Proc. European Microwave Conf.*, 1985, pp. 397–402.

[78] Koster, N. H. L., and R. Jansen, "The Equivalent Circuit of the Asymmetrical Series Gap in Microstrip and Suspended Substrate Lines," *IEEE Trans.*, Vol. MTT-30, No. 8, August 1982, pp. 1273–1279.

[79] Benedek, P., and P. Silvester, "Equivalent Capacitances for Microstrip Gaps and Steps," *IEEE Trans.*, MTT-20, 1972, pp. 729–733.

[80] Gopinath, A., and C. Gupta, "Capacitance Parameters of Discontinuities in Microstriplines," *IEEE Trans.*, Vol. MTT-26, 1978, pp. 831–836.

[81] Kirschning, M., R. H. Jansen, and N. H. L. Koster,"Measurement and Computer-aided Modeling of Microstrip Discontinuites by an Improved Resonator Method," *IEEE GMTT Int. Microwave Symp. Dig.*, 1983.

Chapter 7

Microstrip Loss

There are three types of losses that occur in microstrip lines:

1. Conductor (or ohmic) losses in the strip conductor and the ground plane;
2. Dielectric losses in the substrate;
3. Radiation losses.

We consider only nonmagnetic dielectric substrates. If a magnetic substrate is used, then a magnetic loss component also will be present. Radiation plays an important role in the design of planar antenna elements. However, this subject is beyond the aim of this written material. In Section 6.1.2, we have discussed radiation loss of open-ended microstrip transmission lines in common microstrip circuits. In most conventional microstrip circuit designs with a high substrate dielectric constant, conductor losses in the strip conductor and the ground plane dominate over dielectric and radiation losses.

We assume that the microstrip line has only low loss. In this case, the distortion of the fields penetrating metal walls are negligibly small. Under this assumption, the current density on the conductor can be directly derived from the known distribution of the tangential magnetic field on the conductor surface. Regarding the theory of weakly lossy transmission lines, the voltage wave in the z direction of the transmission line $V(z)$ reads [1]

$$V(z) = V(z=0)\, e^{-\gamma z} = V(z=0)\, e^{-\alpha z} e^{-j\beta z} \tag{7.1}$$

where γ denotes the wave propagation coefficient, and α and β are the attenuation and phase coefficient, respectively. A similar relation can be written for the current wave propagating along the line. For all the line parameters, we prefer the notation “coefficient” instead of “constant,” because in communications these are really frequency-dependent, unlike in power electronics working with a fixed frequency of 50 Hz, and 16 2/3 Hz, respectively. The propagating power $P(z)$, which is merely the product of voltage and current, decreases exponentially with increasing transmission line coordinate z as [2]

$$P(z) = P_0 e^{-2\alpha z} \tag{7.2}$$

P_0 is the transmitted power at $z = 0$. From (7.2), it follows for the power loss $P_v^{\prime}(z)$ per unit length

$$P_v^{\prime}(z) = -\frac{dP(z)}{dz} = 2\alpha P(z) \tag{7.3}$$

Resolving for α gives

$$\alpha = \frac{-dP(z)/dz}{2P(z)} = \frac{P_v^{\prime}(z)}{2P(z)} \tag{7.4}$$

with α in nepers per unit length. A conversion of α into decibels per unit length is given by the relation

$$\alpha_{dB} = 8.686\,\alpha_{Np} \tag{7.5}$$

Regarding α as being the sum of a dielectric attenuation coefficient α_d and a conductive attenuation coefficient α_c,

$$\alpha = \alpha_d + \alpha_c \tag{7.6}$$

(7.4) yields

$$\alpha = -\frac{dP(z)/dz}{2P(z)} = \frac{P_c + P_d}{2P(z)} \tag{7.7}$$

where

$$\alpha_d = \frac{P_d}{2P(z)} \tag{7.8}$$

and

$$\alpha_c = \frac{P_c}{2P(z)} \tag{7.9}$$

Here, P_d and P_c denote the average dielectric power loss and the average conductor power loss per unit length, respectively. The attenuation a over a finite length l of transmission line is defined as

$$a = \alpha\, l \tag{7.10}$$

where a is given in nepers or decibels, and l is the regarded transmission line length. Accordingly, the attenuation per wavelength is obtained from the attenuation per unit length α by multiplying it by the guide wavelength $\lambda = \lambda_0/\sqrt{\epsilon_{eff}}$, where λ_0 is the free space wavelength, and ϵ_{eff} is the effective dielectric constant of the microstrip line,

$$a\,(l=\lambda) \;=\; \alpha\,\lambda \;=\; \alpha\,\frac{\lambda_0}{\sqrt{\epsilon_{eff}}} \;=\; \alpha\,\frac{c_0}{f\sqrt{\epsilon_{eff}}} \tag{7.11}$$

Problem: Derive the neper-decibel conversion relation (7.5) for the attenuation a.

Solution: The definition of the attenuation a reads

$$a/\mathrm{dB} \;=\; 10\log\frac{P_1}{P_2} \;=\; 20\log\frac{V_1}{V_2} \tag{7.12}$$

With (7.2) we get

$$\begin{aligned} a/\mathrm{dB} \;&=\; 20\log\left(e^{a/Np}\right) \;=\; 20\,a/\mathrm{Np}\,\log e \\ &=\; 8.686\,a/\mathrm{Np} \end{aligned} \tag{7.13}$$

The conversion relation (7.13) is often used in a different form. Setting $\log e = x$, taking base 10, $e = 10^x$, and then the natural logarithm $x = (\ln 10)^{-1}$, leads to the equality

$$\log(e) \;=\; 1/\ln 10$$

Thus, (7.13) also can be written as

$$a/\mathrm{dB} \;=\; \frac{20}{\ln 10}\;a/\mathrm{Np} \tag{7.14}$$

7.1 Conductor Loss

In a microstrip line with low loss dielectric substrate, the predominant sources of loss at microwave frequencies are the nonperfect conductors. The current density in the conductors is mainly concentrated in a sheet approximately one skin depth deep inside the conductor surfaces exposed to the electromagnetic field. When the frequency is sufficiently low, the transverse current can be neglected, so that only the longitudinal current essentially contributes to the conductive loss. Under this assumption, the power loss per unit length in the metal walls is calculated as [1]

$$P_c \;=\; \frac{R_{s1}}{2}\int_C |J_1|^2\,dx \;+\; \frac{R_{s2}}{2}\int_{-\infty}^{+\infty} |J_2|^2\,dx \tag{7.15}$$

where indices 1 and 2 denote the strip and the ground plane conductors, respectively. $J(x)$ is the respective surface current density on either electrode. $R_{s_\nu} = \sqrt{\pi f \mu_\nu \rho_\nu}$ ($\nu = 1, 2$) is the surface skin resistivity in ohms per unit square for the strip conductor and ground plane, respectively (Section 7.1.1). The quantities $\mu_{1,2}$ and $\rho_{1,2}$ represent the permeability and the bulk resistivity of the strip and

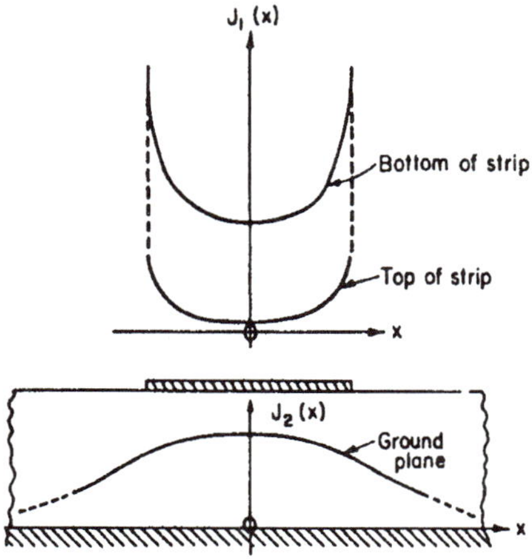

Figure 7.1 Current distribution on microstrip conductors after Pucel, et al. (From: [2], © 1968 IEEE. Reprinted with permission.)

ground conductors, respectively, and f denotes the operating frequency. The first integral of (7.15) is taken over all surfaces of the strip conductor, and the second integral over the infinite ground plate. With the transported power

$$P = \frac{1}{2} Z_0 |I|^2 \tag{7.16}$$

the attenuation per unit length α_c can be written as [2]

$$\alpha_c = \frac{R_{s1}}{2Z_0} \int_C \frac{|J_1|^2 \, dx}{|I|^2} + \frac{R_{s2}}{2Z_0} \int_{-\infty}^{+\infty} \frac{|J_2|^2 \, dx}{|I|^2} \tag{7.17}$$

where Z_0 denotes the characteristic impedance of the microstrip line, and $|I|$ is the magnitude of the total current of the transmission line. The strip conductor thickness t and the ground plane thickness both are assumed to be at least three skin depths.

The evaluation of (7.17) needs knowledge about the current distribution in the ground and upper strip. As is shown in Figure 7.1, for a strip of nonzero thickness t, the current density in the strip conductor and in the ground conductor is not uniform in the transverse plane because of the finite strip width. The strip conductor contributes the major part of the skin loss.

Wiesbeck has analyzed the current distribution in the metal strip and ground plane theoretically and experimentally [3]. Results of the uncovered microstrip line were published in [4], assuming quasi-TEM propagation so that only conductive loss due to the longitudinal current terms have been taken into account. Using the spectral domain method, Schumacher [5] has investigated the two-dimensional current distribution in the ground plane. Very good agreement was found between his simulated results and measurements published in [3].

Wheeler proposed in 1942 his incremental inductance rule [6]. This rule is widely used to calculate the conductive attenuation of a microstrip line dependent on its aspect ratio w/h, conductor thickness t and conductivity σ, and the frequency f. Before we discuss the method in Section 7.1.2, the skin effect is first revised.

7.1.1 Skin Effect

Conduction loss is described by the skin effect. In Figure 7.2, the lower half plane $(0 \le x < \infty,\ y,\ z)$ represents a conductive medium with electrical conductivity σ. The material is assumed to be nonmagnetic ($\mu_r = 1$). We assume a uniform alternating magnetic field directed tangential to the material surface $x = 0$. The magnetic field $\vec{H}$ lies in the y direction. Because of finite conductivity, the continuity condition implies that the magnetic field penetrates into the conductive material and contributes to power loss of the propagating wave. The total current density $\vec{G}$ reads

$$\vec{G} = \vec{J} + j\omega\vec{D} = \sigma\vec{E} + j\omega\epsilon\vec{E} = (\sigma + j\omega\epsilon)\vec{E} \tag{7.18}$$

where $\vec{J}$ and $\vec{D}$ are the volume and displacement current density, respectively. For highly conductive material, (7.18) can be approximated as

$$\vec{G} = \sigma\left(1 + j\frac{\omega\epsilon}{\sigma}\right)\vec{E} \approx \sigma\vec{E} \tag{7.19}$$

with

$$\frac{\omega\epsilon}{\sigma} \ll 1 \tag{7.20}$$

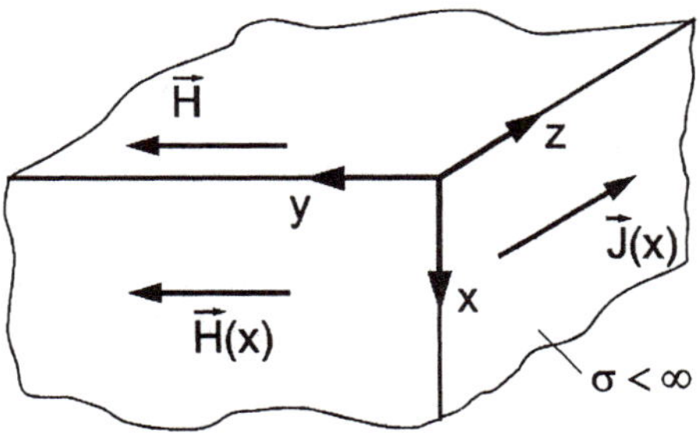

Figure 7.2 Definition of the coordinate system.

We recall Maxwell equations,

$$\text{rot}\,\vec{H} = \vec{\nabla}\times\vec{H} = \vec{J} \tag{7.21}$$

$$\text{rot}\,\vec{E} = \vec{\nabla}\times\vec{E} = -j\omega\mu_0\vec{H} \tag{7.22}$$

and the curl of a vector $\vec{A}$ in Cartesian coordinates,

$$\text{rot}\,\vec{A} = \vec{e}_x\left(\frac{\partial A_z}{\partial y} - \frac{\partial A_y}{\partial z}\right) + \vec{e}_y\left(\frac{\partial A_x}{\partial z} - \frac{\partial A_z}{\partial x}\right) + \vec{e}_z\left(\frac{\partial A_y}{\partial x} - \frac{\partial A_x}{\partial y}\right) \tag{7.23}$$

Then with

$$\vec{H} = H_y\,\vec{e}_y \tag{7.24}$$

it follows from (7.23) with $\partial/\partial z = 0$

$$\text{rot}\,\vec{H} = \vec{e}_z\,\frac{\partial H_y}{\partial x} \tag{7.25}$$

From (7.25), we obtain with (7.21)

$$\frac{\partial H_y}{\partial x} = J_z \tag{7.26}$$

From (7.22) with (7.23) and $\partial/\partial z = 0$, we derive

$$-j\omega\mu_0 H_y = \frac{\partial E_x}{\partial z} - \frac{\partial E_z}{\partial x} = -\frac{1}{\sigma}\frac{\partial J_z}{\partial x}$$

Multiplication by $(-\sigma)$ yields

$$\frac{\partial J_z}{\partial x} = j\omega\mu_0\sigma H_y \tag{7.27}$$

Equation (7.27) differentiated to x and inserted into (7.26) results in

$$\frac{\partial^2 J_z}{\partial x^2} - j\omega\mu_0\sigma J_z = 0 \tag{7.28}$$

with the general solution

$$J_z = e^{-\lambda x} \tag{7.29}$$

With $\partial J_z/\partial x = -\lambda\exp(-\lambda x)$ and $\partial^2 J_z/\partial x^2 = \lambda^2\exp(-\lambda x)$, we obtain from (7.28)

$$\lambda^2 - j\omega\mu\sigma = 0 \tag{7.30}$$

Then λ is obtained as

$$\lambda_{1,2} = \pm\sqrt{j\omega\mu\sigma} \tag{7.31}$$

With (7.31), the solution of the differential equation for $J_z(x)$ in (7.28) reads:

$$J_z(x) = C_1 e^{\lambda_1 x} + C_2 e^{\lambda_2 x} \tag{7.32}$$

The integration constants C_1 and C_2 are determined from the following conditions. The current density vanishes with $x \to \infty$ and becomes J_{z0} at $x = 0$,

$$J_z(x \to \infty) = 0 \quad \to \quad C_1 = 0$$

$$J_z(x = 0) = J_{z0} \quad \to \quad C_2 = J_{z0}$$

Thus, the current density reads

$$J_z(x) = = J_{z0}\, e^{-\frac{x}{\delta}(1 + j)} = J_{z0}\, e^{-\frac{x}{\delta}}\, e^{-j\frac{x}{\delta}} \tag{7.33}$$

with $\sqrt{j} = \exp(j\pi/4) = (1 + j)/\sqrt{2}$ and δ, being defined as

$$\delta = \sqrt{\frac{2}{\omega\mu_0\sigma}} = \left(\pi f \mu_0 \sigma\right)^{-1/2} \tag{7.34}$$

δ denotes the skin depth of a conductor, and it defines the distance to the conductor where the current density drops $1/e$ from a maximum current density of J_{z0}, or 36.7% of its value at the surface of the conductor.

The skin depth is halved if frequency is quadruplicated. Since the current can now flow in only one-half of the original crosssection, the resistance doubles. If magnetic conductors are considered, μ_0 must be replaced by the material permeability μ. The skin depth can be expressed in a different form:

$$\delta = \frac{1}{\sqrt{\pi\sigma}}\frac{1}{\sqrt{f\mu_0}}$$

$$= \sqrt{\frac{\lambda_0}{\eta_0\pi\sigma}} = \frac{1}{2\pi}\sqrt{\frac{\lambda_0 4\pi^2}{\pi\, 120\,\pi\sigma}} \tag{7.35}$$

$$= \frac{1}{2\pi}\sqrt{\frac{\lambda_0}{30\,\sigma}} = 0.0291\sqrt{\frac{\lambda_0}{\sigma}}$$

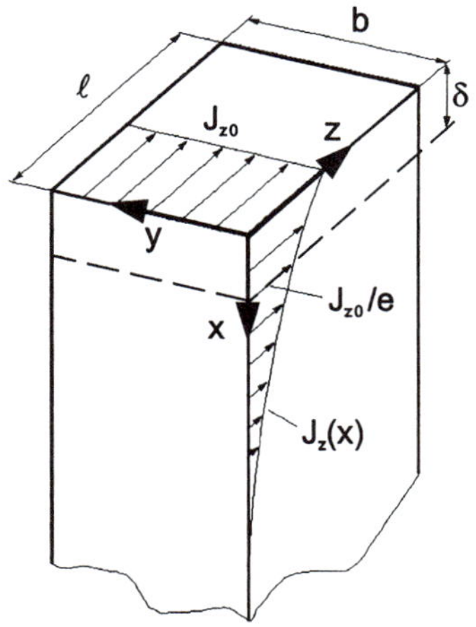

Figure 7.3 Skin effect with exponential decrease of current density in lossy metal.

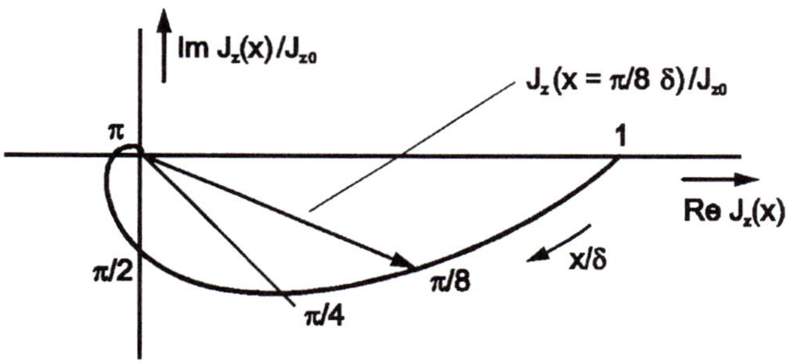

Figure 7.4 Representation of the current density in the complex plane.

With the surface resistance R_s in (7.55), the skin depth also reads

$$\delta = \frac{1}{R_s \sigma} \tag{7.36}$$

With finite conductivity, there is a nonuniform current density starting at the surface and exponentially decaying into the bulk of the conductive metal. At the skin depth of penetration, the current density will be reduced by a factor $1/e$ (or -8.68 dB) and its plane will lag 1 rad. Bearing in mind the exponential decay of the current density it is thus imperative that the conductor thickness should be at least 3 to 5 skin depths.

The amplitude of the current density decreases exponentially with increasing x (Figure 7.3). Furthermore, its phase is shifted with regard to the current density at the surface ($x = 0$) (Figure 7.4).

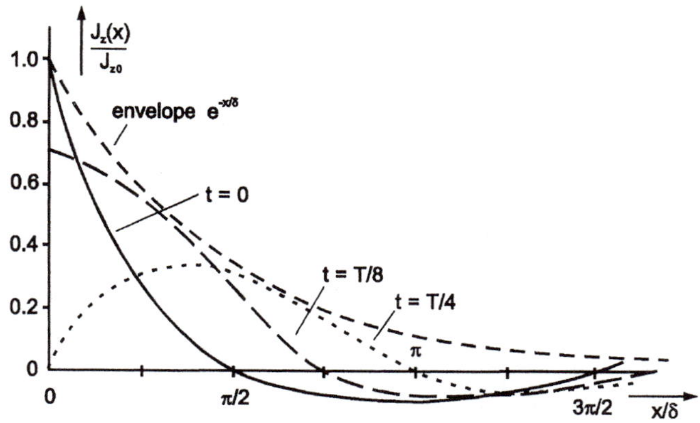

Figure 7.5 Current density along the x direction for different time values.

Time-dependent values $J_z(x,t)$ of the current density follow from (7.33) after multiplication by $e^{j\omega t}$ and taking the real part.

$$J_z(x,t) = \mathrm{Re}\left\{ J_{z0}\, e^{-\frac{x}{\delta}}\, e^{-j\frac{x}{\delta}}\, e^{j\omega t} \right\} = J_{z0}\, e^{-\frac{x}{\delta}} \cos\left(-\frac{x}{\delta} + \omega t \right) \tag{7.37}$$

with J_{z0} real. The current density at different times is calculated as follows:

$$t = 0: \quad J_z(x,0) = J_{z0}\, e^{-\frac{x}{\delta}} \cos\left(\frac{x}{\delta}\right)$$

$$t = \frac{T}{8}: \quad J_z\left(x, \frac{T}{8}\right) = J_{z0}\, e^{-\frac{x}{\delta}} \cos\left(-\frac{x}{\delta} + \frac{\pi}{4} \right)$$

$$t = \frac{T}{4}: \quad J_z\left(x, \frac{T}{4}\right) = J_{z0}\, e^{-\frac{x}{\delta}} \cos\left(-\frac{x}{\delta} + \frac{\pi}{2} \right)$$

The functions for different time points present damped cosine functions with an envelope curve $J_{z0}\exp(-x/\delta)$. As can be seen in Figure 7.5 the current density curves with the time parameter t describe a strongly attenuated wave propagating in the x direction. From (7.27) with (7.33), the magnetic field can be derived,

$$H_y(x) = -\frac{1}{j\omega\mu_0\sigma}\,\frac{1+j}{\delta}\, J_{z0}\, e^{-\frac{x}{\delta}(1+j)} = H_{y0}\, e^{-\frac{x}{\delta}(1+j)} \tag{7.38}$$

The magnetic field principally shows the same behavior as the current density.

In the following discussion, the conductor loss is analyzed. As illustrated in Figure 7.6, we assume a rectangular conductive block with a surface area $b \cdot l$, and infinite extension in the positive x direction.

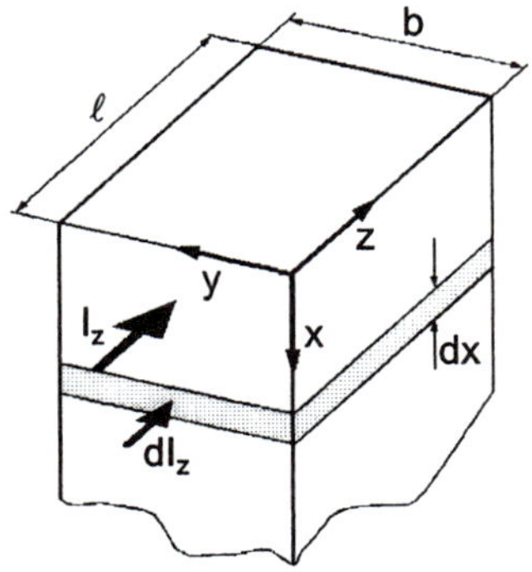

Figure 7.6 Calculation of power loss.

Power loss is calculated as follows. It is assumed that the total current I_z flows in the z direction. The infinitesimal current $dI_z(x)$ can be calculated from the current density $J_z(x)$ in (7.33) as

$$dI_z(x) = J_z(x)\, b\, dx = J_{z_0}\, e^{-\frac{x}{\delta}}\, e^{-j\frac{x}{\delta}}\, b\, dx \tag{7.39}$$

Notice that the phase of the infinitesimal current changes with the variable x. The magnitude then reads

$$|dI_z(x)| = |J_{z_0}|\, e^{-\frac{x}{\delta}}\, b\, dx \tag{7.40}$$

We define a resistance R_{dx} of the infinitesimal sheet,

$$R_{dx} = \frac{l}{\sigma\, b\, dx} \tag{7.41}$$

The conductor loss within the infinitesimal sheet dx is given as

$$\begin{aligned} dP_z &= \frac{R_{dx}|dI_z|^2}{2} = \frac{l}{\sigma\, b\, dx}\, \frac{|J_{z_0}|^2}{2}\, e^{-2\frac{x}{\delta}}\, b^2 (dx)^2 \\ &= \frac{|J_{z_0}|^2}{2}\, \frac{lb}{\sigma}\, e^{-2\frac{x}{\delta}}\, dx \end{aligned} \tag{7.42}$$

The total conductor loss is then calculated as

$$P_z = \int_{x=0}^{\infty} dP = \frac{lb}{\sigma}\,\frac{|J_{z_0}|^2}{2}\int_{x=0}^{\infty} e^{-2\frac{x}{\delta}}\,dx$$
$$= \frac{|J_{z_0}|^2}{4}\,\frac{lb\delta}{\sigma} \tag{7.43}$$

Notice that the skin effect concentrates current near the conductor surface, increasing resistance. Total current I_z reads

$$I_z = \int_{x=0}^{\infty} dI_z = J_{z_0} b \int_{x=0}^{\infty} e^{-\frac{x}{\delta}(1+j)}\,dx = J_{z_0} b\,\frac{\delta}{1+j} \tag{7.44}$$

$$I_z = J_{z_0} b\,\delta\,\frac{e^{-j\frac{\pi}{4}}}{\sqrt{2}} = \frac{J_{z_0} b\,\delta}{\sqrt{2}}\,e^{-j\frac{\pi}{4}} \tag{7.45}$$

It can be seen that the total current I_z lags the current density J_{z_0} at the surface by an angle of 45°. The magnitude of I_z turns out to be

$$|I_z| = \frac{|J_{z_0}|\,b\,\delta}{\sqrt{2}} \tag{7.46}$$

With J_{z_0} from (7.46),

$$|J_{z_0}| = \frac{|I_z|\sqrt{2}}{b\,\delta} \tag{7.47}$$

the total conductor loss P_z in (7.43) can be written as

$$P_z = \frac{|I_z|^2\,2bl\delta}{b^2\delta^2 4\sigma} = \frac{|I_z|^2}{2}\,\frac{l}{\sigma b\delta} \tag{7.48}$$

This is set equal to the power loss calculated from the total current I_z and equivalent total resistance R_z,

$$P_z = \frac{|I_z|^2}{2}\,R_z \tag{7.49}$$

By comparison of (7.48) and (7.49), the total loss resistance R_z is defined as

$$R_z = \frac{l}{\sigma b\delta} \tag{7.50}$$

Table 7.1
Conductivity of Relevant Materials

	Silver	*Copper*	*Gold*	*Aluminium*	*Brass*	*Carbon*
$\sigma/(\mathrm{S/cm})$	$62 \cdot 10^4$	$58 \cdot 10^4$	$41 \cdot 10^4$	$33 \cdot 10^4$	$13 \cdot 10^4$	$0.025 \cdot 10^4$
k_1	1.0	1.03	1.23	1.37	2.2	50

In conclusion, the power loss in a metal wall can be equivalently determined, assuming the total current I_z flowing uniformly in a sheet of thickness δ. Thus, the skin effect can be visualized by an approximation consisting of a uniform current density flowing in a layer near the surface of the metallic elements to a uniform skin depth, δ (Figure 7.7). When the frequency increases, the skin effect layer of current becomes thinner. Then, the current becomes more and more confined to the surface, so that the resistance increases.

The skin depth δ can be simply calculated from the following normalized equation

$$\frac{\delta}{\mu m} = 2.04\, k_1 \frac{1}{\sqrt{f/\mathrm{GHz}}} \tag{7.51}$$

with

$$k_1 = \sqrt{\frac{\sigma_{Ag}}{\sigma}} \tag{7.52}$$

where k_1 is a material parameter referenced to the conductivity of materials listed in Table 7.1 with respect to the conductivity of silver. The frequency is given in gigahertz.

We have already recognized the surface resistance R_s as an important parameter to determine the conductive loss in conductive walls of waveguiding structures. Different notations are used in the literature and will be discussed next.

Surface Resistance R_s

Other terms used are skin effect surface resistivity, sheet resistivity, and high-frequency skin effect resistance per unit square of plane conductor. It is assumed

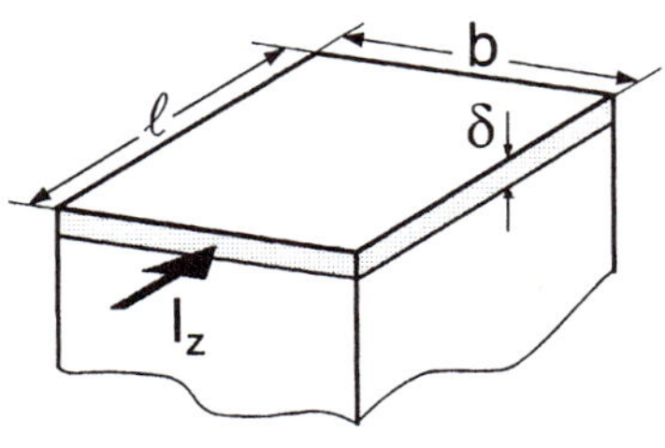

Figure 7.7 Model of skin effect resistance.

that the conductor is sufficiently thick, at least 5δ [1]. Due to the skin effect, the resistance of a conductive metal block with surface area of width b and length l reads

$$R = \frac{1}{\delta\,\sigma}\,\frac{l}{b} = R_s\,\frac{l}{b} \tag{7.53}$$

with

$$R_s = \frac{1}{\sigma\,\delta} = \sqrt{\frac{\pi f \mu}{\sigma}} = \sqrt{\pi f \mu \rho} \tag{7.54}$$

where ρ is the specific resistivity with dimension Ωcm. The resistance per unit square R_s is the resistance of a squared area ($l = b$), and its dimension is Ω per square. It is exactly the same as the dc resistance of a plane conductor of depth δ [1]. The normalized equation for R_s reads

$$R_s = 8.0 \cdot 10^{-3}\, k_1 \sqrt{f/\text{GHz}} \tag{7.55}$$

with k_1 defined in Table 7.1. R_s is also often symbolized as $R_\square$ in thin film technology specifying the sheet resistance of evaporated or sputtered metal layers.

Wave Impedance of Metal

On the surface of metal walls with finite conductivity, both tangential magnetic and electrical field components exist. They result in a Poynting vector perpendicular to the metal surface, meaning power flows into the metal. In Figure 7.5, it was discussed that power loss in conductive material can be thought of as being associated with a strongly damped electromagnetic wave propagating in the metal. Thus, we may derive a characteristic impedance from the transversal electrical and magnetic field components. With respect to the defined coordinate system in Figure 7.2, and taking the opposite sign of the electrical field for inward direction of the wave, the wave impedance reads

$$Z_{F,m} = \frac{-E_z(x)}{H_y(x)} \tag{7.56}$$

With (7.18) and (7.20),

$$E_z(x) = \frac{1}{\sigma}\,J_z(x) \tag{7.57}$$

and with (7.37) and (7.39), it follows for $Z_{F,m}$

$$Z_{F,m} = \frac{1}{\sigma}\,\frac{J_{z0}}{H_{y0}} = +\frac{1}{\sigma}\,\frac{J_{z0}}{(1+j)\,J_{z0}}\,j\omega\mu\sigma\delta = \frac{1+j}{2}\,\omega\mu\delta \tag{7.58}$$

With (7.34), the wave impedance is finally given as

$$Z_{F,m} = \frac{1 + j}{\sigma \delta} \tag{7.59}$$

Complex Internal Surface Impedance
With the notation in Figure 7.8, the complex surface impedance may be defined as

$$Z = \frac{V_z}{I_z} = \frac{E_{z0} l}{J_{z0} b \delta} \tag{7.60}$$

Substituting

$$J_{z0} = \sigma E_{z0} \tag{7.61}$$

in (7.60), we get

$$Z = \frac{(1 + j)}{\sigma \delta} \frac{l}{b} = Z_i \frac{l}{b} \tag{7.62}$$

Assuming a squared element area ($l = b$), Z becomes

$$Z_i = R_i + jX_i = \frac{1 + j}{\sigma \delta} \tag{7.63}$$

The magnitude of the real and imaginary part of the internal impedance per unit square is independent of frequency. The phase of Z_i is 45° for all frequencies. Thus, the surface impedance equals the wave impedance in metal. From the imaginary part of Z_i, the internal inductance can be derived,

$$\text{Im}\{Z_i\} = \omega L_i = \frac{1}{\sigma \delta} \tag{7.64}$$

$$L_i = \frac{1}{\omega \sigma \delta} = \mu \frac{\delta}{2} \tag{7.65}$$

In a similar way as was done with the resistive part of the skin effect impedance, we can also equivalently describe the properties of the reactive part and internal

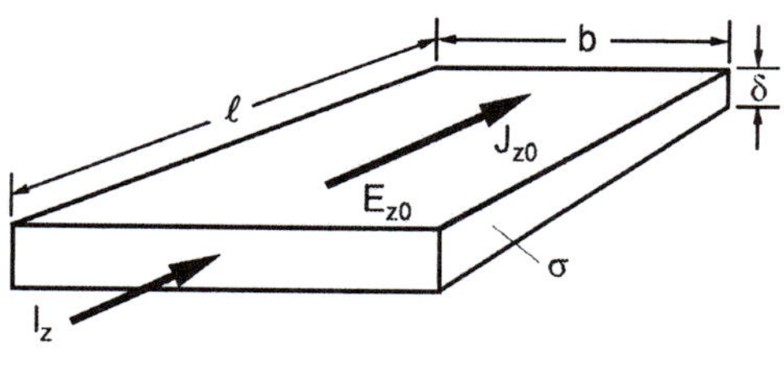

Figure 7.8 Calculation of the complex surface resistance.

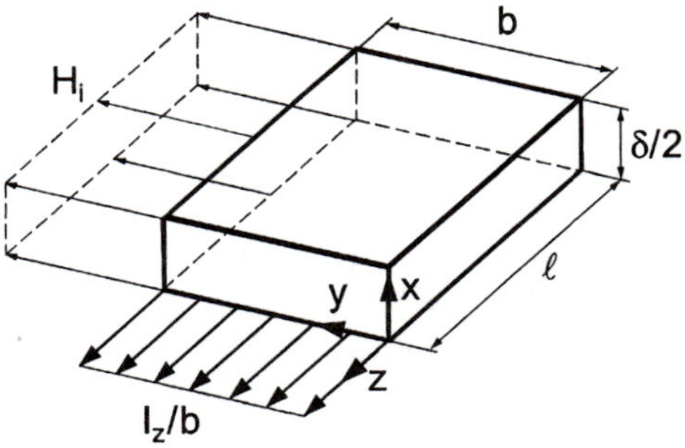

Figure 7.9 Model for the calculation of the internal inductance.

inductance, respectively, by a model shown in Figure 7.9. The model can be understood as a rule to calculate the inner inductance per unit length of a given conductive material from its perfect counterpart. It is assumed that the actual total current I_z occurs as a surface current density $I_{s,z} = I_z/b$ in the perfect model, receded by $\delta/2$ from the actual surface. Then, the internal inductance per unit length in the model is calculated from the surface current density and the associated uniform magnetic field H_i within the volume $bl\delta/2$ caused by the displacement. The inner inductance L_i (compare Section 3.1.5) can be calculated from $\phi_i = \mu_0 H_i \delta/2l$ and $I_{s,z} = H_i b$ as

$$L_i = \frac{\phi_i}{I_{s,z}} = \mu_0 \delta/2 \frac{l}{b}$$

The inner inductance per unit square $(b = l)$ then reads $L_i = \mu_0 \delta/2$, which corresponds to (7.65). Notice that the discussed model also can be interpreted in such a way that the displacement of the surface by $\delta/2$ will increase the external inductance by the respective value of the actual inner inductance of the actual conductive material. This is the essential conclusion in Wheeler's incremental inductance rule [6] discussed in the following section.

7.1.2 Incremental Inductance Rule

As will be shown, the incremental inductance rule permits the determination of the conductive loss of a transmission line from the external inductance per unit length L'. We have shown that the skin effect impedance per unit length, $Z_i' = R_i' + jX_i'$, has a real part R_i' that is equal to its imaginary part X_i', where $X_i' = \omega L_i'$.

Thus, knowing the internal inductance, the resistance per unit length, which is needed for power loss calculation, could be simply determined. Wheeler has pointed out that L_i' can be inferred from L' as the incremental increase in L' caused by an incremental recession of all metallic walls carrying a skin current.

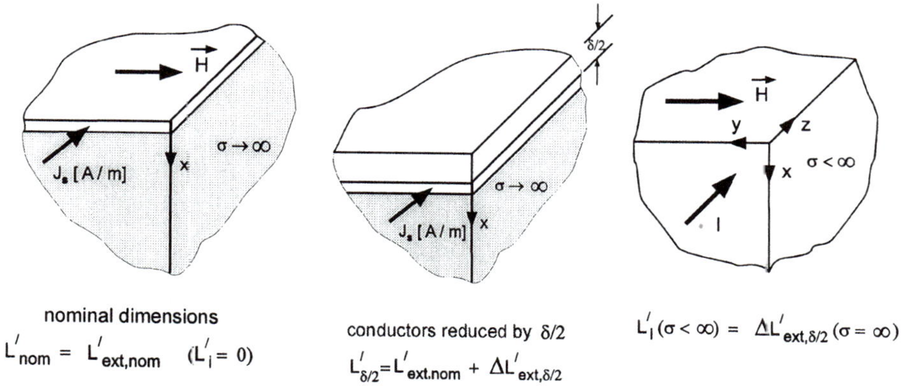

(a) (b) (c)

Figure 7.10 Illustration of Wheeler's incremental inductance rule: (a) lossless conductor with surface current J_s, (b) lossless conductor with receded wall, and (c) lossy conductor.

The amount of recession is equal to one-half the skin depth δ. To be in accordance with the results obtained for the planar skin effect, it is assumed that the radius of curvature and the thickness of the conductors will be several skin depths.

In a TEM transmission line with perfect boundaries and perfect homogeneous dielectric material, the inductance is proportional to the characteristic impedance of an air-filled microstrip line (3.85),

$$L' = \frac{Z_0(\epsilon_r = 1)}{c_0} \tag{7.66}$$

Figure 7.10 illustrates the incremental rule after Wheeler. If the conducting walls were perfect ($\sigma \to \infty$), then the total conductor current would appear as a surface current over the metal walls [Figure 7.10(a)]. Assuming that the surface current is shifted by a distance Δx into the inner part of all conducting walls, then this change would increase the total inductance of the structure by $\Delta L'$, by reason of decreased conductor cross section. If the recession of conducting walls is chosen as $\Delta x = \delta/2$ [Figure 7.10(b)], then the corresponding increase of the external inductance $\Delta L'$ (incremental inductance) is exactly the total internal inductance L_i' of the lossy structure [Figure 7.10(c)]. Thus, the calculation of the resistance per unit length can be formulated as follows

$$\begin{aligned} R' &= \omega L_i' \quad \text{(lossy case)} \\ &= \omega \big[\underset{\text{(lossless)}}{L' \text{ (nominal dimensions)}} - \underset{\text{(lossless)}}{L' \text{(conductors reduced by } \delta/2)} \big] \end{aligned} \tag{7.67}$$

Receding the conductive walls by Δx, the external (total) inductance is increased by $\Delta L'$. Series expansion yields

$$L'(x_j) \approx L'(x_{0_j}) + \left.\frac{\partial L'(x_j)}{\partial x_j}\right|_{x_j = x_{0_j}} \cdot \Delta x_j \qquad (7.68)$$

with the wall index *j*. Generally, the various walls can be of different material, so that the skin depth takes different values for the frequency under consideration. With

$$\Delta x_j = \delta_j / 2 \qquad (7.69)$$

it follows that the increase in inductance per unit length just equals the internal inductance $L'_{i,j}$ per unit length for *j*th wall of the lossy case:

$$L'_{i,j} = \left.\frac{\partial L'(x_j)}{\partial x_j}\right|_{x_j = x_{0_j}} \cdot \frac{\delta_j}{2} \qquad (7.70)$$

The R'_j part of the total loss resistance per unit length is then calculated from the internal inductance as

$$R'_j = 2\pi f\, L'_{i,j} = 2\pi f(\delta_j/2) \cdot \left(\frac{\partial L'}{\partial x_j}\right) \qquad (7.71)$$

With

$$\pi f \delta = \pi f \frac{\delta^2}{\delta}\frac{\sigma}{\sigma} = \pi f R_s \delta^2 \sigma = \pi f R_s \frac{1}{\pi f \mu_0} = \frac{R_s}{\mu_0} \qquad (7.72)$$

we get from (7.71)

$$R'_j = \frac{R_{s_j}}{\mu_0} \cdot \left.\frac{\partial L'}{\partial x_j}\right|_{x_j = x_{0_j}} \qquad (7.73)$$

If the conductor structure comprises *N* conducting walls, then the inner inductance and total loss resistance per unit length R' is derived from

$$L_i = \sum_{j=1}^{N} \frac{\partial L'}{\partial n_j} \frac{\delta_j}{2} \qquad (7.74)$$

$$R' = \sum_{j=1}^{N} \frac{R_{sj}}{\mu_0} \frac{\partial L'}{\partial n_j} \qquad (7.75)$$

where $\partial L'/\partial n_j$ denotes the derivative of the external inductance per unit length with respect to incremental recession of wall j. The normal direction n_j to this wall shows into the conducting region.

For the calculation of the inductance derivatives, it is helpful to refer to Figure 7.11, which shows the surface recessions to be considered.

There is some ambiguity concerning the recession of the bottom surface of the strip conductor. This can be counted towards a recession in t as well as an extension of h. As is pointed out in [2], by counting it as a recession of t, larger values for α_c have been obtained that agreed better with the measured results.

In accordance with these assumptions and Figure 7.11, recessions are [2]

$\delta_{n_j} = \delta h$ (recession of ground plane)

$\delta_{n_j} = \delta h$ (recession of bottom of strip conductor)

$\delta_{n_j} = -\delta w$ (recession of either edge of strip conductor)

$\delta_{n_j} = -\delta t$ (recession of top and bottom of strip conductor)

We obtain with (7.73)

$$R' = \frac{1}{\mu_0}\left[R_{s_1}\left(\frac{\partial L'}{\partial h} - 2\frac{\partial L'}{\partial w} - 2\frac{\partial L'}{\partial t}\right) + R_{s_2}\frac{\partial L'}{\partial h}\right] \tag{7.76}$$

From (7.9) with $P_c = 1/2\,|I|^2 R'$ and (7.16), the conductive attenuation coefficient per unit length α_c is related to the resistance per unit length as

$$\alpha_c = \frac{1}{2}\frac{R'}{Z_0(\epsilon_r)} = \frac{1}{2}\frac{R'}{Z_0(\epsilon_r = 1)}\sqrt{\epsilon_{eff}}$$

$$= \frac{\sqrt{\epsilon_{eff}}}{2\mu_0 Z_0(\epsilon_r = 1)}\sum_j R_{s_j}\frac{\partial L'}{\partial n_j} \tag{7.77}$$

Substituting R' gives [2]

$$\alpha_c = \frac{\sqrt{\epsilon_{eff}}}{2\mu_0 Z_0(\epsilon_r = 1)}\left[R_{s_1}\left(\frac{\partial L'}{\partial h} - 2\frac{\partial L'}{\partial w} - 2\frac{\partial L'}{\partial t}\right) + R_{s_2}\frac{\partial L'}{\partial h}\right] \tag{7.78}$$

Thus, the attenuation of the actual microstrip line with substrate carrier can be calculated from the attenuation of the air-filled line, multiplied by $\sqrt{\varepsilon_{eff}}$. Therefore, in the following, we first apply the incremental inductance rule to the air-filled microstrip line. The geometrical conductor parameters of the microstrip line are width w, height h, and thickness t. Let us assume that the skin resistance of the ground plane is different from the skin resistance of the strip ($R_{s,g} \neq R_{s,s}$), due to different conductor materials.

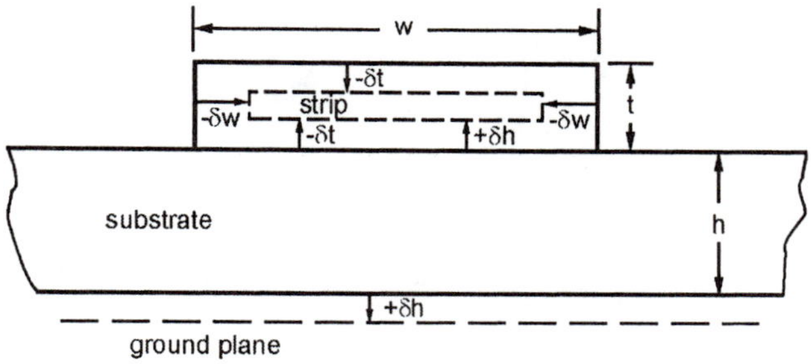

Figure 7.11 Application of the incremental inductance rule to the microstrip transmission line.

Substituting the inductance per unit length $L' = Z_0(\epsilon_r = 1)\sqrt{\epsilon_0\mu_0}$ in (7.73) results in

$$R' = \sqrt{\frac{\epsilon_0}{\mu_0}}\, R_s \frac{\partial Z_0}{\partial x} \tag{7.79}$$

Thus, the attenuation coefficient α_c in (7.77) can be rewritten as

$$\alpha_c = \sqrt{\frac{\epsilon_0}{\mu_0}}\, \frac{R_s}{2Z_0} \frac{\partial Z_0}{\partial x} \tag{7.80}$$

The attenuation $\alpha_{c,g}$ due to the ground plane with a skin resistance $R_{s,g}$ is obtained by receding the metal surface by $\delta n = \delta h$,

$$\alpha_{c,g} = \sqrt{\frac{\epsilon_0}{\mu_0}}\, \frac{R_{s,g}}{2Z_0} \frac{\partial Z_0}{\partial h} \tag{7.81}$$

The strip attenuation $\alpha_{c,s}$ with a skin resistance $R_{s,s}$ is obtained by reducing the strip dimensions by $2\delta w$ and $2\delta t$, and by increasing the ground plane spacing by δh

$$\alpha_{c,s} = \sqrt{\frac{\epsilon_0}{\mu_0}}\, \frac{R_{c,s}}{2Z_0} \left[\frac{\partial Z_0}{\partial h} - 2\frac{\partial Z_0}{\partial w} - 2\frac{\partial Z_0}{\partial t}\right] \tag{7.82}$$

The total attenuation α_c is then

$$\alpha_c = \alpha_{c,g} + \alpha_{c,s} \tag{7.83}$$

If the conductor materials for the ground plane and the strip are the same, we obtain $R_{s,g} = R_{s,s} = R_s$

$$\alpha_c = \sqrt{\frac{\epsilon_0}{\mu_0}} \frac{R_s}{Z_0} \left(\frac{\partial Z_0}{\partial h} - \frac{\partial Z_0}{\partial w} - \frac{\partial Z_0}{\partial t} \right) \tag{7.84}$$

The partial derivations can be written as [7]

$$\frac{\partial Z_0}{\partial w} = \frac{\partial Z_0}{\partial (w/h)} \frac{\partial (w/h)}{\partial w} = +\frac{1}{h} \frac{\partial Z_0}{\partial (w/h)} \tag{7.85}$$

$$\frac{\partial Z_0}{\partial h} = \frac{\partial Z_0}{\partial (w/h)} \frac{\partial (w/h)}{\partial h} = -\frac{w}{h^2} \frac{\partial Z_0}{\partial (w/h)} \tag{7.86}$$

$$\frac{\partial Z_0}{\partial t} = \frac{\partial Z_0(w/h)}{\partial (w/h)} \frac{\partial (w/h)}{\partial t} = +\frac{1}{h} \frac{\partial Z_0(w/h)}{\partial w/h} \frac{\partial w}{\partial t} \tag{7.87}$$

with $\partial w/\partial t$ being the derivative of w with respect to t for constant Z_0. The attenuation α in decibels per unit length is finally, using (7.14) and (7.84) [8]

$$\alpha = -\frac{R_s}{6\pi \ln 10} \frac{\partial Z_0}{\partial (w/h)} \frac{1 + w/h + \partial w/\partial t}{h Z_0} \tag{7.88}$$

The derivative $\partial w/\partial t$ can be calculated from Wheeler's relation [9] for finite strip thickness correction (3.73) which delivers simply

$$\frac{\partial w}{\partial t} = \frac{\partial}{\partial t}(\Delta w) = \frac{1}{\pi} \begin{cases} \ln\left(\dfrac{2h}{t}\right) & \text{for} \quad w/h \geq \dfrac{1}{2\pi} \\ \ln\left(\dfrac{4\pi w}{t}\right) & \text{for} \quad w/h \leq \dfrac{1}{2\pi} \end{cases} \tag{7.89}$$

In the following paragraphs, relations for the calculation of conductive attenuation published successively by Pucel, Schneider, and Hammerstad are briefly discussed. An early approach was published [10] concerning wide microstrip width with $w/h \geq 1$ based on conformal mapping transformation. Pucel et al. [2] calculated the external inductance from $L' = \mu_0 h/w_{eff}$ (3.81) and substituted w_{eff} of (3.44) after Wheeler [11]. Applying (7.74), he found the following formula for the conductive attenuation α_c of an air-filled microstrip line [2, 12]:

$w/h \leq 1/(2\pi)$:

$$\frac{\alpha_c Z_0(\epsilon_r = 1)\, h}{R_s} = \frac{8.68}{2\pi} \left[1 - \left(\frac{w'}{4h}\right)^2\right] \left[1 + \frac{h}{w'} + \frac{h}{\pi w'}\left(\ln\frac{4\pi w}{t} + \frac{t}{w}\right)\right] \tag{7.90}$$

$1/(2\pi) < w/h \le 2$:

$$\frac{\alpha_c Z_0(\epsilon_r = 1)\, h}{R_s} = \frac{8.68}{2\pi}\left[1 - \left(\frac{w'}{4h}\right)^2\right]\left[1 + \frac{h}{w'} + \frac{h}{\pi w'}\left(\ln\frac{2h}{t} - \frac{t}{h}\right)\right] \tag{7.91}$$

$2 \le w/h$:

$$\frac{\alpha_c Z_0(\epsilon_r = 1)\, h}{R_s} = \frac{8.68}{\left\{\frac{w'}{h} + \frac{2}{\pi}\ln\left[2\pi e\left(\frac{w'}{2h} + 0.94\right)\right]\right\}^2}\left[\frac{w'}{h} + \frac{w'/(\pi h)}{\frac{w'}{2h} + 0.94}\right]$$
$$\times\left[1 + \frac{h}{w'} + \frac{h}{\pi w'}\left(\ln\frac{2h}{t} - \frac{t}{h}\right)\right] \tag{7.92}$$

The primed width w' denotes the thickness-corrected width of the microstrip line (see Section 3.1.4).

Schneider applied Wheeler's incremental inductance rule to his own formulas (3.104) and (3.105), and obtained simpler expressions for the derivatives in (7.85) to (7.87) [8]. These read for the air-filled microstrip line:

$w/h \le 1$:

$$\alpha_c = \frac{10\, R_s}{\pi \ln 10}\; \frac{\left(\frac{8h}{w} - \frac{w}{4h}\right)\left(1 + \frac{h}{w} + \frac{h}{w}\frac{\partial w}{\partial t}\right)}{h\, Z_0(\epsilon_r = 1)\exp\left(\frac{Z_0(\epsilon_r = 1)}{60}\right)} \tag{7.93}$$

$w/h \ge 1$:

$$\alpha_c = \frac{R_s Z_0(\epsilon_r = 1)}{720\pi^2 h \ln 10}\left[1 + \frac{0.44h^2}{w^2} + \frac{6h^2}{w^2}\left(1 - \frac{h}{w}\right)^5\right]\left(1 + \frac{w}{h} + \frac{\partial w}{\partial t}\right) \tag{7.94}$$

The formula published by Hammerstad [13] is identical with (7.93) and (7.94), although it is modified in writing. For example, the term $\exp(Z_0/60)$ can be expressed by $8h/w + w/(4h)$ due to (3.117). The relation after Hammerstad reads

$w/h \le 1$:

$$\alpha_c h = \frac{10}{\pi \ln 10}\; \frac{R_s}{Z_0(\epsilon_r = 1)}\; \frac{32 - (w/h)^2}{32 + (w/h)^2}\left[1 + \frac{h}{w}\left(1 + \frac{\partial w}{\partial t}\right)\right] \tag{7.95}$$

$w/h \geq 1$:

$$\alpha_c h = \frac{20}{\ln 10} \frac{Z_0(\epsilon_r = 1) R_s}{\eta_0^2} \left\{ \frac{w}{h} + \frac{6h}{w} \left[\left(1 - \frac{h}{w}\right)^5 + 0.08 \right] \right\} \left\{ 1 + \frac{h}{w} \left(1 + \frac{\partial w}{\partial t} \right) \right\} \tag{7.96}$$

Figure 7.12 shows the normalized conductive attenuation per unit length as a function of w/h and t/h, calculated from (7.95) and (7.96) after Hammerstad. Using (7.90) to (7.92), lower loss values are calculated. This is particularly valid for narrow strips with thick conductive strips. For $w/h = 0.1$ and $t/h = 0.02$, for example, a difference in the order of about 6 dB may be obtained. Notice that regarding dielectric substrates, the relations for the attenuation have to be multiplied by $\sqrt{\epsilon_{eff}}$. Thus in (7.95), $Z_0(\epsilon_r = 1)$ is substituted by $Z_0(\epsilon_r)$, and in (7.96), by $Z_0(\epsilon_r = 1)\sqrt{\epsilon_{eff}} = Z_0(\epsilon_r)\,\epsilon_{eff}$, respectively.

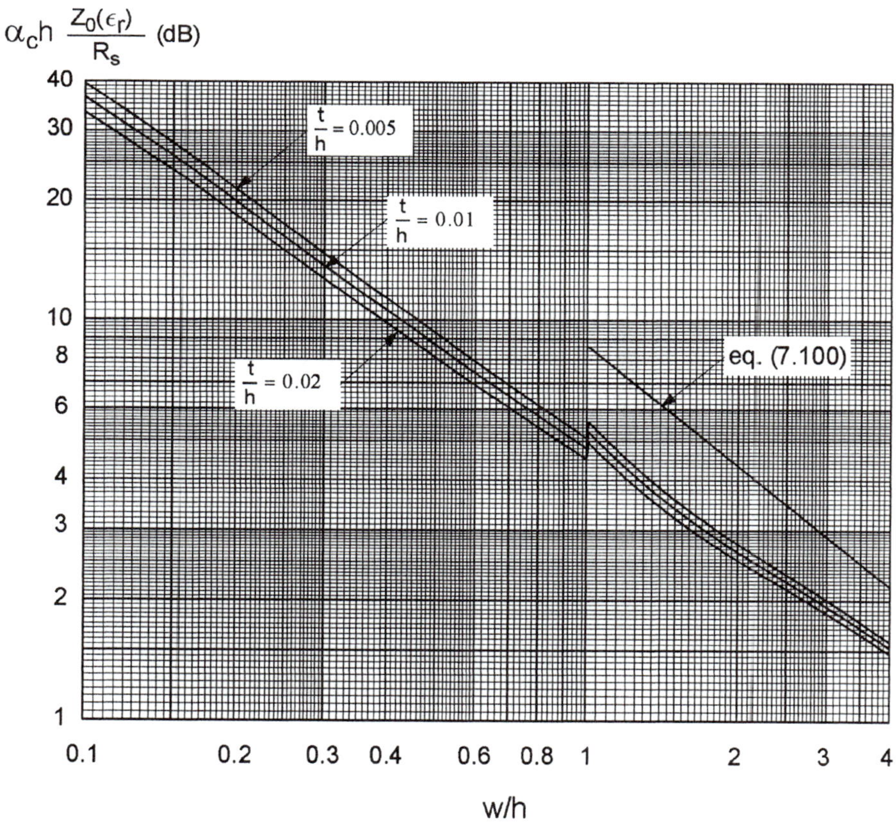

Figure 7.12 Normalized conductive attenuation per unit length as a function of *w/h*, with *t/h* as a parameter after Hammerstad, and approximated formula (7.103).

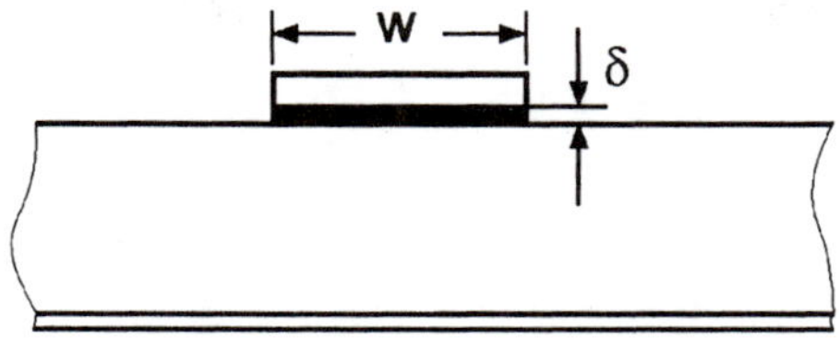

Figure 7.13 Simplified calculation of conductive loss.

With $R_s \sim 1/\delta \sim \sqrt{f}$ from (7.55), the conductor loss increases with the square root of frequency. Notice that the derived formulas are based on the assumption that a TEM wave is propagating along the line. However, this condition is violated more at the higher considered frequency.

In the following paragraph, a strongly simplified formula for the approximated calculation of the conductive attenuation per unit length is presented. We assume that the current distribution is uniform and equal to I/w in both conductors, and confined to the region $|x| < w/2$ (Figure 7.13). Then, the total resistance per unit length reads

$$\alpha_c = 2 \cdot \frac{R'}{2Z_0(\epsilon_r)} \tag{7.97}$$

With (7.53),

$$R' = \frac{R_s}{w} \tag{7.98}$$

we obtain for the attenuation per unit length α_c

$$\alpha_c = 2 \cdot \frac{R_s}{2\,Z_0(\epsilon_r)\,w} \tag{7.99}$$

in nepers per unit length. With (7.14), this reads [6]

$$\alpha_c \approx \frac{8.68\,R_s}{Z_0(\epsilon_r)\,w} \tag{7.100}$$

in decibels per unit length.

This simple expression is valid only for arbitrarily large strip widths, $w/h \to \infty$. Experimental data has shown that the approximation overestimates the skin loss by 80% for $w/h < 2$.

An important figure for a weakly lossy transmission line is its Q factor (2.12). For the series conductive loss, we get

$$Q_c = \frac{\beta}{2\alpha_c} = \frac{\pi}{\lambda\,\alpha_c} \tag{7.101}$$

with

$$\lambda = \frac{\lambda_0}{\sqrt{\epsilon_{eff}}} \tag{7.102}$$

For the approximate calculation of α_c discussed in Figure 7.13, we obtain from (7.100) with (7.54)

$$Q_c = \frac{\pi}{\lambda} \sigma \delta w Z_0(\epsilon_r) \tag{7.103}$$

7.2 Dielectric Loss

Loss in dielectric material originates in the internal polarization of the material and occurs only under ac excitation. Semiconductors may exhibit ohmic loss as well. Losses of either type can be characterized by the dielectric loss factor $\tan\delta_\epsilon$ [14]. Assuming a transmission line entirely embedded in uniform dielectric material with the dielectric constant ϵ_r, the dielectric quality factor Q_d can be written as (2.13)

$$Q_d = \frac{\beta}{2\alpha_d} = \frac{1}{\tan\delta_d} \tag{7.104}$$

Solving for the attenuation per unit length α_d gives

$$\alpha_d = \frac{\beta}{2} \tan\delta_d = \frac{\pi}{\lambda} \tan\delta_d \tag{7.105}$$

in nepers per unit length, and

$$\alpha_d = 27.3 \frac{\tan\delta_d}{\lambda} \tag{7.106}$$

in decibels per unit length, using (7.5) with $\lambda = \lambda_0/\sqrt{\epsilon_r}$. Multiplying the numerator and denominator by $(2f)$, (7.105) reads

$$\alpha_d = \frac{\omega}{2} \sqrt{\epsilon_r \epsilon_0 \mu_0} \tan\delta_d \tag{7.107}$$

If the conductive loss of the dielectric material prevails ($\sigma \gg \omega\epsilon$), then the loss tangent reads [14],

$$\tan\delta_d = \frac{\sigma}{\omega\epsilon} \tag{7.108}$$

and from (7.106), the dielectric attenuation per unit length reads

$$\alpha_{d,\sigma} = \sqrt{\frac{\mu_0}{\epsilon_r \epsilon_0}} \frac{\sigma}{2} \tag{7.109}$$

Regarding a microstrip transmission line with dielectric materials composed of air ($\epsilon_r = 1$) and substrate carrier ($\epsilon_r > 1$), the derived relations for the dielectric attenuation coefficient cannot be directly applied. Due to the partial filling, the effect of the loss tangent of the dielectric layer on the attenuation of the TEM mode must be known. We may first introduce an effective loss tangent $(\tan\delta_d)_{eff}$, so that (7.115) reads

$$\alpha_d = 27.3 \frac{(\tan\delta_d)_{eff}}{\lambda} \tag{7.110}$$

Schneider [15] has introduced a parameter p, which expresses the ratio of the electrical energy stored in the dielectric material W_d over the total electric field energy W,

$$p = \frac{W_d}{W} \tag{7.111}$$

With (7.111), the effective loss tangent can be defined as

$$(\tan\delta_d)_{eff} = p \cdot \tan\delta_d \tag{7.112}$$

It has also been shown that the partial derivative of the total electric field energy W with respect to the dielectric constant ϵ_r can be written as

$$\frac{\partial W}{\partial \epsilon_r} = \frac{W_d}{\epsilon_r} \tag{7.113}$$

The results found in (7.113) also can be applied to the equivalent homogeneous medium with effective dielectric constant, reading

$$\frac{\partial W}{\partial \epsilon_{eff}} = \frac{W}{\epsilon_{eff}} \tag{7.114}$$

Combining (7.113) and (7.114) yields [15]

$$\frac{\partial \epsilon_{eff}}{\partial \epsilon_r} = \frac{W_d}{W} \frac{\epsilon_{eff}}{\epsilon_r} \tag{7.115}$$

With (7.111) and the derivative of the function $\epsilon_{eff}(q, \epsilon_r)$ (3.59), we obtain

$$\frac{\partial \epsilon_{eff}}{\partial \epsilon_r} = q = p\,\frac{\epsilon_{eff}}{\epsilon_r} \tag{7.116}$$

Thus, we have a relation between Wheeler's dielectric filling fraction q and Schneider's dielectric energy fraction p [16],

$$p = \frac{\epsilon_r}{\epsilon_{eff}} \cdot q \tag{7.117}$$

With (7.117) and (7.112) and the filling factor q,

$$q = \frac{\epsilon_{eff} - 1}{\epsilon_r - 1} \tag{7.118}$$

the attenuation coefficient α_d can be written as

$$\alpha_d = 27.3\,\frac{\tan\delta_d}{\lambda_0}\,\frac{\epsilon_r}{\epsilon_r - 1}\,\frac{\epsilon_{eff} - 1}{\sqrt{\epsilon_{eff}}} \tag{7.119}$$

in decibels per unit length. The same relation was published by Welch and Pratt [14].

Applying (7.110) to conductive microstrip transmission lines, Welch and Pratt [14] defined an effective conductivity $\sigma_{d,eff}$ for the mixed dielectric material. In equivalence to the dielectric filling factor after Wheeler, this reads

$$\sigma_{d,eff} = q\,\sigma_d - (1 - q)\,\sigma_0 = q\,\sigma_d \tag{7.120}$$

assuming that the conductivity of air $\sigma_0 = 0$. Thus, for the microstrip line with conductive dielectric loss, we can write

$$\alpha_d = \sqrt{\frac{\mu_0}{\epsilon_{eff}\epsilon_0}}\,\frac{\sigma_{d,eff}}{2} \tag{7.121}$$

$$\alpha_d = \frac{\eta_0}{2}\,\frac{1}{\sqrt{\epsilon_{eff}}}\,\sigma_{d,eff}\,\frac{\epsilon_{eff} - 1}{\epsilon_r - 1} \tag{7.122}$$

in nepers per unit length, and

$$\alpha_d = 4.343\,\eta_0 \frac{1}{\sqrt{\epsilon_{eff}}}\,\sigma_{d,eff}\,\frac{\epsilon_{eff} - 1}{\epsilon_r - 1} \tag{7.123}$$

in decibels per unit length. The dielectric loss increases with frequency ($\sigma_d \sim f$). Both the conductive and relaxation losses contribute to the total dielectric loss. Often ohmic loss can be neglected with high-quality dielectric substrates. Electrical and mechanical properties of substrate materials are discussed in Chapter 9.

7.3 Conductor Roughness

As a consequence of the skin effect, the current in the strip and ground conductor is concentrated near the metal surface next to the substrate. In Section 7.1.2 (see Figure 7.12) it has been assumed that the conductor layer was deposited on a very smooth (polished) substrate surface. However, in practice, substrates feature roughness that should be taken into account for attenuation analysis.

In 1949, Morgan published simulated data about the increase in conductor loss due to periodic grooves of various geometries transverse to the direction of current flow [17]. The theoretical relationship between growing surface roughness and measurement was confirmed in [18, 19]. Hammerstad [13] approximated Morgan's calculations by a simple formula,

$$\alpha_c = \alpha_{c0}\left[1 + \frac{2}{\pi}\arctan 1.4\left(\frac{\Delta}{\delta}\right)^2\right] \tag{7.124}$$

where α_{c0} is the conductor attenuation coefficient at zero frequency, and δ is the skin depth. Δ is the root means square average (RMS) surface roughness and represents the standard deviation of the surface profile heights, taken over a defined evaluation length and area, respectively. Thin-film alumina substrates of various quality are available with RMS values of 0.05 to 0.127 μm. Polished substrates have very low values of 25 to 50 nm [20]. Plastic substrates generally possess higher roughness values of, for example, 1.4 μm for rolled, and from 1.9 to 3.0 μm for electrodeposited (ED) copper foils [21]. Thus, the surface roughness of plastic substrates is much more significant than for ceramic materials. This is also the reason why the simple formula (7.100) gives realistic results for plastic substrates, because it overestimates the theoretical losses by as much as 100% [13].

7.4 Total Loss

For comparison, the total conductive and dielectric loss of a 50Ω microstrip line on different substrates has been calculated versus frequency. The data of the regarded substrate materials is as follows:

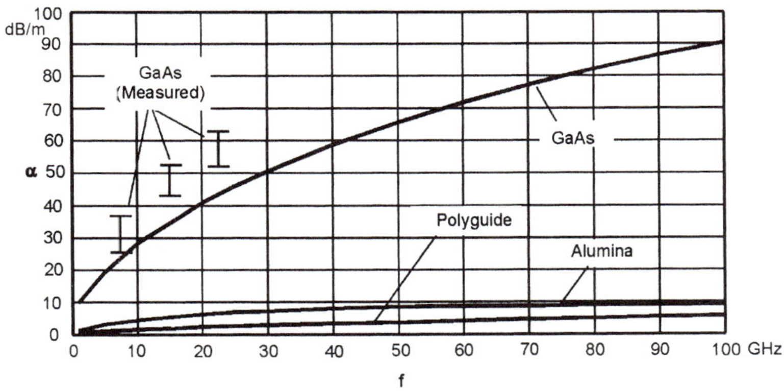

Figure 7.14 Calculated losses in microstrip lines ($Z_0 = 50\Omega$) versus frequency for various substrate materials.

1. Polyguide:
   $\epsilon_r = 2.32$, $h = 0.794$ mm, $t = 35$ μm, $\sigma = 59 \cdot 10^4$ S/cm, $\tan\delta_d = 0.0002$;

2. Al_2O_3:
   $\epsilon_r = 9.8$, $h = 0.635$ mm, $t = 5$ μm, $\sigma = 59 \cdot 10^4$ S/cm, $\tan\delta_d = 0.0001$;

3. GaAs:
   $\epsilon_r = 12.9$, $h = 100$ μm, $t = 3$ μm, $\sigma = 41 \cdot 10^4$ S/cm, $\tan\delta_d = 0$.

The calculation surface roughness of the conductor was not considered. The GaAs data is equal to those published in [22, 23]. As the height of alumina and Polyguide are approximately of similar value, it can be concluded from Figure 7.12 that alumina with $w/h \approx 1$ should have about 4 dB higher loss than Polyguide with $w/h \approx 3$, which is confirmed by the simulated results shown in Figure 7.14. The GaAs microstrip line has by far the highest loss, resulting primarily from the narrow width of about 70 μm. Notice that for given substrate heights, the loss is inversely proportional to the width w in case of very narrow widths, according to (7.95). Shown in Figure 7.14 are loss measurements of a 50Ω line on GaAs substrate material [22], where the bars denote the range of scattered measured data. The measured higher values may result from surface roughness of the conductor.

References

[1] Ramo, S., and J. R. Whinnery, *Fields and Waves in Modern Radio*, 1st ed., New York: John Wiley & Sons, 1944.

[2] Pucel, R. A., D. J. Massé, and C. P. Hartwig, "Losses in Microstrip," *IEEE Trans.*, Vol. MTT-16, 1968, pp. 342–350.

[3] Wiesbeck, W., "Berechnung und Messung der Stromverteilung auf ungeschirmten Streifenleitungen," *Nachrichtentechnische Zeitschrift*, Vol. 25, 1972, pp. 1–6.

[4] Wiesbeck, W., "Berechnung der Dämpfung ungeschirmter Streifenleitungen," *Wissenschaftliche Berichte AEG-TELEFUNKEN*, Vol. 45, 1972, pp. 162–166.

[5] Schumacher, W., "Stromverteilung auf der Grundfläche der Mikrostrip-Leitung und deren Auswirkung auf die ohmsche Leitungsdämpfung," *Arch. Elektr. Übertr.*, Vol. 33, 1979, pp. 207–212.

[6] Wheeler, H. A., "Formulas for the Skin Effect," *Proc. I.R.E.*, 30, No. 9, September, 1942, pp. 412–424.

[7] Schneider, M. V., "Microstrip Lines for Microwave Integrated Circuits," *Bell System Technical Journal*, Vol. 48, 1969, pp. 1421–1444.

[8] Schneider, M. V., B. Glance, and W. F. Bodtmann, "Microwave and Millimeter Wave Hybrid Integrated Circuits for Radio Systems," *Bell System Technical Journal*, Vol. 48, July–August 1969, pp. 1703–1726.

[9] Wheeler, H. A., "Transmission-Line Properties of Parallel Strips Separated by a Dielectric Sheet," *IEEE Trans.*, Vol. MTT-13, 1965, pp. 172–185.

[10] Assadourian, F., and E. Rimai, "Simplified Theory of Microstrip Transmission Systems," *Proc. I.R.E.*, December 1952, pp. 1651–1657.

[11] Wheeler, H. A., "Transmission-Line Properties of Parallel Wide Strips by a Conformal Mapping Approximation," *IEEE Trans.*, Vol. MTT-12, 1964, pp. 280–289.

[12] Pucel, R. A., D. J. Massé, and C. P. Hartwig, "Correction to "Losses in Microstrip"," *IEEE Trans.*, Vol. MTT-16, 1968, p. 1064.

[13] Hammerstad, E. O., and F. Bekkadal, *Microstrip Handbook ELAB Report*, STF 44 A74169, University of Trondheim, Norway, 1975.

[14] Welch, J. D., and H. J. Pratt, "Losses in Microstrip Transmission Systems for Integrated Microwave Circuits," *NEREM RECORD*, 1966, pp. 100–101.

[15] Schneider, M. V., "Dielectric Loss in Integrated Microwave Circuits," *Bell System Technical Journal*, Vol. 48, 1969, pp. 2325–2332.

[16] Wolff, I., *Einführung in die Mikrostrip-Leitungstechnik*, Aachen: Verlag H. Wolff, 2. Auflage, 1978.

[17] Morgan, S. P., "Effect of Surface Roughness on Eddy Current Losses at Microwave Frequencies," *Journ. Appl. Phys.*, 20, 1949, pp. 352–362.

[18] van Heuven, J. H. C., "Conduction and Radiation Losses in Microstrip," *IEEE Trans.*, Vol. MTT-22, 1974, pp. 841–844.

[19] Sobol, H., "Applications of Integrated Circuit Technology to Microwave Frequencies," *Proc. IEEE*, Vol. 59, No. 8, August 1971, pp. 1200–1211.

[20] CoorsTek, *Thin-Film Ceramic Substrates Design Guide*, 2005, www.coorstek.com.

[21] Rogers Corporation, "Copper Foils for High Frequency Circuit Materials," *Publication #92-243*, revised February 6, 2003.

[22] Goldfarb, M. E., and A. Platzker, "Losses in GaAs Microstrip," *IEEE Trans.*, Vol. MTT-38, 1990, pp. 1957–1963.

[23] Gupta, K. C., et al., *Microstrip Lines and Slotlines*, 2nd Edition, Norwood, MA: Artech House, Inc., 1996.

Chapter 8

Applications

In this chapter, some frequently used microstrip circuits are discussed concerning the design of microstrip filters, matching networks for power amplifiers, and directional couplers.

8.1 Microstrip Filter Design

Filter circuits are of great importance in communication systems. They are used, for example, for frequency band selection, channel separation, noise suppression, and matching. Microstrip filters generally include a number of discontinuities and junctions with parasitic capacitive and inductive effects, both of which are frequency-dependent. Figure 8.1 shows frequently used basic filter elements, such as T-junctions, stubs, steps, crossings, and shunt capacitances, which have been analyzed successfully by the mode matching procedure using the microstrip waveguide model (Chapter 5) [1–5].

The intention of this chapter is to consider microstrip discontinuities and junctions in a broader context of microstrip circuit design. Filter components are suitable representatives to study different kinds of parasitic effects, such as

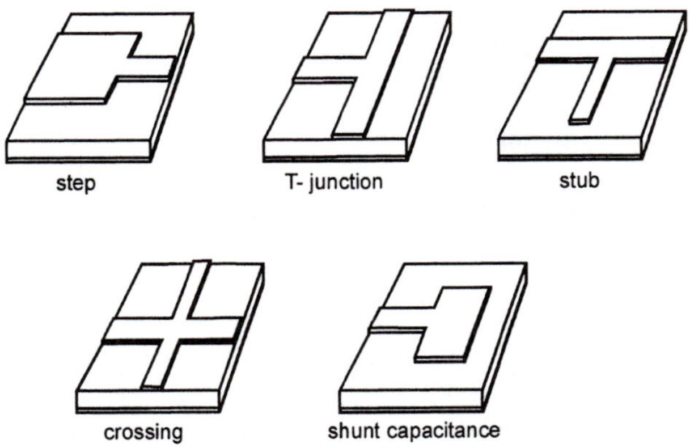

Figure 8.1 Typical discontinuities in microstrip filters.

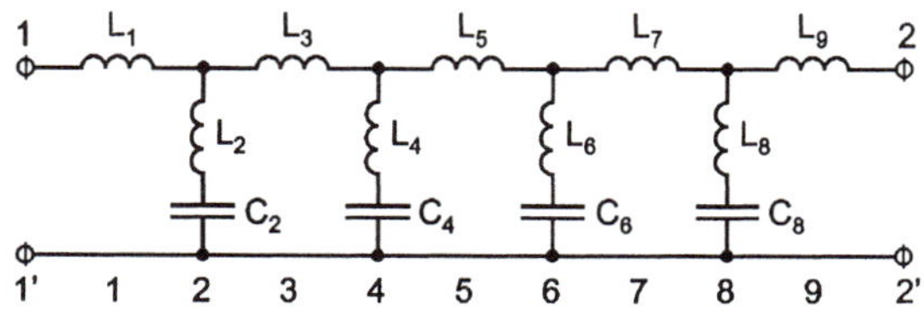

Figure 8.2 LC prototype filter of degree $n = 9$.

electrical stray fields at the transitions that entail shift of reference planes, perturbation of current flow leading to parasitic inductive effects, and at higher frequencies, higher-order mode excitation and radiation. As was already mentioned, the main scope of the book is the fundamental understanding of the physical effects commonly occurring in microstrip circuits. This knowledge will help the designer to analyze the performance of the circuit, and, what is most important, to improve it systematically with the help of commercially available software packages. But the designer should also be always fully aware of the limitation of any model due to frequency, geometry, material, and other parameters. In the following section, we will discuss filter structures that can be simply designed with the knowledge of the previous chapters.

Knowing the scattering coefficients of the structures in Figure 8.1, the design of a group of microstrip filters known as ladder networks [6–9] can be accomplished. Figure 8.2 shows an example of a Cauer parameter filter of ninth order, which will be discussed in more detail in the following sections. Anticipating the final approach, Figure 8.3 shows different layouts of the filter under consideration. In Figure 8.3(a), the series inductances are implemented by electrically short (quasi-lumped) high impedance microstrip lines. The series resonant circuits are modeled by open-ended transmission line stubs. Notice that most of the widths of the stubs are relatively small, so that higher-order modes are not strongly excited in the T-junctions. The impedance step is symmetric, so that higher-order mode excitation on the 50Ω input and output microstrip line is negligible. The larger stub with a width of 11.45 mm (denoted by A) should be regarded carefully, because the cut-off frequency of the first higher-order mode is about 7 GHz (Figure 5.7), which means that above 2.5 GHz, strong influence of the evanescent H_{10} (HE_1) mode is expected. Therefore, in Figure 8.3(b), this critical stub was replaced by two stubs in parallel (denoted by B) having the same characteristic impedance, but of doubled value with respect to Figure 8.3(a). Figure 8.3(c) shows a design solution using only quasi-lumped inductances and capacitances as discussed in Section 6.6.

Computer programs exist to design a filter circuit for given specifications using well-established filter synthesis methods. However, for practical use, filter design catalogues [10–14] that include tabulated data of a large number of filters derived from filter synthesis methods are available. The design tables are very useful to estimate rapidly the necessary circuit complexity to meet given specifications without special knowledge of filter synthesis methods. From tabulated data of

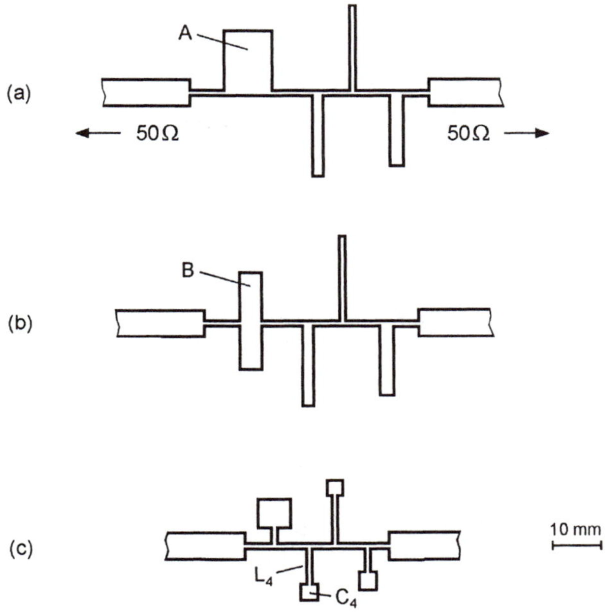

Figure 8.3 (a,b) Mixed quasi-lumped and distributed, and (c) quasi-lumped microstrip ladder filter structures ($\varepsilon_r = 2.32$, $h = 1.56$ mm, narrow lines 0.79 mm) with dimensions given in [5]. (From: [5], © 1980 IEE. Reprinted with permission.)

normalized prototype lowpass filters, a variety of other filter circuits, such as highpass, bandpass, bandstop, frequency-separation filters, and impedance matching networks, can easily be derived by means of network transformations. The design can be started either from the filter transfer function or from the LC realization of the respective tabulated prototype lowpass filter. Using the filter design catalogs, the circuit designer can readily solve most of his filter design problems without becoming involved in the theoretical aspects and special methods of filter synthesis and the time-consuming work of writing extensive computer programs.

8.1.1 Basics in Filter Design

Filter theory has been a widely discussed subject in the literature, and it would go beyond the scope of this book to give a full introduction. Nevertheless, we should deal with some selected points, so that we can use well-known filter catalogs for the design of microstrip filters. The basics can be treated using the low-frequency voltage-current concept. However, at microwave frequencies, the power wave concept is preferred. When useful, we will mention the interrelations between both

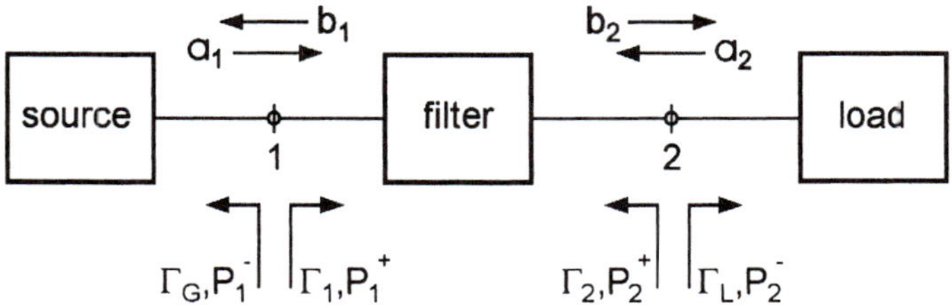

Figure 8.4 Source and load with interconnected filter network.

descriptions. Figure 8.4 shows the circuit under analysis. A filter circuit is connected between the microwave source, characterizedby the source wave b_G and source reflection coefficient Γ_G, and a load with a load reflection coefficient Γ_L. The filter two-port is assumed to be lossless, and it is described by the two-port scattering matrix S. $P_\nu^+ (\nu = 1,2)$ denotes the power traveling toward port 1 and port 2, respectively. Similarly, $P_\nu^- (\nu = 1,2)$ denotes the power reflected at port 1 and port 2, respectively. a_ν and $b_\nu (\nu = 1,2)$ denote the respective incident and backward power waves. The power waves at the ports are related to each other as follows:

Generator:
$$a_1 = b_G + \Gamma_G b_1 \tag{8.1}$$

Two-port:
$$b_1 = S_{11} a_1 + S_{12} a_2 \tag{8.2}$$
$$b_2 = S_{21} a_1 + S_{22} a_2 \tag{8.3}$$

Load:
$$a_2 = b_2 \Gamma_L \tag{8.4}$$

The input reflection coefficient Γ_1 reads

$$b_1 = a_1 \Gamma_1 \tag{8.5}$$

Substituting in (8.1),

$$a_1 = b_G + \Gamma_G \Gamma_1 a_1 \tag{8.6}$$
$$a_1 = \frac{b_G}{1 - \Gamma_1 \Gamma_G} \tag{8.7}$$

Power P_1^+ traveling toward port 1 is

$$P_1^+ = \frac{1}{2} |a_1|^2 = \frac{1}{2} |b_G|^2 \frac{1}{|1 - \Gamma_1 \Gamma_G|^2} \tag{8.8}$$

and power P_1^{abs} absorbed by port 1,

$$P_1^{abs} = P_1^+ - P_1^- = \frac{1}{2}\left(|a_1|^2 - |b_1|^2\right) = \frac{1}{2}|a_1|^2\left(1 - \frac{|b_1|^2}{|a_1|^2}\right) = \frac{1}{2}|a_1|^2\left(1 - |\Gamma_1|^2\right) \tag{8.9}$$

With (8.7),

$$P_1^{abs} = \frac{1}{2}|b_G|^2 \frac{1 - |\Gamma_1|^2}{|1 - \Gamma_1\Gamma_G|^2} \tag{8.10}$$

The power P_2^{abs} absorbed by the load reads

$$P_2^{abs} = P_2^- - P_2^+ = \frac{1}{2}\left(|b_2|^2 - |a_2|^2\right) = \frac{1}{2}|b_2|^2\left(1 - \frac{|a_2|^2}{|b_2|^2}\right) = \frac{1}{2}|b_2|^2\left(1 - |\Gamma_L|^2\right) \tag{8.11}$$

From (8.3), we obtain

$$a_1 = \frac{1}{S_{22}}\left(b_2 - S_{22}a_2\right) = \frac{1 - S_{22}\Gamma_L}{S_{21}}\, b_2 \tag{8.12}$$

Substituting in (8.7) results in

$$b_2 = \frac{S_{21}}{\left(1 - \Gamma_1\Gamma_G\right)\left(1 - S_{22}\Gamma_L\right)}\, b_G \tag{8.13}$$

With (8.13), the absorbed power P_2^{abs} at port 2 reads

$$P_2^{abs} = \frac{1}{2}|b_G|^2 \frac{|S_{21}|^2\left(1 - |\Gamma_L|^2\right)}{|1 - \Gamma_1\Gamma_G|^2\,|1 - S_{22}\Gamma_L|^2} \tag{8.14}$$

Power matching at the input port,

$$\Gamma_G = \Gamma_1^*,\ |\Gamma_G| = |\Gamma_1| \tag{8.15}$$

results in maximum power deliverable by the source (available power). From (8.8), we get

$$P_{1,\max}^+ = P_{1,av} = \frac{1}{2}|b_G|^2 \frac{1}{1 - |\Gamma_G|^2} \tag{8.16}$$

Then, P_2^{abs} in (8.14) related to the available power $P_{1,av}$, gives

$$\frac{P_2^{abs}}{P_{1,av}} = \frac{|S_{21}|^2\left(1 - |\Gamma_L|^2\right)\left(1 - |\Gamma_G|^2\right)}{|1 - \Gamma_1\Gamma_G|^2\,|1 - S_{22}\Gamma_L|^2} \tag{8.17}$$

Assuming a power source with $\Gamma_G = 0$ and matched load termination ($\Gamma_L = 0$), (8.17) simplifies to

$$\frac{P_2^{abs}}{P_{1,av}} = |S_{21}|^2 \tag{8.18}$$

which reflects the power balance relation. In filter theory, the inverse of (8.18) denotes the power transfer function H,

$$|H|^2 = \frac{P_{1,av}}{P_2^{abs}} = \frac{1}{|S_{21}|^2} \tag{8.19}$$

Since we have assumed the filter network to be lossless, the scattering parameters fulfill the well-known unitary property for a reciprocal network

$$|S_{11}|^2 + |S_{21}|^2 = 1 \tag{8.20}$$

In case of output matching, the input reflection coefficient Γ_1 equals the eigen-reflection coefficient S_{11}. With $\Gamma_1 = S_{11}$ and (8.20), it follows from (8.19)

$$|H|^2 = \frac{P_{1,av}}{P_2^{abs}} = \frac{1}{1 - |\Gamma_1|^2} \tag{8.21}$$

The input reflection coefficient Γ_1 is related to the reflected and available power as

$$|\Gamma_1|^2 = \frac{P_1^-}{P_{1,av}} \tag{8.22}$$

In filter theory, the inverse is also defined as parameter T,

$$|T|^2 = \frac{P_{1,av}}{P_1^-} = \frac{1}{|\Gamma_1|^2} \tag{8.23}$$

The power ratio P_1^- / P_2^{abs} is related to the so-called characteristic function K,

$$|K|^2 = \frac{P_1^-}{P_2^{abs}} \tag{8.24}$$

It can be expressed by the power transfer function H and reflection power parameter T,

$$|K|^2 = \frac{P_1^- / P_{1,av}}{P_2^{abs} / P_{1,av}} = \frac{|H|^2}{|T|^2} \tag{8.25}$$

A relation between H and T is obtained from the balance of power analysis. The power incident at port 1, which is the available power $P_{1,av}$, is split into the reflected power P_1^- and the power P_1^{abs} is absorbed by the input port. The latter, however, equals the power absorbed by the load P_2^{abs} because of neutrality of the filter network. Thus, we can write

$$P_{1,av} = P_1^- + P_2^{abs} \tag{8.26}$$

Dividing (8.26) by $P_{1,av}$ results in

$$1 = P_1^-/P_{1,av} + P_2^{abs}/P_{1,av} \tag{8.27}$$

and with (8.21) and (8.23),

$$1 = \frac{1}{|H|^2} + \frac{1}{|T|^2} \tag{8.28}$$

After reforming,

$$|H|^2 = 1 + \frac{|H|^2}{|T|^2} \tag{8.29}$$

and substituting the characteristic function K from (8.25), it follows

$$|H|^2 = 1 + |K|^2 \tag{8.30}$$

The power transfer and the characteristic functions are dependent on the complex frequency variable

$$p = \sigma + j\Omega \tag{8.31}$$

with the normalized frequency Ω, whereby the positive imaginary axis of the complex p plane is correlated with physical frequencies. The real part σ denotes a decreasing $(\sigma < 0)$ or increasing $(\sigma > 0)$ signal in time domain. $\sigma = 0$ characterizes a harmonic signal with constant envelope. Often the frequency is normalized to the passband limit frequency f_p,

$$\Omega = \frac{\omega}{\omega_p} = \frac{f}{f_p} \tag{8.32}$$

The attenuation of the filter network in decibels can be derived from (8.30) as

$$a(\Omega) = 10\log\left(1 + |K(j\Omega)|^2\right) \tag{8.33}$$

Note that $a(\Omega)$ is not a rational function, so that the optimization of the function is focusing on the characteristic function $K(j\Omega)$, which describes a rational function. It is obvious that both functions $a(\Omega)$ and $K(j\Omega)$ have a strong affinity. When $a(\Omega) = 0$, then $|K(j\Omega)|^2 = 0$; and, for $a(\Omega) \to \infty$, $|K(j\Omega)|^2 \to \infty$. Thus, the characteristic function $K(j\Omega)$ reflects the properties of $a(\omega)$ very well. Practically, the transition $a \to K$ implies only a different scaling of the y-axis.

8.1.2 Filter Response Types

The filter specifications needed for the design are illustrated in Figure 8.5. a_p denotes the maximally permissible attenuation in the passband. a_s is the specified minimum attenuation in the stopband. Ω is the normalized frequency f/f_p. Thus, the normalized passband limit frequency (passband cutoff frequency) reads $\Omega_p = 1$, and the normalized stopband limit frequency (stopband cutoff frequency) $\Omega_s = f_s/f_p$. In the following paragraphs, we will discuss the most frequently used filter response types, which are the Butterworth, Chebyshev (also called Tschebyscheff), and Cauer parameter (also called elliptic) filter response.

Butterworth Lowpass Filter

The characteristic function $K(p)$ is defined as a power series reading

$$K(p) = Cp^n \tag{8.34}$$

where $p = j\Omega$ for a lossless filter network. The constant C is related to the maximum input reflection coefficient Γ_1. Starting from the specified filter parameters, passband and stopband attenuation, and cutoff frequency, respectively, and passband reflection coefficient, the filter degree n has to be determined first. This is determined from the conditions for the attenuation a_p and a_s at the passband and stopband cutoff frequencies. From (8.33) with (8.34) we obtain

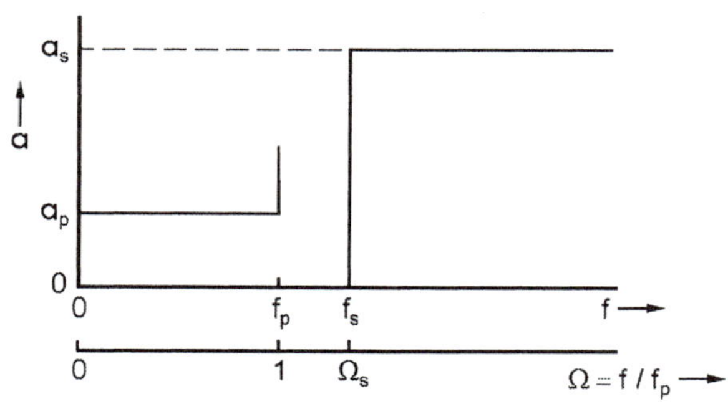

Figure 8.5 Filter design specification.

$$a_p(\Omega_p) = 10\log\left(1 + |C|^2\,|K(j\Omega_p)|^2\right) \tag{8.35}$$

$$a_s(\Omega_s) = 10\log\left(1 + |C|^2\,|K(j\Omega_s)|^2\right) \tag{8.36}$$

where $|K(j\Omega_p)|^2 = 1$. Taking base 10 yields

$$10^{0.1\,a_p(\Omega_p)} = 1 + |C|^2 \tag{8.37}$$

$$10^{0.1\,a_s(\Omega_s)} = 1 + |C|^2\,\Omega_s^{2n} \tag{8.38}$$

From (8.37), the constant C can be determined:

$$|C|^2 = 10^{0.1\,a_p(\Omega_p)} - 1 \tag{8.39}$$

With (8.21), it follows

$$|C|^2 = \frac{1}{1 - |\Gamma_1|^2} - 1 = \frac{1}{|\Gamma_1|^{-2} - 1} \tag{8.40}$$

Thus C reads

$$|C| = \frac{1}{\sqrt{|\Gamma_1|^{-2} - 1}} \tag{8.41}$$

Substituting C in (8.38) gives

$$10^{0.1\,a_s(\Omega_s)} = 1 + \frac{1}{|\Gamma_1(\Omega_p)|^{-2} - 1}\,\Omega_s^{2n} \tag{8.42}$$

and re-sorting,

$$\left(10^{0.1\,a_s(\Omega_s)} - 1\right)\left(|\Gamma_1(\Omega_p)|^{-2} - 1\right) = \Omega_s^{2n} \tag{8.43}$$

Taking the logarithm

$$n\,20\log\Omega_s = 10\log\left[\left(10^{0.1\,a_s(\Omega_s)} - 1\right)\left(|\Gamma_1(\Omega_p)|^{-2} - 1\right)\right] \tag{8.44}$$

and assuming $10^{0.1 a_s(\Omega_s)} \gg 1$ for $a_s \geq 20\,\text{dB}$, we can approximately write

$$n\,20\log\Omega_s \approx 10\log\left[\left(10^{0.1\,a_s(\Omega_s)}\right)\left(|\Gamma_1(\Omega_p)|^{-2} - 1\right)\right] \tag{8.45}$$

which can be simplified to

$$n\,20\log\Omega_s \approx a_s(\Omega_s) + a(|\Gamma_1(\Omega_p)|) \tag{8.46}$$

where

$$a(|\Gamma_1(\Omega_p)|) = 10\log\left(|\Gamma_1(\Omega_p)|^{-2} - 1\right) \tag{8.47}$$

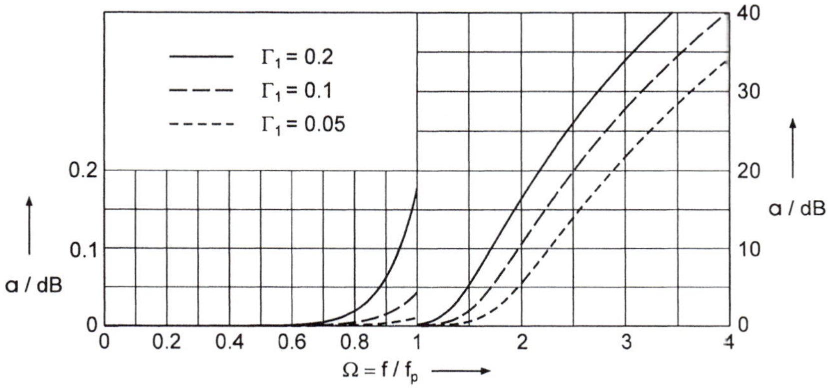

Figure 8.6 Butterworth filter response with $n = 5$ and Γ_1 as a parameter.

$a(|\Gamma_1(\Omega_p)|)$ has been introduced as an auxiliary parameter, and should not be mixed with the passband reflection attenuation $a_p(|\Gamma_1(\Omega_p)|)$. In filter catalogues, (8.46) has been evaluated in the form of diagrams, in which the attenuation $a_s + a$ is plotted versus Ω_s with the filter degree n as a parameter.

Figure 8.6 shows the calculated filter response of a Butterworth filter for a filter degree $n = 5$. The maximum passband reflection attenuation was assumed to be 5%, 10%, and 20%. It can be seen that the attenuation in the passband is smoothly increasing, and reaches its maximum attenuation at the passband cutoff frequency. That is the reason why the Butterworth filter response is also characterized as maximally flat. Regarding the filter curves as a function of reflection coefficient, it is clearly seen that the lower the passband attenuation, the weaker the increase of out-of-passband attenuation.

Chebyshev Lowpass Filter

The Chebyshev function of first kind and order n, termed $T_n(x)$, is defined as

$$T_n(x) = \cos[n(\arccos x)] \qquad (|p| \le 1) \qquad (8.48)$$

and

$$T_n(x) = \cosh[n(\text{areacosh}\, x)] \qquad (|x| \ge 1) \qquad (8.49)$$

The roots of $T_n(x)$ in this nonpolynomial form are given by the closed expression

$$x_k = \cos\left[(2k - 1)\frac{\pi}{2n}\right] \qquad (8.50)$$

with $k = 0, 1, \ldots, n$ and listed in Table 8.1 for $n = 2$ to 7. Let

$$q = \arccos x \qquad (8.51)$$

then $x = \cos q$, and (8.48) becomes

Table 8.1
Roots of the Chebyshev Polynomials

	$\|\Omega_{01}\|$	$\|\Omega_{02}\|$	$\|\Omega_{03}\|$	$\|\Omega_{04}\|$
$n = 2$	0.707107			
3	0.	0.866025		
4	0.382683	0.92388		
5	0.	0.587785	0.951057	
6	0.258819	0.707107	0.965926	
7	0.	0.433884	0.781831	0.974928

$$T_n(x) = \cos(nq) \tag{8.52}$$

From the trigonometric identities

$$\cos[(n+1)q] = \cos(nq)\cos(q) - \sin(nq)\sin(q) \tag{8.53}$$

and

$$\sin(nq)\sin(q) = \cos[(n-1)q] + \cos(nq)\cos(q) \tag{8.54}$$

it follows

$$\cos[(n+1)q] = 2\cos(nq)\cos(q) - \cos[(n-1)q]\sin(q) \tag{8.55}$$

Equation (8.55) reflects the recursion formula for the Chebyshev functions,

$$T_{n+1}(p) = 2\,T_n(p) - T_{n-1}(p) \tag{8.56}$$

With (8.48) and (8.49) it follows for $n = 0$

$$T_0(x) = \left\{\begin{matrix} \cos(0)\,, & |x| \le 1 \\ \cosh(0)\,, & |x| \ge 1 \end{matrix}\right\} = 0 \tag{8.57}$$

and $n = 1$

$$T_1(x) = \left\{\begin{matrix} \cos[\arccos(x)]\,, & |x| \le 1 \\ \cosh[\mathrm{areacosh}(x)]\,, & |x| \ge 1 \end{matrix}\right\} = x \tag{8.58}$$

Using (8.56), it follows for higher order n:

$$T_2(x) = 2x^2 - 1 \tag{8.59}$$

$$T_3(x) = 4x^3 - 3x \tag{8.60}$$

$$T_4(x) = 8x^4 - 8x^2 + 1 \tag{8.61}$$

$$T_5(x) = 16x^5 - 20x^3 + 5x \tag{8.62}$$

$$T_6(x) = 32x^6 - 48x^4 + 18x^2 - 1 \tag{8.63}$$

Thus, (8.59)–(8.63) make clear that the Chebyshev functions are truly polynomials. With $\Omega_{0\nu}$ denoting the roots of the polynomials, the characteristic function can be written for odd n $(n = 3,5,7,...)$

$$K(p) = Cp \prod_{\nu=2}^{(n+1)/2} \left(p^2 + \Omega_{0\nu}^2\right) \tag{8.64}$$

and for even n $(n = 2,4,6,...)$

$$K(p) = C \prod_{\nu=2}^{n/2} \left(p^2 + \Omega_{0\nu}^2\right) \tag{8.65}$$

and

$$K(p) = Cp^2 \prod_{\nu=2}^{n/2} \left(p^2 + \Omega_{0\nu}^2\right) \tag{8.66}$$

respectively. According to (8.35), the attenuation of a Chebyshev filter response of order n at the passband cutoff frequency Ω_p can be written as

$$a_p(\Omega)/\text{dB} = 10\log\left(1 + |C|^2\,|T_n(j\Omega_p)|^2\right) \tag{8.67}$$

where $|T_n(j\Omega_p)|^2 = 1$. Then the constant C is simply determined from (8.67), and agrees with that value already found in (8.41) for Butterworth filters. Recalling

$$T_n^2(j\Omega) = \cosh^2[n \text{ areacosh}(j\Omega)] \tag{8.68}$$

the attenuation in the stopband is given by

$$a_s(\Omega) = 10\log\left(1 + \frac{\cosh^2[n \text{ areacosh}(\Omega)]}{|\Gamma_1(\Omega_p)|^{-2} - 1}\right) \tag{8.69}$$

Taking base 10,

$$\cosh^2[n \text{ areacosh}(\Omega_s)] = \left(10^{a_s(\omega_s)} - 1\right)\left(|\Gamma_1(\Omega_p)|^{-2} - 1\right) \tag{8.70}$$

and solving for the filter degree n, we get

$$n = \frac{\text{areacosh}\sqrt{\left(10^{a_s(\omega_s)} - 1\right)\left(|\Gamma_1(\omega_p)|^{-2} - 1\right)}}{\text{areacosh}\,\Omega_s} \tag{8.71}$$

Thus, n is determined by the stopband attenuation $a_s(\Omega_s)$, the reflection coefficient $\Gamma_1(\Omega_p)$, and the normalized cutoff frequency Ω_s. Making again the approximation $10^{0.1a_s(\omega_s)/\text{dB}} \gg 1$ for $a_s \geq 20$ dB, and taking the logarithm, we can write for (8.70)

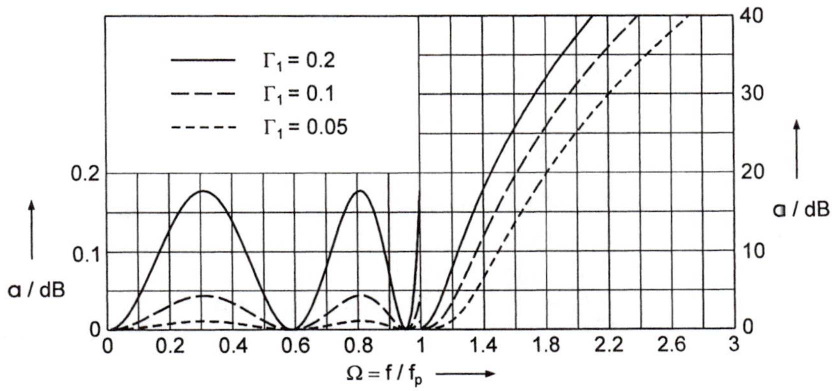

Figure 8.7 Chebyshev filter with $n = 5$ and Γ_1 as a parameter.

$$10\log\left\{\cosh^2\left[n\ \text{areacosh}(\Omega_s)\right]\right\} \approx a_s(\Omega_s) + a(|\Gamma_1(\Omega_p)|) \qquad (8.72)$$

For practical use of (8.72), $a_s(\Omega_s) + a(|\Gamma_1(\Omega_p)|)$ has been plotted as a function of the normalized cutoff frequency Ω_s with the filter degree n as a parameter [12].

Figure 8.7 shows an example for a Chebyshev filter response for $n = 5$ and different reflection coefficient values. As can be seen, this kind of filter exhibits an equiripple filter response in the passband. In the stopband, the attenuation follows a hyperbolic function.

Cauer Parameter Lowpass Filter

Cauer parameter or elliptic lowpass filter responses are characterized by equiripple filter response in both the passband and stopband. The characteristic function $K(p)$ is defined for odd n $(n = 3, 5, 7, \ldots)$,

$$K(p) = C\,\frac{p\prod_{\nu=2}^{(n+1)/2}\left(p^2 + \Omega_{0\nu}^2\right)}{\prod_{\nu=1}^{(n-1)/2}\left(p^2 + \Omega_{\infty 2\nu}^2\right)} \qquad (8.73)$$

and for even n $(n = 2, 4, 6, \ldots)$

$$K(p) = C\,\frac{\prod_{\nu=1}^{n/2}\left(p^2 + \Omega_{0\nu}^2\right)}{\prod_{\nu=1}^{n/2-1}\left(p^2 + \Omega_{\infty 2\nu}^2\right)} \qquad (8.74)$$

and

$$K(p) = C \frac{p^2 \prod_{\nu=2}^{n/2} \left(p^2 + \Omega_{0\nu}^2\right)}{\prod_{\nu=1}^{n/2-1} \left(p^2 + \Omega_{\infty 2\nu}^2\right)} \tag{8.75}$$

respectively. The normalized frequencies Ω_∞ denote the attenuation poles in the stopband frequency region. The characteristic function in (8.74) and (8.75) is also referred to as case b and case c, respectively. The attenuation at $\Omega = 0$ is finite in case b, whereas in case c it is zero. Case a is related, and is derived from the parameters of case b [12]. It denotes a steeper bandpass-bandstop filter response with lowest attenuation pole frequency equal to the bandstop cutoff frequency of case b. Figure 8.8 can be used to determine the filter degree n from the specified attenuation $a_s(\Omega_s) + a(|\Gamma_1(\Omega_p)|)$ and the normalized cutoff frequency Ω_s (see design example in Section 8.3.5).

Figures 8.9 and 8.10 show a reprint of two pages from Saal's filter catalog [12]. It has been shown that Jacobian elliptic functions can be written in polynomial form, as given in (8.73) to (8.75). The modulus m of the elliptic function (compare Section 3.2.1) and the normalized stopband cutoff frequency Ω_s are related to each other by the following equation

$$m = \sin(\theta) = \Omega_s^{-1} \tag{8.76}$$

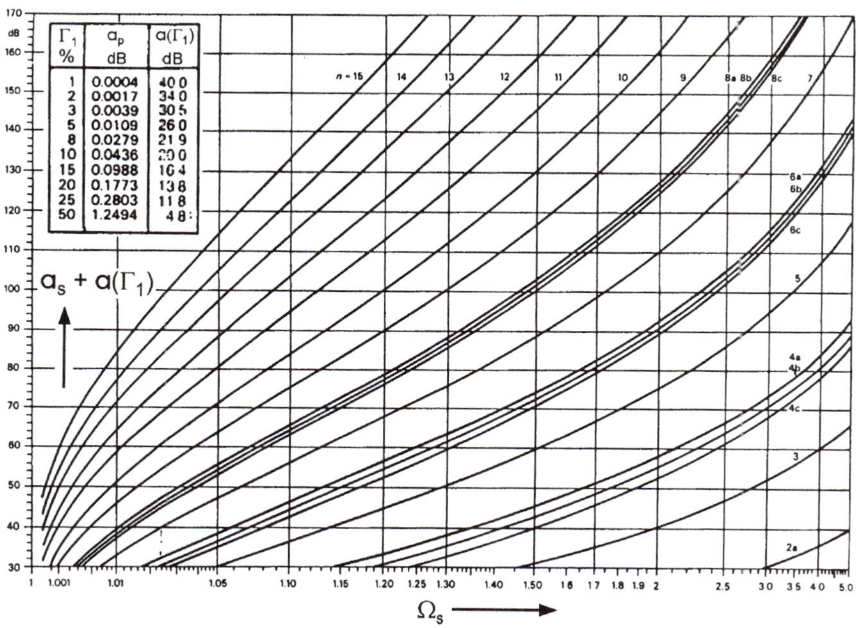

Figure 8.8 Diagram to estimate the degree of Cauer lowpass filters. (From: [12], © 1988 Dr. Alfred Hüthig Verlag. Reprinted with permission.)

C 0905

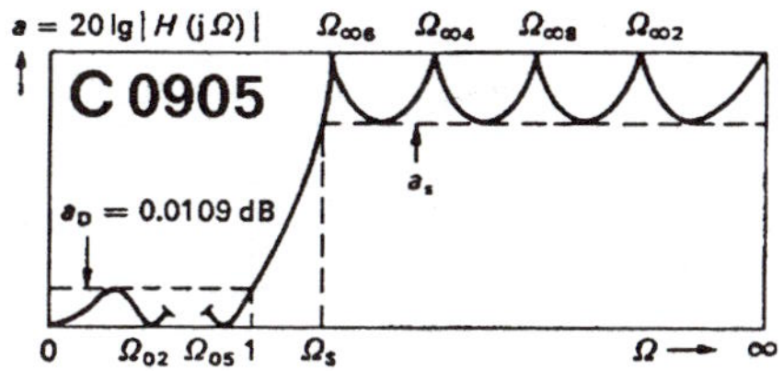

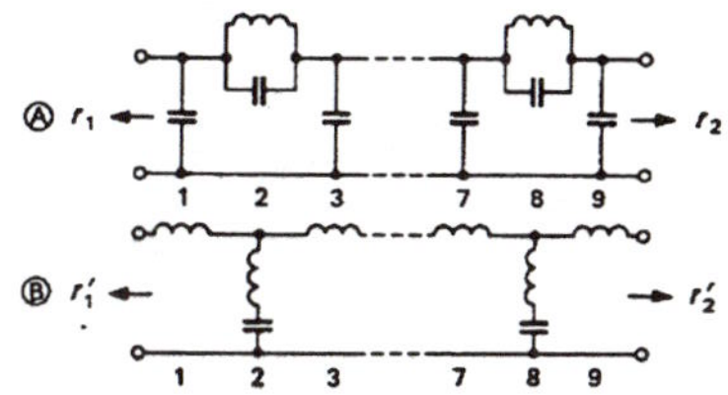

$$H(p) = C\,\frac{(p-\alpha_1)\prod_{\nu=2}^{5}(p^2 - 2\alpha_\nu p + \gamma_\nu)}{\prod_{\nu=1}^{4}(p^2 + \Omega_{\infty 2\nu}^2)}$$

$$\gamma_\nu = \alpha_\nu^2 + \beta_\nu^2$$

Θ	Ω_s	a_s dB	Ⓐ ν	$r_1 = 1$		$r_2 = 1$	$r_1 = \infty$		$r_2 = 1$	$\Omega_{\infty 2\nu}$	$\Omega_{0\nu}$	$-\alpha_\nu$	$\pm\beta_\nu$	C
				$c_{2\nu-1}$	$l_{2\nu}$	$c_{2\nu}$	$c_{2\nu-1}$	$l_{2\nu}$	$c_{2\nu}$					
P			1	0.249001	0.716971		1.120506	1.320957			0.0000000000	1.3947568217	0.0000000000	0.050062617
			2	1.098463	1.347464		1.274193	1.161630			0.0000000000	1.3106426931	0.4770349280	
			3	1.433942	1.347464		1.006436	0.817890			0.0000000000	1.0684457127	0.8965324035	
			4	1.098463	0.716971		0.603235	0.369631			0.0000000000	0.6973784108	1.2078948397	
			5	0.249001			0.124501				0.0000000000	0.2421969804	1.3735873316	
T			1	0.824201	1.430648		1.442001	1.740278			0.0000000000	0.4213732240	0.0000000000	12.816030083
			2	1.812847	1.712374		1.774024	1.749848			0.3420201433	0.3959613092	0.3711439262	
			3	1.913073	1.712374		1.700362	1.594214			0.6427876097	0.3227906167	0.6975224175	
			4	1.812847	1.430648		1.408120	1.045872			0.8660254038	0.2106866120	0.9397694108	
			5	0.824201			0.412101				0.9848077530	0.0731706925	1.0686663437	
44	1.439556540	86.7	1	0.779775	1.363960	0.054541	1.440123	1.711754	0.043460	3.666375663	0.0000000000	0.5090559417	0.0000000000	1958.843927220
			2	1.546246	1.293256	0.301356	1.556307	1.342915	0.290212	1.601834928	0.3926374906	0.4419056358	0.4584808635	
			3	1.402365	1.188714	0.397173	1.232557	1.029421	0.458631	1.455364364	0.7028908188	0.2976075266	0.7879075363	
			4	1.404223	1.167317	0.204235	1.027300	0.694899	0.343081	2.048051420	0.8986921901	0.1584127435	0.9715118838	
			5	0.650106			0.141693				0.9891382362	0.0484381344	1.0489444179	
45	1.414213562	84.7	1	0.777433	1.360451	0.057433	1.439802	1.709991	0.045693	3.577496845	0.0000000000	0.5139603713	0.0000000000	1601.020863265
			2	1.533806	1.274182	0.318011	1.546250	1.324642	0.305897	1.570951885	0.3953081229	0.4441127050	0.4633419774	
			3	1.379828	1.165865	0.419766	1.212332	1.004616	0.487142	1.429459865	0.7058900213	0.2959776446	0.7923094888	
			4	1.386085	1.154559	0.215788	1.011952	0.677978	0.367475	2.003447449	0.9002271654	0.1559050929	0.9728106857	
			5	0.641352			0.126445				0.9893342213	0.0473722683	1.0480488862	
46	1.390163591	82.8	1	0.775004	1.356813	0.060434	1.439447	1.708136	0.048004	3.492207426	0.0000000000	0.5190759818	0.0000000000	1312.794203401
			2	1.521062	1.254694	0.335390	1.535956	1.305990	0.322217	1.541544289	0.3980758934	0.4463736714	0.4684072429	
			3	1.356873	1.142614	0.443435	1.191789	0.979343	0.517363	1.404866589	0.7089802520	0.2942600879	0.7968304199	
			4	1.367598	1.141442	0.227868	0.996528	0.660612	0.393722	1.960793107	0.9017993195	0.1533329780	0.9741221396	
			5	0.632315			0.110487				0.9895342386	0.0462894376	1.0471347793	
47	1.367327461	80.9	1	0.772485	1.353039	0.063550	1.439054	1.706183	0.050396	3.410268222	0.0000000000	0.5244133432	0.0000000000	1079.731317922
			2	1.508008	1.234790	0.353533	1.525424	1.286960	0.339202	1.513520241	0.4009442578	0.4486878841	0.4736860306	
			3	1.333506	1.118966	0.468251	1.170945	0.953604	0.549449	1.381504113	0.7121634171	0.2924501922	0.8014711648	
			4	1.348768	1.127960	0.240503	0.981060	0.642798	0.422027	1.919963071	0.9034087712	0.1506960949	0.9754448066	
			5	0.622987			0.093770				0.9897382484	0.0451901933	1.0482022852	
48	1.345632730	79.0	1	0.769871	1.349125	0.066785	1.438620	1.704128	0.052872	3.331459354	0.0000000000	0.5299839027	0.0000000000	890.584385157
			2	1.494642	1.214471	0.372486	1.514653	1.267551	0.356888	1.486795599	0.4039168985	0.4510544823	0.4791883894	
			3	1.309733	1.094931	0.494292	1.149815	0.927402	0.583582	1.359298837	0.7154415092	0.2905430184	0.8062324792	
			4	1.329597	1.114105	0.253728	0.965586	0.624534	0.452626	1.880642406	0.9050556312	0.1479941688	0.9767765742	
			5	0.613358			0.076242				0.9899462080	0.0440751104	1.0452516081	
Θ	Ω_s	a_s dB	ν Ⓑ	$l_{2\nu-1}$	$c_{2\nu}$	$l_{2\nu}$	$l_{2\nu-1}$	$c_{2\nu}$	$l_{2\nu}$	$\Omega_{\infty 2\nu}$	$\Omega_{0\nu}$	$-\alpha_\nu$	$\pm\beta_\nu$	C
				$r'_1 = 1$		$r'_2 = 1$	$r'_1 = 0$		$r'_2 = 1$					

326

Figure 8.9 Extract of Saal's filter catalog (Cauer lowpass filter C 09 05). (From: [12], © 1988, Dr. Alfred Hüthig Verlag. Reprinted with permission.)

Θ	Ω_s	a_s dB	Ⓐ ν	$r_1 = 1$ $c_{2\nu-1}$	$l_{2\nu}$	$r_2 = 1$ $c_{2\nu}$	$r_1 = \infty$ $c_{2\nu-1}$	$l_{2\nu}$	$r_2 = 1$ $c_{2\nu}$	$\Omega_{\infty 2\nu}$	$\Omega_{0\nu}$	$-\alpha_\nu$	$\pm\beta_\nu$	C
69	1.071144994	43.0	1	0.680660	1.218494	0.179819				2.138096974	0.0000000000	0.7398477482	0.0000000000	23.066526343
			2	1.129684	0.689450	1.136790				1.129557539	0.5009805479	0.5053305285	0.6763158956	
			3	0.733546	0.515240	1.675040				1.076422021	0.8109902409	0.2152052979	0.9331904064	
			4	0.858771	0.716649	0.799886				1.320786539	0.9482872337	0.0764104993	1.0032167386	
			5	0.310340							0.9950976226	0.0183882882	1.0217200730	
70	1.064177772	41.4	1	0.673668	1.206157	0.188994				2.094470158	0.0000000000	0.7677231691	0.0000000000	19.505799984
			2	1.107234	0.659380	1.210484				1.119315818	0.5080892503	0.5070744302	0.6919901204	
			3	0.703665	0.484893	1.804241				1.069127723	0.8171237683	0.2090603016	0.9402500134	
			4	0.833842	0.691081	0.853136				1.302345879	0.9507395103	0.0723631707	1.0041572031	
			5	0.288270							0.9953701034	0.0171491211	1.0204804523	
71	1.057620681	39.7	1	0.666228	1.195162	0.198822				2.051419738	0.0000000000	0.7772279095	0.0000000000	16.482627118
			2	1.084154	0.628763	1.291940				1.109521047	0.5155554768	0.5084960988	0.7086039411	
			3	0.673751	0.454395	1.950356				1.062248344	0.8234408403	0.2025703355	0.9473358063	
			4	0.808806	0.664877	0.911999				1.284391822	0.9532227294	0.0682737049	1.0050376764	
			5	0.264992							0.9956435204	0.0159204836	1.0192376697	
72	1.051462224	38.0	1	0.658296	1.183430	0.209389				2.008868884	0.0000000000	0.7983580716	0.0000000000	13.914386659
			2	1.060405	0.597574	1.382601				1.100160286	0.5234100805	0.5095184200	0.7262376512	
			3	0.643850	0.423767	2.117057				1.055772373	0.8299488831	0.1957151640	0.9544287369	
			4	0.783695	0.637386	0.977491				1.266899981	0.9557354851	0.0641464301	1.0058524710	
			5	0.240367							0.9959175396	0.0147046811	1.0179933815	
73	1.045891756	36.3	1	0.649809	1.170863	0.220801				1.966737792	0.0000000000	0.8213510465	0.0000000000	11.731630256
			2	1.035946	0.565784	1.484305				1.091221749	0.5316884441	0.5100469893	0.7449800481	
			3	0.614016	0.393031	2.309147				1.049689195	0.8366559383	0.1884734174	0.9615064136	
			4	0.758548	0.609154	1.050896				1.249846811	0.9582761314	0.0599863496	1.0065954916	
			5	0.214228							0.9961917888	0.0135042207	1.0167494303	
74	1.040299436	34.6	1	0.640697	1.157343	0.233186				1.924942427	0.0000000000	0.8464975645	0.0000000000	9.875883794
			2	1.010731	0.533361	1.599436				1.082694798	0.5404314545	0.5099655653	0.7649291270	
			3	0.584314	0.362218	2.533016				1.043989048	0.8435707540	0.1808225824	0.9685424677	
			4	0.733415	0.579916	1.133864				1.233209462	0.9608427393	0.0557992574	1.0072601865	
			5	0.186375							0.9964658515	0.0123218383	1.0155078723	
75	1.035276180	32.8	1	0.630875	1.142724	0.246704				1.883392962	0.0000000000	0.8741582598	0.0000000000	8.297833129
			2	0.984707	0.500264	1.731138				1.074569932	0.5496867629	0.5091301079	0.7861926004	
			3	0.554820	0.331363	2.797347				1.038662991	0.8507028954	0.1727390226	0.9755057704	
			4	0.708357	0.549605	1.228549				1.218965624	0.9634330440	0.0515918812	1.0078394943	
			5	0.156560							0.9967392595	0.0111605316	1.0142710119	
76	1.030613629	31.1	1	0.620243	1.126822	0.261559				1.841991777	0.0000000000	0.9047869596	0.0000000000	6.955824549
			2	0.957820	0.466448	1.883645				1.066838803	0.5595104398	0.5073608570	0.8088880130	
			3	0.525628	0.300512	3.114204				1.033702867	0.8580628866	0.1641980416	0.9823594425	
			4	0.683454	0.518143	1.337818				1.201093353	0.9660443794	0.0473720600	1.0083257784	
			5	0.124480							0.9970114842	0.0100235997	1.0130414426	
77	1.026304108	29.3	1	0.608674	1.109402	0.278011				1.800630864	0.0000000000	0.9389641504	0.0000000000	5.814622315
			2	0.930007	0.431862	2.062802				1.059494254	0.5699891860	0.5044317047	0.8331420576	
			3	0.496855	0.269722	3.500799				1.029101282	0.8656623875	0.1551740183	0.9890595824	
			4	0.658809	0.485441	1.465577				1.185570868	0.9686735950	0.0431489721	1.0087107493	
			5	0.089750							0.9972819256	0.0089146912	1.0118220990	
78	1.022340595	27.4	1	0.596013	1.090157	0.296405				1.759188333	0.0000000000	0.9774463444	0.0000000000	4.844375834
			2	0.901205	0.396449	2.276894				1.052530377	0.5811433463	0.5000557941	0.8590884225	
			3	0.468649	0.239071	3.982453				1.024851586	0.8735144261	0.1456406585	0.9955538167	
			4	0.634561	0.451403	1.617279				1.170376315	0.9713169494	0.0389334288	1.0089853713	
			5	0.051878							0.9975498982	0.0078378630	1.0106163211	
Θ	Ω_s	a_s dB	ν Ⓑ	$l_{2\nu-1}$ $r_1' = 1$	$c_{2\nu}$	$l_{2\nu}$ $r_2' = 1$	$l_{2\nu-1}$ $r_1' = 0$	$c_{2\nu}$	$l_{2\nu}$ $r_2' = 1$	$\Omega_{\omega 2\nu}$	$\Omega_{0\nu}$	$-\alpha_\nu$	$\pm\beta_\nu$	C

329

C 0905

Figure 8.10 Continuation of Figure 8.9.

Solving (8.76) for the modular angle θ and converting to degrees results in

$$\theta = \frac{180°}{\pi} \arcsin \frac{1}{\Omega_s} \tag{8.77}$$

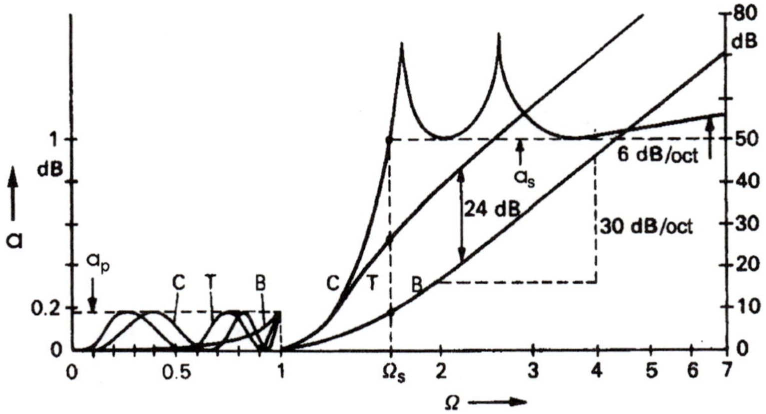

Figure 8.11 Comparison of the attenuation of Butterworth (B), Chebychev (T), and elliptic (C) lowpass filter of degree $n = 5$. (From [12], © 1988, Dr. Alfred Hüthig Verlag. Reprinted with permission.)

In Figure 8.11, the frequency responses of the discussed lowpass filters are compared for a filter degree $n = 5$. The same maximum variation of the passband attenuation a_p and the same steepness $1/\Omega_s$ of the transition region has been assumed. It can be seen that the Cauer filter response (C) yields the highest stopband attenuation a_s. Regarding the Butterworth (B) and the Chebyshev (T) lowpass filter responses, the attenuation in the stopband increases—unlike the Cauer filter—monotonically with $6n$ dB/octave. For $\Omega \gg 1$, the difference between the attenuation of the Chebyshev and the Butterworth lowpass filter response becomes constant $[6(n-1)\ \text{dB}]$.

8.1.3 Design Examples

Problem: A lowpass filter with elliptic filter response shall be designed. In the passband, the maximum magnitude of the reflection coefficient must not exceed the value $|\Gamma_1| = 0.05$. The cutoff frequency of the passband is $f_p = 3$ GHz. At the stopband cutoff frequency $f_s = 3.2$ GHz, the attenuation a_s shall be 39.09 dB (4.5 Np). The input and output terminations R_G and R_L have an impedance of 50Ω.

Solution: The normalized frequency Ω_s, indicating the transition steepness from the passband to the stopband, is calculated as $\Omega_s = f_s / f_p = 3.2/3.0 = 1.066$. The attenuation parameter $a(|\Gamma_1|)$ in (8.47) is calculated as $10\log\left(|\Gamma_1|^{-2} - 1\right) = 26.05\ \text{dB}$. Then, $a_s + a(|\Gamma_1|)$ in (8.71) reads 39.09 dB + 26.05 dB = 65.14 dB.

The needed degree n of the filter response can be taken from Figure 8.8, where $a_s + a(|\Gamma_1|)$ is plotted versus the normalized stopband cutoff frequency Ω_s, which denotes the steepness of the filter response, with the filter degree n as a parameter.

Table 8.2
Normalized Element Values of the Cauer Filter C 09 05 71

c_v	l_{v-1}	Ω_{0v}
$c_2 = 1.1951$	$l_1 = 0.6662$ $l_2 = 0.1988$ $l_3 = 1.0840$	$\Omega_2 = 2.051420$
$c_4 = 0.6288$	$l_4 = 1.2920$ $l_5 = 0.6738$	$\Omega_4 = 1.109521$
$c_6 = 0.4544$	$l_6 = 1.9500$ $l_7 = 0.8088$	$\Omega_6 = 1.062248$
$c_8 = 0.6647$	$l_8 = 0.9120$ $l_9 = 0.2650$	$\Omega_8 = 1.284392$

For $a = 65.14\ \mathrm{dB}$, the required degree $n > 8$ is found. Thus, the smallest possible degree is $n = 9$. The available margin can be used to optimize the filter parameters, for example, to reduce the passband attenuation a_p, to increase the stopband attenuation a_s, or to increase the passband limit and to decrease the stopband limit (Ω_s). If the margin is used to increase the steepness of the attenuation transition, then a new value of 1.054 is read from Figure 8.8 for the normalized frequency Ω_s. The modular angle θ in (8.76) is then calculated as $71.58°$. With $n = 9$, $|\Gamma_1| = 0.05$, and the respective θ value, the circuit C 09 05 71 in [12] can be identified to fulfill all the specifications. Figures 8.10 and 8.11 show the corresponding pages from Saal's filter catalog [12]. The filter notations are: Cauer parameters (C), degree $n = 9\ (09)$, maximum permissible reflection $|\Gamma_1| = 5\%$ (05), and modular angle $\theta = 71°\ (71)$, corresponding to a stopband limit $\Omega_s = f_s/f_p = 1/\sin\theta = 1.0576$. As denoted by the heading of the respective filter page, the filter response can be realized in two ways, as a filter network with parallel resonant circuits in the main (through) line (version A), or as a dual circuit with shunted series resonant circuits (version B). We choose the dual structure B, because it is more appropriate for the implementation in microstrip technique. The normalized element values of the lowpass filter C 09 05 71, listed in Table 8.2, are then found.

The next step is to remove the normalization. For that, we need the passband frequency f_p, which is recalculated with the changed value of Ω_s as $f_p = f_s/\Omega_s = 3.2/1.057621 = 3.025\ \mathrm{GHz}$. Then with $R_G = R_L = R_0 = 50\,\Omega$, the reference inductance L_0 and reference capacitance C_0 are calculated as

$$L_0 = \frac{R_0}{2\pi f_p} = \frac{50\Omega}{2\pi \cdot 3.025 \cdot 10^9\ \mathrm{sec}^{-1}} = 2.63\ \mathrm{nH} \tag{8.78}$$

$$C_0 = \frac{1}{2\pi f_p R_0} = \frac{1}{50\Omega \cdot 2\pi \cdot 3.025 \cdot 10^9\ \mathrm{sec}^{-1}} = 1.052\ \mathrm{pF} \tag{8.79}$$

Then the parameters are denormalized using the relations:

Table 8.3
Denormalized Element Values of the Cauer Filter C 09 05 71

C_ν	$L_{\nu\text{-}1}$	$f_{0\nu}$
C_2 = 1.2571 pF	L_1 = 1.7521 nH L_2 = 0.5228 nH L_3 = 2.8510 nH	f_2 = 6.2068 GHz
C_4 = 0.6615 pF	L_4 = 3.3980 nH L_5 = 1.7721 nH	f_4 = 3.3570 GHz
C_6 = 0.4780 pF	L_6 = 5.1286 nH L_7 = 2.1272 nH	f_6 = 3.2139 GHz
C_8 = 0.6992 pF	L_8 = 2.3986 nH L_9 = 0.6969 nH	f_8 = 3.8861 GHz

$$f_{0\nu} = \Omega_{0\nu}\, f_p, \qquad L_\nu = l_\nu\, L_0, \quad C_\nu = c_\nu\, C_0 \tag{8.80}$$

The denormalized values of the physical components are given in Table 8.3 and Figure 8.12.

The last step is the implementation of the LC prototype filter in microstrip technique. The respective layouts shown in Figure 8.3 have been found by hand-made design on the basis of this compiled design material. Reference plane locations in the junctions and discontinuities have been considered, and microstrip dispersion and open-end effects have been taken into account. Numerical calculations were done on the basis of the microstrip waveguide model [5].

Figure 8.13 shows the measured and calculated magnitude of the transmission coefficient of the filter under analysis. The three versions correspond to the layouts shown in Figure 8.3. The dash-dotted line indicates the specified filter response with a pass-frequency equal to 3.0 GHz, and a stop-frequency equal to 3.2 GHz.

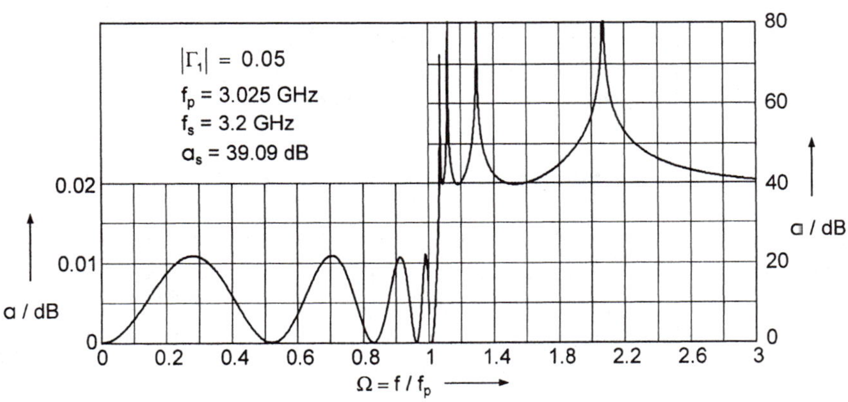

Figure 8.12 Simulated filter response of the designed elliptic filter with the parameters listed in Table 8.3.

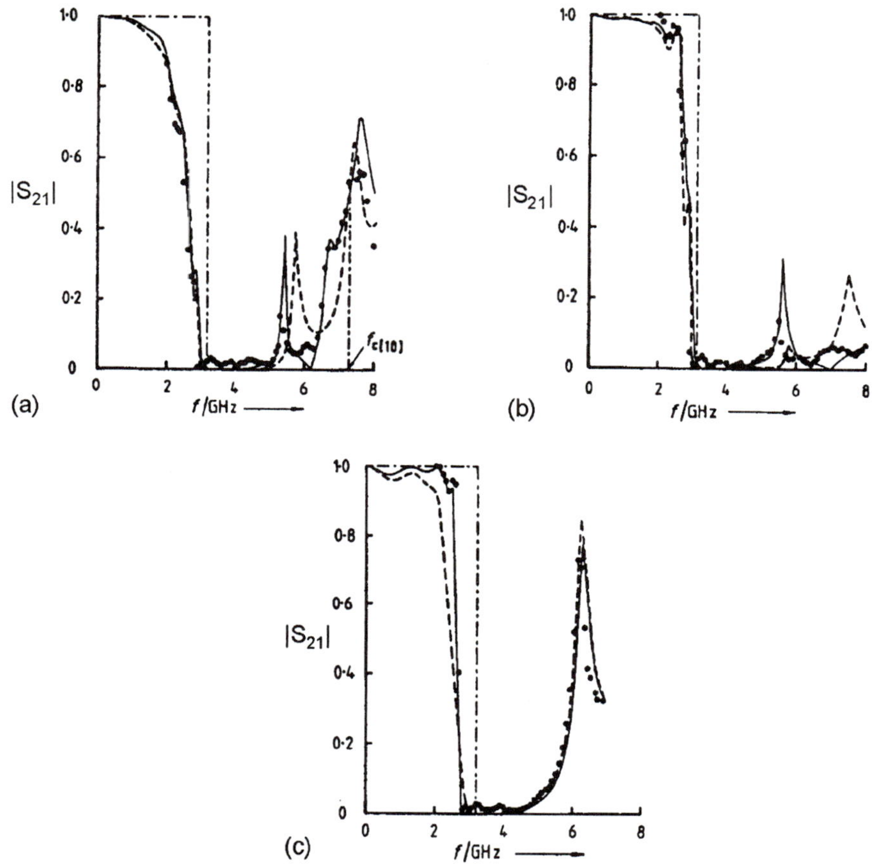

Figure 8.13 Measured and calculated magnitude of the transmission coefficient S_{21} of the lowpass filters, shown in Figure 8.3(a–c), realized on Polyguide substrate (ε_r = 2.32, h = 1.56 mm) (dash-dotted line: LC prototype filter response, solid circles: measurement, dashed line: simulation with junction and discontinuity effects neglected, solid line: simulation using the waveguide model). (From: [5], © 1980 IEE. Reprinted with permission.)

Figure 8.13(a) reveals the weakest transition from passband to stopband. In comparison with the filter layout (b) in Figure 8.3, the impedance value of stub filter element A [Figure 8.3(a)] is very low, which results in larger bandwidth of the respective stub filter response, and thus smoother bandpass-bandstop transition (see Figure 6.101). In addition, the higher-order H_{10} (HE_1) mode is increasingly stimulated by the T-junction with wider width of the stub. From Figure 5.7 we read a cutoff frequency of approximately 7 GHz for the given width w = 11.45 mm. Thus, recalling that a higher-order mode has negligible influence up to one third of its cutoff frequency, the influence can become notable above approximately 2.3 GHz. In addition to the weak transition from passband to stopband, the limit frequency of the stopband f_s is not well met. The filter version in Figure 8.13(b), having replaced the low impedance stub by two shunted higher impedance stubs, show much better performance with respect to steepness of the filter response and agreement with the stopband limit frequency f_s. In the stopband, the transmission coefficient remains rather low.

Figure 8.13(c) shows the transmission coefficient of the filter comprising only quasi-lumped capacitive and inductive elements. It can be seen that this kind of filter realization excellently approximates the specified sharp transition from passband to stopband, although the stopband limit frequency has been measured about 400 MHz lower than specified. The simulation was based on full-wave analysis using the waveguide model, demonstrating excellent agreement with measurement. Optimization was not the primary goal in [5].

In the following paragraphs, some issues on the conversion of an LC network into a microstrip circuit shall be discussed. The design of an open-ended microstrip line with equivalent electrical behavior of a specified lumped LC series resonant circuit has already been treated in Section 6.4. Therefore, we will address the design task illustrated in Figure 1.6, deriving a quasi-lumped microstrip series resonant circuit from a respective lumped LC network. For that we choose the L_4C_4 resonant circuit in Figures 8.2 and 8.3. The prototype filter analysis has delivered the following rounded element values (Table 8.3): $C_4 = 0.66\,\mathrm{pF}$, $L_4 = 3.4\,\mathrm{nH}$, and $f_4 = 3.35\,\mathrm{GHz}$. It is assumed that the single lumped elements C_4 and L_4 are realized by quasi-lumped low- and high-impedance microstrip lines, as discussed in Section 6.6. For the design, we use Polyguide substrate material ($\epsilon_r = 2.32$, $h = 1.58\,\mathrm{mm}$). We will follow three different designs to make the problems more transparent which could occur in the designing process. According to [5], we choose for all narrow lines $w = 0.79\,\mathrm{mm}$, which results in an aspect ratio $w/h = 0.5$. For this value, we read off the table in Appendix F the following line parameters: Characteristic impedance $Z_0(\epsilon_r = 2.32) = 124.25\,\Omega$, and effective dielectric constant $\epsilon_{eff} = 1.8$. The respective effective width can be calculated from (3.82) as

$$w_{eff}(f=0) = \sqrt{\frac{\mu_0}{\epsilon_0}}\,\frac{h}{Z_0(\epsilon_r)\sqrt{\epsilon_{eff}(f=0)}} \qquad (8.81)$$

With the given values, we calculate $w_{eff} = 3.573\,\mathrm{mm}$.

We address ourselves now to the capacitive element. For the first design [Figure 8.14(a)], we choose a microstrip line with $w/h = 2.5$ and $w = 3.95\,\mathrm{mm}$, respectively. From the table in Appendix F, we get $Z_0(\epsilon_r = 2.32) = 56.09\,\Omega$ and $\epsilon_{eff} = 1.94$. The respective effective width is obtained as $w_{eff} = 7.624\,\mathrm{mm}$. The effective length of the capacitive element is derived from (6.144),

$$l_{eff} = \frac{C\,Z_0(\epsilon_r)\,c_0}{\sqrt{\epsilon_{eff}}} \qquad (8.82)$$

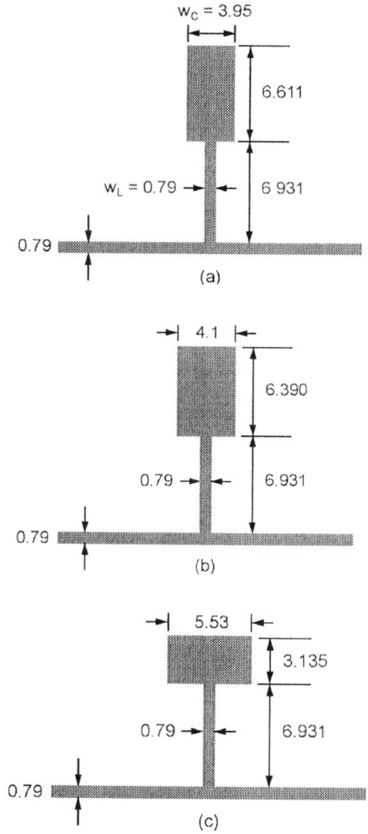

Figure 8.14 Various microstrip layouts of same lumped L_4C_4 series resonant circuit in Figure 8.2, using different ratios of the capacitive and inductive microstrip lines w_C/w_L: (a) 5.0, (b) 5.2, and (c) 7.0.

and reads $l_{eff} = 7.96$ **mm**. This value has to be corrected by the open-end effect, and the capacitive and inductive effects in the step discontinuity. Regarding the open-end effect, Figure 6.9 delivers for $w/h = 2.5$ and the substrate under discussion an effective line extension of $\Delta l/h = 0.48$ and $\Delta l = 0.758$ **mm**, respectively. The open-end effect of the step is calculated from (6.92), and results in $\Delta l_{s,C} = 0.91$ **mm** with $w_2/w_1 = w_L/w_C = 0.2$. The series inductance of the step is obtained from (6.52), and reads $w_1/w_2 = 5$, $L_s/h = 1.127$ **nH/cm**, and $L_s = 0.178$ **nH** respectively. With (3.81), the parasitic inductance can be modeled by a short high impedance line of given characteristic impedance $Z_0 = 124.25\ \Omega$,

$$l_{eff} = \frac{L w_{eff}}{\mu_0 h} \tag{8.83}$$

where $\mu_0 = 0.4\pi$ nH/mm. Substituting the respective values yields $l_{s,L} = 0.32$. The fringing field in the step transition extends the low impedance line beyond the physical step reference plane. Conversely, the step inductance shortens the line, and thus lengthens the narrow line. It can happen that both effects compensate entirely. This can be observed, for example, with ADS, when a circuit comprising a step gives the same result with the step model included and excluded. Thus, whether or not the step parasitics manifest themselves on the whole as inductive or capacitive is dependent on the values of the respective equivalent lengths of the step inductance and capacitance. In this case, the effective reference shift is $\Delta l_s = l_{s,C} - l_{s,L} = 0.91\text{ mm} - 0.32\text{ mm} = 0.59\text{ mm}$. Then, the corrected effective length of the capacitive element reads $l_C = l_{eff,C} - \Delta l - \Delta l_s = (7.96 - 0.7584 - 0.59)\text{ mm} = 6.6116\text{ mm}$.

Next we address the design of the inductance $L_4 = 3.4$ nH. From (8.83), we calculate the effective inductive length as $l_{eff,L} = 6.118$ mm. The reference plane of the T-junction is denoted by the parameter d_2 (see Figure 6.55). For the specified frequency of 3.35 GHz, the wavelength on the narrow line is calculated as $\lambda(w = 0.79\text{ mm}, \epsilon_r = 2.32) = 66.564\text{ mm}$. With same impedance values of the microstrip lines of the T-junction, d_2 is calculated as 1.624 mm. This value can be also directly read from Figure 6.57 with $2w_{eff}/\lambda = 0.107$ and $Z_{01}/Z_{02} = 1$. Then the corrected effective length of the inductive element reads $l_L = l_{eff,L} + d_2 - w/2 = 6.118 + 1.6246 - 0.395 = 7.3476\text{ mm}$. The final layout of the filter element is plotted in Figure 8.14(a). The layout shown in Figure 8.14(b) was designed under the condition that the effective shape of the pad capacitance was a square. Figure 8.14(c) shows a layout where the width of the capacitive element is larger than its length. The design approach is similar. Calculated data is summarized in Table 8.4. Comparing the filters in Figure 8.14(a and b), it is clearly seen that with increasing width ratio w_C/w_L (subscript C indicates capacitive and L as inductive width), the parasitic inductive length $l_{s,L}$ grows faster than the capacitive length $l_{s,C}$, so that with very large impedance steps, compensation of both effects can occur. However, we should also remember that with increasing width of the wider microstrip line, perturbation due to higher-order modes may come up.

Table 8.4

Calculated Parasitic Capacitive and Inductive Lengths $l_{s,C}$ and $l_{s,L}$, the Corrected Physical Capacitive Filter Element Length l_C, and Simulated Center Frequency $f_{0,s}$ and $f_{0,m}$, Using ADS and Momentum, Respectively

filter	w_C/w_L	$l_{eff,C}$ (mm)	Δl (mm)	$l_{s,C}$ (mm)	$L_s/$n H	$l_{s,L}$ (mm)	$\Delta l = l_{s,C} - l_{s,L}$ (mm)	l_c (mm)	$f_{0,s}$ (GHz)	$f_{0,m}$ (GHz)
(a)	5.0	7.96	0.758	0.91	0.178	0.32	0.59	6.61	2.89	2.917
(b)	5.2	7.75	0.758	0.93	0.188	0.34	0.59	6.39	2.90	2.927
(c)	7.0	4.48	0.821	1.05	0.295	0.53	0.52	3.13	3.45	3.500

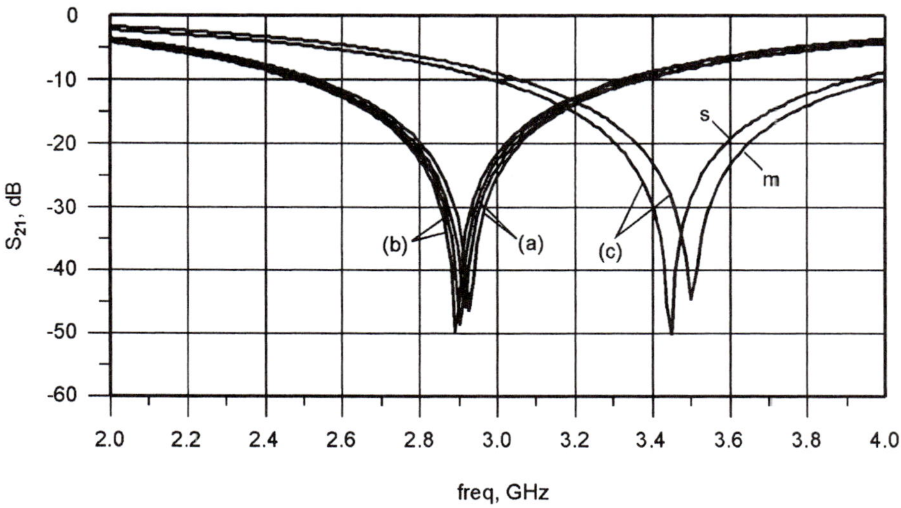

Figure 8.15 Simulated transmission coefficient of analyzed resonant circuits in Figure 8.14 [ADS schematic (s) and Momentum (m) simulation].

Simulation of the designed filter elements in Figure 8.14 has been performed with an ADS schematic simulator and Momentum [15]. As shown in Figure 8.15, for the smaller width ratio, both simulations agree relatively well, whereas for the larger ratio, a higher discrepancy is observed, which may originate from coupling of the open-end discontinuity and the fringing field in the step transition. Such coupling effects cannot be treated with a schematic simulator.

A strong frequency shift of 0.3 GHz is observed with an increase of width ratio from 5 to 7 with a respective decrease of the physical length of the capacitive element. Recalling the conditions for the realization of quasi-lumped L and C elements in Section 6.6, $Z_0^L \gg Z_0^C$ and $l_{eff,\nu} \le \lambda_\nu/10$ $(\nu = C, L)$ must be fulfilled. The effective electrical length of the capacitive element of the filter in Figure 8.14(a) (see also Table 8.4) is calculated as $l_L/\lambda = 0.09$, whereas the capacitive element yields $l_C/\lambda = 0.123$, which is beyond the limit value of 0.1. Thus, also recalling the moderate width ratio of 5, the conditions for quasi-lumped circuit design is less fulfilled. It seems that this is also the reason why a damping pole at 6.2 GHz has not been realized in Figure 8.13(c). Instead, a passband spike was measured and calculated at the specified frequency. Thus, the quasi-lumped element conditions must be carefully taken into account in microstrip circuit design, otherwise strong disagreement can occur between the designed layout and simulated results.

8.1.4 Frequency Transformations

The lowpass filter may be considered as a reference filter, because the filter response of highpass, bandpass, and bandstop filters can be easily derived from the tabulated data using frequency transformation. This is briefly discussed in the following sections [12].

Highpass Filter

To obtain a highpass filter response as shown in Figure 8.16, we consider the frequency transformation

$$\tilde{\Omega} = \frac{1}{\Omega} \tag{8.84}$$

The passband and stopband cutoff frequencies read

$$\tilde{\Omega}_p = \frac{1}{\Omega_p} = 1 \tag{8.85}$$

and

$$\tilde{\Omega}_s = \frac{1}{\Omega_s} \tag{8.86}$$

respectively. As illustrated in Figure 8.17, capacitances are converted to inductances, and vice versa.

Bandpass Filter

The bandpass filter response (Figure 8.18) is obtained by using the frequency transformation

$$\Omega = a\left(\tilde{\Omega} - \frac{1}{\tilde{\Omega}}\right) \tag{8.87}$$

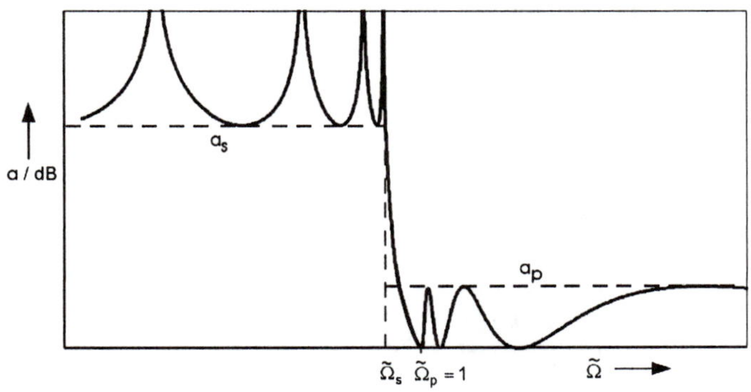

Figure 8.16 Highpass filter response.

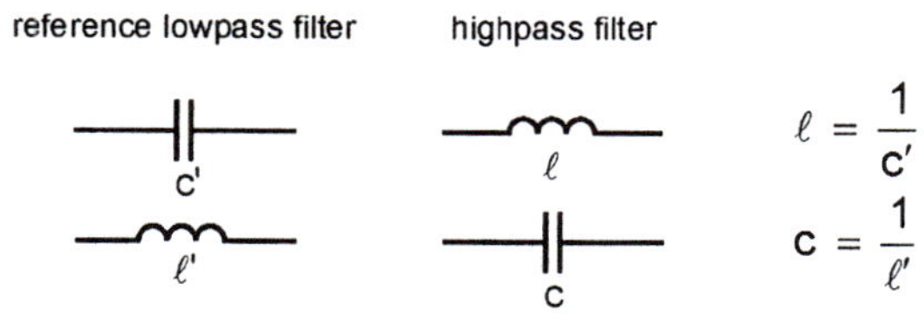

Figure 8.17 Lowpass-to-highpass filter element transformation.

where a is a transformation coefficient. Equation (8.87) solved for $\tilde{\Omega}$ yields

$$\tilde{\Omega}_{\pm} = \sqrt{\left(\frac{\Omega}{2a}\right)^2 + 1} \pm \frac{\Omega}{2a} \tag{8.88}$$

The lowpass filter response corresponds to the bandpass filter with respective positive frequencies (see Figure 8.19). Frequency points $\tilde{\Omega}_+ > 1$ and $\tilde{\Omega}_- < 1$ derived from (8.88) are geometrically symmetric to 1,

$$\tilde{\Omega}_+ \tilde{\Omega}_- = 1 \tag{8.89}$$

From (8.87) and (8.89), we obtain for the passband cutoff frequency

$$\Omega_p = 1 = a\left(\tilde{\Omega}_{+p} - \tilde{\Omega}_{-p}\right) \tag{8.90}$$

and for the stopband cutoff, respectively,

$$\Omega_s = a\left(\tilde{\Omega}_{+s} - \tilde{\Omega}_{-s}\right) \tag{8.91}$$

From (8.87), the transformation coefficient is obtained as

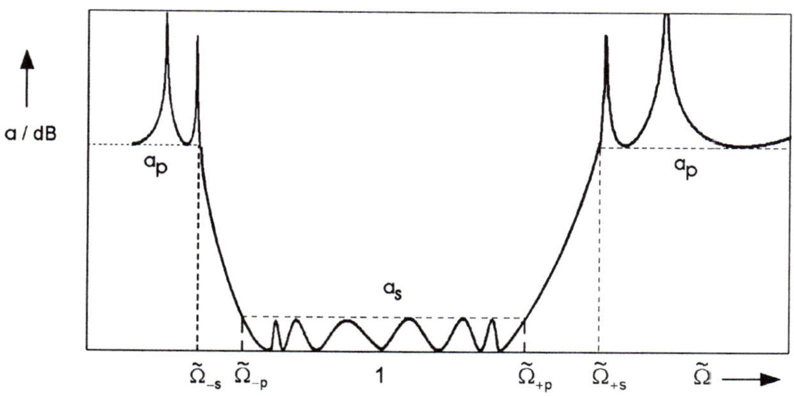

Figure 8.18 Bandpass filter response.

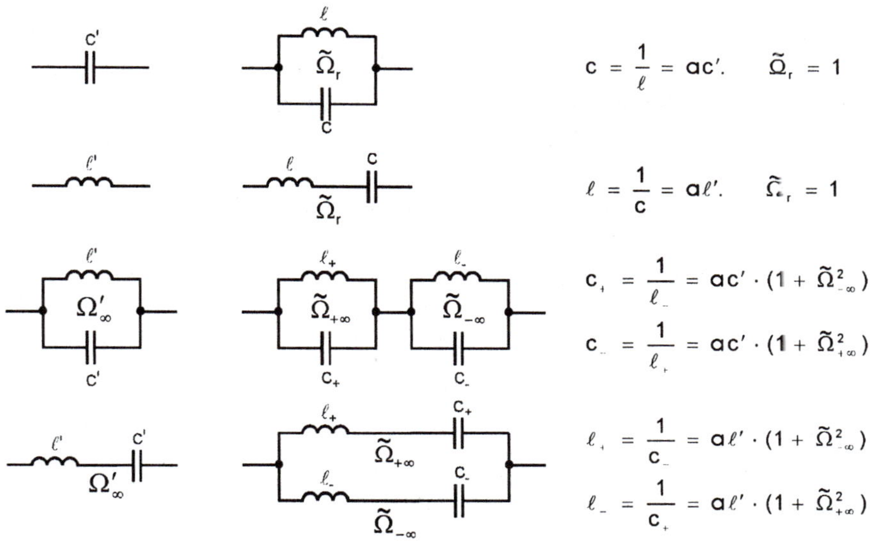

Figure 8.19 Lowpass-to-bandpass filter element transformation.

$$a = \frac{1}{\left(\tilde{\Omega}_{+p} - \tilde{\Omega}_{-p}\right)} \tag{8.92}$$

and substituting a in (8.91) gives

$$\Omega_s = \frac{\tilde{\Omega}_{+s} - \tilde{\Omega}_{-s}}{\tilde{\Omega}_{+p} - \tilde{\Omega}_{-p}} \tag{8.93}$$

Figure 8.19 illustrates the transformation of lowpass to respective bandpass filter elements.

8.2 Impedance Matching Networks

Impedance matching networks are inserted between source and load to optimize the power delivered from the source to the load. In this section, we will treat a special matching problem existing with bipolar and LDMOS power devices in the 10W to 125W region with a real part of the output impedance of only 1Ω to 4Ω. The matching network usually includes very wide microstrip transmission lines with large impedance step ratio [16]. The practical problem is well discussed in [17], and we will use the given example to discuss the various problems in connection with the design of the output matching network. The optimum load impedance for a power device can be found by load-pull measurements [18–21]. Assuming a transistor output impedance of $Z_1 = (2 + j4)\Omega$, maximum power is transferred when the transistor is terminated by its conjugate complex value $Z_1^* = (2 - j4)\,\Omega$. We transform Z_1 into a load impedance of 50Ω using the Smith Chart, which is treated in Appendix B. The circuit shall be realized on Polytetrafluorethylene (PTFE) substrate ($\epsilon_r = 2.5$, $h = 0.381$ mm) [22] for an operating frequency $f = 2.14$ GHz. We further assume that the lead pad width of the power device is $w_1 = 5.6$ mm [23–25]. To avoid any risk of shortening with the grounded package, a gap of $l_1 = 1.0$ mm is maintained between the power device and metallization pattern of the microstrip circuit (Figure 8.20).

The gap with the lead pad above the substrate acts as a microstrip line, which transforms Z_1 into Z_2. With $w/h = 5.6\,\text{mm}/0.381\,\text{mm} = 14.7$, the characteristic impedance is obtained from Figure 3.19 as $Z_{01} = 13.75\Omega$ and $\sqrt{\epsilon_{eff}} = 1.4875$, respectively. Normalizing Z_1 to Z_{01} results in $z_1 = Z_1/Z_{01} = (2 + j4)\Omega/13.75\Omega = 0.145 + j0.291$ (point A in the Smith Chart of Figure 8.21). The corresponding reflection coefficient $\Gamma_1 = 0.775\angle 147°$, and the electrical length $l_A/\lambda = 0.046$. The wavelength on the line with Z_{01} reads $\lambda_1 = c_0/\left(f\sqrt{\epsilon_{eff_1}}\right) = 94.180$ mm. Then the electrical length l_1 of the gap is obtained as $\beta l_1 = 2\pi l_1/\lambda = 0.0667$ and $l_1/\lambda = 0.0106$, respectively. Rotating Γ_1 by $l_1/\lambda = 0.0106$, we find z_2 with respective l_B/λ value, $l_B/\lambda = 0.046 + 0.0106 = 0.0566$ (point B). The denormalized impedance reads $Z_2 = z_2 \cdot Z_{01} = (0.148 + j0.366) \cdot 13.75\Omega = (2.035 + j5.0325)\,\Omega$.

Following the concept of $\lambda/4$-transformation in Figure B.20 (Appendix B), it is necessary that the termination is a real impedance. Therefore, we had to rotate Γ_2 (point B) clockwise by $l/\lambda = 0.25 - 0.0566 = 0.1934$ to obtain a purely real impedance prior to $\lambda/4$ transformation. This, however, would unnecessarily extend the lateral dimension of the output matching circuit.

An alternative approach is now described, which results in a relatively short overall length. The idea is to regard the imaginary part of $Im\{Z_2\} = 5.0325\Omega$ as

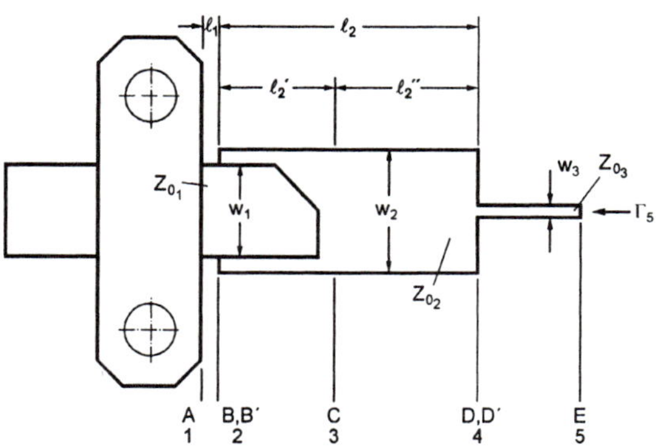

Figure 8.20 Power device package and designed output matching network with defined reference planes and dimensions.

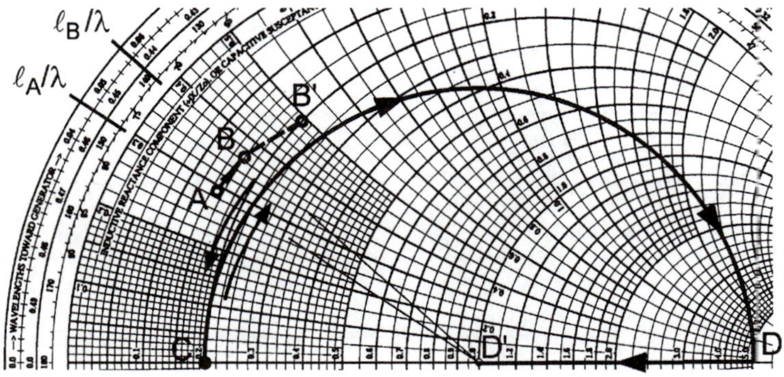

Figure 8.21 Graphical representation of the matching operation.

part of a single step quarter-wave transformer. This is approximately possible in case of very low impedance (approximately short-circuited) termination.

We may define a characteristic impedance of a $\lambda/4$-transformer calculated from the geometric mean of the real part of Z_2 and the impedance $Z_3 = 50\,\Omega$, thus $Z_{02} = (50\,\Omega \cdot 2.035)^{1/2} = 10.087\,\Omega$. Since the real part of the complex impedance termination Z_2 is much lower than the characteristic impedance Z_{02} (quasi short-circuited termination of line), we can approximately describe the imaginary part of $Im\{Z_2\} = 5.0325\,\Omega$ by a respective electrically short length of transmission line. The normalized value of Z_2 reads $z_2 = Z_2/Z_{02} = 0.2017 + j0.499$. Using the periphery scale l/λ, the imaginary value of 0.499 is equivalent to a short-circuited transmission line of electrical line length $l_2'/\lambda = 0.074$. Thus, Z_2 can be approximated by a line with a characteristic impedance $Z_{02} = 10.087\,\Omega$ and electrical length l_2'/λ, which is terminated by a resistance $R = 2.035\,\Omega$. The respective line length can be regarded as an integral part of the single step $\lambda/4$-transformer of the matching problem under consideration. Thus, for the remaining length of the transformer l_2'', it follows $l_2''/\lambda = 0.25 - 0.074 = 0.176$. To determine the length l_2'', we need the effective dielectric constant of the line. For the microstrip line of $Z_{02} = 10.087\,\Omega$ a $w/h = 20.75$ ratio is obtained from Figure 3.19, which yields a respective width $w = 7.9057$ mm. Figure 3.18 delivers an effective dielectric constant $\sqrt{\epsilon_{eff}} = 1.55$, so that the wavelength is calculated as $\lambda = 90.383$ mm. Thus, the effective length of the $\lambda/4$ transformer is obtained as $l_2 = 0.176 \cdot 90.383\text{ mm} = 16.178$ mm. l_2' is found to be 6.688 mm. The simulated reflection coefficient Γ_5 (Figure 8.20) (named S_{11} in ADS plot) is shown in Figure 8.22. In the ADS simulation, the different microstrip lines were directly connected with each other without implementation of the step model. As shown by curve 1, the center frequency is really near the specified one, regarding limit design accuracy using the Smith Chart.

Table 8.5
Parasitic Capacitive and Inductive Lengths $l_{s,C}$ and $l_{s,L}$ for the Impedance Steps of the Analyzed Transformer

Step	w_2/w_1	$l_{s,C}$ (mm)	L_s/nH	$l_{s,L}$ (mm)	$\Delta l = l_{s,C} - l_{s,L}$ (mm)
Left	1.411	0.1066	0.002	0.029	0.0776
Right	7.545	0.317	0.079	0.316	0.001

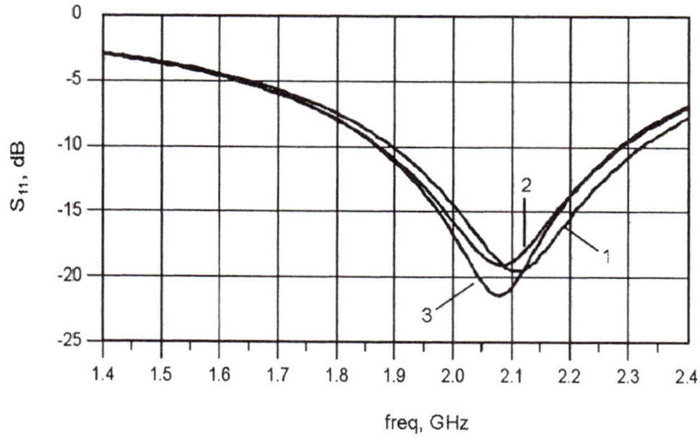

Figure 8.22 Simulated reflection coefficient of the matching network under analysis with ADS. (1) Schematic simulation of the layout with effective capacitive and inductive lengths (step models not included in the simulation). (2) Schematic simulation of the parasitics corrected layout (simulation with step models included). (3) Momentum simulation of parasitics corrected layout.

In the next step, we analyze the parasitics in the matching network. The approach is similar to that in Section 8.1. The parasitic inductive and capacitive lengths for the impedance steps are summarized in Table 8.5. The length l_2, taking into account the step and open-end parasitics, then reads $l_2 = l_{2,eff} - \Delta_{s_1} - \Delta_{s_2} = 16.178 - 0.0776 - 0.001 = 16.1$. Notice that the equivalent lengths due to the inductive and capacitive effects compensate each other for the large step. Although correction lengths due to parasitics may be negligible in some cases, they were maintained for educational reasons.

Figure 8.23 shows the final parasitics corrected layout of the transformer. Curve 2 in Figure 8.22 shows the respective schematic simulation. The shift is marginal, showing the applicability of the equations of the discontinuities existing with the transformer. Simulation with Momentum shows a further slight shift to lower frequencies.

Regarding higher-order mode excitation in the step discontinuities, it can be concluded from Figure 5.7 that the cutoff frequency of the first even higher-order mode $H_{[20]}$ (HE_2) is above X band frequencies. Thus, dynamic parasitic effects are essentially excluded. This can be different with other substrate materials. Using, for

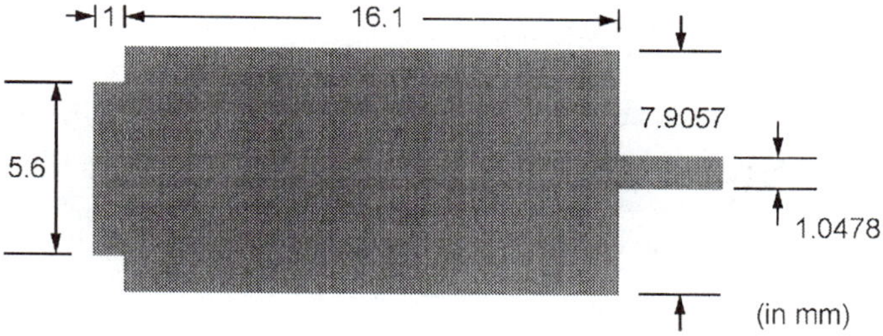

Figure 8.23 Layout under consideration of step discontinuities.

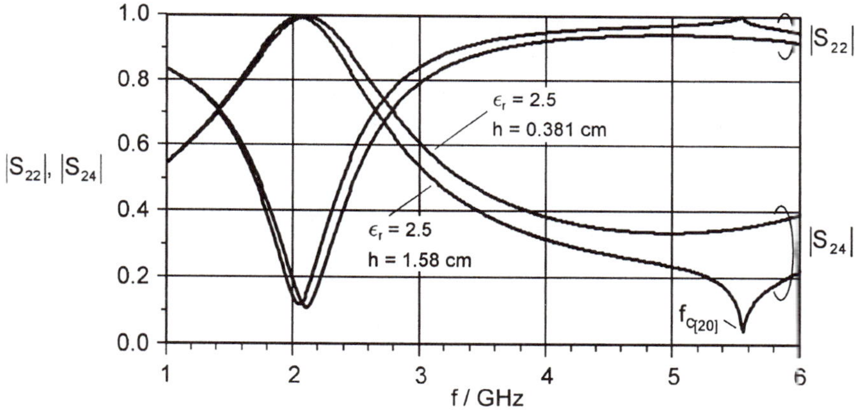

Figure 8.24 ADS simulation of the reflection and transmission coefficients of the discussed matching problem for TLT substrate with different heights.

example, a thickness of 1.58 mm of the used substrate material, the transformer width would increase to approximately 32.8 mm. In this case, indeed, troubles in the design can occur as discussed in [17], due to the relatively low cutoff frequency of the H_{20} mode of approximately 5.5 GHz. This is confirmed with ADS simulation, as illustrated in Figure 8.24 using the same substrate material TLT [22], but different substrate heights. It can be clearly seen that in the case of the thicker substrate, which results in respective wider strip conductors of the microstrip circuit, the higher-order mode interferes already far below the cutoff frequency of the first even higher-order mode $H_{[20]}$ (HE_2) (see Section 6.2.4, and Figures 5.7 and 5.8, respectively). The increasing influence of the higher-order mode with growing frequency results in a shift of the matching network center frequency to lower frequency and a reduction of the 3-dB bandwidth.

8.3 Coupled-Line Directional Coupler

The microstrip directional coupler has specific properties that can be exploited for many applications in the microwave and millimeter wave region. They can probe a signal for frequency and power control, and they can analyze independently and simultaneously forward and backward traveling waves on a transmission line. They also can be used as a power splitter.

Figure 8.25 shows a typical layout of a directional coupler. It comprises a short section of coupled microstrip lines of length L, which are perfectly matched at all four ports. Along the coupled lines, an even and odd mode is propagating (Figure 2.32). Coupling is provided by the strong electrical field in the gap spacing between the lines, which is due to the odd mode, and the magnetic field lines surrounding the microstrips due to the even mode. The notation of the four ports is arbitrary. Assuming that the lines have the same width and spacing between the lines does not change along its entire length (i.e., the coupled part is uniform and symmetrical), then the scattering matrix S, taking into account reciprocal properties of the device and provided the coupler is perfectly matched at all four ports, reads:

$$(S) = \begin{pmatrix} 0 & S_{12} & S_{13} & S_{14} \\ S_{12} & 0 & S_{14} & S_{13} \\ S_{13} & S_{14} & 0 & S_{12} \\ S_{14} & S_{13} & S_{12} & 0 \end{pmatrix} \tag{8.94}$$

For an ideal coupler, the isolated port is entirely decoupled, so that $S_{31} = 0$. In this case, (8.81) simplfies to

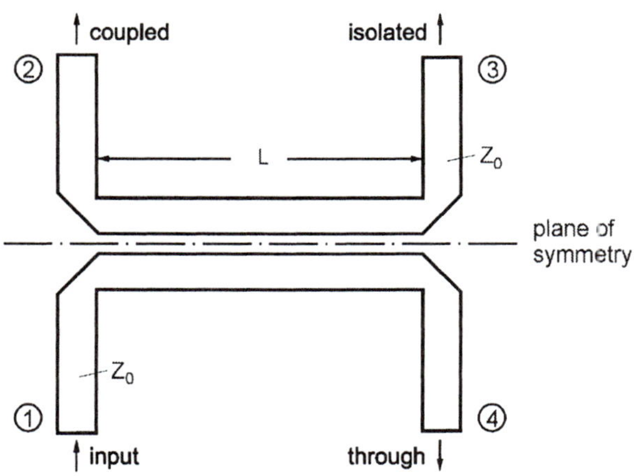

Figure 8.25 Layout of a microstrip directional coupler.

$$(S) = \begin{pmatrix} 0 & S_{12} & 0 & S_{14} \\ S_{12} & 0 & S_{14} & 0 \\ 0 & S_{14} & 0 & S_{12} \\ S_{14} & 0 & S_{12} & 0 \end{pmatrix} \tag{8.95}$$

In this case, part of the incident power is coupled to the coupled port, and the difference of incident and coupled power is traveling to the through port.

Using the given port definition, the performance of the coupler is described with the help of the following four main parameters:

Coupling:

$$C/\text{dB} = 10 \log \frac{P_1^+}{P_2^-} = -20 \log |S_{12}| \tag{8.96}$$

Throughput:

$$G = 10 \log \frac{P_1^+}{P_4^-} = -20 \log |S_{41}| \tag{8.97}$$

Isolation:

$$I = 10 \log \frac{P_1^+}{P_3^-} = -20 \log |S_{31}| \tag{8.98}$$

Directivity:

$$D = 10 \log \frac{P_2^-}{P_3^-} = 10 \log \frac{P_2^- / P_1^+}{P_3^- / P_1^+} = 20 \log \frac{|S_{21}|}{|S_{31}|} \tag{8.99}$$

where P_1^+ is the incident power at port 1, and P_2^-, P_3^-, and P_4^- are the transmitted power at the coupled, isolated, and direct ports, respectively. In the following section, we consider the electrical properties of coupled microstrip lines and derive the impedance and scattering matrix for a section of length L.

8.3.1 Impedance Matrix of Coupled Lines

Figure 8.26 shows the equivalent circuit of two coupled lines. Stimulation of both lines with I_{01} at port $1-1'$ and port $2-2'$, or with I_{03} at port $3-3'$ and port $4-4'$, respectively, will excite an even mode. In this case, the coupled line parameters are denoted by the even phase coefficient β_e and even characteristic impedance Z_{0_e}. On the other hand, source current I_{02} between electrodes 1 and 2, or I_{04} between electrodes 3 and 4, will excite an odd mode. In this case, the coupled line parameters are denoted by the odd phase coefficient β_o and odd characteristic

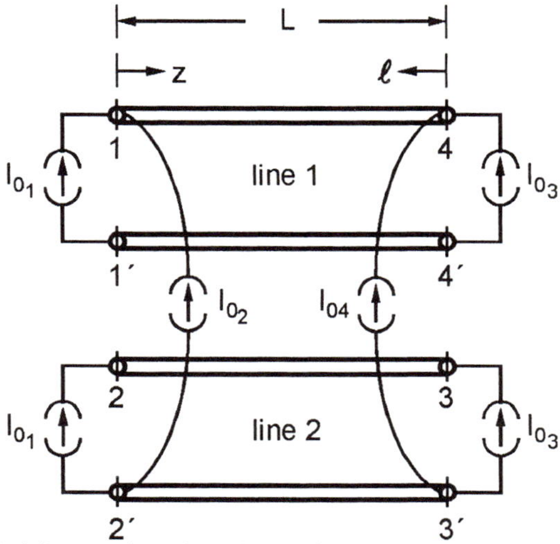

Figure 8.26 Coupled lines and modes of stimulation.

impedance Z_{0_0}. By virtue of the superposition principle, we can confine our analysis separately to I_{0_1}, I_{0_2}, I_{0_3}, and I_{0_4}, as shown in Figure 8.27.

We recall the transmission line equations, which describe complex voltage and current along the line as a function of voltage and current at the output at $l = 0$,

$$V(l) = V_0 \cos \beta l + jZ_0 I_0 \sin \beta l \tag{8.100}$$

$$I(l) = I_0 \cos \beta l + j\frac{1}{Z_0} V_0 \sin \beta l \tag{8.101}$$

where V_0 and I_0 are voltage and current at the output of the line. We first consider Figure 8.27(a). Since the lines terminate in an open circuit, (8.100) and (8.101) simplify with $I_0 = 0$ to

$$V_1(l) = V_0 \cos \beta_e l \tag{8.102}$$

$$I_1(l) = j\frac{V_0}{Z_{0_e}} \sin \beta_e l \tag{8.103}$$

From (8.103), we get for $l = L$

$$I(l = L) = j\frac{V_0}{Z_{0_e}} \sin \beta L = I_1 = I_{0_1} \tag{8.104}$$

V_0 substituted in (8.102) yields

$$V_1(l) = -j Z_{0_e} I_{0_1} \frac{\cos \beta l}{\sin \beta L} \tag{8.105}$$

for both lines.

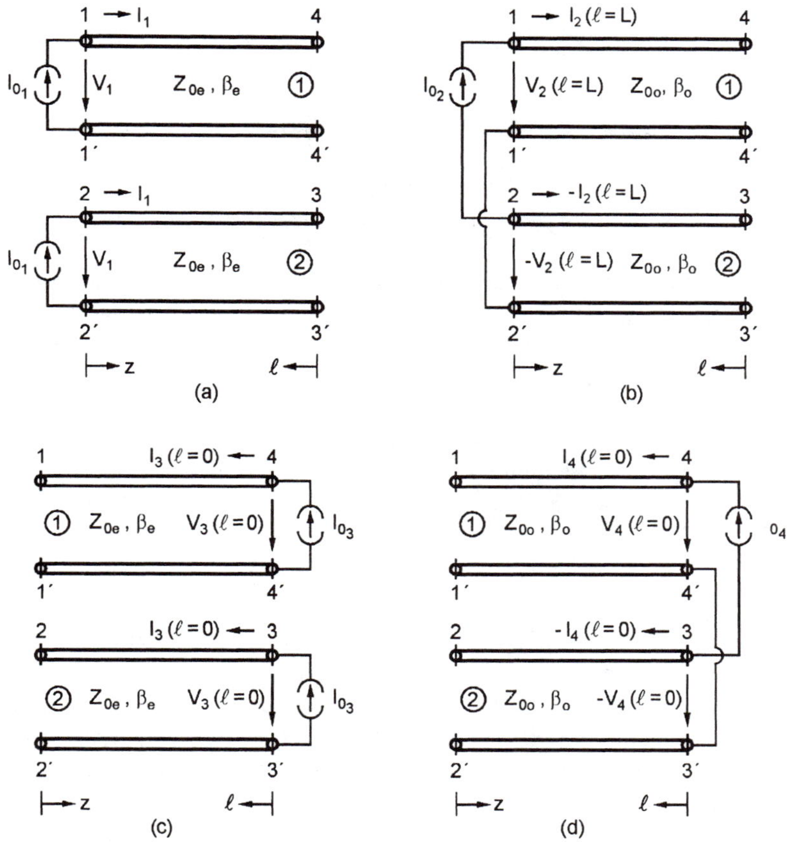

Figure 8.27 Derivation of the impedance matrix on the basis of even and odd current source stimulation at the input and ouput of the section of coupled line.

By interchanging the respective indices, we can directly derive from (8.105) voltage $V_3(z)$ on both lines when the even mode is stimulated by I_{0_3} at the opposite side [Figure 8.27(c)],

$$V_3(z) = -j Z_{0_e} I_{03} \frac{\cos \beta z}{\sin \beta L} = V_3(l) = -j Z_{0_e} I_{03} \frac{\cos [\beta_e (L - l)]}{\sin(\beta_e l)} \qquad (8.106)$$

Similarly, the voltages V_2 and V_4 can be calculated in the case of odd stimulus with I_{02} and I_{04}, respectively. The line parameters are now denoted by the odd phase coefficient β_o and odd characteristic impedance Z_{0_o}. Taking into account the opposite sign of current and voltage on the respective lines, we can derive from (8.105)

$$V_2(l) = \mp j Z_{0_o} I_{0_2} \frac{\cos \beta_o l}{\sin \beta_o L} \tag{8.107}$$

for stimulating current I_{02}. For I_{04}, we obtain corresponding to (8.106) for V_4

$$V_4(l) = \mp j Z_{0_o} I_{04} \frac{\cos [\beta_o (L - l)]}{\sin(\beta_o L)} \tag{8.108}$$

The upper negative sign relates to line 1 and the lower positive sign to line 2.

The impedance matrix relates port voltages to port currents. Regarding even and odd mode operation, this results in a 4×4 impedance matrix,

$$\begin{pmatrix} V_1 \\ V_2 \\ V_3 \\ V_4 \end{pmatrix} = \begin{pmatrix} Z_{11} & Z_{12} & Z_{13} & Z_{14} \\ Z_{21} & Z_{22} & Z_{23} & Z_{24} \\ Z_{31} & Z_{32} & Z_{33} & Z_{34} \\ Z_{41} & Z_{42} & Z_{43} & Z_{44} \end{pmatrix} \begin{pmatrix} I_1 \\ I_2 \\ I_3 \\ I_4 \end{pmatrix} \tag{8.109}$$

As can be seen from Figures 8.26 and 8.27, the port currents I_1, I_2, I_3, and I_4 (directed inward at the ports in Figure 8.25) are obtained by superposing the respective source currents,

$$I_1 = I_{01} + I_{02}, \quad I_2 = I_{01} - I_{02}, \quad I_3 = I_{03} - I_{04}, \quad I_4 = I_{03} + I_{04} \tag{8.110}$$

Resolved for the source currents, (8.110) reads

$$I_{01} = \frac{1}{2}(I_1 + I_2), \qquad I_{02} = \frac{1}{2}(I_1 - I_2)$$
$$I_{03} = \frac{1}{2}(I_4 + I_3), \qquad I_{04} = \frac{1}{2}(I_4 - I_3) \tag{8.111}$$

The total port voltages V_1, V_2, V_3, and V_4 arise from the superposition of the voltages stimulated by the current sources, as discussed before. Note that total voltage is characterized without any line variable l.

$$V_1 = V_1(l = L) + V_2(l = L) + V_3(z = 0) + V_4(z = 0) \tag{8.112}$$
$$V_2 = V_1(l = L) + V_2(l = L) + V_3(z = 0) + V_4(z = 0) \tag{8.113}$$

for lines 1 and 2, respectively. Further on, the relations for V_3 and V_4 read

$$V_3 = V_1(l=0) + V_2(l=0) + V_3(l=0) + V_4(l=0) \tag{8.114}$$

$$V_4 = V_1(l=0) + V_2(l=0) + V_3(l=0) + V_4(l=0) \tag{8.115}$$

for lines 2 and 1, respectively. Substituting the respective voltages and currents, we get a relationship between total port voltages and currents. This reads, for example, for V_1 [26]:

$$V_1 = -\frac{j}{2}(I_1 + I_2)Z_{0_e}\frac{\cos\beta_e L}{\sin\beta_e L} - \frac{j}{2}(I_1 - I_2)Z_{0_o}\frac{\cos(\beta_o L)}{\sin(\beta_o L)} \tag{8.116}$$

Taking into account symmetry properties of the analyzed line section, the impedance parameters in (8.109) read:

$$Z_{11} = Z_{22} = Z_{33} = Z_{44} = -jZ_{0_e}\frac{\cot\theta_e}{2} - jZ_{0_o}\frac{\cot\theta_o}{2} \tag{8.117}$$

$$Z_{12} = Z_{21} = Z_{34} = Z_{43} = -jZ_{0_e}\frac{\cot\theta_e}{2} + jZ_{0_o}\frac{\cot\theta_o}{2} \tag{8.118}$$

$$Z_{13} = Z_{31} = Z_{24} = Z_{42} = -jZ_{0_e}\frac{\csc\theta_e}{2} - jZ_{0_o}\frac{\csc\theta_o}{2} \tag{8.119}$$

$$Z_{14} = Z_{41} = Z_{23} = Z_{32} = -jZ_{0_e}\frac{\csc\theta_e}{2} - jZ_{0_o}\frac{\csc\theta_o}{2} \tag{8.120}$$

where $\theta_{e,o} = \beta_{e,o}L$ is the normalized electrical length of the line section for the even and odd modes, respectively.

8.3.2 Scattering Matrix of Coupled Lines

The microstrip coupler shown in Figure 8.25 is symmetric with respect to ports 1 and 2, and 3 and 4, respectively. In this case, the analysis of the four-port is simplified by applying the method proposed in [27]. Even excitation of the four-port (Figure 8.28) at ports 1 and 2 results in the electromagnetic field of an even mode, which is symmetric with respect to the plane of symmetry. Thus, the plane of symmetry represents a magnetic wall. Similarly, choosing an odd excitation, the plane of symmetry becomes an electric wall (Figure 8.28). Then the four-port scattering matrix can be derived from the eigenreflection and transmission coefficients of the remaining two-ports with perfect electric and magnetic walls, as shown in Figure 8.29. The scattering matrix $\mathbf{S}_{14}^{e,o}$ of the respective two-port is defined as

$$\mathbf{S}_{14}^{\nu} = \begin{pmatrix} S_{11}^{\nu} & S_{14}^{\nu} \\ S_{14}^{\nu} & S_{11}^{\nu} \end{pmatrix} \tag{8.121}$$

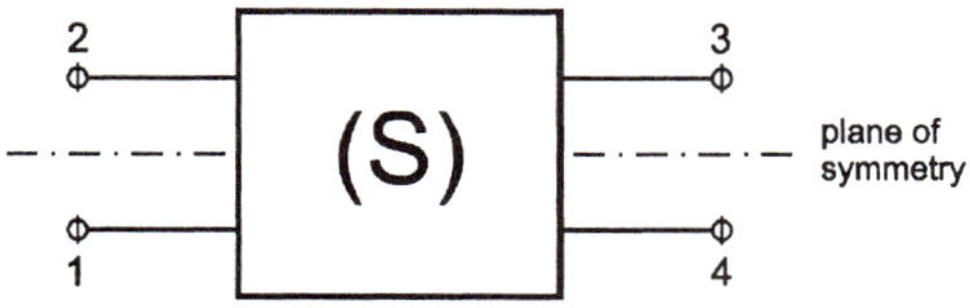

Figure 8.28 Linear passive four-port with plane of symmetry indicated.

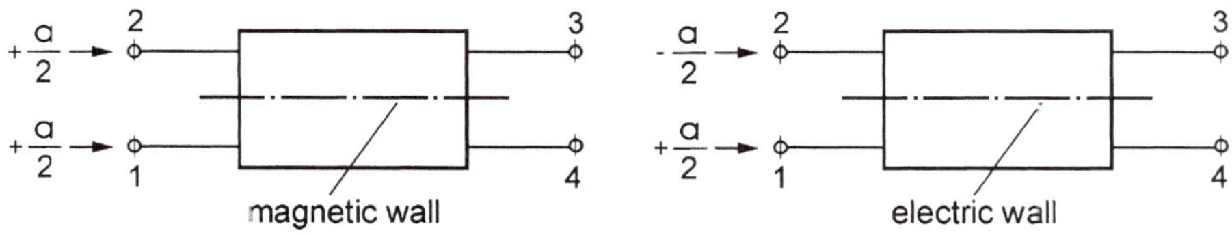

Figure 8.29 Even (left) and odd (right) excitation of the symmetric four-port.

Figure 8.30 Partitioning of the symmetrical four-port into two-ports with electric and magnetic walls (respective two-ports with ports 2 and 3 not shown).

where $\nu = (e, o)$. If two power waves a of amplitude $+\frac{1}{2}$ and in-phase are applied at ports 1 and 2, then the plane of symmetry represents a magnetic wall (Figure 8.29). Similarly, if two power waves of amplitude $+\frac{1}{2}$ and out-of-phase are applied at ports 1 and 2, then the plane of symmetry becomes an electric wall. For the reduced two-port with ports 1 and 4 (Figure 8.30), we can write

$$b_1 = S_{11}^{\nu} a_1 + S_{14}^{\nu} a_4 \tag{8.122}$$

$$b_4 = S_{41}^{\nu} a_1 + S_{44}^{\nu} a_4 \tag{8.123}$$

where $\nu = (e, o)$. Then, with the given half-amplitude excitation, we get from (8.122) and (8.123) for even and odd case with $a_4 = 0$ (see Figure 8.30)

$$b_1^e = S_{11}^e \frac{a_1}{2} \qquad b_4^e = S_{41}^e \frac{a_1}{2}$$

$$b_1^o = S_{11}^o \frac{a_1}{2} \qquad b_4^o = S_{41}^o \frac{a_1}{2} \tag{8.124}$$

By superposition, we obtain

$$b_1 = b_1^e + b_1^o = \frac{1}{2}\left(S_{11}^e + S_{11}^o\right)a_1 = S_{11}\,a_1 \tag{8.125}$$

$$b_4 = b_4^e + b_4^o = \frac{1}{2}\left(S_{41}^e + S_{41}^o\right)a_1 = S_{41}\,a_1 \tag{8.126}$$

Now, consider ports 2 and 3 under given excitation. The respective scattering matrix relations read

$$b_2 = S_{22}^{\nu}\,a_2 + S_{23}^{\nu}\,a_3 \tag{8.127}$$

$$b_3 = S_{32}^{\nu}\,a_2 + S_{33}^{\nu}\,a_3 \tag{8.128}$$

where $\nu = (e,o)$. Then, with the given half-amplitude excitation we get from (8.127) and (8.128) for even and odd case with $a_3 = 0$

$$b_2^e = S_{22}^e\,\frac{a_1}{2} \qquad b_3^e = S_{32}^e\,\frac{a_1}{2}$$

$$b_2^o = -S_{22}^o\,\frac{a_1}{2} \qquad b_3^o = -S_{32}^o\,\frac{a_1}{2} \tag{8.129}$$

By superposition, we obtain

$$b_2 = b_2^e + b_2^o = \frac{1}{2}\left(S_{22}^e - S_{22}^o\right)a_1 = S_{21}\,a_1 \tag{8.130}$$

$$b_3 = b_3^e + b_3^o = \frac{1}{2}\left(S_{32}^e - S_{32}^o\right)a_1 = S_{31}\,a_1 \tag{8.131}$$

Thus, taking into account symmetry, the four-port scattering coefficients read [28]:

$$S_{11} = \frac{1}{2}\left(S_{11}^e + S_{11}^o\right) \qquad S_{12} = \frac{1}{2}\left(S_{11}^e - S_{11}^o\right)$$

$$S_{13} = \frac{1}{2}\left(S_{13}^e - S_{13}^o\right) \qquad S_{14} = \frac{1}{2}\left(S_{14}^e + S_{14}^o\right) \tag{8.132}$$

With the commutative relation

$$S = (Z_N + E)^{-1}(Z_N - E) = (Z_N - E)(Z_N + E)^{-1} \tag{8.133}$$

the scattering coefficients can be obtained from (8.117) to (8.120). For the two-port scattering coefficients in (8.121), the following relations are given:

$$S_{11}^e = j\,\frac{\left[\left(Z_{0_e}/Z_0\right) - \left(Z_0/Z_{0_e}\right)\right]\sin\theta_e}{2\cos\theta_e + j\left[\left(Z_{0_e}/Z_0\right) + \left(Z_0/Z_{0_e}\right)\right]\sin\theta_e} \tag{8.134}$$

$$S_{14}^e = \frac{2}{2\cos\theta_e + j\left[\left(Z_{0_e}/Z_0\right) + \left(Z_0/Z_{0_e}\right)\right]\sin\theta_e} \tag{8.135}$$

$$S_{11}^{o} = j\frac{\left[\left(Z_{0_o}/Z_0\right) - \left(Z_0/Z_{0_o}\right)\right]\sin\theta_o}{2\cos\theta_o + j\left[\left(Z_{0_o}/Z_0\right) + \left(Z_0/Z_{0_o}\right)\right]\sin\theta_o} \tag{8.136}$$

$$S_{14}^{o} = \frac{2}{2\cos\theta_o + j\left[\left(Z_{0_o}/Z_0\right) + \left(Z_0/Z_{0_o}\right)\right]\sin\theta_o} \tag{8.137}$$

8.3.3 TEM Directional Coupler

In the following section, we assume that the electrical length of the coupled microstrip lines are equal for the even and odd mode to simplify further calculations,

$$\theta_e = \theta_o = \beta L = \frac{2\pi f}{c_0}\sqrt{\epsilon_r}\,L \tag{8.138}$$

The condition is generally fulfilled for coupled lines embedded in a homogeneous medium, such as a directional triplate coupler. Then, it follows for the eigen-reflection coefficient in (8.132)

$$S_{11} = \frac{1}{2}(S_{11}^{e} + S_{11}^{o}) = \frac{1}{2}j\frac{\left[\left(Z_{0_e}/Z_0\right) - \left(Z_0/Z_{0_e}\right)\right]\sin\theta}{2\cos\theta + j\left[\left(Z_{0_e}/Z_0\right) + \left(Z_0/Z_{0_e}\right)\right]\sin\theta}$$

$$+ j\frac{\left[\left(Z_{0_o}/Z_0\right) - \left(Z_0/Z_{0_o}\right)\right]\sin\theta}{2\cos\theta + j\left[\left(Z_{0_o}/Z_0\right) + \left(Z_0/Z_{0_o}\right)\right]\sin\theta} \tag{8.139}$$

Under assumption

$$Z_0^2 = Z_{0_e}Z_{0_o},\; Z_0 = \sqrt{Z_{0_e}Z_{0_o}} \tag{8.140}$$

the relations $S_{11}^{e} = -S_{11}^{o}$ and $S_{12}^{e} = S_{12}^{o}$ are valid. Then, we obtain from (8.132)

$$S_{11} = 0, \qquad S_{12} = S_{11}^{e}, \qquad S_{13} = 0, \qquad S_{14} = S_{12}^{e} \tag{8.141}$$

Thus, the circuit has the properties of a directional coupler. A wave incident at port 1 will be transfered to port 2 and port 4. Port 3 is isolated, and port 1 is matched. The scattering coefficients S_{12} and S_{14} read:

$$S_{12} = j\frac{\left[\left(Z_{0_e}/Z_{0_o}\right) - \left(Z_{0_o}/Z_{0_e}\right)\right]\sin\theta}{2\cos\theta + j\left[\left(Z_{0_e}/Z_{0_o}\right) + \left(Z_{0_o}/Z_{0_e}\right)\right]\sin\theta} \tag{8.142}$$

$$S_{14} = \frac{2}{2\cos\theta + j\left[\left(Z_{0_e}/Z_{0_o}\right) + \left(Z_{0_o}/Z_{0_e}\right)\right]\sin\theta} \tag{8.143}$$

The transmission coefficients become maximum if $\theta = \beta L = \pi/2$, which means $L = \lambda/4$. Under this assumption, (8.142) and (8.143) read

$$S_{12} = \frac{\left(Z_{0_e}/Z_0\right)^{1/2} - \left(Z_0/Z_{0_e}\right)^{1/2}}{\left(Z_{0_e}/Z_0\right)^{1/2} + \left(Z_0/Z_{0_e}\right)^{1/2}} = \frac{Z_{0_e}/Z_{0_o} - 1}{Z_{0_e}/Z_{0_o} + 1} = k \tag{8.144}$$

$$S_{14} = -j\frac{2}{\left(Z_{0_e}/Z_0\right)^{1/2} + \left(Z_0/Z_{0_e}\right)^{1/2}} \tag{8.145}$$

For design purposes, the relevant relations for the TEM directional coupler are summarized as follows:

Coupling coefficient k:

$$k = \frac{Z_{0_e}/Z_{0_o} - 1}{Z_{0_e}/Z_{0_o} + 1} \tag{8.146}$$

Characteristic impedance Z_0 of terminating microstrip lines:

$$Z_0 = \sqrt{Z_{0_e}Z_{0_o}} \tag{8.147}$$

Coupling length L:

$$L = \lambda/4 \tag{8.148}$$

Center frequency f_0:

$$f_0 = \frac{c_0}{4l\sqrt{\epsilon_r}} \tag{8.149}$$

Scattering coefficients at f_0:

$$S_{12}(f_0) = k \tag{8.150}$$

$$S_{14}(f_0) = -j\sqrt{1 - k^2} \tag{8.151}$$

$$S_{11} = S_{22} = S_{33} = S_{44} = 0 \tag{8.152}$$

Furthermore, the reciprocity and symmetry conditions reflected in (8.95) hold.

8.3.4 Quasi-TEM Directional Coupler

In the last section, it was pointed out that a TEM coupler is based on the assumption that the phase velocity of the even and odd modes are equal (8.138), and that the matching condition (8.147) is fulfilled. The matching condition can be accomplished, whereas due to the inhomogeneous medium and different field distribution of the even and odd modes, the phase velocity matching can only approximately be achieved:

$$\frac{\Delta\epsilon_{eff}}{\epsilon_{eff_{av}}} = \frac{\epsilon_{eff_e} - \epsilon_{eff_o}}{(\epsilon_{eff_e} + \epsilon_{eff_o})/2} = 2\,\frac{\epsilon_{eff_e} - \epsilon_{eff_o}}{\epsilon_{eff_e} + \epsilon_{eff_o}} \ll 1 \tag{8.153}$$

The calculation of the scattering parameters with the different line parameters for the even and odd mode delivers similar expressions as for the TEM mode assumption. Defining a center frequency

$$f_0 = \frac{f_{0_e} + f_{0_o}}{2} = \frac{f_{0_e}}{2}\left(1 + \sqrt{\epsilon_{eff_e}/\epsilon_{eff_o}}\right) \tag{8.154}$$

delivers the following results, found for the scattering parameters of a quasi-TEM coupler [29–31]:

$$S_{11}(f_0) = -j\frac{\pi}{8}k\sqrt{1 - k^2}\,\frac{\Delta\epsilon_{eff}}{\epsilon_{eff_{av}}} \tag{8.155}$$

$$S_{14}(f_0) = -j\sqrt{1 - k^2} \tag{8.156}$$

$$S_{21}(f_0) = k \tag{8.157}$$

$$S_{31}(f_0) = -\frac{\pi}{8}(1 - k^2)\,\frac{\Delta\epsilon_{eff}}{\epsilon_{eff_{av}}} \tag{8.158}$$

The directivity D reads

$$D(f_0) = -20\log\left[\frac{\pi}{8}\left(\frac{1}{k} - k\right)\frac{\Delta\epsilon_{eff}}{\epsilon_{eff_{av}}}\right] \tag{8.159}$$

8.3.5 Design Example

A directional coupler shall be designed on alumina substrate ($\epsilon_r = 9.6$, $h = 0.635$ mm). The coupling factor is specified as 10 dB and the midband frequency as 9.375 GHz. The coupler is matched with 50 Ω on all ports.

We first calculate the coupling coefficient k from the coupling factor C,

$$C = -20\log(k) \tag{8.160}$$

Then we get for k:

$$k = 10^{-\frac{C/\text{dB}}{20}} = 0.316 \tag{8.161}$$

With (8.146) and (8.147), the even-mode and odd-mode characteristic impedances are related to the coupling coefficient k as

$$Z_{0_e} = Z_0\sqrt{(1+k)/(1-k)} = 69.353\,\Omega \tag{8.162}$$

$$Z_{0_o} = Z_0\sqrt{(1-k)/(1+k)} = Z_0^2/Z_{0_e} = 36.047\,\Omega \tag{8.163}$$

From the graph in Figure J.2, we determine for the given values of characteristic impedance of even and odd mode,

$$w/h = 0.83, \quad s/h = 0.28 \tag{8.164}$$

From the graphs in Figure J.3, we then read the effective dielectric constants ϵ_{eff_e} and ϵ_{eff_o},

$$\epsilon_{eff_e} = 6.8, \; \epsilon_{eff_o} = 5.46 \tag{8.165}$$

Next, we specify the length L of the coupled lines. For that, we need the averaged center frequency f_0,

$$f_0 = \frac{f_{0_e} + f_{0_o}}{2} \tag{8.166}$$

With (8.149), it follows

$$f_0 = \frac{1}{2}\frac{c_0}{4l}\left(\frac{1}{\sqrt{\epsilon_{eff_e}}} + \frac{1}{\sqrt{\epsilon_{eff_o}}}\right) \tag{8.167}$$

Then, it follows for the coupler length L:

$$L = \frac{c_0}{8f_0}\left(\frac{1}{\sqrt{\epsilon_{eff_e}}} + \frac{1}{\sqrt{\epsilon_{eff_o}}}\right) \tag{8.168}$$

Using (8.166) the length is calculated as $L = 3.2$ mm.

Figure 8.31 shows the scattering parameters simulated with ADS (version 2003C) [15]. With $\Delta\epsilon_{eff}/\epsilon_{eff_{av}} = 0.218$, the magnitude of $S_{11}(f_0)$ is calculated from (8.155) as 0.0257. Expressed in decibels it gives -31.78 **dB**, which agrees satisfactorily with the simulated result of -26.4 **dB**, taking into account that the on-hand design is based on quasi-static assumption. S_{14} is calculated from (8.156) as -0.22 **dB** in comparison with simulated -0.63 **dB**. The transmission coefficient of the isolated

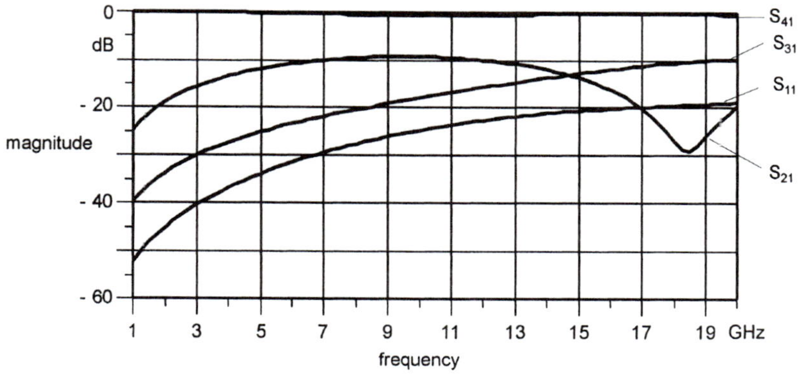

Figure 8.31 Magnitude response of 9.375-GHz directional coupler.

path S_{31} is calculated from (8.149) as –22.2 dB. Simulation yields –18.5 dB. Using (8.146), the directivity is calculated as $D = 12.25$ **dB**, whereas simulation delivers a value of $D = 9.2$ **dB**.

It should be recognized that the coupler response is periodic with frequency as can be seen from Figure 8.31. Coupling becomes worse at approximately 18.5 GHz. At this frequency, the electrical length of the coupled line becomes a half-wavelength, which does not meet the condition (8.148).

References

[1] Kompa, G., *Die frequenzabhängigen Übertragungseigenschaften von Streifenleitungs-Wellenwiderstandssprüngen, Filtern und Stichleitungen in Mikrostrip-Technik*, Doctoral Thesis, Aachen: RWTH, 1975.

[2] Mehran, R., *Grundelemente des rechnergestützten Entwurfs von Mikrostreifenleitungs-Schaltungen*, Aachen: Verlag H. Wolff, 1983.

[3] Kompa, G., and R. Mehran, "Planar Waveguide Model for Calculating Microstrip Components," *Electron. Lett.*, Vol. 11, September 1975, pp. 459–460.

[4] Kompa, G., "Reduced Coupling Aperture of Microstrip Stubs Provides New Aspects in Stub-Filter Design," *Proc. 6th European Microwave Conf.*, Rome 1976, pp. 39–43.

[5] Kompa, G., and R. Mehran, "Microstrip Filter Analysis Using a Microstrip Waveguide Model," *The Radio and Electronic Engineer*, Vol. 50, No. 1/2, January/February 1980, pp. 54–58.

[6] Skwirzynski, J. K., and J. Zdunek, "Note on Calculation of Ladder Coefficients for Symmetrical and Inverse Impedance Filters on a Digital Computer," *IRE Trans. Circuit Theory*, December 1958, pp. 328–347.

[7] Meinguet, J., and V. Belevitch, "On the Realizability of Ladder Filters," *IRE Trans. Circuit Theory*, December 1958, pp. 253–255.

[8] Watanabe, H., "Synthesis of Band-Pass Ladder Network," *IRE Trans. Circuit Theory*, December 1958, pp. 256–264.

[9] Belevitch, B., "Recent Developments in Filter Theory," *IRE Trans. Circuit Theory*, December 1958, pp. 236–252.

[10] Saal, R., and E. Ulbrich, "On the Design of Filters by Synthesis," *IRE Trans. Circuit Theory CT 5*, December 1958, pp. 284–327.

[11] Saal, R., *Der Entwurf von Filtern mit Hilfe des Kataloges normierter Tiefpässe*, Telefunken, 1961.

[12] Saal, R., *Handbuch zum Filterentwurf - Handbook of Filter Design*, AEG-Fachbücher, 2. Auflage, Dr. Alfred Hüthig Verlag, 1988.

[13] Levy, R., "Tables of Element Values for the Distributed Low-Pass Prototype Filter," *IEEE Trans.*, Vol. MTT-13, 1965, pp. 514–534.

[14] Matthaei, G. L., L. Young, and E. M. T. Jones, *Microwave Filters, Impedance-Matching Networks, and Coupling Structures*, New York: McGraw-Hill Book Company, 1964.

[15] Agilent Technologies, *Advanced Design System (ADS)*, Version 2003(C), http://eesof.tm.agilent.com.

[16] Cripps, S. C., *RF Power Amplifiers for Wireless Communications*, Norwood, MA: Artech House, 1999.

[17] Arntz, B., "Analyzing Abrupt Microstrip Transitions," *rfdesign*, March 2002, pp. 22–32 (www.rfdesign.com).

[18] Bunz, B., and G. Kompa, "Active Load Pull with Fourth Harmonic Tuning on an IQ Modulator Concept," *Proc. European Microwave Conf.*, 2003, pp. 359–361.

[19] Benedikt, J., et al., "High-Power Time-Domain Measurement System with Active Harmonic-Load Pull for High-Efficiency Base-Station Amplifier Design," *IEEE MTT-S Symposium Digest*, Boston, 2000, pp. 2617–2624.

[20] Teyssier, J.-P., et al., "40-GHz/150-ns Versitile Pulsed Measurement System for Microwave Transistor Isothermal Characterization," *IEEE Trans.*, Vol. MTT-46, 1998, pp. 2043–2052.

[21] Teyssier, J.-P., et al., "Large Signal Characterization of Microwave Power Devices," *Proc. 12th GaAs Syposium*, Amsterdam, 2004, pp. 187–190.

[22] Taconic, *High-Performance Laminates Material Guide*, www.taconic-add.com/en-products.php.

[23] Motorola, *RF Power Field Effect Transistors*, Document MRF21010/D, 2003, www.freescale.com.

[24] Motorola, *Using Data Sheet Impedances for RF LDMOS Devices*, Document EB212/D, 2004, www.freescale.com.

[25] Motorola, *Mounting Method with Mechanical Fasteners for the MRF19090 and Similar Packages*, Document AN1923/D, 2001, www.freescale.com.

[26] Wolff, I., *Einführung in die Mikrostrip-Leitungstechnik*, Aachen: Verlag H. Wolff, 2. Auflage, 1978.

[27] Reed, J., and G. J. Wheeler, "A Method of Analysis of Symmetrical Four-Port Networks," *IEEE Trans.*, Vol. MTT-4, 1956, pp. 246–252.

[28] K. C. Gupta, R. Garg, and R. Chadha, *Computer-aided Design of Microwave Circuits*, Norwood, MA: Artech House, Inc., 1981.

[29] Hoffmann, R. K., *Integrierte Mikrowellenschaltungen*, Berlin: Springer-Verlag, 1983.

[30] Brenner, H. E., "Perturbations of the Critical Parameters of Quarter-Wave Directional Couplers," *IEEE Trans.*, Vol. MTT-15, 1967, pp. 384–385.

[31] Brenner, H., "λ/4-Richtkoppler in inhomogenem Medium mit Abweichungen der Gleich- und Gegentaktparameter vom Idealwert," *Frequenz*, Vol. 26, 1972, pp. 156–165.

Chapter 9

Technology

The fabrication of microwave integrated circuits needs knowledge, skill, and experience in standard photolithography and chemical etching techniques [1]. Regarding widely used hybrid technology, this includes preparation of the mask of the circuit layout, cleaning the substrate surface, covering the surface with photo resist, and illuminating it with the mask attached. After exposure, the photo resist is developed, and the substrate etched and cleaned. Then, if needed, lumped elements are placed onto the circuit. Finally, for practical usage, the complete circuit is mounted in a package with microstrip-to-coaxial adapters and feedthroughs for the bias leads.

The photolithographic process for the realization of microstrip circuits has been widely discussed in the literature (e.g., [1–7]), so that the basic processing steps do not need to be described further. Nevertheless, technology has to be developed given specific conditions and specifications with respect to substrate materials, instruments, and chemicals. To specify the various process parameters, partially extensive test runs have to be performed. Process-related underetching effects or attainable minimum width of a microstrip line are important design constraints. Especially with coupled line circuits, underetching may be significant [8].

9.1 Substrate

There exists a wide range of substrate materials with various material parameters. With respect to easy mechanical processing, soft plastic substrates are frequently prefered to rigid hard substrates, such as alumina or, in special cases, ferrite materials. Soft materials have advantages in grounding devices by cutting or drilling holes through the substrate and shortening by conductive copper foil. In a similar way, short-circuited stubs can be realized. Nevertheless, soldering on soft substrates necessitates cautiousness, and reproducibility may become questionable. This is, in particular, valid for circuits operating at higher frequencies.

9.1.1 Rigid Substrates

Alumina Substrate

Aluminium oxide (Al_2O_3), commonly referred to as alumina, is a widely used substrate material. It has high hardness and good thermal conductivity, comparable with GaAs. Initially, the raw material of ceramic powder of proper particle size (grain size $<1.0\,\mu m$) is mixed with organic binders and solvents to achieve slip rheology. Next, the material is formed to a ceramic tape by a continuous casting machine. During this process, the material is gradually heated. Then the tape is punched to produce the desired substrate configurations. The parts are sintered at a high temperature. The sintering process brings about several changes in the ceramic: total surface area is reduced, bulk volume is reduced, and strength is increased [9]. A surface finish of less than 26 μm can be obtained. Al_2O_3 is available in purity ranges from 94% to 99.5%, depending on the organic binder content of the sintered ceramic material. Aluminium oxide substrate material is offered as standard substrate material. The lower purity material is commonly used for thickfilm technology.

High-purity alumina substrate is commonly used for thinfilm technology production. It exhibits excellent dielectric properties from dc to gigahertz frequencies, and the loss tangent is $\tan\delta_\epsilon = 0.0001$. It has a relative dielectric constant of 9.7. However, the dielectric constant may vary from substrate to substrate.

Silicon Substrate

Low resistivity Silicon wafers used in industry for the realization of silicon devices [metal-oxide-semiconductor field-effect transistor (MOSFET), hetero-bipolar-transistor (HBT)] have a specific resistivity less than $50\,\Omega cm$. Material with a lower value of $20\,\Omega cm$ is commonly processed [10].

HRS is defined as monocrystalline silicon having a bulk resistivity larger than $1{,}000\,\Omega cm$. High resistivity polished silicon wafer can be obtained at reasonable price [11, 12]. It can be micromachined using wet or dry etching to fabricate micro-electro-mechanical systems (MEMS) [13]. Micromachining is also applied to fabricate high Q planar transmission lines, as shown in Figure 2.29. It has also been used for a new microintegration technique termed quasi-monolithic integration technology, which is discussed in Section 9.2.3.

The thermal conductivity of HRS material is $\sigma_{th} = 1.5\,W cm^{-1} K^{-1}$, and thus much larger than that of GaAs with $0.46\,W cm^{-1} K^{-1}$. This is important for high power device application. The dielectric constant at 300 K is $\epsilon_r = 11.7$, and the specific resistivity ρ varies between 8 to $30{,}000\,\Omega\,cm$, depending on the doping level of the material (p or n type) [14]. The loss tangent is approximately $\tan\delta_d = 0.004$ for a Si wafer, having bulk resistivity of $1{,}000\,\Omega\,cm$ at 35 GHz.

Gallium Arsenide Substrate

Single crystal gallium arsenide is used for MMICs. Semi-insolating (s.i.) 6″ GaAs wafers are commercially available. Larger GaAs wafer sizes are still under development, to consistently drop down production cost per millimeter [15]. The dielectric constant is $\epsilon_r = 12.9$ and higher than that of alumina. The dielectric loss tangent is very low ($\tan\delta_d = 0.016$). The coefficient of thermal expansion (CTE) is $3.8 \cdot 10^{-6}\mathrm{K}^{-1}$. GaAs substrates can be micromachined. Compared with silicon, GaAs material has poor thermal conductivity σ_{th} ($\sigma_{th,GaAs} = 0.46\,\mathrm{Wcm^{-1}K^{-1}}$).

Because of high material cost, GaAs has not found wide use in common microstrip circuit design.

Gallium Nitride Substrate

Gallium nitride (GaN) is a wide bandgap material (bandgap of 3.4 eV in comparison with 1.43 eV for GaAs and 1.11 eV for Si) with high breakdown voltage ($35 \cdot 10^5$ V/cm in comparison with $6.5 \cdot 10^5$ V/cm for GaAs and $6 \cdot 10^5$ V/cm for Si) [16], and therefore very promising for high power devices. However, there are still issues with crystal quality, and with growing a crystal or ingot of large enough size to provide the economic advantages of large wafer processing. Therefore, early GaN thin films were grown on silicon carbide (SiC) or sapphire substrates, which are lattice mismatched to GaN. Therefore, a thin film of aluminum nitride layer is growing on top of the sapphire substrate before laying the GaN film onto the surface [17]. Nevertheless, the risk of stress and defects induced by growing GaN epitaxial layers on nonnitride substrates, remains inherent, and device processing becomes more expensive.

Meanwhile, much progress has been done in GaN wafer fabrication. Today, three-inch wafers are already commercially available, and four-inch wafers are under development [18].

The dielectric constant is $\epsilon_r = 9.5$, the thermal conductivity $\sigma_{th} = 1.7\,\mathrm{Wcm^{-1}K^{-1}}$ [20, 21], and the coefficient of thermal expansion $\mathrm{CTE} = 5.11 \cdot 10^{-6}\mathrm{K}^{-1}$ (a-axis), and $4.47 \cdot 10^{-6}\mathrm{K}^{-1}$ (c-axis), respectively [19].

Silicon Carbide Substrate

SiC has very high thermal conductivity, and therefore is especially used for the design of microwave devices dissipating very high power. Highest thermal conductivity has been achieved in pure material, and in the direction perpendicular to the c-axis being $4.8\,\mathrm{Wcm^{-1}K^{-1}}$ at room temperature. The dielectric constant is $\epsilon_r = 9.5$ [20], the thermal conductivity $\sigma_{th} = 3.3\,\mathrm{Wcm^{-1}K^{-1}}$ [21] and the coefficient of thermal expansion $\mathrm{CTE} = 4.34 \cdot 10^{-6}\mathrm{K}^{-1}$ (a-axis), and $3.74 \cdot 10^{-6}\mathrm{K}^{-1}$ (c-axis), respectively [19].

Sapphire

Sapphire is a single-crystal material with the chemical formula Al_2O_3. It is used when very good surface finish is required. It serves also as a carrier for processing GaAs and other semiconductors. The conductivity is $\sigma_{th} = 0.35\,\mathrm{Wcm^{-1}K^{-1}}$.

Table 9.1
Properties of Rigid Substrate Materials

Substrate	ϵ_r	$\tan\delta_d$	σ_{th} $(\mathrm{Wcm^{-1}K^{-1}})$	CTE $(10^{-6}\mathrm{K^{-1}})$
Al_2O3 (99.5%)	9.7	0.0001	0.37	6.3
Al_2O3 (96%)	9.4	0.001	0.35	6.4
Si (HRS)	11.7	0.004	1.5	2.5
GaAs	12.9	0.0016	0.46	3.8
GaN	9.5		1.6	5.11, 4.47
SiC	9.5		3.3	4.34, 3.74
Sapphire	9.4, 11.6	0.0001	0.35	6.0
BeO	6.6	0.006	3.0	6.1

Beryllium Oxide
Beryllium oxide (BeO), also referred to as Beryllia, exhibits high thermal conductivity. However, machining is very dangerous because of the high toxicity, and therefore, reasonable precautions must be taken for working. The cost of the raw material is high. Fabrication of substrates is similar to alumina. Purity levels of greater than 99% are commercially available.

Thermal conductivity is extremely high and comparable with copper. At room temperature typical value is $\sigma_{th} = 3.0\,\mathrm{Wcm^{-1}K^{-1}}$. Due to its high electrical resistivity, BeO is classed as an electrical insulator. The material is often exploited as an effective heat sink for high-power applications. The dielectric constant is $\epsilon_r = 6.6$.

Table 9.1 gives a summary of relevant material parameters of rigid substrates. Reference [1] is recommended for additional material.

9.1.2 Soft Substrates

The relative dielectric constant of plastic substrates in common use is in the range from 2.2 to 2.6. The main group constitutes the fluorinated polymer PTFE, which is frequently called Teflon. Other names are Duroid, Polyflon, RT/Duroid, and TFE [22]. Pure PTFE is soft, easily deformed, and very prone to creep. Its dielectric constant is $\epsilon_r = 2.0$ to 2.1 at 1 MHz, and the dissipation factor $\tan\delta_d = 0.0003$ to 0.0007. The coefficient of thermal expansion is from 100 to $160\cdot10^{-6}\mathrm{K^{-1}}$ and the thermal conductivity $0.0025\,\mathrm{Wcm^{-1}K^{-1}}$. Upper working temperature is from 180°C to 260°C.

PTFE material is commonly glass reinforced for better mechanical stability [23]. Two types of glass reinforced Teflon are available, one with uniformly distributed glass microfibers, and the other with a layer of woven glass. This leads to low thermal expansion coefficient parallel to the surface plane of woven glass substrate material, but perpendicular to the surface, the CTE value remains relatively high.

Polyolefin [24] is a mixture of olefin polymers and copolymers plus additives. Important polyolefins are, for example, polyethylene and polypropylene. These basic raw materials have a typical dielectric constant of $\epsilon_r = 2.3$, and dissipation factor $\tan\delta_d = 0.0002$ to 0.00035.

The standard FR-4 substrate material has been mostly used as low-cost base material for analog and digital circuit design for lower frequencies because of relative high dissipation loss and dielectric constant variation. Resin is combined with high-quality core materials like fine glass fabrics or aramid paper. Meanwhile, resin-based substrates with enhanced electrical properties are commercially available for high frequency applications ensuring low dissipation loss over a broad frequency range.

Copper cladding of the plastic substrates commonly consists of 35 μm and 70 μm. Substrates are also available with a thicker copper layer of 70 μm with the advantage of better adhesion of the copper to the substrate. This advantage, however, may be offset by the decreasing accuracy due to underetching [8].

Higher value of dielectric constant can be obtained by adding ceramic to the plastic material, termed TFE-based composite substrates. Epsilam 10 was one of the first ceramic-filled high dielectric materials with a dielectric constant $\epsilon_r = 10.3 \pm 0.5$, but with only moderate ϵ_r tolerance over the sheet. The substrate was very flexible, and the electrical properties were similar to alumina. To the knowledge of the writer, the material is no longer available. Today, improved woven fiberglass, and reinforced ceramic-filled PTFE-based composite materials are commercially available [23]. Table 9.2 gives an overview about soft substrates and its material properties for 1 GHz [24–28].

In the following sections, some comments are given to plastic substrate materials, which have been predominantly experienced by the writer. Detailed information is found in [1].

Table 9.2
Properties of Soft Substrate Materials

Substrate	ϵ_r	$\tan\delta_\epsilon$	σ_{th} $(\mathrm{W cm^{-1} K^{-1}})$	CTE $(10^{-6}\mathrm{K^{-1}})$
Teflon	2.0–2.1	0.0003–0.007	0.0025	100–160
Polyguide	2.325	0.0002	0.005	108
TLT	2.4–2.6	0.006	0.0034	9–12/130–145
CuClad	2.4–2.6	0.001	0.00254	
RT/Duroid	2.33	0.001		
AD1000	10.2	0.0018	0.813	7.5/9.5/19.9
FR-4	4.6–4.9	0.02		
IS640	3.5–3.7	0.0045–0.0048		

Soft substrates may suffer from localized pressure, which appears by multiple mounting a microstrip-to-coaxial adapter to a microstrip circuit, for example. This happens particularly in connection with in-fixture calibration procedures when using microstrip transmission line standards. Therefore, in order to circumvent this reproducibility issue, an in-fixture calibration procedure is described in [29], where mounting and demounting of the adapter during calibration and for final DUT measurement is no longer necessary, so that high reliability in the test-fixture calibration and measurement is obtained.

Wire-bonding of chip transistors, MIM capacitors, or diodes onto various soft substrates, using ultrasonic bonding machines, can be excellently performed. However, most important is a flat support to avoid any vertical movement of the substrate under processing. This may become a problem after having already soldered shunt devices, yielding an uneven surface due to the solder joint. A remedy is found by providing a respective recess in the circuit support.

Frequently, shunted devices need an electrical connection to ground. This is done with thin plastic substrates, such as 0.381 μm TLT (also termed TLX) substrate material [26]. A repeated cut through the substrate with a scalpel opens a narrow slit, which is used to feed through a conductive copper strip, before soldering it to ground and the device. More difficult is the planar integration of small active devices that need a perfect squared hole for grounding. Though this fabrication technology is very useful for laboratory work, it may be less suitable for volume production with respect to cost and required narrow fabrication tolerances, especially at higher frequencies. A summary of the soft substrate materials is given in Table 9.2. The data refers to room temperature (300K).

9.1.3 Substrate Height

The substrate height determines the width of the microstrip line for a given characteristic impedance. The design of low impedance microstrip lines can result in very wide strips with cutoff frequencies of higher-order modes within or near the operating frequency range. As has been shown in Section 8.2, this can be avoided by using respective thin substrate materials. With narrow microstrip lines on thick substrates, the risk of undesired parasitic effects occur. When the substrate thickness h reaches a quarter-wavelength, then the dielectric-cladded metal plate forms an open-ended resonator in normal direction with respect to the substrate surface, which would be prone to radiating electromagnetic energy. Notice that for a high dielectric constant, the dielectric-air interface acts like a magnetic wall for the resonating field.

With increasing substrate thickness, the excitation of surface waves also becomes more severe (see Sections 3.4.2.3 and 6.1.2). Thus, in any case, the substrate thickness should be chosen as small as possible, at least

$$\lambda/10 \geq h \tag{9.1}$$

Table 9.3
Alumina Substrate Heights and Recommended Frequency Range

Frequency Range (*GHz*)	*h* (*mil*)	*h* (*mm*)
1–15	25	0.635
8–18	20	0.508
13–22	15	0.381
18–35	10	0.254

Table 9.4
Estimated Maximum Operating Frequency for Different Values of Dielectric Constant of Alumina Material with Substrate Height as Parameter

h	f_{max}/GHz	
(*mm*)	$\varepsilon_r = 9.6$	$\varepsilon_r = 10.7$
0.635	15.2	14.4
0.508	19.1	12.1
0.381	25.4	24.1
0.254	38.0	36.0

as illustrated in Figure 9.1. With $\lambda = \lambda_0/\sqrt{\epsilon_r} = c_0/(f\sqrt{\epsilon_r})$, it then follows for the applicable frequency range

$$f \leq \frac{0.1\,c_0}{\sqrt{\epsilon_r}\,h} \tag{9.2}$$

Alumina (Al_2O_3) is frequently used as substrate. In Table 9.3, standard substrate thicknesses are listed with the recommended frequency ranges. Applying (9.2) to a substrate with $\epsilon_r = 9.6$ and $\epsilon_r = 10.7$, respectively, the values obtained in Table 9.4 are in good agreement with those in Table 9.3.

9.2 Chip Devices

There is a wide range of chip devices, including resistors, capacitors, inductors, diodes, transistors, and others. Manifold package constructions have been developed. Packaging originates in additional parasitics, lowering the practical frequency range of application. The size of chip resistors and capacitors are defined by a four-digit code that approximates its footprint. Thickness is not relevant in the size code. In the United States and most parts of Europe, chip size codes are defined in inches. In Japan and some places in the Orient, chip size codes are

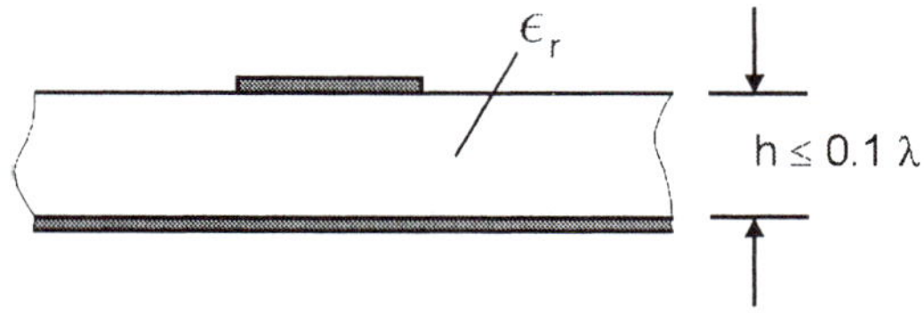

Figure 9.1 Recommended substrate height range.

defined in millimeters [30]. For example, size 1206 means that the length of the device is $0.12'' = 3.048$ mm, and the width is $0.06'' = 1.524$ mm. The corresponding chip in metric size code reads 3216 and denotes the size $3.2\,\text{mm} \times 1.6\,\text{mm}$.

Chips are commonly soldered or glued to the microstrip circuit. Conductive silver paint is often used as an alternative. However, the latter is more convenient for use in the laboratory, where it is not exposed to a harsh industrial environment with mechanical vibrations and shocks.

9.2.1 Resistor

Thinfilm and thickfilm resistors have been widely used both for hybrid thickfilm and thinfilm circuit design and fabrication. They are available in a large range of resistive values with selectable tight tolerances. Resistors are by nature suited to broadband application [31, 32]. They are used as matched attenuators, ring or star power dividers, Wheatstone bridges used in network analyzers, and others [33]. The manufacturers of chip devices partly offer CAD models for their products. Present-day CAD programs also contain libraries with surface mount devices. In particular, ADS contains a huge number of specified devices from known companies. However, as reported in [33], experience has shown that simulation can differ severely from measured data. Therefore, for more accurate circuit design, it seems that model confirmation is needed by one's own measurements.

The respective assembling, the placement of the chip, its alignment to the microstrip line, and the kind of contact by soldering, gluing or mere fixing with conductive paint will have influence on the device model parameters.

Low impedance chip resistors are frequently used as a sensing resistor for current control. This is a particular problem with pulsed high-power laser diodes used in laser radar systems. The peak value of the avalanche discharging current may exceed the specified dissipation power, leading to degradation of the resistive device. Because of the high current amplitude, impedance values less than 1 Ω are desired. To solve this problem, the resistance of conductive paints has been investigated. A microstrip gap of 10 mm was filled with various conductive paints, and dc measured for different strip width. The sheet resistance $R_\square$ (see Section 7.1) can be derived from measurement with $w = l = 1\,\text{cm}$.

Figure 9.2 shows the measurement result. The paint was applied evenly with a scalpel. The scattering points of the EPO-TEK H20E [34], which is a two-component silver-filled epoxy designed specifically for chip bonding, may indicate that the epoxy and hardener were not sufficiently mixed. The coating thickness, the

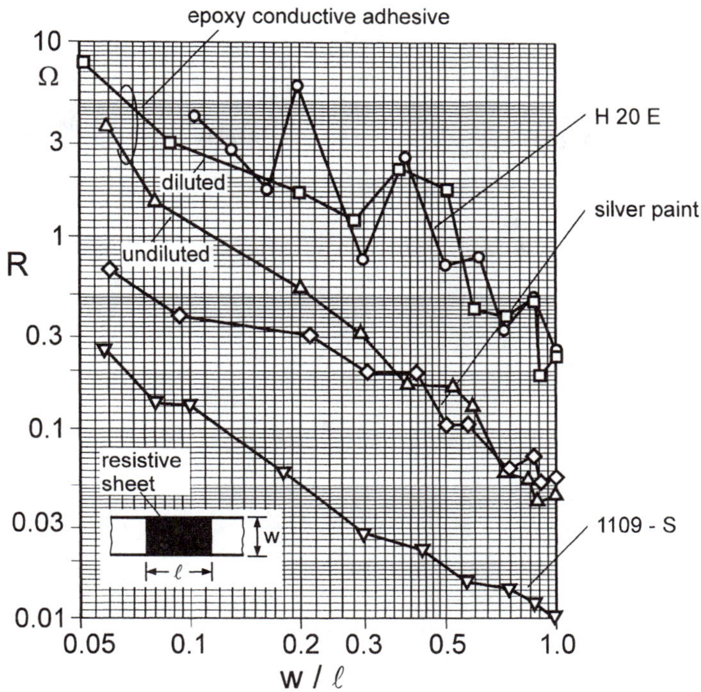

Figure 9.2 Measured rectangular patch resistance value versus patch width *w* (length *l* = 10 mm) for various conductive paints.

curing conditions, and the evaluated sheet resistance values are summarized in Table 9.5. It has been found that conductive silver paint usually recommended for electronic circuit repair (here, Leitsilber 200 from Demetron was used, which is, to the authors knowledge, no longer commercially available) is less suited for post-tuning with a scalpel. The sheet crumbles. Tuning of 1109-S [35], which is a silver-filled thermosetting phenolic resin material, is, due to its excellent hardening, non-tunable. H20E can be tuned easily after hardening at moderate temperature.

9.2.2 Capacitor

Chip capacitors are available in a wide range of sizes and capacitance values. They are frequently used in active devices as a dc break in the gate and drain line of a microwave field-effect transistor. The insertion loss should be low. Typical values are **≤0.1dB**, up to 18 GHz. The precise capacitance value is of less importance. Since the geometry is like a parallel plate capacitor, the chip must be wire-bonded from the top plate to the microstrip line. Maximum operating frequency may be specified as 40 GHz. High breakdown voltages up to 300V are available

Rectangular capacitor chip geometries with connecting pads wrapped around opposing sides can be easily mounted by soldering or gluing. Operating at higher

Table 9.5
Conductive Coating Materials Used for the Realization of Low-Valued Resistors

Material	*Hardening*	*Sheet Thickness* (μm)	*Sheet Resistance* (*ohm per square*)
1109-S Electro-Science Labs	220°C @ 2 hours	65	0.01
Epoxy Conductive Glue Demetron	180°C @ 2 hours	115	0.05
Silver Paint Leitsilber/Demetron	Room Temperature	18	0.05
Epoxy Conductive Glue Demetron	180°C @ 2 hours	115	0.05
H20-E EPO-TEK	60°C @ 11 hours	55	0.3

frequencies, the distributed nature of the chip must be taken into account. Regarding the bias networks shown in Figures 6.117 to 6.120, the radial stubs could principally be replaced by chip capacitors to obtain electrical shorts. Thickfilm capacitors are only usable at lower frequencies because of the occurrence of eigenresonances in the lower gigahertz range. More useful are thinfilm capacitors. Regarding a chip size of 0603, which corresponds to the dimension of $1.524\ \text{mm} \times 0.762\ \text{mm}$, the capacitor can be regarded as a piece of planar transmission line (sandwich line). In the case of a thin plastic substrate of $\epsilon_r = 2.5$, the phase of the capacitive load reflection coefficient is estimated to be approximately 40° to 50° for a frequency of approximately 18 GHz. With an estimated effective width of 1 mm, the characteristic impedance value is approximately 30Ω.

9.3 Circuit Fabrication

The choice of circuit fabrication depends on various parameters, mainly the volume of fabricated circuits [36]. Monolithic technology is attractive for mass production, and the domain for hybrid technology is a moderate volume of fabrication. Recently, a novel microintegration technology, also termed quasi-monolithic integration technology (QMIT), has been developed, which may become economically attractive for mass production. However, competition with MMICs will depend on various targets to be reached, for example, first pass success design, high fabrication yield, low-cost batch processing, and heat transfer management for high-power devices. An historical sketch of microwave integrated circuits, monolithic microwave integrated circuits, and their application, is presented in [37].

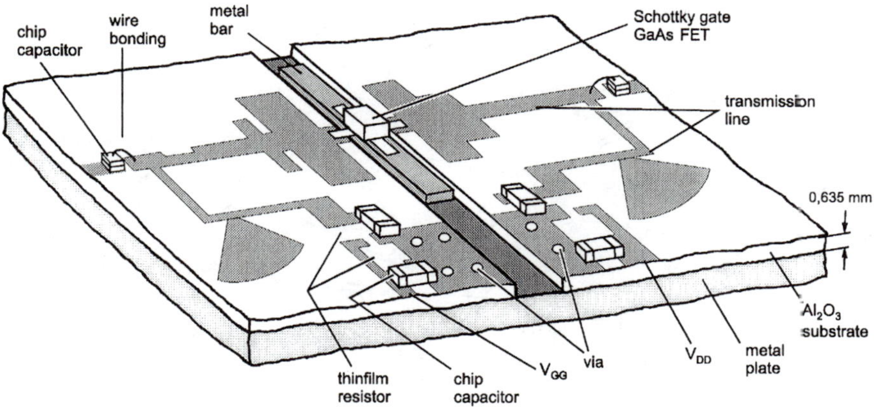

Figure 9.3 Example of a hybrid microwave circuit.

9.3.1 Hybrid Microwave Circuits

Hybrid integration of circuits is essential for many applications in order to achieve small overall size, minimum weight, and low production cost. Hybrid microwave circuits are realized similar to printed circuit fabrication, and include two separate processing steps. In the first step, the planar transmission line construction and integrated planar elements like resistors, low-valued inductors, and capacitors are realized based on photolithography. Plastic, ceramic, and semiconductor material can be used as substrate. Metallization of the substrate is done by evaporation or sputtering (thinfilm). In the thickfilm technique, metallization is performed by using conductive pastes that are burnt at high temperature. The copper sheet (foil) on plastic substrates is fixed under heat and pressure (or stuck with glue, or deposited electrochemically).

In the second processing step, hybrid chip elements, such as transistors, diodes, resistors, capacitors, and inductors are bonded onto the substrate with silver epoxy or are eutactically soldering. Chip connection technologies, such as wire bonding, tape automated bonding (TAB), and flip-chip solder bump, are treated in [38]. Figure 9.3 shows a typical example of a microwave amplifier in hybrid technology. A final important processing step is the packaging of the circuit, including the dc and RF connectors. The construction of the external RF connections is commonly realized in conventional coaxial technique.

9.3.2 Monolithic Microwave Integrated Circuits

The origin of the MMIC approach can be traced back to the 1960s. A summary about the historical development of MMICs is found in [37]. MMICs had become increasingly important since the 1970s, for many reasons. The purity of GaAs wafer material could be enhanced. Large-area low-loss GaAs material (semi-

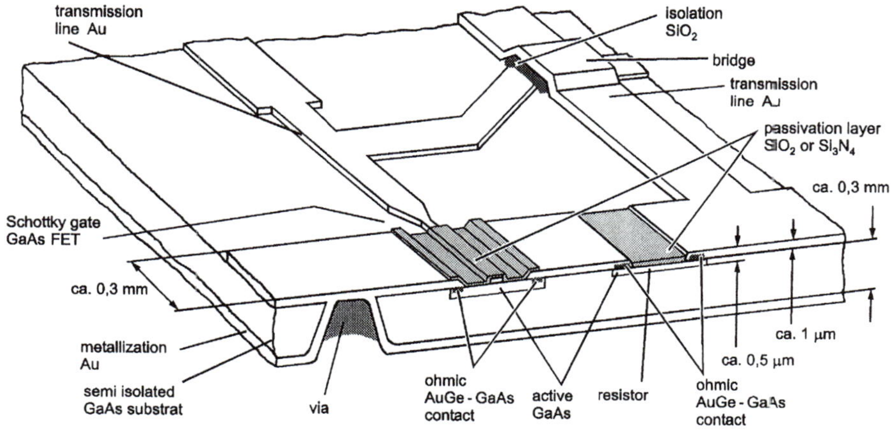

Figure 9.4 Example of a monolithic microwave integrated circuit.

isolated) had become available. High yield and large-scale production had become possible. The metal-semiconductor field-effect transistor (MESFET) had become mature. It could be used for low-noise (receiver), and power applications (transmitter) as well, up to microwave frequencies. This opened up a new perspective for future mass production, such as direct satellite TV [direct broadcast satellite (DBS)] and phased arrays.

Today, technology based on GaAs and related III-V compound semiconductors such as indium phosphide (InP), has matured, opening up frequency ranges above 100 GHz. Silicon-based technology [silicon germanium (SiGe)] is competing and consistently improving.

In the case of MMIC, all passive and active devices and connections are fabricated on a semiconductor substrate in the same process run. Processing includes the common semiconductor processing steps, such as epitaxy, ion implantation for the realization of active (doped) regions; and the known thinfilm processing steps such as evaporation, sputtering, electroplating, and photolithography. Tuning of the finished circuit is generally not possible. Therefore, the design must be done very carefully and take into account process tolerances. This can be done with the help of CAD software used for circuit analysis and optimization. Figure 9.4 shows an example of a microwave monolithic analog circuit.

9.3.3 Quasi-Monolithic Microwave Integrated Circuits

Multichip module (MCM) technology for the fabrication of low-cost microwave and millimeterwave circuits is of growing interest. Devices of different technologies can be combined and integrated on low-cost dielectric substrate material. Recently, high-resistivity silicon substrate material has received much attention [13]. On the

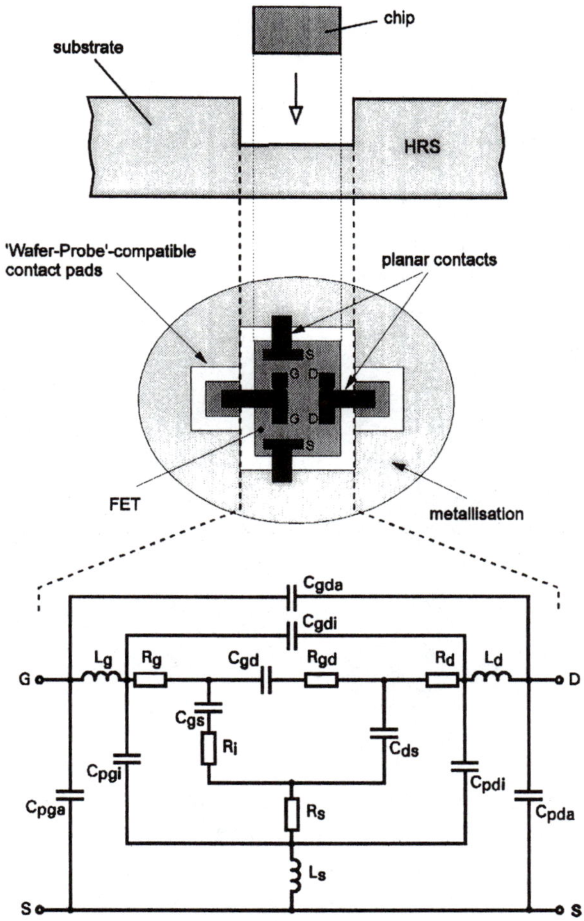

Figure 9.5 Concept of micromachined test-fixture for accurate device modelling in silicon environment.

one hand, this has been due to the reasonable wafer price. On the other hand, silicon can be excellently micromachined using wet or dry etching. Whereas in conventional hybrid technology, chips are attached onto the surface of the substrate carrier and wire-bonded to the circuit, micromachining can be used to provide a recess in the silicon substrate, so that the upper side of the chip is fully planarized with the surface of the substrate. This results in short interconnection lengths, and hence minimization of the associated parasitic inductances.

QMIT also addresses a fundamental design problem in hybrid technology, as illustrated in Figure 9.5. Optimum performance of the finished circuit depends very much on the accurate characterization and modeling of the active devices. Because in most cases the circuit designer has insufficient information about the technology for physical modeling, experimental modeling is best suited for commercial

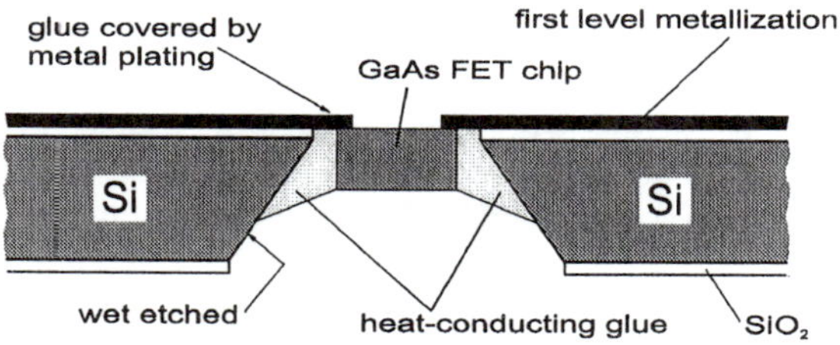

Figure 9.6 Quasi-monolithic chip embedding.

Figure 9.7 Complete air-bridge construction.

devices. The usual characterization procedure of microwave transistor chips requires the use of test-fixtures, in which the interconnections to the transistors are made using bond wires. Efficient device characterization methods and model parameter extraction procedures have been developed recently to derive models that are capable of accurately reproducing measured microwave data [39–42]. These models are employed in circuit design, simulation, and optimization. However, in the realization of the final circuit, it is usual practice to use another transistor of the same type instead of the one previously characterized. The reasons for this practice are twofold. First, one can perform a statistical analysis of the design to minimize the effects of device tolerances. Second, it is difficult if not impossible to remove bond-wires from a device, and the device from the test-fixture, without damaging it. For MMIC designs, the only option is the use of statistical methods.

Figure 9.6 shows a QMIT arrangement where the chip is precisely placed in a wet-etched hole and fixed with heat-conducting glue. Figure 9.7 shows a microphoto of an air-bridge construction from the transistor electrode to the coplanar connection line. The different thermal expansion coefficients of the mixed GaAs-Si-Epoxy material system has yielded enormous fabrication problems [43–45]. Figure 9.8 shows a general view of an embedded and contacted GaAs FET.

Under the conditions of high planarity ($<2\ \mu m$) of the surface at the chip-silicon

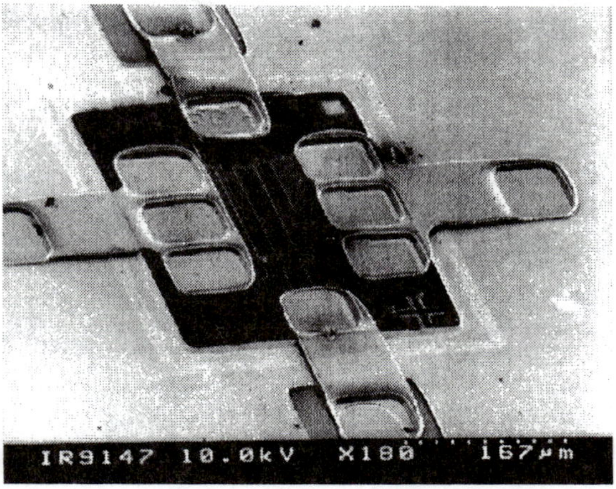

Figure 9.8 Micromachined test-fixture with wafer-probe compatible coplanar contacts for accurate GaAs FET characterization. (From: [48], © 2003 IEEE. Reprinted with permission.)

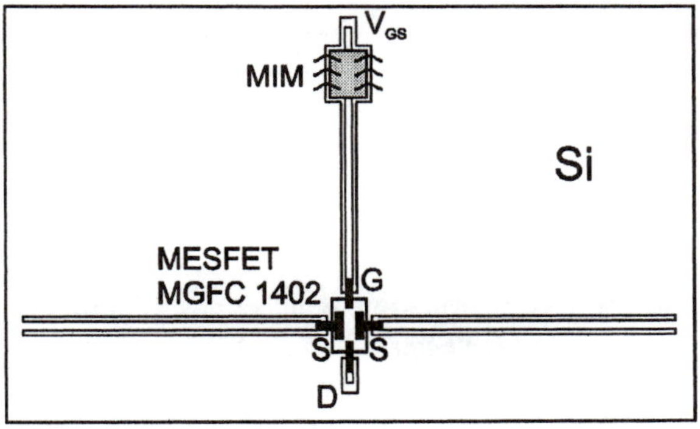

Figure 9.9 MESFET oscillator with 50Ω coplanar lines (frequency tuning by V_{GS} control).

interface, minimum space between embedded chip and opening sidewalls (<10 μm), and usage of low-temperature thinfilm processing, a first coplanar QMIT MESFET oscillator for X band application could be realized and measured [46, 47]. Figure 9.9 shows the layout, and Figure 9.10 shows the output power as a function of tuned gate-source voltage V_{GS} of the FET transistor. Maximum output power of approximately 5 mW at a frequency of 11.35 GHz was obtained.

As described in [11, 48], the QMIT integration process has been improved substantially by processing an amorphous silicon membrane to fix the chip device in the wet-etched hole, thus avoiding the temperature-sensitive glue. The new fabrication process yields very low thermal resistance and less thermomechanical stress. The embedding of the chip needs no tight geometry tolerances. The contacting of the device electrodes is no longer done by elaborate air-bridge technology. The planar-embedded device is covered with polyimide ($\epsilon_r = 2.9$),

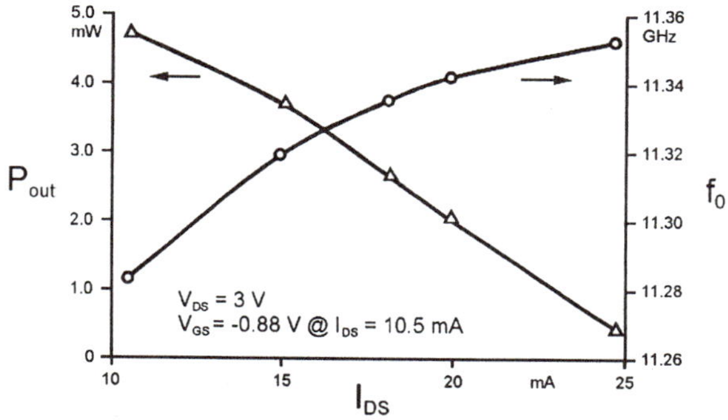

Figure 9.10 Output power and oscillating frequency versus V_{GS} controlled drain current I_{DS}.

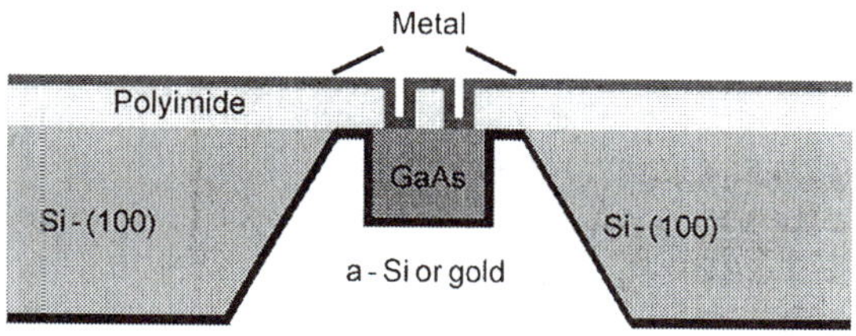

Figure 9.11 Chip device microintegration using enhanced QMIT. (From: [48], © 2003 IEEE. Reprinted with permission.)

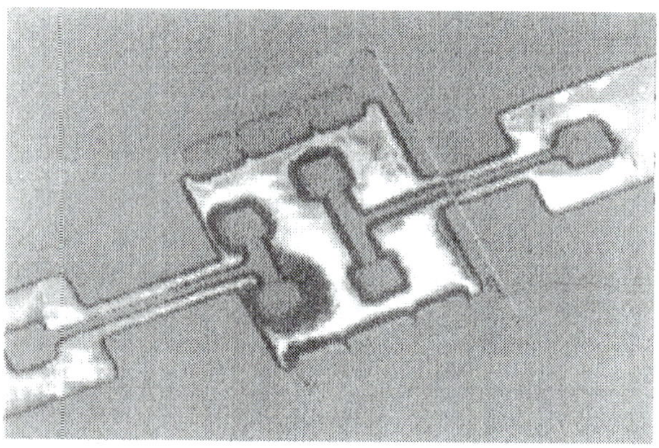

Figure 9.12 Coplanar microintegration of a GaAs device using enhanced QMIT. (From [48], © 2003 IEEE. Reprinted with permission.)

which serves as a substrate carrier for the planar circuit. Oxygen plasma etching is used to prepare via holes for contacting the device by metal deposition and patterning. It has been confirmed by thermal management analysis that the QMIT concept is usable for the integration of power devices. Figure 9.11 shows a cross-

and contacted GaAs device, fixed by amorphous silicon for coplanar circuit fabrication. In the case of microstrip circuit realization, a gold layer is electroplated to keep the chip in place.

References

[1] Laverghetta, S. T., *Microwave Materials and Fabrication Techniques*, Dedham, MA: Artech House, 1964.

[2] Wolff, I., *Einführung in die Mikrostrip-Leitungstechnik*, Aachen: Verlag H. Wolff, 2. Auflage, 1978.

[3] Sweet, A. A., *MIC & MMIC Amplifier and Oscillator Ciruit Design*, Norwoood, MA: Artech House, 1990.

[4] Young, L., and H. Sobol, *Advances in Microwaves*, Vol. 8, Academic Press, 1974.

[5] Edwards, T., *Foundations for Microstrip Circuit Design*, 2nd ed., New York: John Wiley & Sons, 1992.

[6] Maloratsky, L. G., *Passive RF & Microwave Integrated Circuits*, Elsevier Inc., 2004.

[7] Hoffmann, R. K., *Integrierte Mikrowellenschaltungen*, Berlin: Springer-Verlag, 1983.

[8] Hammerstad, E. O., and F. Bekkadal, *Microstrip Handbook ELAB Report*, STF 44 A74169, University of Trondheim, Norway, 1975.

[9] CoorsTek, *Design Guidelines for As-Fired, Lasered, and Polished Substrates*, 8510-1041 Rev A (2002), www.coorstek.com/coorstek/pdf/8510-1164ThinFilm.pdf.

[10] Joodaki, M., private communication.

[11] Rieh, J.-S., et al., "X- and Ku-Band Amplifiers Based on Si/SiGe HBT's and Micromachined Lumped Elements," *IEEE Trans.*, Vol. MTT-46, 1998, pp. 685–693.

[12] Crystec, *Silicon for Research and Development*, Datasheet, http://www.crystec.de.

[13] Joodaki, M., *An Enhanced Quasi-Monolithic Integration Technology for Microwave and Millimeterwave Applications*, Doctoral Thesis, University of Kassel, 2003.

[14] TOPSIL Semiconductor Materials A/S, *HiRes™ Silicon for GHz Technology*, Application Note, http://www.toipsil.com.

[15] Freiberger Compound Materials GmbH, *Semi-insulating Gallium Arsenide Wafers*, http://www.fcm-semicon.com.

[16] Weitzel, C. E., "RF Power Devices for Wireless Communications," *2002 Radio Frequency Integrated Circuits Symposium (RFIC)*, Seattle, WA, USA, 2002, pp. 369–372.

[17] Kramer, D. A., "Gallium," *GALLIUM*, 1998, pp. 29.1–29.7.

[18] Kyma, "Kyma Implements Key Initiatives and Expands Product Portfolio," Press Release 3/07/05, http://www.kymatech.com.

[19] Wang K., and R. Reeber, "Thermal Residual Stress Modeling in AIN and GaN Multilayer Samples," MRS Int. J. Nitride Semicond. Res. 4S1, G3.18, 1999.

[20] Mishra, U. K., P. Parikh, and Y. Wu, "AlGaN/GaN HEMTs—An Overview of Device Operation and Applications," *Proc. IEEE*, Vol. 90, 2002, pp. 1022–1031.

[21] Park, J., M. W. Shin, and C. C. Lee, "Thermal Modeling and Measurement of AlGaN-GaN HFETs Built on Sapphire and SiC Substrates," *IEEE Trans. Electron Devices*, Vol. 51, 2004, pp. 1753–1759.

[22] Nist, *Chemistry WebBook*, http://webbook.nist.gov.

[23] Rogers Corporation, *About our Products—Laminates and Flexible Circuit Materials*, http://www.rogerscorporation.com/acm/about_our_products.htm.

[24] Polyflon Company, *Microwave Substrates*, http://www.polyflon.com/ microwavesubstrates.htm.

[25] ARLON, *Microwave Materials*, http://www.arlonmed.com/micro.html.

[26] Taconic, *High Performance Laminates Material Guide*, http://www.taconic-add.com/en-products.php.

[27] Rogers Corporation, "Device Attachment Methods and Wirebonding Notes for RT/duroid® and RO4000® Series High Frequency Laminates," *Fabrication Guidelines RT4.1.3*, http://www.rogerscorporation.com.

[28] isola, *Resin/Core Materials*, http://www.isola-group.com.

[29] Kompa, G., M. Schlechtweg, and F. van Raay, "Precisely Calibrated Coaxial-to-Microstrip Transitions Yield Improved Performance in GaAs FET Characterization," *IEEE Trans.*, Vol. MTT-38, 1990, pp. 62–68.

[30] TopLine, *Dummy Class 101*, http://www.TopLineDummy.com.

[31] Yin, W.-Y., et al., "Experimental Characterisation and Modelling of On-Chip Capacitors and Resistors on GaAs Substrates," *IEE Proc. Microw. Antennas and Propag.*, Vol. 149, No. 1, February, 2002, pp. 50–52.

[32] Chramiec, J., and M. Kitlinski, "Comment—Experimental Characterisation and Modelling of On-Chip Capacitors and Resistors on GaAs Substrates," *IEE Proc. Microw. Antennas and Propag.*, Vol. 151, No. 6, December, 2004, p. 544.

[33] Chramiec, J., M. E. Adamski, and M. Kitlinski, "Evaluation of Chip Resistor CAD Models Used in Microwave Circuit Design Programs," http://www.ise.pw.edu.pl/~madamski/c26.pdf.

[34] Epoxy Technology, Inc., *H20E - Data Sheet*, http://www.isola-group.com.

[35] Electro-Science Laboratories, Inc., Polymer Conductive Silver 1109-S, http://www.electroscience.com/pdf/1109-S.pdf.

[36] Kompa, G., E. Wasige, and M. Joodaki, "Quasi-Monolithic Hybrid Technology Based on Si Micromachining and Low-Temperature Thin-Film Processing," *Conf. Proc. MICRO.tec 2000*, VDE World Microtechnologies Congress, September 25–27, 2000, Expo 2000, Hannover, Germany, Vol. 2, pp. 109–114.

[37] Niehenke, E. C., R. A. Pucel, and I. J. Bahl, "Microwave and Millimeter-Wave Integrated Circuits," *IEEE Trans.*, Vol. MTT-50, 2002, pp. 846–857.

[38] Doane, D. A., and P. D. Franzon, *Multichip Module Technologies and Alternatives: The Basics*, New York: Van Nostrand Reinhold, 1993.

[39] Root, D. E., S. Fan, and J. Meyer, "Technology Independent Large-Signal Nonquasistatic FET Models by Direct Construction from Automatically Characterized Device Data," *Proc. 21st European Microwave Conf.*, September 1991, pp. 927–932.

[40] Kompa, G., and M. Novotny, "Frequency-Dependent Measurement Error Analysis and Refined FET Model Parameter Extraction Including Bias-Dependent Series Resistors," *Int. IEEE Workshop on "Experimentally Based FET Device Modelling & Related Nonlinear Circuit Design,"* University of Kassel, July 17-18, 1997, Paper 6.1–16.

[41] Kompa, G., "Modeling of Dispersive Microwave FET Devices Using a Quasi-Static Approach," *International Journal of Microwave and Millimeter-Wave Computer-Aided Engineering*, Vol. 5, No. 3, 1995, pp. 173–194.

[42] Kompa, G., "Reliable Extraction of Small-Signal Elements of a Generalized Distributed FET Model," *IEEE MTT-S Int. Microwave Symp. Dig.*, Baltimore, 1998, Paper TUA-8, pp. 291–294.

[43] Wasige, E., *Quasi-monolithische Si-GaAs-Mischtechnologie für Mikrowellen- und Millimeter-wellenschaltungen*, Doctoral Thesis, University of Kassel, 1999, published by Düsseldorf: VDI Verlag, Fortschritt-Berichte VDI, Reihe 9 (Elektronik), Nr. 299, 1999.

[44] Wasige, E., et al., "Air Bridge Based Planar Hybrid Technology for Microwave and Millimeterwave Applications," *IEEE MTT-S Digest*, Denver, 1997, pp. 925–928.

[45] Kompa, G., and E. Wasige, "Quasi-Monolithic Si-GaAs Hybrid Technology for Microwave and Millimeter-Wave Application," *Proc. of First Joint Symposium on Opto- and Microelectronic Devices and Circuits (SODC 2000)*, Nanjing, China, April 10–15, 2000, pp. 33–36.

[46] Wasige, E., et al., "GaAs FET Characterization in a Quasi-Monolithic Si Environment," *IEEE MTT-S Int. Microwave Symp. Dig.*, Anaheim, 1999, pp. 1889–1891.

[47] Wasige, E., et al., "12 GHz Coplanar Quasi-Monolithic Oscillator," *IEEE MTT-S Int. Microwave Symp. Digest*, 1999, pp. 227–228.

[48] Joodaki, M., G. Kompa, and H. Hillmer, "An Enhanced Quasi-Monolithic Integration Technology for Microwave and Millimeter Wave Applications," *IEEE Trans. on Advanced Packaging*, Vol. 26, November 2003, pp. 402–409.

Appendix A

Method of Conformal Transformation

The purpose of this appendix is to give a review of the theory of functions of a complex variable that we shall need in the discussion of conformal mapping of a microstrip transmission line onto an ideal parallel plate guiding structure, which has less complicated boundary conditions, and therefore makes transmission line analysis much simpler.

A.1 Theory of Functions of a Complex Variable

The theory of functions with a complex variable is well described in numerous publications [1–6]. In the theory of complex variables, we use the complex variable

$$z = x + jy \tag{A.1}$$

where both x and y are real variables. The values of the independent variables x and y cover the total plane spanned by the coordinates. Any given value of z is associated with a point in the x-y-plane, usually termed complex the z-plane. The complex number z also may be expressed by polar coordinates in terms of r and θ:

$$r = \sqrt{x^2 + y^2}, \quad \theta = \tan^{-1}(y/x) \tag{A.2}$$

with the magnitude of z, denoted by $|z|$. θ is called the argument of z, and denoted by $\arg z$. Hence, we can write z in polar form

$$z = x + jy = r(\cos\theta + j\sin\theta) = re^{j\theta} \tag{A.3}$$

Note, however, that for a given z, $\arg z$ is not unique. Clearly, we can add any integer multiple of 2π to θ without affecting the value of the function $z = x + jy$. If θ satisfies the equation $-\pi < \theta \leq \pi$ it is called the principal argument of z. The principal argument of z, then is usually denoted by $\operatorname{Arg} z$. Assume a different complex variable w, where

$$w = f(z) = u + jv \tag{A.4}$$

This means that w is some function of z. Equation (A.4) indicates a rule that specifies for each assigned value of z a corresponding value of w. This assignment means that both the real and imaginary parts of the complex variable w are dependent on the (x, y) coordinates,

$$u = u(x,y), \quad v = v(x,y) \tag{A.5}$$

In Figure A.1, curve C illustrates a continuous variation of the variable z. Applying (A.4) may result in the corresponding trace C'. To avoid confusion, the values of w are usually shown on a separate graph, called the complex w-plane. Thus, (A.4) transforms the z-plane (or some part of it) onto the w-plane (or some part of it). In order to prove whether or not the transformation is unique, we consider the derivative of the function $w = f(z)$, which is defined as the limit of the ratio $\Delta w/\Delta z$ as the element Δz becomes infinitesimal, for every point in the z-plane:

$$\lim_{|\Delta z| \to 0} \frac{\Delta w}{\Delta z} = \lim_{|\Delta z| \to 0} \frac{(w + \Delta w) - w}{(z + \Delta z) - z} \tag{A.6}$$

with $w = f(z) = u + jv$, $u = u(x,y)$ and $v = v(x,y)$. Δw is the change of the complex function w if the variable z is changed by Δz. A complex function $f(z)$ is said to be *analytic* or *regular* whenever it is differentiable at every z in some neighborhood of the considered point. The function $f(z)$ is *conformal* at every point where it is analytic and has a nonzero derivative. The derivative may fail to exist at certain (*singular*) points where it may be infinite or undetermined. For the derivative to be unique in (A.6), it is necessary that the limit value of the ratio $\Delta w/\Delta z$ should be independent of the direction Δz. This means that we should obtain the same result if z is changed alone in the x direction, or only in the

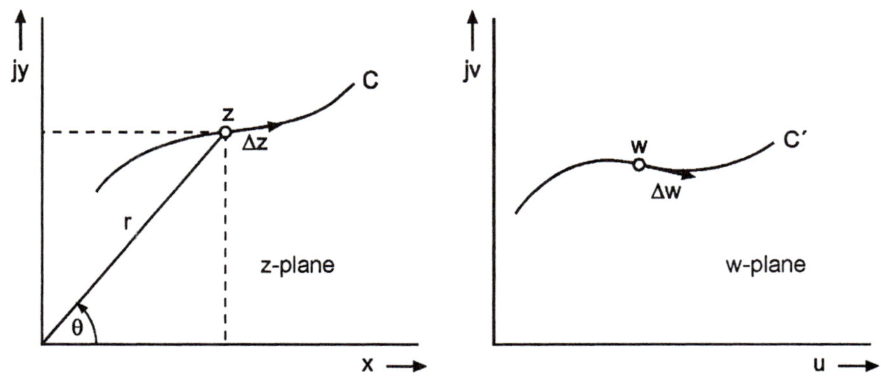

Figure A.1 Transformation of curve C in the z-plane on the curve C' in the w-plane.

y direction. Assuming that the functions u and v are continuous and differentiable functions of the x and y coordinates, the limit value for $\Delta z = \Delta x$ becomes

$$\lim_{\Delta z = \Delta x \to 0} \frac{\Delta w}{\Delta z} = \frac{dw}{dz} = \frac{\partial w}{\partial x} = \frac{\partial (u + jv)}{\partial x} = \frac{\partial u}{\partial x} + j\frac{\partial v}{\partial x} \tag{A.7}$$

For a differential shift in the y direction, $\Delta z = j\Delta y$, we get

$$\lim_{\Delta z = j\Delta y \to 0} \frac{\Delta w}{\Delta z} = \frac{dw}{dz} = \frac{\partial w}{\partial (jy)} = \frac{\partial (u + jv)}{j\,\partial y} = \frac{\partial v}{\partial y} - j\frac{\partial u}{\partial y} \tag{A.8}$$

Assuming that the limit value is independent of the direction Δz, (A.7) and (A.8) are equal. Two complex quantities are equal if and only if their real and imaginary parts are separately equal. Hence, from (A.7) and (A.8), the following necessary conditions can be derived,

$$\frac{\partial w}{\partial x} = -j\frac{\partial w}{\partial y} \tag{A.9}$$

Equating real and imaginary parts in (A.7) and (A.8),

$$\frac{\partial u}{\partial x} = \frac{\partial v}{\partial y} \tag{A.10}$$

$$\frac{\partial v}{\partial x} = -\frac{\partial u}{\partial y} \tag{A.11}$$

leads to a pair of partial differential equations. Differentiability in complex analysis always leads to a pair of partial differential equations. Equations (A.10) and (A.11) are called *Cauchy-Riemann* differential equations. They ensure that the derivative dw/dz is unique at a point, and that the function $w = f(z)$ is analytic there. It can be shown that (A.10) and (A.11) are obtained for any arbitrary direction of the change Δz, so that they are also sufficient conditions [3]. From (A.6), we obtain

$$\lim_{|\Delta z| \to 0} \frac{\Delta w}{\Delta z} = \lim_{|\Delta z| \to 0} \frac{\Delta u + j\Delta v}{\Delta x + j\Delta y} \tag{A.12}$$

Series expansion for $u(x,y)$ and $v(x,y)$ and truncation after the first linear term delivers

$$\Delta u(x,y) = \frac{\partial u(x,y)}{\partial x}\Delta x + \frac{\partial u(x,y)}{\partial y}\Delta y \tag{A.13}$$

$$\Delta v(x,y) = \frac{\partial v(x,y)}{\partial x}\Delta x + \frac{\partial v(x,y)}{\partial y}\Delta y \tag{A.14}$$

Δu and Δv substituted in (A.12),

$$\frac{dw}{dz} = \lim_{|\Delta z| \to 0} \frac{\left(\frac{\partial u}{\partial x} + j\frac{\partial v}{\partial x}\right)\Delta x + \left(\frac{\partial u}{\partial y} + j\frac{\partial v}{\partial y}\right)\Delta y}{\Delta x + j\Delta y} \tag{A.15}$$

After division by Δx, we get

$$\frac{dw}{dz} = \lim_{|\Delta z| \to 0} \frac{\frac{\partial w}{\partial x} + \frac{\partial w}{\partial y}\frac{\Delta y}{\Delta x}}{1 + j\frac{\Delta y}{\Delta x}} \tag{A.16}$$

The limit value in (A.16) is independent of the direction if the quotient is independent of the ratio $\Delta y/\Delta x$. To analyze the respective condition, we consider the bilinear function

$$y = \frac{ax + b}{cx + d} = \frac{a}{c}\,\frac{x + b/a}{x + d/c} \tag{A.17}$$

Adding and subtracting the term a/c gives

$$y = \frac{a}{c} + \frac{a}{c}\left(-1 + \frac{x + b/a}{x + d/c}\right) = \frac{a}{c} + \frac{a}{c}\,\frac{b/a - d/c}{x + d/c} = \frac{a}{c} \tag{A.18}$$

if the condition

$$bc - ad = 0 \tag{A.19}$$

is fulfilled; that is, with (A.19), the function y is independent of the coordinate x. Comparing (A.16) and (A.18), it follows that the limit value is independent of $\Delta y/\Delta x$ if

$$j\frac{\partial w}{\partial x} - \frac{\partial w}{\partial y} = 0 \tag{A.20}$$

is satisfied, which agrees with (A.9). Thus, only with the assumption, made by (A.20), the limit value

$$\frac{dw}{dz} = \lim_{|\Delta z| \to 0} \frac{\Delta w}{\Delta z} \tag{A.21}$$

is uniquely defined, and denoted as the derivative of the function w with respect to the complex variable z. The complex function is then denoted as complex differentiable, holomorph, or regular. Substituting (A.20) in (A.16) gives

$$\frac{dw}{dz} = \lim_{|\Delta z| \to 0} \frac{\dfrac{\partial w}{\partial x} + j\dfrac{\partial w}{\partial x}\dfrac{\Delta y}{\Delta x}}{1 + j\dfrac{\Delta y}{\Delta x}} = \frac{\partial w}{\partial x} \tag{A.22}$$

It is seen that the derivative of the complex function w with respect to the complex function z equals the partial derivative of this function with respect to the coordinate x. Thus, we can write

$$f'(z) = \frac{dw}{dz} = \lim_{|\Delta z| \to 0} \frac{\Delta w}{\Delta z} = \frac{\partial w}{\partial x} \tag{A.23}$$

The differentiability of $f(z)$ implies continuity at the regarded point. However, partial differential equations are seldom of interest at a single point, but rather in a region. Differentiating (A.10) with respect to x and (A.11) with respect to y, and adding the resulting equations, yields the relation

$$\frac{\partial^2 u}{\partial x^2} + \frac{\partial^2 u}{\partial y^2} = 0 \tag{A.24}$$

Similarly, if the differentiation is reversed, we obtain

$$\frac{\partial^2 v}{\partial x^2} + \frac{\partial^2 v}{\partial y^2} = 0 \tag{A.25}$$

These differential equations are recognized as potential equations or Laplace's differential equations in two dimensions. Thus, both the real and imaginary parts of an analytic function of a complex variable satisfy Laplace's equations and would be suitable for use as the potential functions for two-dimensional electrostatic problems.

When the function $f(z)$ is analytic, the derivative $f'(z) = dw/dz$ at a point is independent of the direction of the change dz from the point. In general, the differential quotient is a complex quantity. It may be written in terms of magnitude and phase. If the difference Δz is sufficiently small (Figure A.2), then we get the linear relation between the difference Δw and Δz,

$$\frac{\Delta w}{\Delta z} = \frac{f(z_0 + \Delta z) - f(z_0)}{\Delta z} \approx f'(z_0) \tag{A.26}$$

or

$$\Delta w \approx f'(z)\,\Delta z \tag{A.27}$$

which expresses that the transformation is isogonal and faithful in length. The differential quotient is commonly a complex number,

$$f'(z) = |f'(z)| \measuredangle f'(z) \tag{A.28}$$

Thus,

$$|\Delta w| = |f'(z)|\,|\Delta z| \tag{A.29}$$

$$\arg(\Delta w) = \arg[f'(z)] + arg(\Delta z) \tag{A.30}$$

$\arg(\Delta z)$ is the angle of the line from the shifted point $(z_0 + \Delta z)$ to the reference point (z_0) with the positive x-axis. We can interpret this geometrically in the following way. By the rule for the product of complex quantities, the magnitude of Δw is $|f'(x)|$ times the magnitude of Δz, and the angle of Δw is $\arg[f'(z)]$ plus the angle of Δz. Thus, Δz is magnified by a scale factor $|f'(x)|$ and rotated by an angle $\arg[f'(z)]$. The derivative $f'(z)$ determines the extension or shrinking of a line element, or its rotation as well. Figure A.3 illustrates the transformation of two arcs $\widehat{AB}$ and $\widehat{AC}$ that intersect at an angle α in the z-plane. Since both transformed curves are rotated through the same angle in the w-plane, the angle α between both in point A of the z-plane will also appear in the w-plane. Moreover, the triangle ABC is transformed to a similar triangle, $A'B'C'$. A transformation with these properties is called a conformal transformation. Conformality is considered a local property of analytic functions. $|f'(z)|$ is a local scaling factor at z_0 and is independent of the direction of z from z_0. Conformality is illustrated in Figure A.4, where an equidistant squared grid in the w-plane is mapped to a curvilinear squared grid in the z-plane. With continuous refinement of the grid, the grid meshes approach a squared shape. Regarding infinitesimal area, geometric shapes are mapped into similar figures. Shapes of finite dimensions will be distorted; however, the angle between the two curves remain.

The lines $u = const$ and the lines $v = const$ in the w-plane (Figure A.4) intersect at right angles. We may interpret the constant u lines as equipotentials. Then the constant v lines would represent electric field lines, and vice versa. From this point of view, the conformal transformation may be thought of as one that takes a uniform field in the w-plane (represented by the equispaced constant u and constant v lines) and transforms it so that it fits the given boundary conditions in the z-plane (Figure A.4).

Frequently, the transformation is done in several steps. For example, the uniform field is transformed first into some intermediate complex plane by $z_1 = f(w)$, and then into a plane $z_2 = g(z_1)$, in which the boundary conditions are satisfied. Intensive work has been done to analyze the conformal mapping of

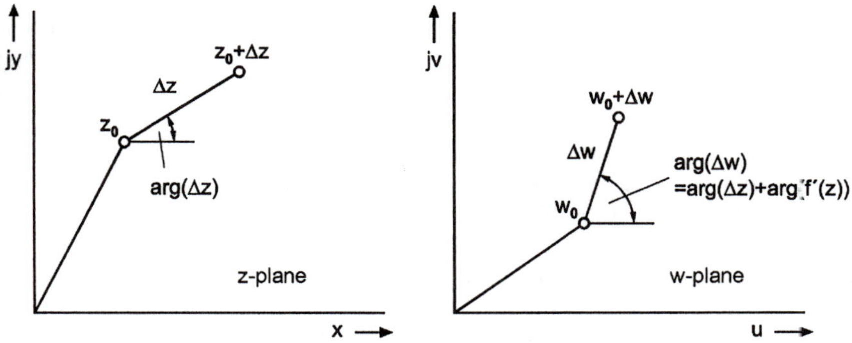

Figure A.2 Transformation of two points in small distance.

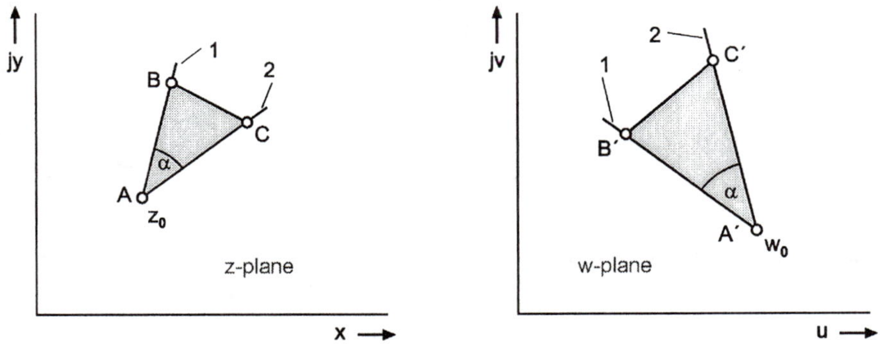

Figure A.3 Intersection of two curves.

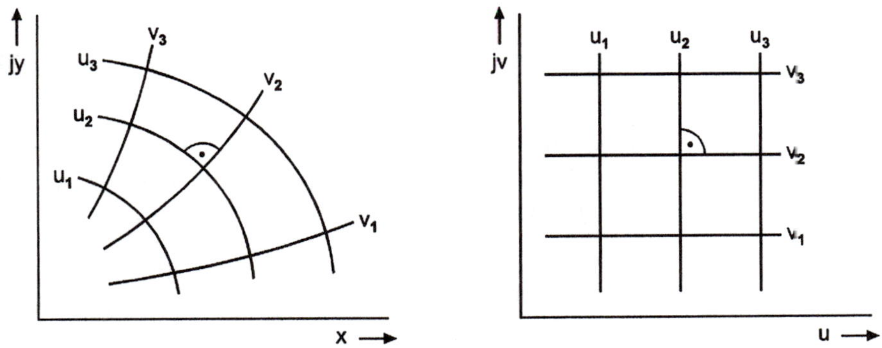

Figure A.4 A mapping of coordinate lines of the *w*-plane in the *z*-plane.

numerous analytic functions $w = f(z)$, that is, to describe the mapping of curves or regions in one plane on certain curves and regions in the other plane [7]. However, from the practical point of view, physical boundary problems generally do not comply with such analytic solutions. Therefore, in practice, the inverse problem is important to find the conformal transformation function for a given boundary problem comprising complicated electrode contours. We will see that the method of Schwarz-Christoffel transformation can tackle this difficult problem.

A1.1 Application to Electrostatic Problems

After having discussed the mathematical properties of complex functions, we will now focus on the application—finding the solution of complicated two-dimensional electrostatic boundary problems. For the following consideration, we make the assumptions that the transversal electrical and magnetic field is curl- and divergence-free, and that the medium in which the field exists is homogeneous, time independent, isotropic, and linear. These conditions describe an electrostatic field problem. The electrical field strength obeys the relations

$$\mathrm{rot}\,\vec{E}\,(x',y',z') = 0 \tag{A.31}$$

$$\mathrm{div}\,\vec{E}\,(x',y',z') = 0 \tag{A.32}$$

The primed quantities x', y', and z' are physical variables. We may derive the electric vector field $\vec{E}$ from a two-dimensional scalar potential function $\varphi(x',y')$,

$$\vec{E}(x',y') = -\mathbf{grad}\,\varphi(x',y') \tag{A.33}$$

which fulfills (A.31). The substitution of $\vec{E}$ into (A.32) leads to the well-known two-dimensional Laplace differential equation,

$$\nabla^2\varphi(x',y') = 0 \tag{A.34}$$

To relate the physical scalar potential $\varphi\,(x',y')$ to the complex functions z and w, and complex variables (x,y) and (u,v), respectively, we normalize the physical quantities by a normalizing voltage V_N and length l_N.

$$u = \frac{\varphi}{V_N}, \quad x = \frac{x'}{l_N}, \quad y = \frac{y'}{l_N} \tag{A.35}$$

We have chosen the normalized scalar potential function as the real part of the complex function $w = u + jv$ with $u = (x,y)$ and $v = (x,y)$. It is clear that the

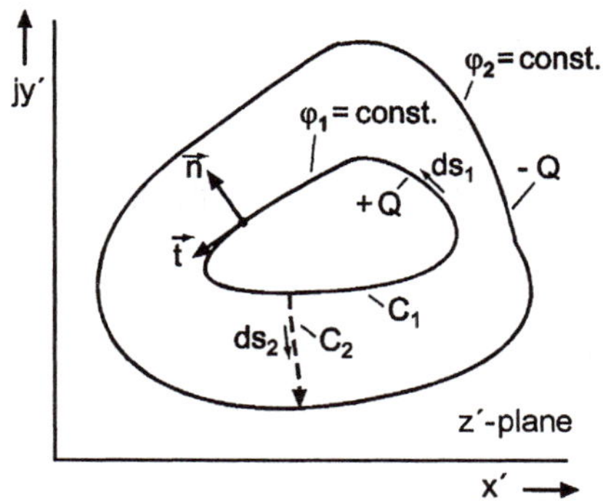

Figure A.5 Calculation of the capacitance per unit length of a cylindrical capacitor (physical plane).

normalized quantities are dimensionless. If the function u is continuous and its derivative exists at any point, then a function v can be found using the Cauchy-Riemann differential equations, such that $w = u + jv$ is analytic. As will be shown next, the results can be used to calculate the capacitance of a cylindrical capacitor configuration of length h with irregular shape [3]. Figure A.5 illustrates the contours of the inner and outer electrodes of the capacitor structure under analysis in the physical z'-plane. The capacitor per unit length $C' = C/h$ is defined as

$$C' = \frac{C}{h} = \frac{1}{h}\frac{Q}{V} \tag{A.36}$$

with the stored charge Q and voltage V between the electrodes. The charge is calculated as

$$Q = \oint_A \vec{D}\cdot d\vec{A} \tag{A.37}$$

where $\vec{D}$ is the electrical displacement and $d\vec{A}$ is the vector of an infinitesimal area of the capacitor surface. With $d\vec{s}_1$ as a line element along curve C_1, $d\vec{A}$ can be written as $d\vec{A} = \vec{n}\,dA = \vec{n}\,h\,ds_1$ with the normal unit vector $\vec{n}$, which can be expressed by the tangent unit vector $\vec{t}$ and $d\vec{s}_1$, respectively,

$$\vec{n} = \vec{t}\times\vec{e}_z = d\vec{s}_1/|d\vec{s}_1| \times \vec{e}_{z'} \tag{A.38}$$

where

$$\vec{ds} \times \vec{e}_{z'} = \begin{vmatrix} \vec{e}_{x'} & \vec{e}_{y'} & \vec{e}_{z'} \\ dx' & dy' & 0 \\ 0 & 0 & 1 \end{vmatrix} = \vec{e}_{x'} dy' - \vec{e}_{y'} dx' = \begin{pmatrix} dy' \\ -dx' \\ 0 \end{pmatrix} \quad \text{(A.39)}$$

With (A.39) and (A.38), $\vec{dA}$ is written as

$$\vec{dA} = (\vec{ds}_1 \times \vec{e}_z) h = (\vec{e}_{x'} dy' - \vec{e}_{y'} dx') h \quad \text{(A.40)}$$

Then the charge in (A.37) reads

$$Q = \epsilon h \oint_C (E_{x'} \vec{e}_{x'} + E_{y'} \vec{e}_{y'}) (\vec{e}_{x'} dy' - \vec{e}_{y'} dx') = \epsilon h \oint_C (E_{x'} dy' - E_{y'} dx') \quad \text{(A.41)}$$

With

$$\vec{E} = -\operatorname{grad} \varphi = -\frac{\partial \varphi}{\partial x'} \vec{e}_{x'} - \frac{\partial \varphi}{\partial y'} \vec{e}_{y'} = E_x \vec{e}_{x'} + E_y \vec{e}_{y'} \quad \text{(A.42)}$$

(A.41) yields

$$Q = -\epsilon h \oint_C \left(\frac{\partial \varphi}{\partial x'} dy' - \frac{\partial \varphi}{\partial y'} dx' \right) \quad \text{(A.43)}$$

After normalizing according to (A.35), the charge Q reads

$$Q = -\epsilon h \oint_C \left(\frac{\partial u}{\partial x} dy - \frac{\partial u}{\partial y} dx \right) \quad \text{(A.44)}$$

Taking into account the Cauchy-Riemann differential equations (A.10) and (A.11), we get

$$Q = \epsilon h \oint_C \left(\frac{\partial v}{\partial y} dy + \frac{\partial v}{\partial x} dx \right) \quad \text{(A.45)}$$

With the total differential of $v(x,y)$,

$$dv(x,y) = \frac{\partial v(x,y)}{\partial x} dx + \frac{\partial v(x,y)}{\partial y} dy \quad \text{(A.46)}$$

the charge is obtained as

$$Q = \epsilon h \oint_C dv \tag{A.47}$$

The voltage V is calculated from the line integral over the electrical field strength between the inner and outer electrode 1 and 2, and the gradient of the electrical potential, respectively,

$$\begin{aligned} V &= \int_1^2 \vec{E}\cdot d\vec{s}_2^{\,\prime} = \int_1^2 \left(E_{x'}\vec{e}_{x'} + E_{y'}\vec{e}_{y'}\right)\left(dx'\vec{e}_{x'} + dy'\vec{e}_{y'}\right) = \int_1^2 \left(E_{x'}dx' + E_{y'}dy'\right) \\ &= \int_1^2 \left(-\frac{\partial\varphi}{\partial x'}dx' - \frac{\partial\varphi}{\partial y'}dy'\right) \end{aligned} \tag{A.48}$$

After normalizing according to (A.35), the voltage V reads

$$V = \int_1^2 \left(-\frac{\partial u}{\partial x}dx - \frac{\partial u}{\partial y}dy\right) \tag{A.49}$$

With the total differential of $u(x,y)$,

$$du(x,y) = \frac{\partial u(x,y)}{\partial x}dx + \frac{\partial u(x,y)}{\partial y}dy \tag{A.50}$$

(A.49) yields,

$$V = \int_1^2 du = u_2 - u_1 \tag{A.51}$$

Then the capacitance per unit length is obtained as

$$C' = \frac{1}{h}\frac{Q}{V} = \frac{\epsilon \oint_C dv}{u_2 - u_1} \tag{A.52}$$

The scalar potential φ in the physical x'-y'-plane, and the respective normalized potential $\varphi(x,y)/V_N = u(x,y)$ in the complex z-plane, are constant on the surface of the electrodes (equipotential lines). Thus, equipotential lines in the x'-y'-plane correspond to lines $u(x,y) = const.$ in the w-plane (Figure A.6). The electrical field lines are orthogonal to the equipotential lines, and therefore are identified by the lines $v(x,y) = const.$ in the w-plane. Then (A.47) simply reads

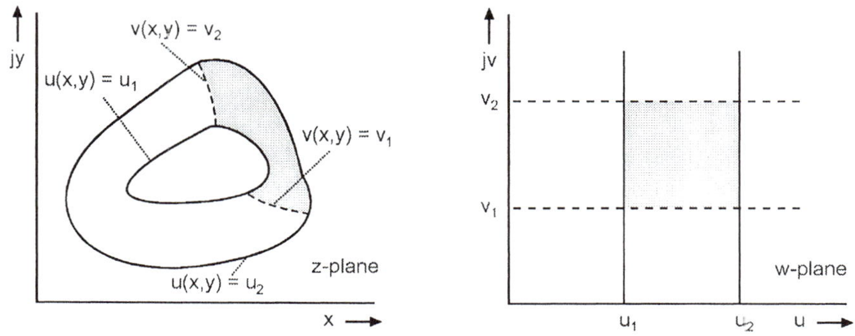

Figure A.6 Conformal transformation of a cylindrical capacitor in the *z*-plane to a parallel plate configuration in the *w*-plane.

$$Q = \epsilon h (v_2 - v_1) \tag{A.53}$$

and (A.52)

$$C' = \frac{\epsilon (v_2 - v_1)}{u_2 - u_1} \tag{A.54}$$

Equation (A.54) can be interpreted as the capacitance of a parallel plate capacitor $C = \epsilon A/d$ with a plate area $A = h(v_2 - v_1)$ and electrode spacing $d = u_2 - u_1$. Note that in case of a circular coaxial crosssection, $v_2 - v_1$ would yield the value of 2π. The electric field vector

$$\vec{E}(x', y') = E_{x'} \vec{e}_{x'} + E_{y'} \vec{e}_{y'} \tag{A.55}$$

can be described by a complex quantity

$$E(x,y) = E_x(x,y) + jE_y(x,y) \tag{A.56}$$

in the complex z-plane. The electric field can be obtained from the potential function $\varphi(x',y')$, and hence from the real part $u(x,y)$ of the complex function $w = u(x,y) + jv(x,y)$.

$$\vec{E} = -\operatorname{grad} \varphi = -\frac{\partial \varphi}{\partial x'} \vec{e}_{x'} - \frac{\partial \varphi}{\partial y'} \vec{e}_{y'} \tag{A.57}$$

$$E_{x'}(x',y') = -\frac{\partial \varphi(x',y')}{\partial x'} = -\frac{V_N}{l_N} \frac{\partial u(x,y)}{\partial x} = \frac{V_N}{l_N} E_x(x,y) \tag{A.58}$$

$$E_{y'}(x',y') = -\frac{\partial\varphi(x',y')}{\partial y'} = -\frac{V_N}{l_N}\frac{\partial u(x,y)}{\partial y} = \frac{V_N}{l_N}E_y(x,y) \qquad \text{(A.59)}$$

From Cauchy-Riemann differential equations, (A.56) yields with (A.11), (A.23), (A.58), and (A.59)

$$E(x,y) = -\frac{\partial u}{\partial x} - j\frac{\partial u}{\partial y} = -\left(\frac{\partial u}{\partial x} - j\frac{\partial v}{\partial x}\right) = -\left(\frac{\partial w}{\partial x}\right)^* = -\left(\frac{dw}{dz}\right)^* \qquad \text{(A.60)}$$

This means that if a function $w = u(x,y) + jv(x,y)$ does exit, so that the real part of this function $u(x,y)$ becomes constant in the z-plane on a given electrode structure, then the electrical field distribution can be calculated from (A.60). The amplitude of E reads

$$|E| = \sqrt{\left(\frac{\partial u}{\partial x}\right)^2 + \left(\frac{\partial v}{\partial x}\right)^2} = \left|\frac{\partial u}{\partial x} + j\frac{\partial v}{\partial x}\right| = \left|\frac{dw}{dz}\right| \qquad \text{(A.61)}$$

Equation (A.61) means that the amplitude of the electrical field equals the amplitude of the derivative of the complex function w at any point.

Specific complex functions have been analyzed by several authors regarding the mapping of $u(x,y) = const.$ and $v(x,y) = const.$ onto the z-plane, and their properties are identified with related electrostatic problems. However, regarding practical problems, the reverse solution is usually required. That is, the equipotential conducting boundaries are given, and the complex function has to be found to solve the problem. It must be conceded that there is no straightforward procedure that, for generally shaped boundaries, will lead to the transformation for a given physical problem. However, a solution can be derived if the boundaries consist of straightline sides with angle intersections. The procedure is known as the Schwarz-Christoffel transformation.

A.2 The Schwarz-Christoffel Mapping Theorem

A frequent problem in electrical engineering is the analysis of the electrical field distribution and capacitors with irregular polygonal electrodes. The Schwarz-Christoffel mapping transformation is well-suited to treat such boundary problems. The transformation maps an arbitrary polygon with straight line segments in the z-plane to a series of segments along the real axis in a w-plane, as shown in Figure A.7. The segments correspond to the sides of the polygon. Thus, the upper half of the complex w-plane is mapped onto the interior of a given polygon with vertices P_ν ($\nu = 1, 2, 3, \ldots$) in the z-plane. Suppose that the vertices of the polygon P are given by $P_1, \ldots, P_2$ in the counterclockwise direction, then the interior region lies to the left of the straight line segment contour. The same situation is given in the w-plane; the real axis is run from negative to positive values so that the upper half-

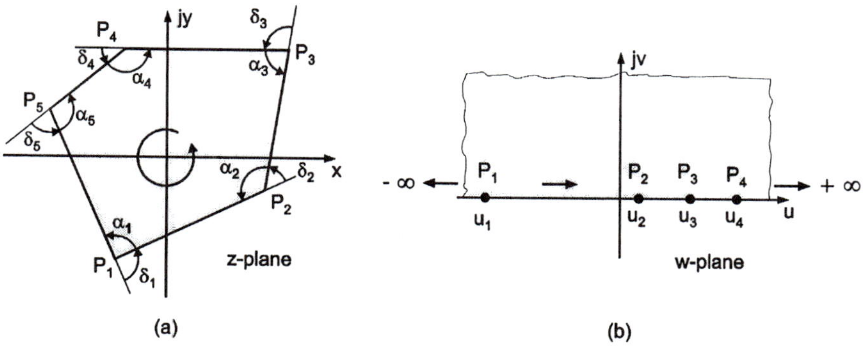

Figure A.7 (a) General polygon in the *z*-plane, and (b) polygon of figure transferred into a straight line in the *w*-plane.

plane lies to the left side of the contour. $\delta_r = -\mu_r \pi$ $(r = 1, 2, ..., n)$ is the external angle of vertex P_r $(r = 1, 2, ..., n)$ and $\alpha_r = \pi(\mu_r + 1)$ $(r = 1, 2, ..., n)$ is the corresponding internal angle $(r = 1, 2, ..., n)$. The internal angle of a vertex is measured between the sides of the given angle and is always positive for the given path direction. That is, the area to be transformed lies on the left side of the assumed path direction. The real quantities u_r $(r = 1, 2, ..., n)$ are determined by the placement of the vertices in the w-plane. At the vertices of the polygon, the mapping function is not conformal.

A.2.1 Schwarz-Christoffel Theorem

Assume that the system of real numbers, $u_1, u_2, u_3, ..., u_n$, with $u_1 < u_2 < ... < u_n$, is given on the real axis in the w-plane. Then the transformation

$$z = A \int (w - u_1)^{\mu_1} (w - u_2)^{\mu_2} (w - u_3)^{\mu_3} ... (w - u_n)^{\mu_n} dw + B \quad \text{(A.62)}$$

maps the upper half of the w-plane $[Im(w) > 0]$ onto the polygonal region in the z-plane. The points $u_1, u_2, u_3, ..., u_n$ and $w = \infty$, respectively, correspond to the coordinates of the vertices in the z-plane, $z_1, z_2, z_3, ..., z_n$ of the polygon. Equation (A.62) is also called the Schwarz-Christoffel integral. The differential quotient (mapping gradient between the z-plane and the w-plane) is

$$\frac{dz}{dw} = A (w - u_1)^{\mu_1} (w - u_2)^{\mu_2} (w - u_3)^{\mu_3} ... (w - u_n)^{\mu_n} \quad \text{(A.63)}$$

The angle of dz can be written as

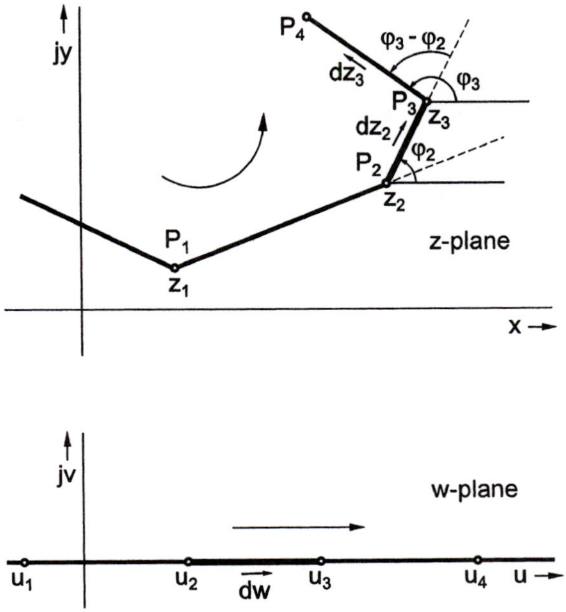

Figure A.8 Graphical illustration of the Schwarz-Christoffel relation.

$$\arg(dz) = \arg(A) + \mu_1 \arg(w - u_1) + \mu_2 \arg(w - u_2) + \dots + \mu_n \arg(w - u_n) \quad \text{(A.64)}$$

with $\arg(dw) = 0$, if the circulation direction is from $-\infty$ to $+\infty$ in the w-plane. The argument of the difference $(w - u_r)$, $1 \le r \le n$, is zero for all w values lying to the left of the respective point u_r, because in this case the difference is a positive number, so that the argument is zero. However, if the point w lies to the right side of u_r, then $(w - u_r)$ becomes a negative value, so that the argument becomes π. Thus we can write [2]

$$\arg(w-u_r) = \begin{cases} 0 & \text{for} \quad w > u_r \\ \pi & \text{for} \quad w < u_r \end{cases} \qquad (1 \le r \le n) \qquad \text{(A.65)}$$

We denote the argument of dz by φ_r when the point on the polygonal curve is within the line interval $u_r < w < u_{r+1}$. Then, as shown in Figure A.8, the following relation is obtained for $\varphi_2 = arg(dz_2)$:

$$\varphi_2 = \arg(A) + \pi(\mu_3 + \mu_4 + \dots \mu_n) \qquad \text{(A.66)}$$

All terms of $(w - u_i)$ vanish, where u_i is lying to the left of the considered point. The slope φ_2 is maintained between P_2 and P_3. Regarding φ_3 describing the

The slope φ_2 is maintained between P_2 and P_3. Regarding φ_3 describing the argument of dz_3 between the points P_4 and P_3, the relation for φ_3 is given by

$$\varphi_3 = \arg(A) + \pi(\mu_4 + \mu_5 + \dots \mu_n) \quad \text{(A.67)}$$

The difference $\varphi_3 - \varphi_2$ is then calculated as

$$\varphi_3 - \varphi_2 = -\pi\mu_3 \quad \text{(A.68)}$$

which corresponds to the external angle of the polygon curve in point P_3. Thus, we have seen that the transformation (A.62) maps the singular intervals (u_r, u_{r+1}) in the w-plane into pieces of straight lines in the z-plane with an external intersection angle of $-\mu_r\pi$.

Under the assumption that the polygon is closed then the sum of the external angles becomes 2π,

$$-\mu_1\pi - \mu_2\pi - \dots -\mu_n\pi = 2\pi \quad \text{(A.69)}$$

or

$$-\sum_{i=1}^{n} \mu_i = 2 \quad \text{(A.70)}$$

The number of constants to be determined is as follows [2]. Supposing the number of u_i is n and the number μ is $n - 1$ [due to the relation (A.70)], and taking into account the complex scaling factor A and complex integration constant B, we get in total $n + (n - 1) + 4 = 2n + 3$ constants. The uniqueness of the transformation permits the choice of three constants. The remaining $2n$ independent quantities are sufficient to determine the location of an n-sided polygon in the z-plane.

In practice, the determination of the constants includes problems. The vertices of the polygon are given in the z-plane, and from the external angles, we can derive the value of μ_i. However, at most three values of the u_i parameters, referred to the vertices, can be freely chosen. The other constants, including the constants A and B, are then determined automatically. However, as experience has shown, determination of the latter constants sometimes includes extreme difficulties. The setting down of (A.63) for a specific problem is usually easy, but the difficulties come in its integration [1].

The real quantities u_ν $(\nu = 1,2,\dots)$ are determined by the location of the corner points in the w-plane. The real quantities μ_ν $(\nu = 1,2,\dots)$ can be derived from the external angles of the polygon curve. A is a complex quantity; its amplitude is undetermined. It acts as a transformation scaling factor, which is mostly defined as 1.

As illustrated in Figure A.9, the sign of δ_r depends on the change of path direction in the vertices. The external angle is positive if the rotation is counterclockwise [Figure A.9(a)], and negative if the rotation is clockwise [Figure A.9(b)]. In the given examples, the parameters are calculated as follows:

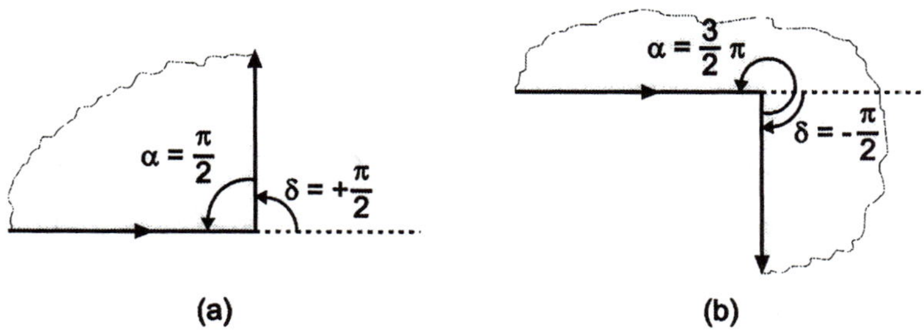

Figure A.9 Sign of external angle.

a) $\delta = +\pi/2 = -\mu\pi \rightarrow \mu = -1/2$

$\alpha = \pi(\mu + 1) = \pi(-1/2 + 1) = \pi/2$

b) $\delta = -\pi/2 = -\mu\pi \rightarrow \mu = +1/2$

$\alpha = \pi(\mu + 1) = \pi(1/2 + 1) = (3/2)\pi$

In practical application, one or more of the vertices may be at infinity, and part of the boundary may be at a different potential from the remaining part. In this case, the real axis in the w-plane consists of two parts at different potentials. A further transformation is then needed to map the single plane capacitor to a parallel plate configuration.

Problem A.1: Figure A.10(a) illustrates the edge of a parallel-plate capacitor with an infinite bottom electrode ($\varphi = V_0$). The upper electrode with the potential $\varphi = 0$V is unlimited for positive x' values. By using the Schwarz-Christoffel theorem, the edge contour described by the polygon A-B-C-D-E is to be mapped to an infinite parallel-plate configuration with homogeneous electrostatic field, as shown in Figure A.10(b).
Hint: Solve the problem by combining two transformations: (1) mapping of the capacitor edge into a single plane, and (2) mapping of a parallel-plate capacitor to the transformed single plane. Note that the quantities in Figure 4.10 are physical ones.

Solution: First, the physical quantities are normalized according to (A.35): $x = x'/l_N$, $y = y'/l_N$, and $v = \varphi/V_N$. As normalizing parameters, we choose $l_N = h/\pi$ and $V_N = V_0$. Figure A.11 shows the normalized configurations in the complex z- and w'-planes. Mapping is done in two transformation steps: (1) Mapping of the capacitor edge (Figure A.12) into a single plane in the w-plane, (2) mapping of the parallel-plate capacitor (Figure A.13) to the transformed single plane in the w-plane.

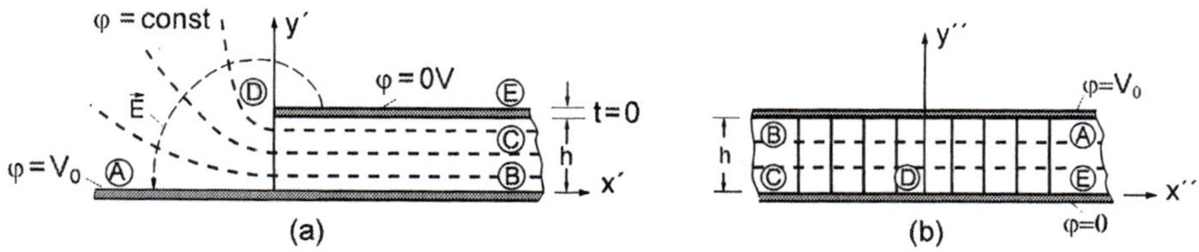

Figure A.10 (a) Edge field of a capacitor, mapped (b) into a homogeneous field of a parallel-plate structure.

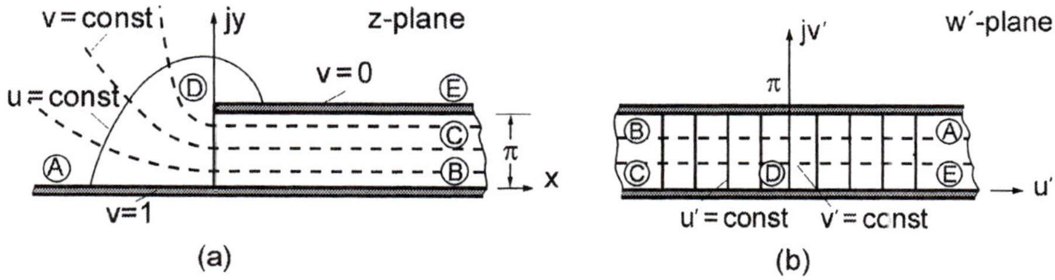

Figure A.11 Normalized structures in the complex (a) *z*-plane and (b) *w*′-plane.

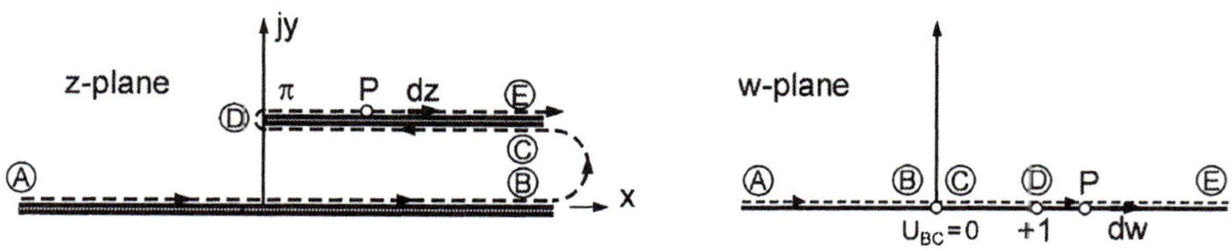

Figure A.12 Mapping path in the *z*-plane and *w*-plane.

Step I: Figure A.12 illustrates the mapping path in the complex planes under consideration. The exponents μ in (A.62) are obtained from the external angles $2\delta = -\mu\pi$, identifying the reversals of the mapping path in the *z*-plane. There are two reversals: $\delta_{BC} = +\pi \rightarrow \mu_{BC} = -1$ and $\delta_D = -\pi \rightarrow \mu_D = +1$. We place the reversal B-C and point *D* in the *z*-plane to $w = 0$, and the point $u = +1$ in the *w*-plane. Thus, (A.63) reads

$$\frac{dz}{dw} = A\,(w)^{-1}\,(w - 1)^{+1} = A\left(1 - \frac{1}{w}\right) \tag{A.71}$$

Equation (A.71) can simply be integrated, and yields

$$z = A\,(w - \ln w) + B \tag{A.72}$$

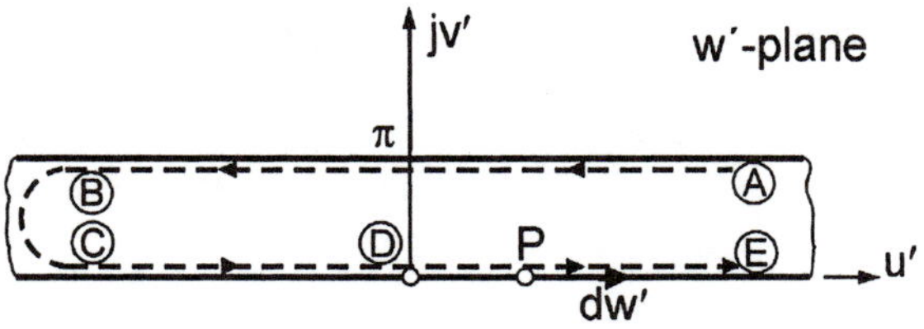

Figure A.13 Mapping path in the complex w'-plane.

z, w, and A in (A.72) are complex quantities. The arguments obey the relation (A.64),

$$\arg(dz) = \arg(dw) + \arg(A) + \arg(w) + \arg(w - 1) \tag{A.73}$$

Regarding point P in Figure A.12(b), the arguments read

$$\arg(dz) = \arg(dw) = \arg(w) = \arg(w - 1) = 0 \tag{A.74}$$

Then, it follows from (A.73) that $\arg(A) = 0$. We choose $|A| = 1$, and (A.72) reads

$$z = w - \ln w + B \tag{A.75}$$

For the determination of the integration constant B, we consider point D, which delivers the corresponding relationship between the variables in the z- and w-planes. For $z = j\pi$ and $w = +1$, we obtain from (A.75) $B = -1 + j\pi$. Thus, (A.75) simplifies to the mapping relation

$$z = (w - 1) - \ln w + j\pi \tag{A.76}$$

Step II: Figure A.13 shows the mapping path in the w'-plane. With only one reversal B-C, $\delta_{BC} = +\pi \rightarrow \mu_{BC} = -1$, and placing the reversal to $w = 0$ in the w-plane, the mapping gradient simply reads

$$\frac{dw'}{dw} = A' w^{-1} \tag{A.77}$$

We consider point P and apply (A.64),

$$\arg(dw') = \arg(dw) + \arg(A') - \arg(w) \tag{A.78}$$

With $\arg(dw') = \arg(dw) = \arg(w) = 0$, it follows from (A.78) that $\arg(A') = 0$. We choose $|A'| = 1$, and (A.77) reads after integration

$$w = \ln w + B' \tag{A.79}$$

Regarding point D with the coordinates $w' = 0$ and $w = +1$, the integration constant is determined to be $B' = 0$. Thus, the mapping transformation reads

$$w' = \ln w \tag{A.80}$$

or

$$w = e^{w'} \tag{A.81}$$

Substituting (A.81) in (A.76) results in

$$z = e^{w'} - 1 - w' + j\pi \tag{A.82}$$

References

[1] Ramo, S., J. R. Whinnery, and T. van Duzer, *Fields and Waves in Communication Electronics*, 2nd ed., New York: John Wiley & Sons, 1984.

[2] Simonyi, K., *Theoretische Elektrotechnik*, Berlin: VEB Deutscher Verlag der Wissenschaften, 1966.

[3] Wolff, I., *Einführung in die Mikrostrip-Leitungstechnik*, Aachen: Verlag H. Wolff, 2. Auflage, 1978.

[4] Magnusson, P. C., et al., *Transmission Lines and Wave Propagation*, 4th ed., CRC Press, 2001.

[5] Kuipers, L., and R. Timman, *Handbuch der Mathematik*, Berlin: Walter de Gruyter & Co., 1968.

[6] Chen, W. W. L., "Introduction to Complex Analysis," Lecture notes, updated 2002, www.maths.mq.edu.au/~wchen/lnicafolder/lnica.html.

[7] Kober, H., *Dictionary of Conformal Representations*, Dover Publications, Inc., 2nd ed. with Correction, 1957.

Appendix B

Smith Chart

The Smith Chart is a graphical tool which is very useful when solving transmission line and network problems [1]. It was published in 1939 by Phillip H. Smith at the Bell Telephone Laboratories [2]. An enhanced version followed in 1944 [3], which may have become in the course of time the most widely used impedance and reflection coefficient chart. The objective of most transmission line problems is to match a line to a load [4]. Although matching problems can be tackled by using current CAD software, the Smith Chart is still very useful in finding solutions step-by-step. This may lead to a deeper understanding of transmission line transformation properties. Therefore, it is also supposed that the application of the Smith Chart is of high pedagogical importance. In the context of this book, we will learn to find lumped and distributed (or mixed) elements of the matching network and to convert these elements into a microstrip circuit. The construction of the Smith Chart is based on a bilinear transformation. The z-plane is mapped onto the Γ-plane due to the relation

$$\Gamma = \frac{z - 1}{z + 1} \tag{B.1}$$

where z denotes the normalized impedance and Γ the reflection coefficient. z and Γ are complex numbers and can be written as $z = Re(z) + jIm(z)$, and $\Gamma = |\Gamma| e^{j arg \Gamma} = Re(\Gamma) + jIm(\Gamma) = u + jv$, respectively. The inverse transformation reads

$$z = \frac{1 + \Gamma}{1 - \Gamma} = \frac{(1 + \Gamma)(1 - \Gamma^*)}{|1 - \Gamma|^2} = \frac{1 - |\Gamma|^2 + \Gamma - \Gamma^*}{|1 - \Gamma|^2} \tag{B.2}$$

As can be seen from Figure B.1, the following relations are valid,

$$\Gamma + \Gamma^* = 2Re(\Gamma) \tag{B.3}$$

$$\Gamma - \Gamma^* = 2jIm(\Gamma) \tag{B.4}$$

Regarding Figure B.2, the law of cosine delivers,

$$|1 - \Gamma|^2 = |\Gamma|^2 + 1 - 2|\Gamma| \cos\alpha \tag{B.5}$$

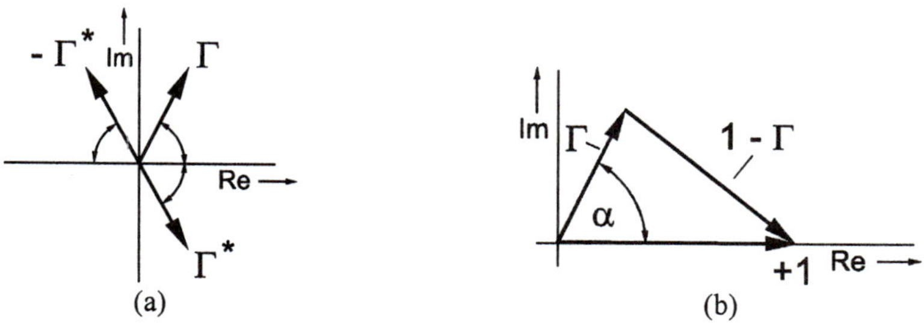

Figure B.1 Graphical illustration of equations (a) (B.4) and (b) (B.5).

where $\cos\alpha = Re(\Gamma)/|\Gamma|$ and $|\Gamma|^2 = u^2 + v^2 = \Gamma \cdot \Gamma^*$. With (B.4) and (B.5), (B.2) becomes

$$z = \frac{1 - |\Gamma|^2 + 2jIm(\Gamma)}{1 + |\Gamma|^2 - 2Re(\Gamma)} = \frac{1 - u^2 - v^2 + 2jv}{1 + u^2 + v^2 - 2u} \tag{B.6}$$

The real and imaginary parts of z are identified as

$$Re(z) = \frac{1 - u^2 - v^2}{1 + u^2 + v^2 - 2u} \tag{B.7}$$

$$Im(z) = \frac{2v}{1 + u^2 + v^2 - 2u} \tag{B.8}$$

Equations (B.7) and (B.8) are used to study special properties of the bilinear transformation. We analyze the mapping of constant $Im(z)$- and $Re(z)$-lines, circles of constant amplitude of z, and lines of constant phase of z onto the complex Γ-plane.

B.1 Construction of the Smith Chart

B.1.1 Analysis of *Im*(*z*) = *const.*

a) $Im(z) = 0 \quad \rightarrow \quad v = Im(\Gamma) = 0$

The real axis of the z-plane is mapped onto the real axis of the Γ-plane within the unit circle. Only the region within the unit circle is of interest, because this is the mapped region of the z-plane with $Re(z) \geq 0$. Thus, we limit our considerations solely to impedances with a positive resistive part.

b) $Im(z) \neq 0$

From (B.8), we get

$$(u - 1)^2 + v^2 - 2v/Im(z) = 0 \tag{B.9}$$

and after rearranging,

$$(u - 1)^2 + [v - 1/Im(z)]^2 = [1/Im(z)]^2 \tag{B.10}$$

With $v_0 = 1/Im(z)$ and $R = 1/Im(z)$ (B.10) reads

$$(u - 1)^2 + (v - v_0)^2 = R^2 \tag{B.11}$$

The ratio of v_0 and R gives $v_0/R = 1$. Thus, the relation $Im(z) \neq 0$ represents circuits in the Γ-plane, whose centers are all lying on the straight line $Re(\Gamma) = u = 1$, and all of which are going through the point $(u = 1, v = 0)$.

B.1.2 Analysis of $Re(z) = const.$

After multiplying the expression for the real part of z in (B.7) with the denominator,

$$(1 + u^2 + v^2 - 2u)\,Re(z) = 1 - u^2 - v^2 \tag{B.12}$$

and sorting,

$$u^2[Re(z) + 1] - 2u\,Re(z) + v^2[1 + Re(z)] + Re(z) = 1 \tag{B.13}$$

we find the following circle equation for u and v,

$$u^2 - \frac{2\,Re(z)}{1 + Re(z)}u + v^2 + \frac{Re(z)}{1 + Re(z)} = \frac{1}{1 + Re(z)} \tag{B.14}$$

$$\left(u - \frac{Re(z)}{1 + Re(z)}\right)^2 + v^2 = \left(\frac{Re(z)}{1 + Re(z)}\right)^2 + \frac{1 - Re(z)}{1 + Re(z)}$$

$$= \frac{Re(z)^2 + 1 - Re(z)^2}{[1 + Re(z)]^2} = \frac{1}{[1 + Re(z)]^2} \tag{B.15}$$

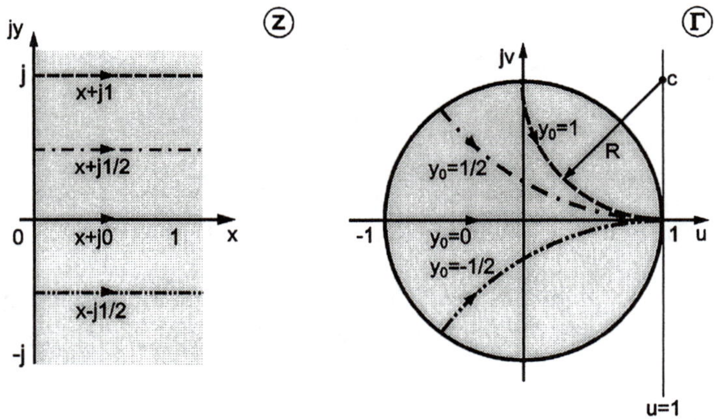

Figure B.2 Loci of constant y lines [$Im(z)$=*const.*] mapped in the Γ-plane.

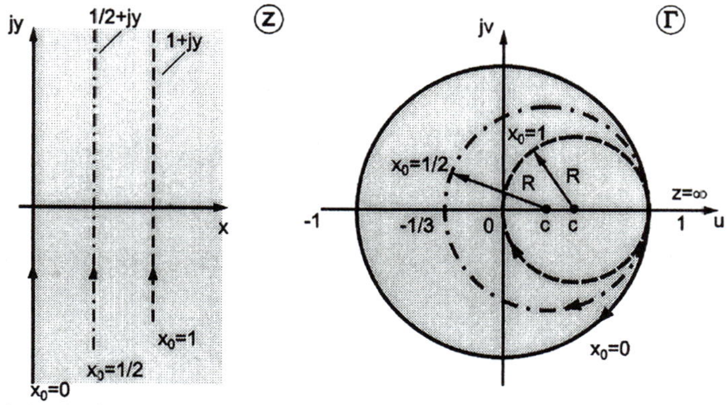

Figure B.3 Loci of constant x lines [$Re(z)=const.$] mapped in the Γ-plane.

Substituting $u_0 = Re(z)/[1 + Re(z)]$ and $R = 1/[1 + Re(z)]$, (B.15) is written as

$$(u - u_0)^2 + v^2 = R^2 \tag{B.16}$$

The sum of u_0 and R yields

$$u_0 + R = 1 \tag{B.17}$$

With u_0 being the center point, all circuits go through the point $(u = 1, v = 0)$ (Figure B.3).

B.1.3 Analysis of $|z| = const.$

The modulus of squared z is given by

$$|z|^2 = z \cdot z^* = \frac{(1 + \Gamma)(1 + \Gamma^*)}{(1 - \Gamma)(1 - \Gamma^*)} = \frac{1 + \Gamma + \Gamma^* + |\Gamma|^2}{1 - \Gamma - \Gamma^* + |\Gamma|^2} \tag{B.18}$$

Putting the value of Γ,

$$|z|^2 = \frac{1 + u^2 + v^2 + 2u}{1 + u^2 + v^2 - 2u} \tag{B.19}$$

and sorting, we get

$$|z|^2(1 + u^2 + v^2 - 2u) = 1 + u^2 + v^2 + 2u \tag{B.20}$$

$$(1 + u^2 + v^2)(|z|^2 - 1) - 2u(|z|^2 + 1) = 0 \tag{B.21}$$

$$u^2(|z|^2 - 1) - 2u(|z|^2 + 1) + v^2(|z|^2 - 1) + |z|^2 - 1 = 0 \tag{B.22}$$

Assuming $|z| = 1$, we obtain $u = 0$. Thus, the unit circle in the z-plane degenerates to a straight line $Re(\Gamma) = 0$ in the Γ-plane. Further modification of (B.22),

$$u^2 - 2\frac{|z|^2 + 1}{|z|^2 - 1}u + v^2 = -1 \tag{B.23}$$

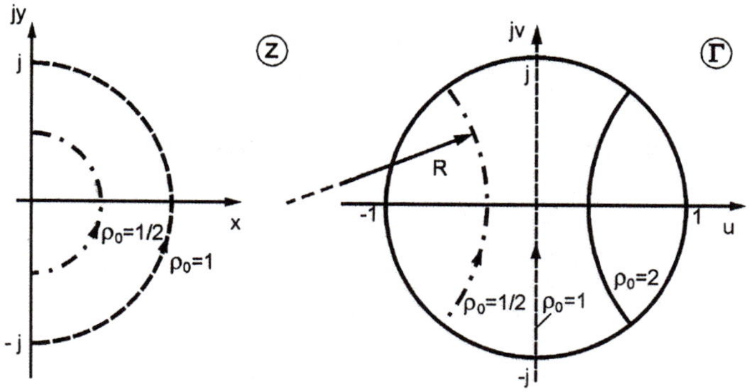

Figure B.4 Loci of constant amplitude of z in the Γ-plane.

leads to the following circle equation,

$$\left(u - \frac{|z|^2 + 1}{|z|^2 - 1}\right)^2 + v^2 = \left(\frac{|z|^2+1}{|z|^2-1}\right)^2 - 1 = \frac{4|z|^2}{(|z|^2 - 1)^2} \tag{B.24}$$

or

$$(u - u_0)^2 + v^2 = R^2 \tag{B.25}$$

with the center point $u_0 = (|z|^2 + 1)/(|z|^2 - 1)$ and radius $R = 2|z|/(||z|^2 - 1|)$ Figure B.4). Equation (B.25) shows that circles in the z-plane yield circles in the Γ-plane. u_0 denotes the shift of the center of the circle for a given value of $|z|$. Regarding the inverse of $|z|$, that is $1/|z|$, the following circle parameters (u_0', R') are obtained:

$$u_0' = \frac{1/|z|^2 + 1}{1/|z|^2 - 1} = \frac{|z|^2 + 1}{-|z|^2 + 1} = -u_0$$

$$R' = \frac{2/|z|}{|1/|z|^2 - 1|} = \frac{2|z|}{|1 - |z|^2|} = R \tag{B.26}$$

B.1.4 Analysis of φ = *const.*

Expressing z by its modulus and phase, which is $z = |z|\, e^{j\varphi}$, we consider the ratio of z and its complex conjugate value z^*,

$$\frac{z}{z^*} = e^{j2\varphi} = \cos 2\varphi + j\sin 2\varphi \tag{B.27}$$

Substituting Γ for z in (B.27) and taking into account (B.4) and (B.5) gives

$$\frac{z}{z^*} = \frac{(1+\Gamma)}{(1-\Gamma)}\,\frac{(1-\Gamma^*)}{(1+\Gamma^*)} = \frac{1+\Gamma-\Gamma^*-|\Gamma|^2}{1-\Gamma+\Gamma^*-|\Gamma|^2}$$

$$= \frac{1-|\Gamma|^2+2jIm(\Gamma)}{1-|\Gamma|^2-2jIm(\Gamma)} = \frac{(1-|\Gamma|^2+2jIm(\Gamma))^2}{(1-|\Gamma|^2)^2+4Im^2(\Gamma)}$$

$$= \frac{(1-|\Gamma|^2)+2\cdot 2jIm(\Gamma)(1-|\Gamma|^2) - 4Im^2(\Gamma)}{(1-|\Gamma|^2)^2+4Im^2(\Gamma)}$$

$$= \frac{(1-|\Gamma|^2)^2-4\,Im^2(\Gamma)}{(1-|\Gamma|^2)^2+4Im^2(\Gamma)} + j\,\frac{4\,Im(\Gamma)(1-|\Gamma|^2)}{(1-\Gamma^2)^2+4\,Im^2(\Gamma)} \tag{B.28}$$

By comparison of the real and imaginary parts in (B.27) and (B.28), we find the following relations:

$$\cos 2\varphi = \frac{(1-u^2-v^2)^2-4v^2}{(1-u^2-v^2)^2+4v^2}, \quad \sin 2\varphi = \frac{4v(1-u^2-v^2)}{(1-u^2-v^2)^2+4v^2} \tag{B.29}$$

With the relation $\cot\varphi = (1 + \cos 2\varphi)/\sin 2\varphi$ [5], and substituting the sine and cosine in (B.29), we can write

$$\cot\varphi = \frac{1+\left[(1-u^2-v^2)^2-4v^2\right]/N}{\left[4v(1-u^2-v^2)\right]/N}, \quad N = (1-u^2-v^2)^2+4v^2 \tag{B.30}$$

Multiplying numerator and denominator by N leads to

$$\cot\varphi = \frac{(1-u^2-v^2)^2+4v^2+(1-u^2-v^2)^2-4v^2}{4v(1-u^2-v^2)} = \frac{1-u^2-v^2}{2v} \tag{B.31}$$

Eliminating the denominator,

$$u^2+v^2+2v\cot\varphi = 1 \tag{B.32}$$

and writing the expression in the form of a circle equation, we get

$$u^2+(v+\cot\varphi)^2 = \cot^2\varphi+1 = \frac{1}{\sin^2\varphi} \tag{B.33}$$

or

$$u^2+(v+v_0)^2 = R^2 \tag{B.34}$$

where $v_0 = \cot\varphi$ and $R = 1/|\sin\varphi|$ (Figure B.5).

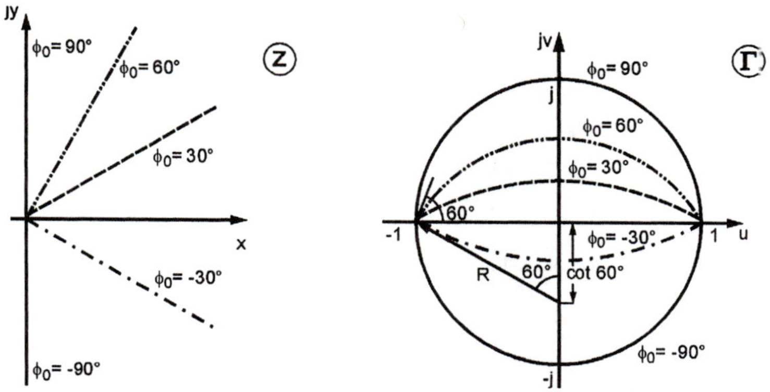

Figure B.5 Loci of constant phase of z in the Γ-plane.

B.2 Smith Chart as a Graphic Aid for the Solution of Transmission Line and Network Problems

In the following section we will recognize the Smith Chart as being a useful graphic aid for the solution of transmission line and network problems. Figure B.6 shows a typical situation in a distributed microwave network. A transmission line that is supposed to be lossless is terminated by a load with load impedance Z_L. The transmission line is characterized by its characteristic impedance Z_0 and phase coefficient β. The coordinate variable l is introduced to signify a given reference plane at a given distance from the load. At the load terminal, the load reflection coefficient $\Gamma(l=0) = \Gamma(0)$ reads

$$\Gamma(l = 0) = \frac{Z_L - Z_0}{Z_L + Z_0} \tag{B.35}$$

normalized with respect to the characteristic impedance Z_0, (B.35) reads

$$\Gamma(l = 0) = \frac{Z_L/Z_0 - 1}{Z_L/Z_0 + 1} = \frac{z_L - 1}{z_L + 1} \tag{B.36}$$

where z_L denotes the normalized load impedance. Equation (B.35) solved for the load impedance Z_L yields

$$Z_L = Z_0 \frac{1 + \Gamma(0)}{1 - \Gamma(0)} \tag{B.37}$$

or in normalized writing

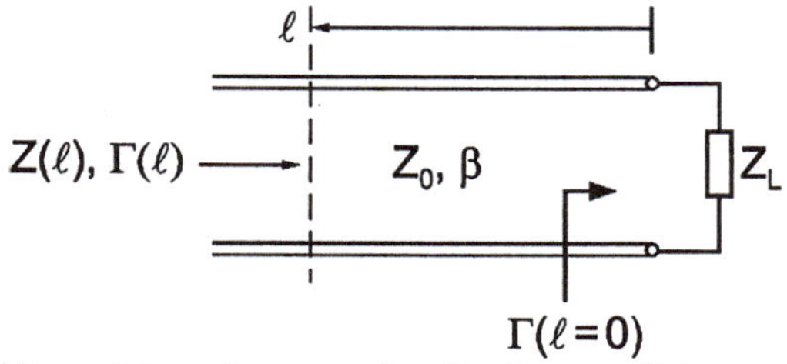

Figure B.6 Definition of impedances and reflection coefficients of transmission line terminated by a load.

$$z_L = \frac{Z_L}{Z_0} = \frac{1 + \Gamma(0)}{1 - \Gamma(0)} \tag{B.38}$$

Moving toward the generator by a length l, the load impedance Z_L is transformed into the line input impedance $Z(l)$ by the well-known transformation equation

$$Z(l) = \frac{Z_L + jZ_0 \tan\beta l}{1 + jZ_L/Z_0 \tan\beta l} \tag{B.39}$$

The normalized input impedance $z(l) = Z(l)/Z_0$ then reads

$$z(l) = \frac{z_L + j\tan\beta l}{1 + jz_L \tan\beta l} \tag{B.40}$$

Because the line parameters are complex, the mathematical evaluation is a little cumbersome. Instead of that, we regard the reflection coefficient Γ, which obeys the much simpler relation

$$\Gamma(l) = \Gamma(l=0)\, e^{-j2\beta l} \tag{B.41}$$

Thus, as we move along the line, the change of Γ manifests itself as a vector having constant magnitude in the Γ-plane, but being rotated by an angle of $(-2\beta l)$. The input impedance is obtained according to (B.38) as

$$z_{in} = \frac{Z_{in}}{Z_0} = \frac{1 + \Gamma(l)}{1 - \Gamma(l)} \tag{B.42}$$

The Smith Chart also can be used as a pure calculation tool if the physical problem does not contain any transmission lines. In this case, Γ has only purely mathematical meaning. The characteristic impedance Z_0 is then replaced by a normalizing impedance Z_N; therefore

$$z = Z/Z_N, \quad y = 1/z = Z_N/Z = YZ_N \tag{B.43}$$

In the following, we will discuss the interrelation of impedances

$$z = Z/Z_N = R/Z_N + jX/Z_N \tag{B.44}$$

and admittances, respectively,

$$y = YZ_N = GZ_N + jBZ_N \tag{B.45}$$

and the line parameter terms

$$\Gamma(l=0), \qquad \Gamma(l) = \Gamma(0)e^{-j2\beta l}, \qquad \mathrm{VSWR} = s = \frac{1 + |\Gamma|}{1 - |\Gamma|} \tag{B.46}$$

and their graphical representation in the Smith Chart. VSWR is called the standing wave ratio. The inverse value is denoted as the matching coefficient m,

$$m = \frac{1}{s} = \frac{1 - |\Gamma|}{1 + |\Gamma|} = \frac{|V|_{\min}}{|V|_{\max}} \tag{B.47}$$

m can take values between 0 and 1, whereas VSWR covers the range from 1 to ∞. A graphical representation of the Smith Chart is shown in Figure B.7 At the bottom, there are additional scales for radial parameters, such as magnitude of reflection coefficient and its squared value. The latter denotes the power of the reflected wave normalized to the power in the incident wave. Furthermore, VSWR can be directly read from the respective scale.

The reflection coefficient is directly related to the corresponding normalized impedance z, and its real and imaginary parts can be taken from the curves of constant values R/Z_0 and X/Z_0. A given point on the Γ-plane signifies the amplitude and phase of the reflection coefficient Γ. The magnitude of the reflection coefficient Γ is the distance from the center of the Smith Chart to the respective impedance or admittance. Reading the angle of the reflection coefficient is done by using the scale $\arg\Gamma$ on the circumference of the unit circle. The outer scale on the circumference with l/λ as a parameter is used to rotate the reflection coefficient Γ according to the relationship

$$\Gamma(l) = \Gamma(0)e^{-j2\beta l} = \Gamma(0)e^{-j4\pi l/\lambda} \tag{B.48}$$

which describes the transformation of the reflection coefficient along the line of electrical length l/λ. Thus, clockwise rotation on the Smith Chart plane is movement toward the generator.

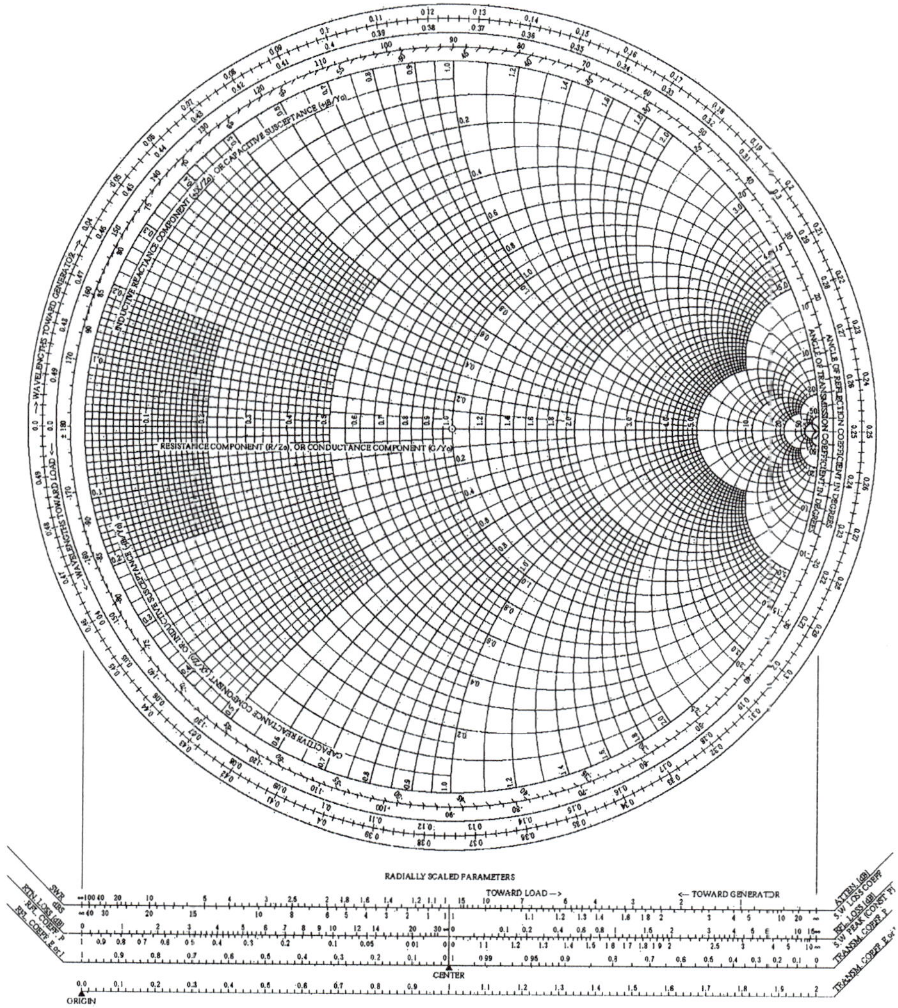

Figure B.7 Smith Chart.

B.3 Using the Smith Chart

The Smith Chart can be used to determine the reflection coefficient Γ of a circuit and other quantities that are in relation to the reflection coefficient. Moreover, the Smith Chart can be used for the determination of an input impedance and admittance. It is a useful tool for the designing of circuits, the evaluation of loci, the determination of the quality factor of a resonator, and the solving of matching problems.

B.3.1 Series and Parallel Combination of Resistances, Capacitances, and Inductances

For the calculation of the input impedance or admittance of an RLC network comprising both series and parallel branches of circuit elements, addition and subtraction operations must be performed in the z- or y-plane. Arbitrary mixed series and shunted interconnections of two-ports can be treated by the operation $y = 1/z$. Figure B.8 shows the transformation of a load resistance R_L into the input impedance Z_{in} with an LC two-port transformation.

The input impedance Z_{in} reads

$$Z_{in} = \frac{(R_L + jX_2)jX_1}{R_L + jX_2 + jX_1} \tag{B.49}$$

After normalizing the series circuit $R_L + jX_2$,

$$z_2 = R_L/Z_N + j\,X_2/Z_N \tag{B.50}$$

and taking the inverse

$$y_2 = 1/z_1 \tag{B.51}$$

the total admittance reads

$$y_{in} = y_1 + y_2 \tag{B.52}$$

where $y_1 = Z_N/jX_1$. The normalized input impedance then is

$$z_{in} = 1/y_{in} \tag{B.53}$$

The denormalized value is given as

$$Z_{in} = z_{in}Z_N = R_{in} + jX_{in}$$

The corresponding graphical representation on the Smith Chart is shown in Figure B.9. The mathematical operations must be graphically performed in the given sequential steps. Since the Γ-plane is covered with a net of curves of constant values R/Z_N

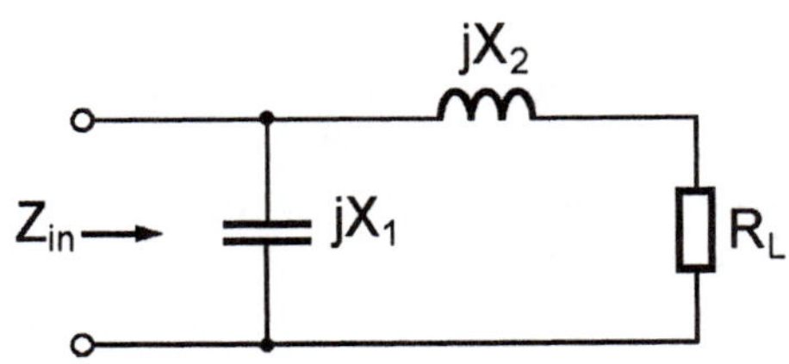

Figure B.8 Transformation of R_L into Z_{in} by an LC transformation two-port.

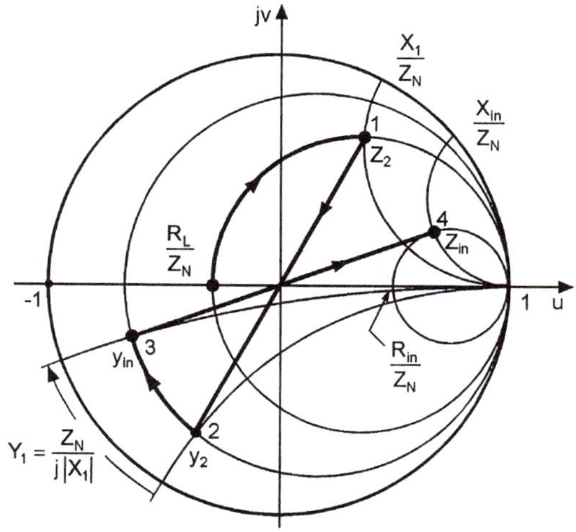

Figure B.9 Transformation of R_L into Z_{in} in Figure B.8.

and X/Z_N, addition and subtraction can simply be done as in the z-plane itself. The normalized value R_L/Z_N is found on the real axis $v = 0$. We move on the circle $R_L/Z_N = const.$ and reach the impedance $z_2 = R_L/Z_N + jX_2/Z_N$. For summing up the admittances of the shunted branches, we convert z_2 into y_2 by rotating the corresponding reflection coefficient by $180°$. Next, we have to add the admittance $y_{in} = 1/jX_1$, moving on the circle $Im(y_2) = const$. Then, z_{in} is found by inversion, $z_{in} = 1/y_{in}$.

Impedance matching normally involves usage of lossless matching elements to move from a point on the impedance (or admittance) plane to the center of the Smith Chart, where $Z/Z_0 = Y/Y_0 = 1$. This movement should be done with as few lossless elements as possible (usually two), which should have the minimum length of microstrip lines.

B.3.2 Power Matching with Lumped Elements (Resonance Transformation)

The problem of power matching at a given frequency is commonly solved by inserting a lossless two-port transformation network between the generator and load, as shown in Figure B.10. The power-matching network comprises two reactances, which make it possible to match any arbitrary passive load Z_L at a given frequency. Smith [1] showed that eight matching network topologies, illustrated in Figure B.11, are possible, which comprise specific L-L, L-C, C-L, or C-C element combinations, with each combination having either a series-shunt or shunt-series sequence. The selection of one of the eight possible matching network topologies is simplified if the values of Z_L are known, for which a matching can be achieved for a given topology. In order to identify this region in the Smith

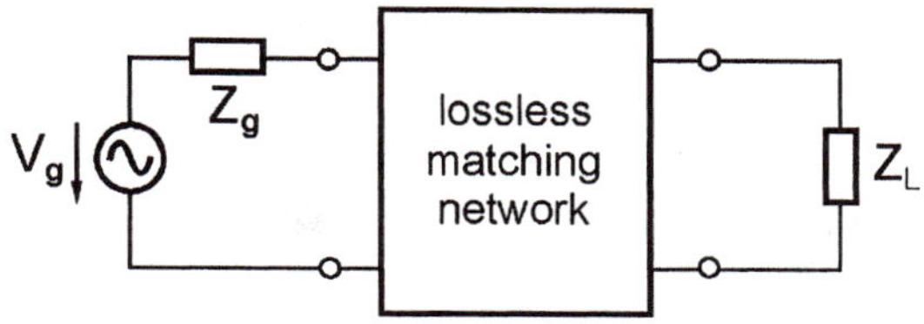

Figure B.10 Power matching network for maximum power transfer.

Chart, the normalization impedance Z_N must be known. In the following section, the transformation paths and the regions in which matching cannot be achieved (shaded) are presented, under the assumption that the reference impedance is real; that is, $Z_N = Z_g$ (real). The eight different L-L, L-C, C-L, or C-C circuit combinations, and their associated allowed areas in which a perfect single-frequency match can be accomplished, are illustrated in Figure B.11. The respective normalized L-type circuit is also shown. Normalization has been done as follows: $Z_g/Z_N = 1$, $Z_L/Z_N = z_L$, X_1/Z_N, X_2/Z_N.

B.3.3 Matching Network Comprising Sections of Transmission Line

The usual impedance matching problem is to match Z_L to the center of the Smith Chart, which gives a zero reflection coefficient and maximum power transfer from the source to the load. Regarding very high frequencies, the use of lumped element circuits becomes problematic. As is known, a short- or open-circuited transmission line of adjustable length can equivalently describe any reactive impedance value. The matching circuit also could be a combination of lumped and distributed microstrip elements.

B.3.3.1 Transmission Line of Adjustable Length with Adjustable Stub (Single Stub)

With the help of the line of length l_1 in Figure B.12, the load impedance z_L will be transformed to z_L' (see Figure B.13). Inversion leads to the admittance y_L'. By addition of the input admittance b_2 of the stub, the point $y = z = 1$ is reached. One can read the length l_1/λ from the periphery of the Smith Chart. The input admittance jb_2 can be realized by a short-circuited line ($y = \infty$), and the corresponding electrical length l_2/λ can be taken from the peripheral scale. The characteristic impedance of the stub is usually equal to Z_0; otherwise, the computation is more complicated. With free choice of l_1 and l_2 matching for all Z_L is possible.

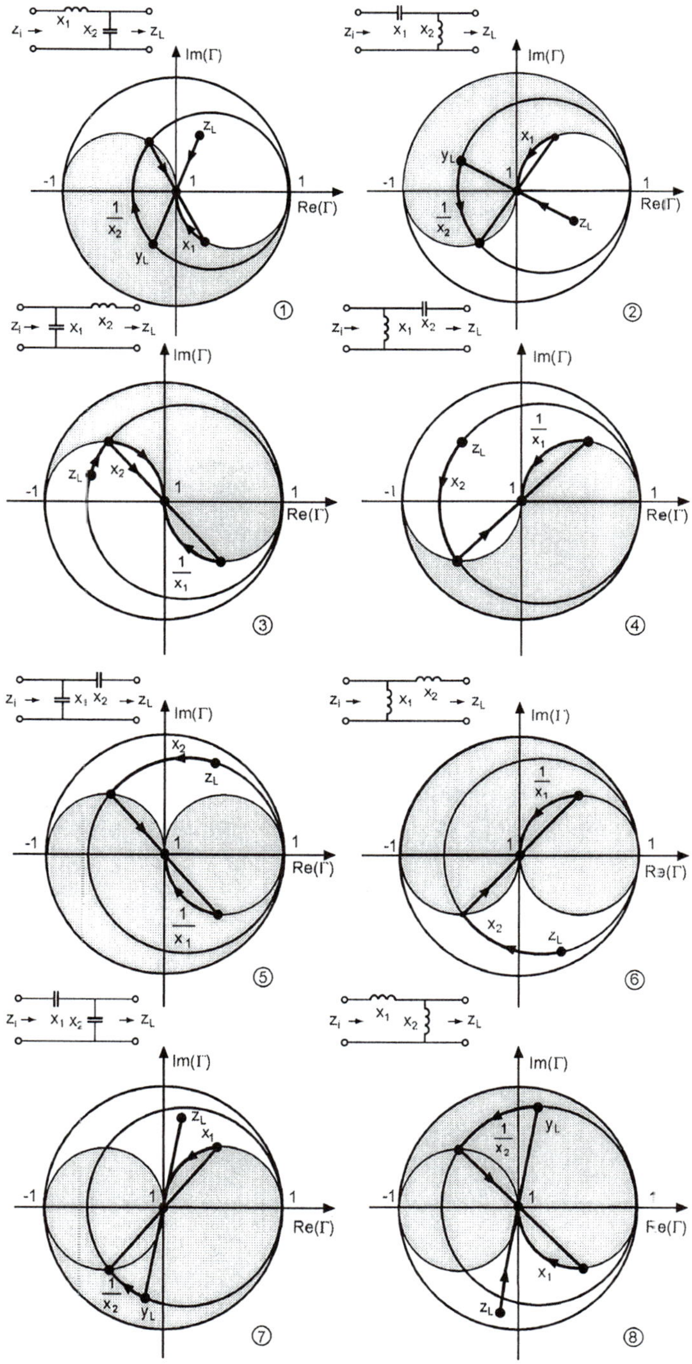

Figure B.11 The eight possible matching network topologies with two reactances.

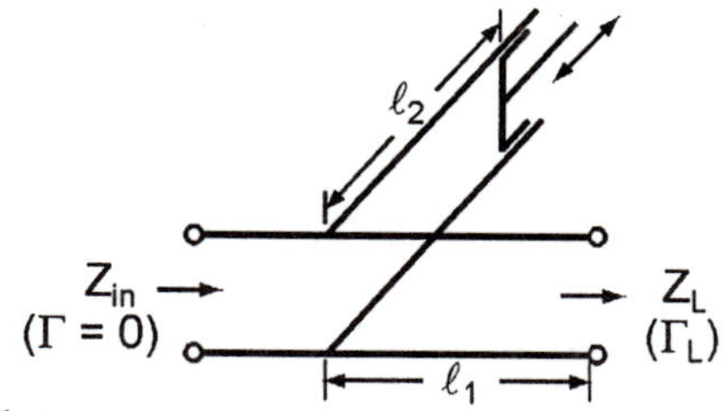

Figure B.12 Single stub tuner.

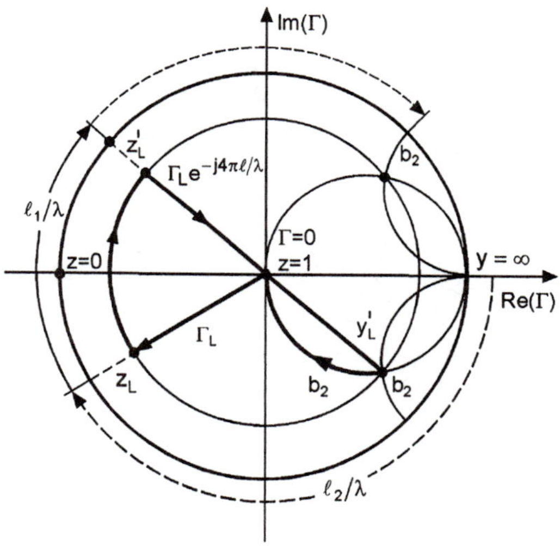

Figure B.13 Transformation path in the Smith Chart (matching with single stub).

B.3.3.2 Matching Network with Two Stubs (Double Stub Tuner)

Stubs with variable length are simpler and usually adopted for circuit realization rather than for a tunable main transmission line. A double stub configuration, shown in Figure B.14, has great practical importance. The distance between the two stubs is usually between $\lambda/16$ and $\lambda/4$, and determines the range of load impedances in the Smith Chart for which matching can be obtained.

The input susceptances b_1 and b_2 of the stubs are determined as follows. From y_L, one moves along the circle $Re(y_L) = const.$ in the mathematically positive sense, up to the intersection point P with the circle $y = 1 + jb$ rotated by l/λ in the mathematically positive sense, and thereby receives b_1. From point P, one reaches the circle $y = 1 + jb$ by rotating over a length l/λ. Matching ($\Gamma = 0$) is obtained by adding b_2. If y_L and/or z_L lies in the region of forbidden admittances and/or impedances (shaded areas in Figure B.15), then no intersection P occurs. The corresponding line length of the stubs to realize b_1 and b_2 can be read from the periphery of the Smith Chart.

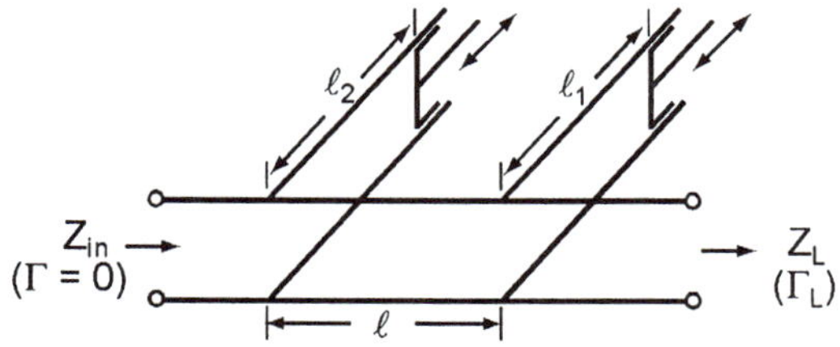

Figure B.14 Matching with double stub tuner.

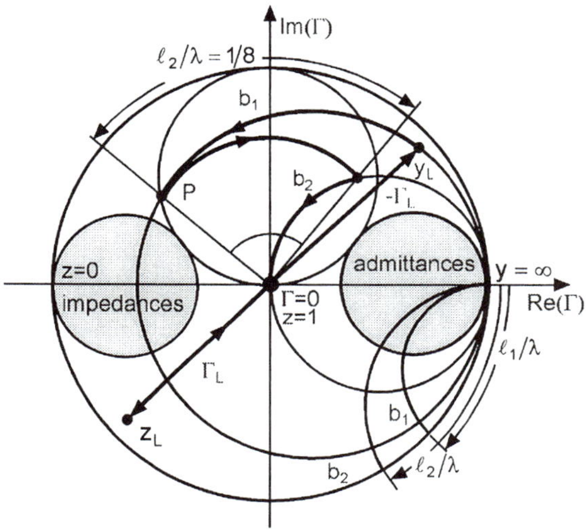

Figure B.15 Transformation path in the Smith Chart (matching with double stub).

B.3.3.3 λ/4-Transformer

The $\lambda/4$-transformer provides an easy solution for impedance matching at one frequency, using a piece of transmission line of length $l = \lambda/4$. The input impedance of a transmission line of length $l = \lambda/4$ can be evaluated from (B.39), dividing numerator and denominator by $\tan \beta l$, and taking the limit value $\beta l = (2\pi/\lambda)(\lambda/4) = \pi/2$. The result obtained is

$$Z_{in}(l = \lambda/4) = \frac{Z_0^2}{Z_L} \tag{B.54}$$

After normalization to the characteristic impedance Z_0 of the transmission line, the relation (B.54) reads

$$Z_{in}/Z_0 = \frac{1}{Z_L/Z_0} = z_{in} = 1/z_l \tag{B.55}$$

Choosing the characteristic impedance of the $\lambda/4$-transmission line as the geometric mean of the characteristic impedance values of the input and output transmission lines Z_{01} and Z_{03}, respectively (Figure B.16),

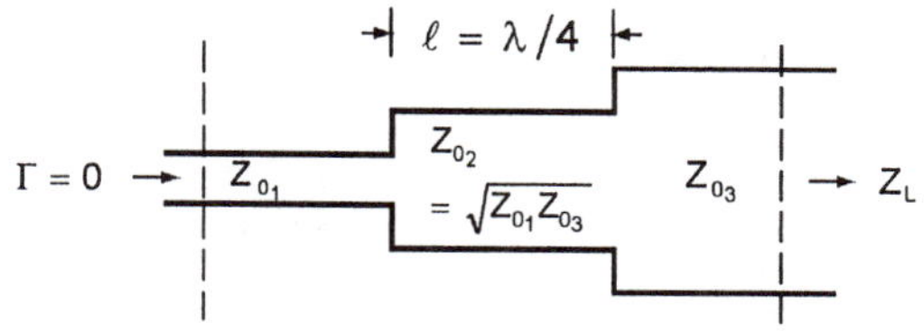

Figure B.16 The quarter-wave matching transformer.

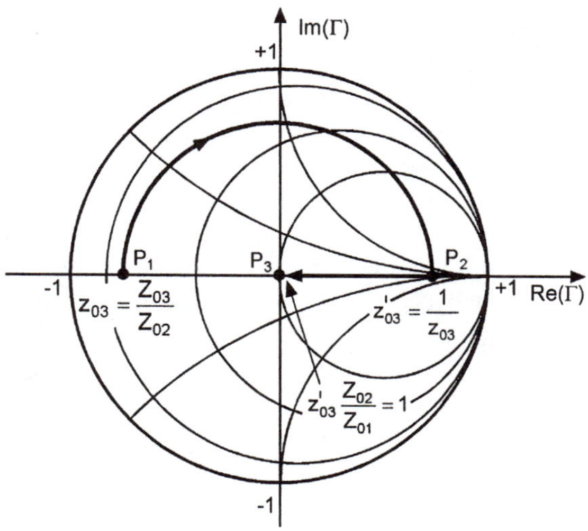

Figure B.17 Transformation path in the Smith Chart of a quarter-wave transformer.

$$Z_{02} = \sqrt{Z_{01}Z_{03}} \tag{B.56}$$

a resistive load impedance $Z_L = Z_{03}$ will be transformed by the $\lambda/4$-transformer, so that the input reflection coefficient Γ becomes zero. Figure B.17 illustrates the transformation in the Smith Chart. The load is first normalized to Z_{02} $(= \sqrt{Z_{01}Z_{03}})$, so that $z_{03} = Z_{03}/Z_{02} = Z_{03}/\sqrt{Z_{01}Z_{03}} = \sqrt{Z_{03}/Z_{01}}$ (point P_1). Point P_2 is reached after clockwise rotation by 180°. The normalized impedance now reads $z_{03}^{\prime} = 1/z_{03} = \sqrt{Z_{01}/Z_{03}}$. Denormalizing (i.e., multiplication with Z_{02}) and normalizing to the characteristic impedance Z_{01} of the input transmission line yield $z_{03}^{\prime} Z_{02}/Z_{01} = 1$ (point P_3). The described $\lambda/4$-transformation also can be applied to complex load impedances after transforming into real values through an appropriate length of transmission line.

B.3.4 Reading of Voltage (and/or Current) from Smith Chart

The amplitudes of forward and reverse propagating voltage and current waves on the transmission line read

$$V(l) = V^i(l) + V^r(l) = V^i(l)(1 + \Gamma) \tag{B.57}$$

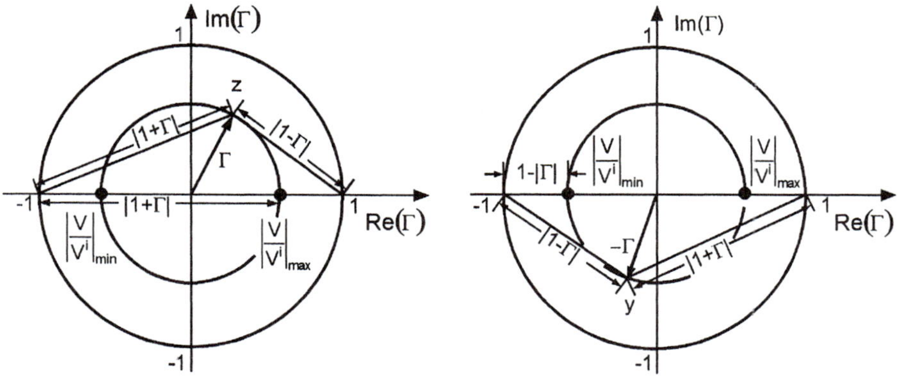

Figure B.18 Reading amplitudes of total voltage and current from the Smith Chart.

$$Z_0 I(l) = V^i(l) - V^r(l) = V^i(l)(1 - \Gamma) \tag{B.58}$$

where V and I denote the total line voltage and line current, respectively. V^i and V^r are the forward and backward traveling voltage waves. Normalizing (B.57) and (B.58) with respect to V^i yields

$$\left|\frac{V(l)}{V^i}\right| = |1 + \Gamma|, \qquad \left|\frac{Z_0 I(l)}{V^i}\right| = |1 - \Gamma| \tag{B.59}$$

These relations can be directly read from the Smith Chart, as shown in Figure B.18. It is seen that the reading of the normalized line voltage and current is different for impedances and admittances in the Γ-plane. The magnitude of the amplitude V^i is constant along the line, and reads $|V^i| = |V_g|/2$ if the internal resistance of the generator is equal to the characteristic impedance of the attached line. If $|V_g|$ is not known, then the normalized voltage $|V^i|$ can be derived from the transferred power. In any case, one also can calculate the voltage distribution along the transmission line with normalization to $|V|_{max}$ or $|V|_{min}$ without knowledge of V_g.

References

[1] Smith, P. H., *Electronic Applications of the Smith Chart*, New York: McGraw-Hill Book Company, 1969.

[2] Smith, P. H., "Transmission Line Calculator," *Electronics*, January 1939, p. 29.

[3] Smith, P. H., "An Improved Transmission Line Calculator," *Electronics*, January 1944, pp. 130–133, 318, 320, 322, 324–325.

[4] Kraus, J. D., and D. A. Fleisch, *Electromagnetics with Applications*, 5th ed., New York: McGraw-Hill, 1999.

[5] Bronstein, L. N., and K. A. Semendjajew, *Taschenbuch der Mathematik*, 8. Auflage, Zürich und Frankfurt/M: Verlag Harri Deutsch, 1968.

Appendix C

Pulse Shaping with Coupled Microstrip and Slot Line

In [1], the design of a broadband thin film signal sampler is described using slot lines to generate balanced sampling pulses with opposite polarity. Equal amplitudes of balanced sampling pulses are essential because good decoupling is needed to keep the gate pulse transients off the signal line. In [2, 3], a balanced avalanche discharging pulse generator was used with two coupled microstrip lines as electrical charge storage. Experience had shown that the same amplitude of the generated pulses only could be obtained after additional tuning of the load impedances. Such difficulties were not observed with sampling pulse generation using a microstrip-slot line combination. Therefore, some basic experimental results are described in this Appendix.

Slot lines can simply be incorporated in microstrip circuits by etching the slot line circuit in the ground plane of the microstrip circuit. Figure C.1 shows a short-circuited slot line of length l and slot width s, positioned perpendicular and symmetrical to a microstrip drive line. The microstrip line is terminated in its matched load, and stimulated by an electrical pulse of 47V peak value, 1.7 ns pulse width (FWHM), and 440 ps rise-time. A miniature, semi-rigid coaxial cable is placed at the center of the short-circuited slot line and used for signal detection.

In Figure C.2, the electric and magnetic field distribution both in the microstrip and slot line are illustrated to explain the coupling of the microstrip and slot line

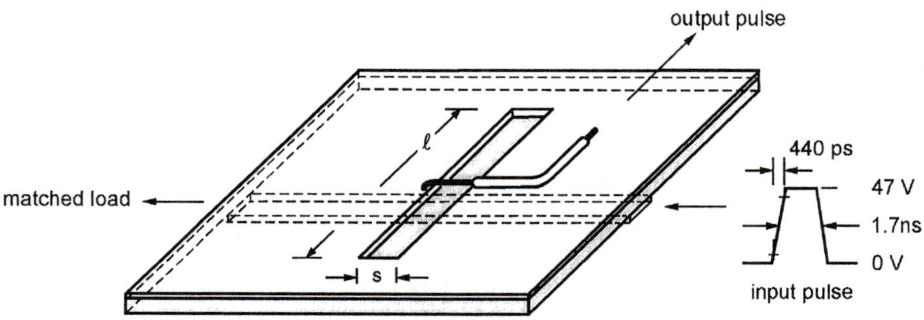

Figure C.1 Electrical pulse shaping using a short-circuited slot line.

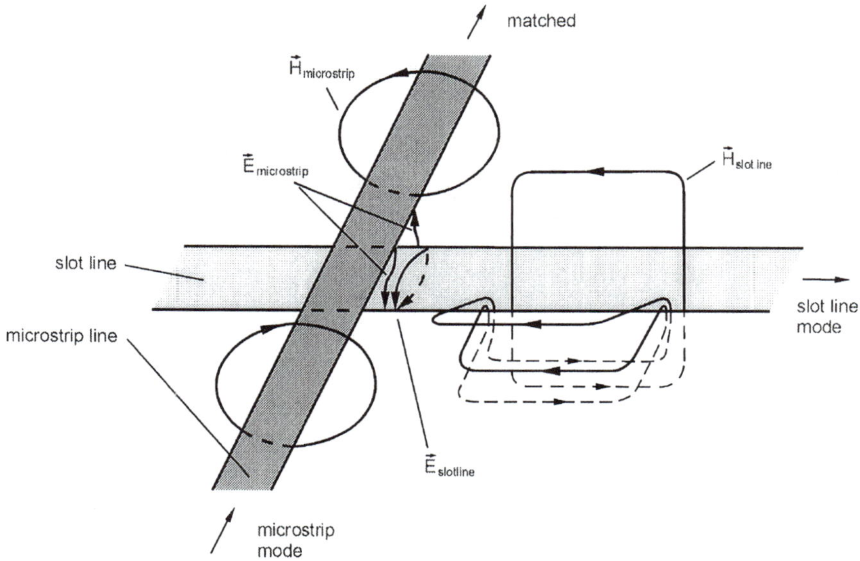

Figure C.2 Illustrative representation of the excitation of a slot line wave by a quasi-TEM mode on a microstrip line.

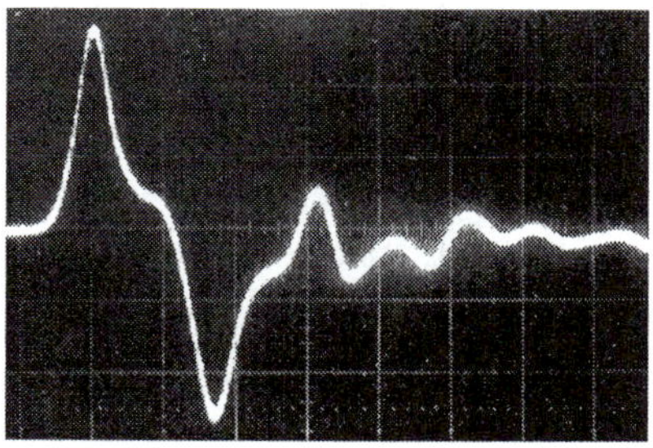

Figure C.3 Measured electrical pulse after differentiation with a short-circuited slot line (5V/division, 1 ns/division).

mode. This can be explained qualitatively by considering the $\vec{E}$-field distribution associated with the microstrip-to-slot line transition. Referring to Figure C.2, we note that the $\vec{E}$-field, passing the slot, will change its polarity and will excite a strong electrical field between the half-conductive planes of the slot line. Regarding the $\vec{H}$-field, the transversal magnetic field lines of the microstrip mode produce a strong longitudinal magnetic field component in the slot line. The direction of the excited transverse electric and magnetic field originates in a Poynting vector, indicating wave propagation from both sides of the coupling point. Using a voltage step function as the drive signal, the slot wave propagates along the line and is

reflected

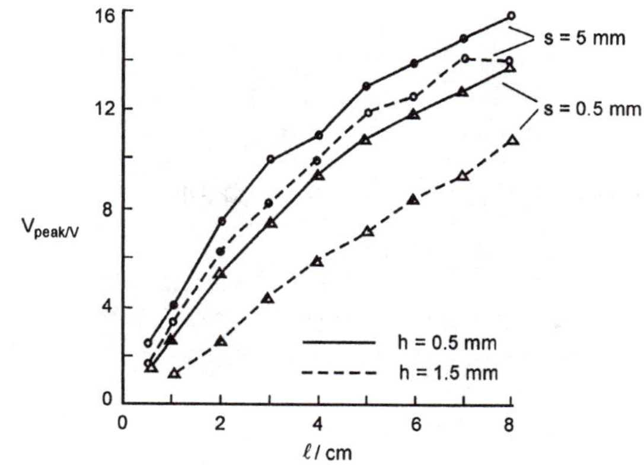

Figure C.4 Measured peak voltage of the differentiated pulse.

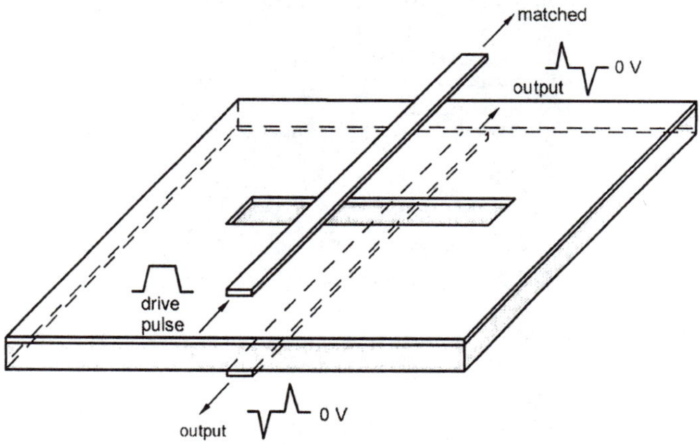

Figure C.5 Short-circuited slot line coupled to two microstrip lines.

at the short with opposite polarity. Therefore, depending on the delay time of the pulse, which is $\tau_d = l/v_{slot}$, a differentiated drive pulse shape is measured as the output signal.

Figure C.3 shows the transient behavior of the differentiated pulse coupled out at the center of the short-circuited slot line. The substrate used was low-cost Duraver-E-Cu (FR-4) (Section 9.1.2) with a dielectric constant $\epsilon_r = 4.7$ and height 0.5 mm and 1.5 mm, respectively. As shown in Figure C.4, the amplitude increases up to a maximum value of 16V. The peak value depends strongly on the substrate height and slot width. The pulse width increases linearly from 400 ps to approximately 600 ps with growing slot length. The rise time is comparable with that of the stimulating pulse.

Differentiated pulses with opposite polarity can be obtained with the circuit configuration shown in Figure C.5. In this case, two microstrip lines are placed on both sides of the slot line oriented at 90° to the microstrip lines. The excitation of

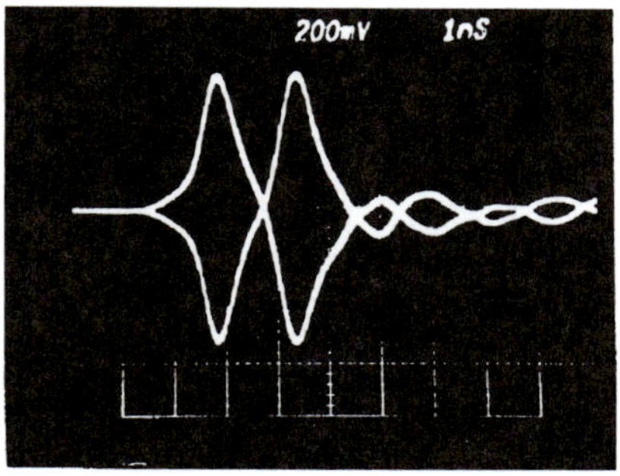

Figure C.6 Measured slot line differentiated pulses on Duraver-E-Cu (2V/division, 1 ns/division).

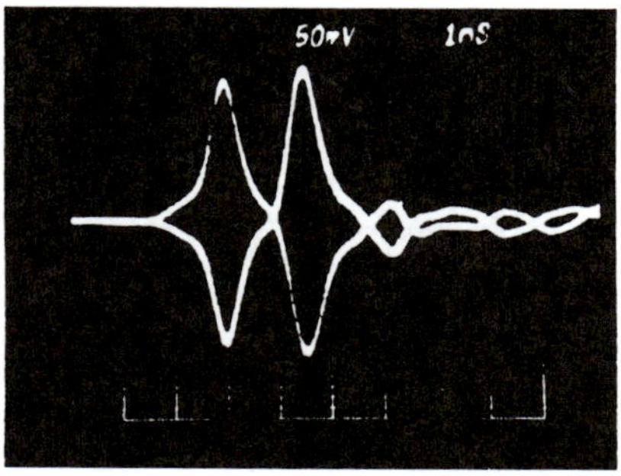

Figure C.7 Measured slot line differentiated pulses on Al_2O_3 ceramics (0.5V/division, 1 ns/division).

the microstrip pulses of opposite polarity by the slot line mode can be understood with Figure C.2. Figure C.6 shows typical waveforms on Duraver-E-Cu 1(h = 1.5 mm, ϵ_r = 4.7) with a slot width of 3 mm and slot length of 35 mm. A voltage peak value 5V was obtained. Figure C.7 shows typical waveforms on alumina (h = 0.635 mm, ϵ_r = 9.7) with a slot width of 1 mm and slot length of 20 mm. In this case the voltage peak value was measured as 1.3V. It can be seen from the recorded traces that the balanced signals have equal amplitude and are opposite in sign, with some ringing that is damped out by the diode conduction in a normal sampler [1].

References

[1] Merkkelo, J., and R. D. Hall, "Broad-Band Thin-Film Signal Sampler," *IEEE Trans.*, Vol. SC-7, 1972, pp. 50–54.

[2] Schwarte, R., *Zur Messung der Kurvenform einmaliger, zugleich im Subnanosekunden- und Millivoltbereich ablaufender Vorgänge mittels Samplingtechnik*, Doctoral Thesis, Aachen, Germany: RWTH, 1972.

[3] Kompa, G., "Entfernungsmeßgerät nach dem Impulslaufzeitverfahren," European Patent, No. 0015566, granted 1984.

Appendix D

Modes in Waveguiding Structures

D.1 Maxwell Equations

The analysis of waveguiding structures requires the solution of the complete set of Maxwell equations, which relate the field vectors $\vec{E}$, $\vec{D}$, $\vec{J}$, $\vec{H}$, and $\vec{B}$ and the scalar free electric space charge density ρ to each other. $\vec{B}$ is called the magnetic flux density, magnetic field strength, or the magnetic induction, whereas $\vec{H}$ is called the magnetic field. Often the distinction is ignored, and both symbols are frequently referred to as the magnetic field [1]. $\vec{E}$ is the electric field, $\vec{D}$ is the electric displacement field, and $\vec{J}$ is the current density. In general, these quantities are space- and time-dependent. The Maxwell equations [2, 3], originally derived from experience, can be written both in integral and differential form. Not all the knowledge gained from experience is reflected by the Maxwell equations; for example, the quantization of charge, or the fact that the energy content of any system has to be finite. In particular cases, additional conditions must be defined. However, classical waveguiding problems can be solved with the standard Maxwell equations. The Maxwell equations in integral form read:

$$\oint_{C(A)} \vec{H} \cdot d\vec{s} = \oiint \vec{G} \cdot d\vec{A} = \oiint (\vec{J} + \dot{\vec{D}}) \cdot d\vec{A} \tag{D.1}$$

$$\oint_{C} \vec{E} \cdot d\vec{s} = - \oiint \dot{\vec{B}} \cdot d\vec{A} \tag{D.2}$$

$$\oiint \vec{B} \cdot d\vec{A} = 0 \tag{D.3}$$

$$\oiint \vec{D} \cdot d\vec{A} = \iiint \rho \, dV \tag{D.4}$$

Note that the dot on a vector or scalar denotes a time derivative. When the integration boundaries are invariant with time (i.e., when the integration area does not change in a stationary material), the integration and the time differentiation can be interchanged,

$$\frac{\partial}{\partial t} \oiint \vec{F} \cdot d\vec{A} = \oiint \frac{\partial}{\partial t} \vec{F} \cdot d\vec{A} = \oiint \dot{\vec{F}} \cdot d\vec{A} \tag{D.5}$$

Maxwell equations (D.1) and (D.2) relate an integral over an arbitrary surface A

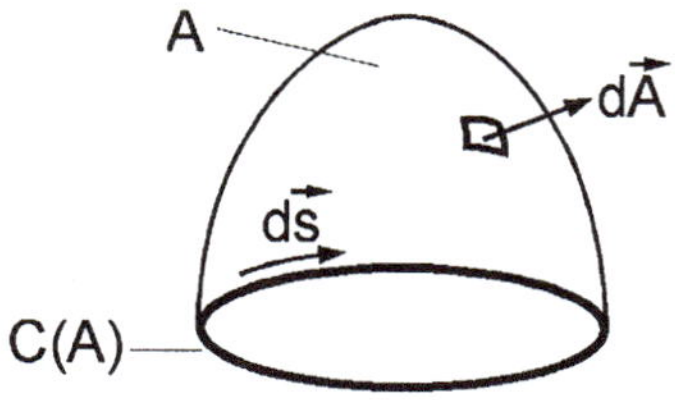

Figure D.1 Related surface and contour integrals.

to a contour integral along the closed-loop boundary of this surface (see Figure D.1). The infinitesimal contour element $d\vec{s}$ along the boundary curve $C(A)$ and the infinitesimal area vector $d\vec{A}$ are assigned to each other by a right-hand system. The surface integral of the magnetic flux density

$$\psi_m = \oiint \vec{B} \cdot d\vec{A} \tag{D.6}$$

is called magnetic flux. In analogy, the electric flux ψ_e is defined as

$$\psi_e = \oiint \vec{D} \cdot d\vec{A} \tag{D.7}$$

For field quantities with harmonic time dependency, defined as

$$\vec{F}(\vec{r}, t) = Re \left\{ \underline{\vec{F}}\ (\vec{r})\ e^{+j\omega t} \right\} \tag{D.8}$$

the Maxwell equations can be written in the following complex notation:

$$\oint_{C(A)} \underline{\vec{H}} \cdot d\vec{s} = \oiint \underline{\vec{G}} \cdot d\vec{A} = \oiint (\underline{\vec{J}} + j\omega \underline{\vec{D}}) \cdot d\vec{A} \tag{D.9}$$

$$\oint_C \underline{\vec{E}} \cdot d\vec{s} = -j\omega \oiint \underline{\vec{B}} \cdot d\vec{A} \tag{D.10}$$

$$\oiint \underline{\vec{B}} \cdot d\vec{A} = 0 \tag{D.11}$$

$$\oiint \underline{\vec{D}} \cdot d\vec{A} = \iiint \underline{\rho}\, dV \tag{D.12}$$

Note that the underlining of symbols denotes complex quantities, but will be used only if necessary. Maxwell equation (D.9) tells us that the surface integral over the total current density is equal to the contour integral over the magnetic field strength along the boundary curve of the surface A. Maxwell equation (D.10) indicates that the decrease of the magnetic flux over the area A versus time is equal to the contour integral over the electric field strength along the boundary curve of the surface A. Maxwell equations (D.11) and (D.12) relate a volume integral to a surface integral over the closed envelope surface A of the volume V (Figure D.2).

Equation (D.11) means that the magnetic flux through a closed surface envelope must vanish. By reason of the nonexistence of magnetic charges, the magnetic field lines are closed loops. Equation (D.12) expresses the fact that the electric flux

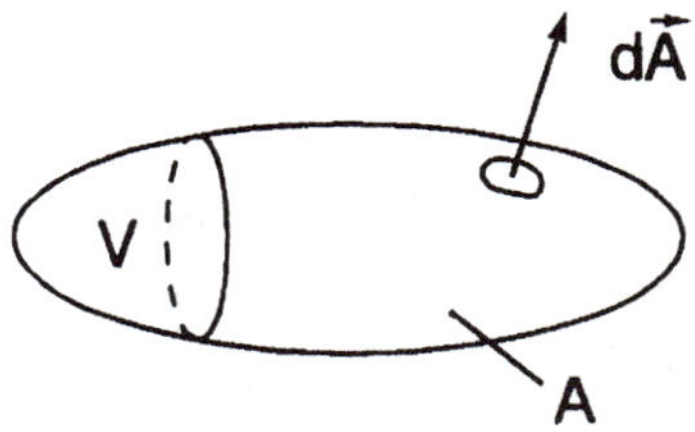

Figure D.2 Related surface and volume integrals.

through a closed surface area is equivalent to the total charge that is included therein. With application of the vector analysis integral theorems of Gauß and Stokes, the Maxwell equations are obtained in differential form. After Stokes, we write

$$\oint \vec{F} \cdot d\vec{s} = \oiint rot\ \vec{F} \cdot d\vec{A} \tag{D.13}$$

Applying (D.13) to (D.1), it reads

$$\oiint \left[rot\ \vec{H} - (\vec{J} + \dot{\vec{D}}) \right] \cdot d\vec{A} = 0$$

Because the integral vanishes for arbitrary areas, we find:

$$rot\ \vec{H} - (\vec{J} + \dot{\vec{D}}) = \vec{0} \tag{D.14}$$

Similarly, we obtain from (D.2) in differential form

$$rot\ \vec{E} + \dot{\vec{B}} = \vec{0} \tag{D.15}$$

With the application of Gaussian theorem,

$$\oiint \vec{F} \cdot d\vec{A} = \iiint div\, \vec{F} dV \tag{D.16}$$

we find from (D.11)

$$\oiint \vec{B} \cdot d\vec{A} = \iiint div\, \vec{B} dV = 0$$

The integral relationship is fulfilled for arbitrary volumes V only if

$$div\ \vec{B} = 0 \tag{D.17}$$

holds. Similarly, one obtains from (D.12)

$$div\ \vec{D} = \rho \tag{D.18}$$

The Maxwell equations in differential form read in time domain

$$rot\, \vec{H} = \vec{G} = \vec{J} + \dot{\vec{D}} \tag{D.19}$$

$$rot\, \vec{E} = -\dot{\vec{B}} \tag{D.20}$$

$$div\,\vec{B} = 0 \tag{D.21}$$

$$div\,\vec{D} = \rho \tag{D.22}$$

and in frequency domain

$$rot\,\underline{\vec{H}} = \underline{\vec{G}} = \underline{\vec{J}} + j\omega\underline{\vec{D}} \tag{D.23}$$

$$rot\,\underline{\vec{E}} = -j\omega\,\underline{\vec{B}} \tag{D.24}$$

$$div\,\underline{\vec{B}} = 0 \tag{D.25}$$

$$div\,\underline{\vec{D}} = \underline{\rho} \tag{D.26}$$

The field vectors $\vec{E}$ and $\vec{D}$, and $\vec{H}$ and $\vec{B}$, respectively, are linked with each other by the material equations.

Material Equations:

$$\vec{D} = \epsilon\vec{E} \tag{D.27}$$

$$\vec{B} = \mu\vec{H} \tag{D.28}$$

$$\vec{J} = \sigma\vec{E} \tag{D.29}$$

where ϵ is the electrical permittivity and μ the magnetic permeability [1]. The material equations, also termed constitutive relations [4], describe the interaction of $\vec{E}$ and $\vec{H}$ in a given material. For the determination of the 15 components of the 5 field quantities $\vec{E}$, $\vec{D}$, $\vec{H}$, $\vec{B}$, and $\vec{J}$, we have 15 scalar differential equations with the relationships (D.23), (D.24), and (D.27) to (D.29), as well. Equation (D.21) is a true additional condition, which must be fulfilled by the solution of the set of Maxwell equations. Finally, the space charge density ρ can be determined from (D.22). The Maxwell equations in differential form are complete only if the boundary conditions (see Sections D.1.1 and D.1.2) are added, which describes the behavior of the field quantities at interfaces of electrically different materials.

D.1.1 Boundary Conditions

It will be shown that the tangential magnetic and electric field strengths at dielectric interfaces are continuous. As shown in Figure D.3, a small rectangular area ΔA penetrates the interface in normal direction. The area vector $d\vec{A}$ lies parallel to the interface. We calculate the limit value of the contour integral with respect to Δh,

$$\lim_{\Delta h \to 0} \oint \vec{H}\cdot d\vec{s} = \int_{\Delta s} \vec{H}\cdot d\vec{s}_1 + \int_{\Delta s} \vec{H}\cdot d\vec{s}_2 = \int_{\Delta s} \left(\vec{H}_1 - \vec{H}_2\right)\cdot d\vec{s} \tag{D.30}$$

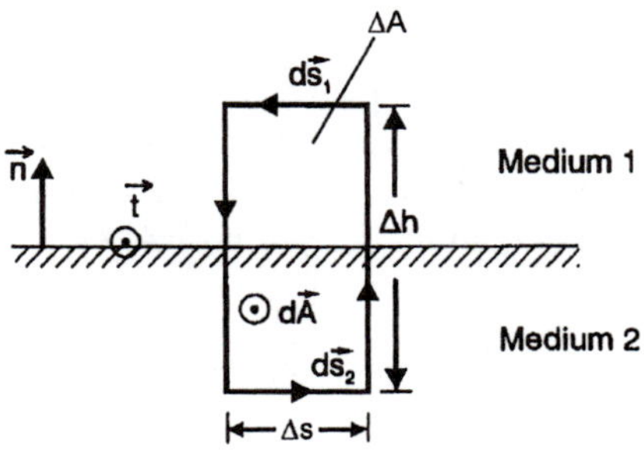

Figure D.3 Analysis of electromagnetic field at dielectric interface.

where $d\vec{s} = d\vec{s}_1 = -d\vec{s}_2$, and where $d\vec{s}$ is tangential to the boundary plane. In the limit case $\Delta h \to 0$, it follows for the total current density that

$$\lim_{\Delta h \to 0} \iint \vec{G} \cdot d\vec{A} = \lim_{\Delta h \to 0} \iint \left(\vec{J} + \dot{\vec{D}}\right) \cdot \vec{t}\, dA = 0 \tag{D.31}$$

because $\dot{\vec{D}}$ is always finite, and $\vec{J}$ is assumed to remain finite. Because (D.30) and (D.31) must be equal, a condition for the tangential components of the $\vec{H}$ field is obtained. The tangential component $H_{\tan}$ of the magnetic field strength is continuous, if no forced surface current is present. Thus, we obtain

$$\vec{H}^{(1)}_{\tan} - \vec{H}^{(2)}_{\tan} = \vec{0}$$

or

$$\vec{n} \times \left(\vec{H}_1 - \vec{H}_2\right) = \vec{0} \tag{D.32}$$

where subscripts 1 and 2 specify the media, and $\vec{n}$ indicates the normal unit vector at the boundary. In the special case when a surface current exists in the boundary plane, then (D.31) changes to read

$$\lim_{\Delta h \to 0} \iint \left(\vec{J} + \dot{\vec{D}}\right) \cdot d\vec{A} = \int \vec{J}_s \cdot \vec{t}\, ds \tag{D.33}$$

with the surface current density $\vec{J}_s$ ($[\vec{J}_s] = A/m$) and tangential unity vector $\vec{t}$ ($\vec{t} = \vec{n} \times d\vec{s}/ds$). If a surface current is present, the tangential magnetic field changes discontinuously at the transition from medium 1 to 2,

$$H^{(1)}_{\tan} - H^{(2)}_{\tan} = J_s$$

$$\vec{n} \times \left(\vec{H}_1 - \vec{H}_2\right) = \vec{J}_s \tag{D.34}$$

Similarly, from (D.10), it follows for the electric field $\vec{E}$ at the interface

$$E_{1_{\tan}} = E_{2_{\tan}}$$

$$\vec{n} \times \left(\vec{E}_1 - \vec{E}_2\right) = \vec{0} \tag{D.35}$$

because $\dot{\vec{B}}$ is always finite for real fields. That means that the tangential components of the magnetic and electric fields are continuous at the boundary of two media. The interface conditions for the normal electric displacement $\vec{D}$ and magnetic field strength $\vec{B}$ read

$$\vec{n} \cdot \left(\vec{D}_1 - \vec{D}_2\right) = \rho_s \tag{D.36}$$

Problem D.1: For the dielectric-coated metal plate (surface waveguide) shown in Figure D.4, write down the boundary conditions at the air-dielectric interface. The dielectric material is assumed to be lossless.

Solution: There are neither surface currents in the air-dielectric interface nor surface charges. Therefore, $H_{\tan}$ and $E_{\tan}$ are continuous at the interface at $y = 0$, and we can write: $H_{\tan}^{(1)} = H_{\tan}^{(2)}$ and $E_{\tan}^{(1)} = E_{\tan}^{(2)}$ $(0 \le |x| < \infty,\ y = 0,\ 0 \le |z| < \infty)$.

Problem D.2: Assume a metal plane with infinite conductivitiy (Figure D.5). Discuss the boundary conditions.

Solution: Because of $\sigma \to \infty$, the external electromagnetic field in region 1 does not penetrate into region 2, which means $\vec{H}_{\tan}^{(2)} = \vec{0}$. Therefore, (D.34) reads $\vec{n} \times \vec{H}_{\tan}^{(1)} = \vec{J}_s$ and $H_{\tan}^{(1)} = J_s$, respectively.

Problem D.3: Figure D.6 shows a microstrip transmission line. Let us suppose that the metallization is infinitely conductive ($\sigma = \infty$) and has negligible thickness. Formulate the boundary conditions along the air-dielectric interface (including the strip).

Solution: For the interface at $y = 0$, the boundary conditions for the electric field components read:

$$\begin{aligned} E_{2x} - E_{1x} &= 0 && 0 \le |x| < \infty \\ \epsilon_2 E_{2y} - \epsilon_1 E_{1y} &= 0 && w/2 \le |x| < \infty \\ \epsilon_2 E_{2y} - \epsilon_1 E_{1y} &= \rho_s && 0 \le |x| < w/2 \\ E_{2z} - E_{1z} &= 0 && 0 \le |x| < \infty \end{aligned} \tag{D.37}$$

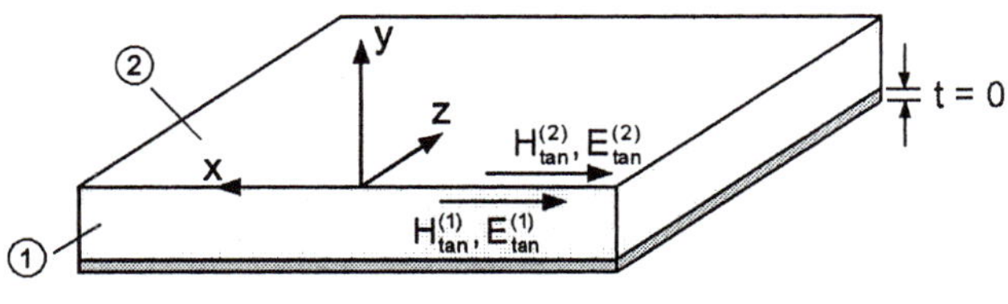

Figure D.4 Air-dielectric interface conditions of grounded dielectric plate.

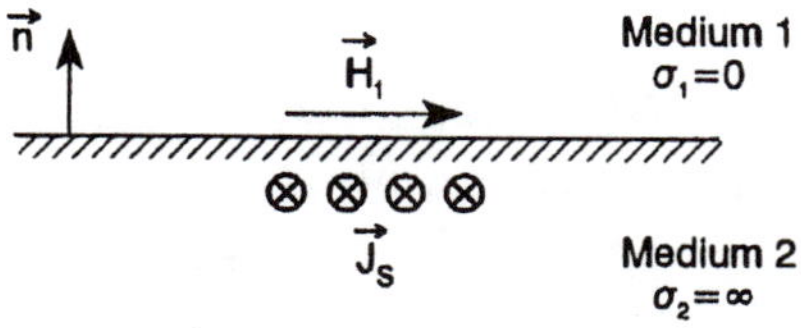

Figure D.5 Boundary properties at a metal surface of infinite conductivity.

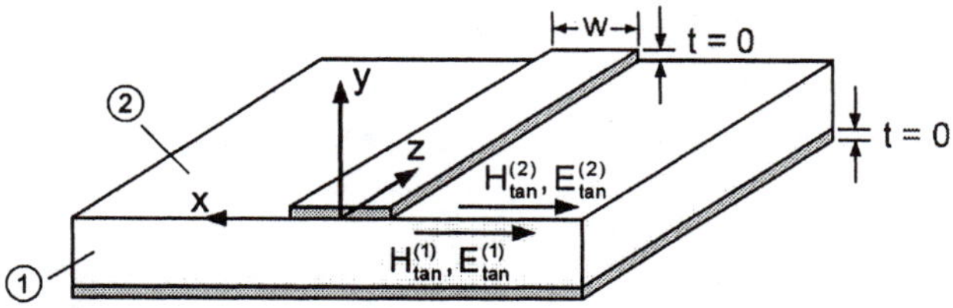

Figure D.6 Microstrip interface boundaries.

For the magnetic field components, we get:

$$
\begin{aligned}
H_{2x} - H_{1x} &= 0 && w/2 \le |x| < \infty \\
H_{2x} - H_{1x} &= -J_z && 0 \le |x| < w/2 \\
H_{2y} - H_{1y} &= 0 && 0 \le |x| < \infty \\
H_{2z} - H_{1z} &= 0 && w/2 \le |x| < \infty \\
H_{2z} - H_{1z} &= J_x && 0 \le |x| < w/2
\end{aligned}
\tag{D.38}
$$

At low frequencies, we have $J_x \ll J_z$. Thus $J_s \approx J_z \approx I_z/w$.

D.1.2 Particular Boundary Conditions

For given particular boundary value problems, different mathematical solutions can be found from Maxwell equations, but only one solution is physically meaningful. Typical examples are the analysis of open waveguiding structures with the electromagnetic field extended to infinity, or the consideration of geometrical singularities, such as sharp conductive edges. In such cases, additional physical boundary conditions have to be introduced to attain unique solutions.

Edge Condition:

For sharp edges or tips, the so-called edge condition must be fulfilled. The edge condition says that the electric and magnetic energy that is stored in the vicinity of an edge must be finite. For steady-state sinusoidal time-varying fields, it reads

$$\lim_{V \to 0} \iiint \left(\frac{1}{4} \epsilon \vec{E} \cdot \vec{E}^* + \frac{1}{4} \mu \vec{H} \cdot \vec{H}^* \right) dV = 0 \tag{D.39}$$

Sommerfeld's Radiation Condition:
Regarding open waveguiding problems the boundary conditions have to be fulfilled at infinity. If the material is lossy, then the electromagnetic field must vanish at infinity. If we assume that the material is lossless and isotropic, then the radiation condition after Sommerfeld signifies that each transversal component ψ of the electromagnetic field has to fulfill the following relationship:

$$\lim_{r \to \infty} r \left(\frac{\partial}{\partial r} \psi(\vec{r}) + jk\psi(\vec{r}) \right) = 0 \qquad \text{(D.40a)}$$

It can be easily shown that a spherical wave with

$$\psi(\vec{r}) = \frac{1}{r} e^{-jkr} \qquad \text{(D.40b)}$$

fulfills the radiation condition. Therefore, at any point in space that is sufficiently far away from the point of origin of the spherical coordinate system, the fields have to behave like planar waves:

$$\vec{E} = -\sqrt{\frac{\mu}{\epsilon}} \left(\frac{\vec{r}}{r} \times \vec{H} \right) + o\left(\frac{1}{r}\right)$$

$$\vec{H} = \sqrt{\frac{\epsilon}{\mu}} \left(\frac{\vec{r}}{r} \times \vec{E} \right) + o\left(\frac{1}{r}\right) \qquad \text{(D.41)}$$

The second term in (D.41) vanishes for large values of r. The function $f(r) = o(1/r)$ converges to zero more strongly than $1/r$ with increasing r. The radiation condition does not allow a wave to impinge from infinity toward the point of origin.

D.2 Modified Maxwell Equations for Solving Waveguiding Boundary Problems

The Maxwell equations are coupled differential (integral) equations. For mode analysis in waveguiding structures, modified versions of the equations are usually applied. We will first decouple the Maxwell equations, which will lead to wave equations for the electrical field and magnetic field, respectively, in time and frequency domain. The wave equation then can be solved in different ways: (1) reducing the field analysis problem to the determination of the longitudinal electric and magnetic field components, and (2) introducing an electric and magnetic vector potential as an auxiliary quantity (electrodynamic potentials).

First, we modify the Maxwell equations to obtain decoupled differential equations for the electric field and magnetic field (wave equations). We assume that the considered material is linear, homogeneous, and isotropic. Furthermore, its properties do not change with time, so it can be described with the three scalar material parameters μ, ϵ, and σ. We assume further that the sources of excitation of the electromagnetic field on the waveguiding structures are outside the field region under consideration. We assume homogeneous lines, so that mode conversion or excitation cannot occur. We start with Maxwell equations in time domain (D.19) and (D.20), using the divergence operator $\vec{\nabla}\cdot$ and the curl operator $\vec{\nabla}\times$,

$$\vec{\nabla} \times \vec{H} = \vec{J} + \dot{\vec{D}} \tag{D.42a}$$
$$\vec{\nabla} \times \vec{E} = -\dot{\vec{B}} \tag{D.42b}$$

Application of the curl operator ($\vec{\nabla}\times$) to (D.42a) with consideration of the material equation $\vec{D} = \epsilon\vec{E}$ results in

$$\vec{\nabla} \times (\vec{\nabla} \times \vec{H}) = \vec{\nabla} \times (\vec{J} + \dot{\vec{D}}) = \vec{\nabla} \times \vec{J} + \epsilon\vec{\nabla} \times \dot{\vec{E}} \tag{D.43}$$

With $\vec{\nabla}\times(\vec{\nabla}\times\vec{H}) = \vec{\nabla}(\vec{\nabla}\cdot\vec{H}) - \vec{\nabla}^2\vec{H}$ and $\vec{\nabla}\cdot\vec{H} = 0$, it follows that

$$-\vec{\nabla}^2\vec{H} = \vec{\nabla}\times\vec{J} + \epsilon\,\vec{\nabla}\times\dot{\vec{E}} \tag{D.44}$$

From (D.42b), after partial differentiation with respect to time and interchanging space and time derivatives under assumption of stationary (not moving) material, we obtain:

$$\vec{\nabla}\times\dot{\vec{E}} = -\mu\ddot{\vec{H}} \tag{D.45}$$

Inserting (D.45) into (D.44) and re-sorting leads to the vector differential equation,

$$\vec{\nabla}^2\vec{H} - \epsilon\mu\ddot{\vec{H}} = -\vec{\nabla}\times\vec{J} \tag{D.46}$$

Similarly, we gain a vector differential equation for the electrical field strength

$$\vec{\nabla}\times(\vec{\nabla}\times\vec{E}) = -\mu\vec{\nabla}\times\dot{\vec{H}} \tag{D.47}$$

With $\vec{\nabla}\times(\vec{\nabla}\times\vec{E}) = \vec{\nabla}(\vec{\nabla}\cdot\vec{E}) - \vec{\nabla}^2\vec{E}$ and $\vec{\nabla}\cdot\vec{E} = \rho/\epsilon$, it follows that

$$\vec{\nabla}^2\vec{E} = \frac{1}{\epsilon}\vec{\nabla}\rho + \mu\,\vec{\nabla}\times\dot{\vec{H}} \tag{D.48}$$

Partial differentiation of the first Maxwell equation with respect to time leads to

$$\vec{\nabla}\times\dot{\vec{H}} = \dot{\vec{J}} + \epsilon\ddot{\vec{E}} \tag{D.49}$$

and insertion into (D.48) results in

$$\vec{\nabla}^2\vec{E} - \epsilon\mu\ddot{\vec{E}} = \frac{1}{\epsilon}\vec{\nabla}\rho + \mu\dot{\vec{J}} \tag{D.50}$$

Assuming that the field region is free of charge and conductive current, two homogeneous vector differential equations for $\vec{E}$ and $\vec{H}$ follow from (D.46) and (D.50),

$$\vec{\nabla}^2\vec{E} - \epsilon\mu\ddot{\vec{E}} = \vec{0} \tag{D.51}$$

$$\vec{\nabla}^2\vec{H} - \epsilon\mu\ddot{\vec{H}} = \vec{0} \tag{D.52}$$

With the inhomogeneous wave equations (D.46) and (D.50), or the homogeneous wave equations (D.51) and (D.52) for $\vec{E}$ and $\vec{H}$, Maxwell equations are decoupled. It must be noted that the individual pairs of equations cannot be solved totally independent of each other, because the electromagnetic field strengths $\vec{E}$ and $\vec{H}$ always remain coupled via the Maxwell equations. By experience, the standard procedure is to determine one of the field quantities from the corresponding wave equation, and to obtain the (dual) field from the corresponding Maxwell equation in differential form. If the conductive current density $\vec{J}$ is decomposed in a forced current density $\vec{J}_0$ and a second part that is proportional to the electric field strength in conductive material, then the relations can be described in more detail. With

$$\vec{J} = \sigma\vec{E} + \vec{J}_0 \tag{D.53}$$

it follows from (D.42a) and (D.42b)

$$\vec{\nabla}\times\vec{\nabla}\times\vec{E} = -\mu\left(\dot{\vec{J}} + \ddot{\vec{D}}\right) = -\mu\left(\sigma\dot{\vec{E}} + \dot{\vec{J}}_0 + \epsilon\ddot{\vec{E}}\right)$$

$$= \vec{\nabla}\left(\vec{\nabla}\cdot\vec{E}\right) - \vec{\nabla}^2\vec{E} = \vec{\nabla}\left(\frac{\rho}{\epsilon}\right) - \vec{\nabla}^2\vec{E}$$

Thus,

$$\left(\vec{\nabla}^2 - \mu\sigma\frac{\partial}{\partial t} - \mu\epsilon\frac{\partial^2}{\partial t^2}\right)\vec{E} = \mu\frac{\partial}{\partial t}\dot{\vec{J}}_0 + \vec{\nabla}\left(\frac{\rho}{\epsilon}\right) \tag{D.54}$$

Similarly, we find the following inhomogeneous wave equation for $\vec{H}$,

$$\left(\vec{\nabla}^2 - \mu\sigma\frac{\partial}{\partial t} - \mu\epsilon\frac{\partial^2}{\partial t^2}\right)\vec{H} = -\vec{\nabla}\times\vec{J}_0 \tag{D.55}$$

For complex notation, the time derivative $\partial/\partial t$ is substituted by $j\omega$.

$$\left(\vec{\nabla}^2 + \underline{k}^2\right)\underline{\vec{E}}(\vec{r}) = j\omega\mu\,\underline{\vec{J}}_0 + \frac{1}{\epsilon}\vec{\nabla}\underline{\rho}(\vec{r}) \tag{D.56}$$

$$\left(\vec{\nabla}^2 + \underline{k}^2\right)\underline{\vec{H}}(\vec{r}) = -\vec{\nabla}\times\underline{\vec{J}}_0 \tag{D.57}$$

with the complex material wavenumber $\underline{k}$. From

$$\underline{k}^2 = \omega^2\epsilon\mu - j\omega\mu\sigma = \omega^2\underline{\epsilon}\mu \tag{D.58}$$

we find

$$\underline{k} = \underset{(-)}{+}\sqrt{\omega^2\underline{\epsilon}\mu} = \underset{(-)}{+}\sqrt{\omega^2\epsilon\mu - j\omega\mu\sigma} \tag{D.59}$$

We choose the positive sign. Thus, $\underline{k}$ is identified as being limited tc the second quadrant of the complex plane. The equations for the complex amplitudes (D.56) and (D.57) are the so-called inhomogeneous Helmholtz equations. Assuming that the field region is free of space charge [$\vec{\rho}(\vec{r}) = 0$], and assuming the nonexistence of any forced current density ($\vec{\underline{J}}_0 = \vec{0}$), the wave equations for lossy material read

$$\left(\vec{\nabla}^2 + \underline{k}^2\right) \left\{ \begin{matrix} \vec{\underline{H}}(\vec{r}) \\ \vec{\underline{E}}(\vec{r}) \end{matrix} \right\} = \vec{0} \tag{D.60}$$

If the material is a perfect isolator ($\sigma = 0$), then we find for the wavenumber $\underline{k}$

$$\underline{k} = k = \omega\sqrt{\mu\epsilon} = \frac{\omega}{c_0}\sqrt{\mu_r\epsilon_r} \tag{D.61}$$

and (D.60) turns out to be

$$\left(\vec{\nabla}^2 + k^2\right) \left\{ \begin{matrix} \vec{\underline{H}}(\vec{r}) \\ \vec{\underline{E}}(\vec{r}) \end{matrix} \right\} = \vec{0} \tag{D.62}$$

D.2.1 Wave Equation for the Longitudinal Field Components

In the following section, it will be shown that, in the case of lossless cylindrical waveguides, the solution of Maxwell equations can be reduced to the solution of two-dimensional partial differential equations for the longitudinal magnetic and electric field components E_z and H_z with given boundary conditions. Then, from known E_z and H_z, the transversal components of the electromagnetic field can be determined. To find the relations between the transversal and longitudinal field components, we follow the approach described in [5]. First, we decompose the $\vec{E}$ and $\vec{H}$ fields into a transversal and a longitudinal term,

$$\vec{E} = \vec{E}_t + \vec{E}_z = \vec{E}_t + E_z\vec{e}_z \tag{D.63}$$

$$\vec{H} = \vec{H}_t + \vec{H}_z = \vec{H}_t + H_z\vec{e}_z \tag{D.64}$$

Similarly, the Nabla operator can be decomposed into two parts. A first part describes the differentiation with respect to the coordinates of the transversal (u,v)-plane, and the other part describes the differentiation with respect to the longitudinal or axial coordinate z,

$$\vec{\nabla} = \vec{\nabla}_t + \vec{\nabla}_z = \vec{\nabla}_t + \frac{\partial}{\partial_z}\vec{e}_z \tag{D.65}$$

In the Cartesian coordinate system, the variables u and v correspond to the coordinates x and y, and in the polar coordinate system, they correspond to the variables r and φ. The vector term $\vec{\nabla} \times \vec{E}$ can be decomposed into transversal and longitudinal terms as follows:

$$\begin{aligned} \vec{\nabla} \times \vec{E} &= \left(\vec{\nabla}_t + \vec{\nabla}_z\right) \times \left(\vec{E}_t + \vec{E}_z\right) \\ &= \vec{\nabla}_t \times \vec{E}_t + \vec{\nabla}_t \times \vec{E}_z + \vec{\nabla}_z \times \vec{E}_t + \vec{\nabla}_z \times \vec{E}_z \end{aligned} \tag{D.66}$$

Thus, we have

$$(\vec{\nabla} \times \vec{E})_t = \vec{\nabla}_t \times \vec{E}_z + \vec{\nabla}_z \times \vec{E}_t \tag{D.67}$$

$$(\vec{\nabla} \times \vec{E})_z = \vec{\nabla}_t \times \vec{E}_t \tag{D.68}$$

From (D.42b), the transversal term of $\vec{\nabla} \times \vec{E}$ is obtained

$$(\vec{\nabla} \times \vec{E})_t = -j\omega\mu\vec{H}_t = \vec{\nabla}_z \times \vec{E}_t + \vec{\nabla}_t \times \vec{E}_z \tag{D.69}$$

or

$$\vec{\nabla}_z \times \vec{E}_t + \vec{\nabla}_t \times \vec{E}_z = -j\omega\mu\vec{H}_t \tag{D.70}$$

Similarly, one obtains from (D.42a):

$$(\vec{\nabla} \times \vec{H})_t = j\omega\varepsilon\vec{E}_t = \vec{\nabla}_z \times \vec{H}_t + \vec{\nabla}_t \times \vec{H}_z \tag{D.71}$$

or

$$\vec{\nabla}_z \times \vec{H}_t + \vec{\nabla}_t \times \vec{H}_z = j\omega\epsilon\vec{E}_t \tag{D.72}$$

Application of the longitudinal curl operator ($\vec{\nabla}_z \times$) to (D.72) results in

$$\vec{\nabla}_z \times (\vec{\nabla}_z \times \vec{H}_t) + \vec{\nabla}_z \times (\vec{\nabla}_t \times \vec{H}_z) = j\omega\varepsilon\vec{\nabla}_z \times \vec{E}_t$$

$$\vec{\nabla}_z \underbrace{(\vec{\nabla}_z \cdot \vec{H}_t)}_{=0} - \vec{H}_t(\vec{\nabla}_z \cdot \vec{\nabla}_z) + \vec{\nabla}_t(\vec{\nabla}_z \cdot \vec{H}_z) - \vec{H}_z \underbrace{(\vec{\nabla}_z \cdot \vec{\nabla}_t)}_{=0} = j\omega\varepsilon\,\vec{\nabla}_z \times \vec{E}_t \tag{D.73}$$

$$-\frac{\partial^2}{\partial z^2}\vec{H}_t + \vec{\nabla}_t(\vec{\nabla}_z \cdot \vec{H}_z) = j\omega\epsilon\vec{\nabla}_z \times \vec{E}_t \tag{D.74}$$

Next, we consider the wave properties in the direction of propagation. This will lead to a separation of the z dependency of the electromagnetic field quantities. After having found a solution for the z dependency, we will return to (D.74) again. The separation of the z dependency of the electromagnetic field on cylindrical lines can be easily achieved by the separation of variables:

$$\vec{E}(u,v,z) = \vec{g}(u,v) \cdot G(z) \tag{D.75}$$

$$\vec{H}(u,v,z) = \vec{f}(u,v) \cdot F(z) \tag{D.76}$$

with

$$\vec{g}(u,v) = \vec{g}_t(u,v) + \vec{e}_z g_z(u,v) \tag{D.77}$$

$$\vec{f}(u,v) = \vec{f}_t(u,v) + \vec{e}_z f_z(u,v) \tag{D.78}$$

The electrical vector function $\vec{g}(u,v)$ and the magnetic vector function $\vec{f}(u,v)$ depend only on the tranversal coordinates u and v. The scalar functions $G(z)$ and $F(z)$ are only dependent on the longitudinal coordinate z. From the homogeneous complex vectorial wave equation for the electric field $\vec{E}$,

$$(\vec{\nabla}^2 + k^2)\vec{E} = \vec{0} \tag{D.79}$$

it follows with (D.75), and

$$\vec{\nabla}^2 = \vec{\nabla}_t^2 + \vec{\nabla}_z^2 \tag{D.80}$$

$$\left(\vec{\nabla}_t^2 + \vec{\nabla}_z^2 + k^2\right) \vec{g}\, G = \vec{0} \tag{D.81}$$

Regarding the particular properties of the defined Nabla operators,

$$G(z)\ \vec{\nabla}_t^2\ \vec{g}(u,v) + \vec{g}(u,v)\ \vec{\nabla}_z^2\ G(z) + k^2\ \vec{g}(u,v)\ G(z) = \vec{0} \tag{D.82}$$

Division by $G(z)$, with $G(z) \neq 0$, results in

$$\vec{\nabla}_t^2\, \vec{g}(u,v) + \left(\frac{1}{G(z)} \vec{\nabla}_z^2 G(z) + k^2\right) \vec{g}(u,v) = \vec{0} \tag{D.83}$$

To ensure that the differential equation for $\vec{g}(u,v)$ can be fulfilled for any value of z, the bracket term in (D.83) must be a constant with respect to z. We therefore choose

$$\frac{1}{G(z)}\ \vec{\nabla}_z^2\ G(z) = -k_z^2 \tag{D.84}$$

and multiplied by $G(z)$,

$$\vec{\nabla}_z^2\, G(z) + k_z^2\, G(z) = 0 \tag{D.85}$$

The quantity k_z is formally denoted as a separation constant with respect to the z direction. k_z is also related to the guide wavelength λ_g,

$$\lambda_g = \frac{2\pi}{k_z} \tag{D.86}$$

Therefore, k_z is also termed a longitudinal wavenumber. Then (D.83) reads

$$\vec{\nabla}_t^2\, \vec{g}(u,v) + \left(k^2 - k_z^2\right) \vec{g}(u,v) = 0 \tag{D.87}$$

With the transversal wavenumber k_t,

$$k_t^2 = k^2 - k_z^2 \tag{D.88}$$

we obtain

$$\vec{\nabla}_t^2\, \vec{g}(u,v) + k_t^2\, \vec{g}(u,v) = 0 \tag{D.89}$$

From (D.88), the transversal wavenumber cannot be uniquely determined,

$$k_t = \overset{+}{(-)} \sqrt{k^2 - k_z^2} \tag{D.90}$$

The sign of k_t can be chosen arbitrarily without a loss in generality. The solution of the differential equation (D.85) is

$$G(z) = C^+ e^{-jk_z z} + C^- e^{+jk_z z} \tag{D.91}$$

Therefore, the electric field $\vec{E}(u,v,z)$ can be written as

$$\vec{E}(u,v,z) = \vec{g}(u,v)\left(C^{+}e^{-jk_z z} + C^{-}e^{+jk_z z}\right) = \vec{g}^{+}(u,v)e^{-jk_z z} + \vec{g}^{-}(u,v)e^{+jk_z z} \quad \text{(D.92)}$$

The first term in (D.92) represents an electrical vector field, dependent on the transversal variables u and v, which propagates in positive z direction, and the second term describes an electrical vector field that propagates in negative z direction. For the longitudinal $\vec{\nabla}$ operator, it follows for the positive partial wave with (D.91) that

$$\vec{\nabla}_z = \vec{e}_z \frac{\partial}{\partial z} = -jk_z \vec{e}_z \quad \text{(D.93)}$$

We return now to (D.74), and with (D.93), it follows

$$\begin{aligned} k_z^2 \vec{H}_t - jk_z \vec{\nabla}_t H_z &= j\omega\epsilon \vec{\nabla}_z \times \vec{E}_t = j\omega\epsilon(-j\omega\mu\vec{H}_t - \vec{\nabla}_t \times \vec{E}_z) \\ &= \omega^2\epsilon\mu\vec{H}_t + j\omega\epsilon(\vec{e}_z \times \vec{\nabla}_t E_z) = k^2\vec{H}_t + j\omega\epsilon(\vec{e}_z \times \vec{\nabla}_t E_z) \end{aligned} \quad \text{(D.94)}$$

Similarly, we can derive the following relationship from (D.70):

$$k_z^2 \vec{E}_t - jk_z \vec{\nabla}_t E_z = k^2 \vec{E}_t - j\omega\mu(\vec{e}_z \times \vec{\nabla}_t H_z) \quad \text{(D.95)}$$

Sorting of (D.94) and (D.95) results in

$$(k^2 - k_z^2)\vec{H}_t = -j\omega\epsilon(\vec{e}_z \times \vec{\nabla}_t E_z) - jk_z \vec{\nabla}_t H_z \quad \text{(D.96)}$$

$$(k^2 - k_z^2)\vec{E}_t = j\omega\mu(\vec{e}_z \times \vec{\nabla}_t H_z) - jk_z \vec{\nabla}_t E_z \quad \text{(D.97)}$$

With $k^2 = k_t^2 + k_z^2$ (D.88), we find

$$k_t^2 \vec{H}_t = -j\omega\epsilon(\vec{e}_z \times \vec{\nabla}_t E_z) - jk_z \vec{\nabla}_t H_z \quad \text{(D.98)}$$

$$k_t^2 \vec{E}_t = j\omega\mu(\vec{e}_z \times \vec{\nabla}_t H_z) - jk_z \vec{\nabla}_t E_z \quad \text{(D.99)}$$

In matrix notation,

$$\begin{pmatrix} \vec{E}_t \\ \vec{H}_t \end{pmatrix} = \frac{1}{k_t^2} \begin{pmatrix} -jk_z & j\omega\mu\vec{e}_z \times \\ -j\omega\epsilon\vec{e}_z \times & -jk_z \end{pmatrix} \vec{\nabla}_t \begin{pmatrix} E_z \\ H_z \end{pmatrix} \quad \text{(D.100)}$$

Thus, the determination of the transversal components of the electric and magnetic fields can be calculated from the longitudinal field components. Therefore, we search for a differential equation for E_z and H_z. These are the so-called two-dimensional Helmholtz equations for E_z and H_z. We start with (D.100). With

$$V_t = \begin{pmatrix} \vec{E}_t \\ \vec{H}_t \end{pmatrix} \quad \text{(D.101)}$$

$$V_z = \begin{pmatrix} E_z \\ H_z \end{pmatrix} \quad \text{(D.102)}$$

$$M_1 = \begin{pmatrix} -jk_z & j\omega\mu\vec{e}_z\times \\ -j\omega\varepsilon\vec{e}_z\times & -jk_z \end{pmatrix} \tag{D.103}$$

(D.100) reads

$$V_t = \frac{1}{k_t^2} M_1 \vec{\nabla}_t V_z \tag{D.104}$$

Analogous to (D.69) and (D.71), we obtain from Maxwell equations for the longitudinal components of the magnetic and electric fields

$$(\vec{\nabla} \times \vec{H})_z = \vec{\nabla}_t \times \vec{H}_t = j\omega\epsilon\vec{E}_z \tag{D.105}$$

$$(\vec{\nabla} \times \vec{E})_z = \vec{\nabla}_t \times \vec{E}_t = -j\omega\mu\vec{H}_z \tag{D.106}$$

Thus,

$$E_z = -\frac{j\omega\mu}{k^2}\vec{e}_z\cdot(\vec{\nabla}_t \times \vec{H}_t) \tag{D.107}$$

$$H_z = \frac{j\omega\epsilon}{k^2}\vec{e}_z\cdot(\vec{\nabla}_t \times \vec{E}_t) \tag{D.108}$$

In matrix notation,

$$\begin{pmatrix} \vec{E}_z \\ \vec{H}_z \end{pmatrix} = \frac{1}{k^2}\begin{pmatrix} 0 & -j\omega\mu \\ j\omega\epsilon & 0 \end{pmatrix} \vec{e}_z \cdot \left[\vec{\nabla}_t \times \begin{pmatrix} E_z \\ H_z \end{pmatrix}\right] \tag{D.109}$$

With

$$M_2 = \begin{pmatrix} 0 & -j\omega\mu \\ j\omega\epsilon & 0 \end{pmatrix} \tag{D.110}$$

we write in abbreviated form

$$V_z = \frac{1}{k^2} M_2\, \vec{e}_z \cdot (\vec{\nabla}_t \times V_t) \tag{D.111}$$

Decomposing M_1 in (D.103), and substituting coefficient matrix M_2 (D.110) and a two-row unity matrix, leads to

$$M_1 = \begin{pmatrix} -jk_z & j\omega\mu\vec{e}_z\times \\ -j\omega\epsilon\vec{e}_z\times & -jk_z \end{pmatrix} = \begin{pmatrix} 0 & j\omega\mu \\ -j\omega\varepsilon & 0 \end{pmatrix}\vec{e}_z \times + \begin{pmatrix} -jk_z & 0 \\ 0 & -jk_z \end{pmatrix} = -M_2\vec{e}_z \times -jk_z E \tag{D.112}$$

M_1 substituted in (D.104) yields

$$V_t = -\frac{1}{k_t^2}(jk_z E + M_2\vec{e}_z\times)\vec{\nabla}_t V_z = -j\frac{k_z}{k_t^2} E\,\vec{\nabla}_t V_z - \frac{1}{k_t^2} M_2\vec{e}_z \times \vec{\nabla}_t V_z \tag{D.113}$$

Therefore, (D.111) reads

$$
\begin{aligned}
V_z &= \frac{1}{k^2} M_2 \vec{e}_z \cdot \left[\vec{\nabla}_t \times \left(-j \frac{k_z}{k_t^2} E \vec{\nabla}_t V_z - \frac{1}{k_t^2} M_2 \vec{e}_z \times \vec{\nabla}_t V_z \right) \right] \\
&= -\frac{1}{k^2 k_t^2} M_2^2 \vec{e}_z \cdot \left[\vec{\nabla}_t \times \left(\vec{e}_z \times \vec{\nabla}_t V_z \right) \right]
\end{aligned} \tag{D.114}
$$

With

$$
\vec{\nabla}_t \times (\vec{e}_z \times \vec{\nabla}_t V_z) = \vec{e}_z (\vec{\nabla}_t \cdot \vec{\nabla}_t V_z) - \vec{\nabla}_t V_z \underbrace{(\vec{\nabla}_t \cdot \vec{e}_z)}_{=0}
$$

it is

$$
V_z = -\frac{1}{k^2 k_t^2} M_2^2 \vec{e}_z \cdot \left[\vec{e}_z \cdot \left(\vec{\nabla}_t \cdot \vec{\nabla}_t V_z \right) \right]
$$

or

$$
V_z = -\frac{1}{k^2 k_t^2} M_2^2 \vec{\nabla}_t^{\,2} V_z \tag{D.115}
$$

Calculating M_2^2,

$$
M_2^2 = \begin{pmatrix} 0 & -j\omega\mu \\ j\omega\varepsilon & 0 \end{pmatrix} \begin{pmatrix} 0 & -j\omega\mu \\ j\omega\varepsilon & 0 \end{pmatrix} = \begin{pmatrix} \omega^2\varepsilon\mu & 0 \\ 0 & \omega^2\varepsilon\mu \end{pmatrix} = k^2 E \tag{D.116}
$$

and inserting into (D.115) results in

$$
\vec{\nabla}_t^{\,2} V_z + k_t^2 V_z = 0 \tag{D.117}
$$

With V_z from (D.102), we can determine the longitudinal components of the electric and magnetic fields using the two-dimensional partial scalar differential equations as follows:

$$
\vec{\nabla}_t^{\,2} E_z + k_t^2 E_z = 0 \tag{D.118}
$$

$$
\vec{\nabla}_t^{\,2} H_z + k_t^2 H_z = 0 \tag{D.119}
$$

With

$$
\begin{aligned}
\vec{\nabla}_t^{\,2} &= \vec{\nabla}^2 - \vec{\nabla}_z^{\,2} \\
k_t^2 &= k^2 - k_z^2 = k^2 + \vec{\nabla}_z^{\,2}
\end{aligned}
$$

(D.118) and (D.119) can be reformulated:

$$
(\vec{\nabla}^2 - k_z^2) E_z + (k^2 + \vec{\nabla}_z^2) E_z = (\vec{\nabla}^2 + k^2) E_z = 0
$$

$$
\vec{\nabla}^2 E_z + k^2 E_z = 0 \tag{D.120}
$$

$$
\vec{\nabla}^2 H_z + k^2 H_z = 0 \tag{D.121}
$$

The boundary conditions for E_z and H_z are:

$$
\left.
\begin{aligned}
E_z &= 0 \\
\frac{\partial H_z}{\partial n} &= 0
\end{aligned}
\right\} \quad \text{on the metallic boundary} \tag{D.122}
$$

E Modes ($H_z = 0$):
The transversal electric field $\vec{E}_t$ reads (D.100):

$$\vec{E}_t = \frac{1}{k_t^2}\left[j\omega\mu\left(\vec{e}_z \times \vec{\nabla}_t H_z\right) - jk_z\vec{\nabla}_t E_z\right] = -j\frac{k_z}{k_t^2}\vec{\nabla}_t E_z = -j\frac{k_z}{k_t^2}\left(\frac{\partial}{\partial x}\vec{e}_x + \frac{\partial}{\partial y}\vec{e}_y\right)E_z \tag{D.123}$$

and the transversal magnetic field:

$$\vec{H}_t = \frac{1}{k_t^2}\left[-j\omega\varepsilon\left(\vec{e}_z \times \vec{\nabla}_t E_z\right) - jk_z\vec{\nabla}_t H_z\right] = -j\frac{\omega\varepsilon}{k_t^2}(\vec{e}_z \times \vec{\nabla}_t E_z) \tag{D.124}$$

With

$$\vec{\nabla}_t \vec{E}_z = \left(\frac{\partial}{\partial x}\vec{e}_x + \frac{\partial}{\partial y}\vec{e}_y\right)E_z$$

it follows from (D.124)

$$\vec{H}_t = -j\frac{\omega\varepsilon}{k_t^2}\left[\vec{e}_z \times \left(\frac{\partial}{\partial x}\vec{e}_x + \frac{\partial}{\partial y}\vec{e}_y\right)E_z\right] = -j\frac{\omega\varepsilon}{k_t^2}\left(\frac{\partial}{\partial x}\vec{e}_y - \frac{\partial}{\partial y}\vec{e}_x\right)E_z \tag{D.125}$$

Thus, the transversal field components of *E* (or *TM*) modes are given as

$$E_x = -j\frac{k_z}{k_t^2}\frac{\partial E_z}{\partial x} \qquad H_x = j\frac{\omega\epsilon}{k_t^2}\frac{\partial E_z}{\partial y} = -\frac{1}{Z_F}E_y$$

$$E_y = -j\frac{k_z}{k_t^2}\frac{\partial E_z}{\partial y} \qquad H_y = -j\frac{\omega\epsilon}{k_t^2}\frac{\partial E_z}{\partial x} = \frac{1}{Z_F}E_x \tag{D.126}$$

with the wave impedance of *E* (or *TM*) modes

$$Z_F = \frac{k_z}{\omega\epsilon} \tag{D.127}$$

The transversal magnetic field components also can be directly derived using the relationship

$$\vec{H}_t = \frac{1}{Z_F}\left(\vec{e}_z \times \vec{E}_t\right) \tag{D.128}$$

Introducing the transversal Cartesian components,

$$\vec{H}_t = \frac{1}{Z_F}\left[\vec{e}_z \times (E_x\vec{e}_x + E_y\vec{e}_y)\right] = \frac{1}{Z_F}(E_x\vec{e}_y - E_y\vec{e}_x)$$

again delivers $H_x = -E_y/Z_F$ and $H_y = E_x/Z_F$.

H Modes: $H_z \neq 0,\ E_z = 0$
From (D.100) with $E_z = 0$, we find for the transversal electric field:

$$\vec{E}_t = \frac{1}{k_t^2} j\omega\mu \left(\vec{e}_z \times \vec{\nabla}_t H_z \right) = j \frac{\omega\mu}{k_t^2} \left(\frac{\partial}{\partial x}\vec{e}_y - \frac{\partial}{\partial y}\vec{e}_x \right) H_z \qquad \text{(D.129)}$$

and for the transversal magnetic field:

$$\vec{H}_t = -j \frac{k_z}{k_t^2} \vec{\nabla}_t H_z = -j \frac{k_z}{k_t^2} \left(\frac{\partial}{\partial x}\vec{e}_x + \frac{\partial}{\partial y}\vec{e}_y \right) H_z \qquad \text{(D.130)}$$

Thus, the transversal field components of *H* (or *TE*) modes are given as

$$E_x = -j \frac{\omega\mu}{k_t^2} \frac{\partial H_z}{\partial y} \qquad H_x = -j \frac{k_z}{k_t^2} \frac{\partial H_z}{\partial x} = -\frac{1}{Z_F} E_y$$

$$E_y = j \frac{\omega\mu}{k_t^2} \frac{\partial H_z}{\partial x} \qquad H_y = -j \frac{k_z}{k_t^2} \frac{\partial H_z}{\partial y} = \frac{1}{Z_F} E_x \qquad \text{(D.131)}$$

with the wave impedance of *H* (or *TE*) modes

$$Z_F = \frac{\omega\mu}{k_z} \qquad \text{(D.132)}$$

D.2.2 Electric and Magnetic Vector Potentials

In Section D.2.1, the transversal electric and magnetic field components were derived from the longitudinal electric field in the case of an *E* (*TM*) mode, and from the longitudinal magnetic field in the case of an *H* (*TE*) mode. Now we will follow a frequently used alternative procedure, namely, to derive the field vectors $\vec{H}$ and $\vec{E}$ from an electric and magnetic vector potential $\vec{A}^e$ and $\vec{A}^m$, respectively. We suppose that any source exciting the electromagnetic field lies outside the field region under analysis. In this case, the vector potentials only can be considered as auxiliary quantities, and their introduction essentially reflects a different formulation of Maxwell equations leading to wave equations for $\vec{A}^e$ and $\vec{A}^m$, in analogy to the wave equations for $\vec{E}$ and $\vec{H}$.

We assume a forced electric current density and charge distribution, which excite the electric and magnetic fields $\vec{E}^e(\vec{r},t)$ and $\vec{H}^e(\vec{r},t)$, and in dual sense, forced magnetic current density and magnetic charges, which stimulate the electric and magnetic fields $\vec{E}^m(\vec{r},t)$ and $\vec{H}^m(\vec{r},t)$. Then, a total electromagnetic field can be described as a superposition of both these partial fields:

$$\vec{E}(\vec{r},t) = \vec{E}^e(\vec{r},t) + \vec{E}^m(\vec{r},t) \qquad \text{(D.133)}$$

$$\vec{H}(\vec{r},t) = \vec{H}^e(\vec{r},t) + \vec{H}^m(\vec{r},t) \qquad \text{(D.134)}$$

The superscript *e* denotes excitation by electric sources, and the superscript *m* denotes excitation by magnetic sources.

Partial Problem I
Assuming electrical current density $\vec{J}^e$ and electrical charge density ρ as the electric sources for the stimulation of $\vec{E}^e$ and $\vec{H}^e$, the Maxwell equations (see Section D.1) read in the frequency domain

$$rot\vec{H}^e - j\omega\vec{D}^e = \vec{J}^e \tag{D.135}$$

$$rot\vec{E}^e + j\omega\vec{B}^e = \vec{0} \tag{D.136}$$

$$div\vec{B}^e = \vec{0} \tag{D.137}$$

$$div\vec{D}^e = \varrho \tag{D.138}$$

Partial Problem II
Similarly, assuming magnetic current density $\vec{J}^m$ and magnetic charge density ρ^m as the magnetic sources for the stimulation of $\vec{E}^m$ and $\vec{H}^m$, the Maxwell equations can be written as

$$rot\vec{H}^m - j\omega\vec{D}^m = \vec{0} \tag{D.139}$$

$$rot\vec{E}^m + j\omega\vec{B}^m = -\vec{J}^m \tag{D.140}$$

$$div\vec{B}^m = \varrho^m \tag{D.141}$$

$$div\vec{D}^m = \vec{0} \tag{D.142}$$

As is shown in Sections D.2.2.1 and D.2.2.2, the partial field $\vec{E}^e$ and $\vec{H}^e$ can be derived from an electrical vector potential $\vec{A}^e$, and the partial field $\vec{E}^m$ and $\vec{H}^m$ can be derived from a magnetic vector potential $\vec{A}^m$. As a final result, the total electric and magnetic fields $\vec{E}$ and $\vec{H}$ are calculated from the vector potentials $\vec{A}^e$ and $\vec{A}^m$, satisfying the following relations:

$$\vec{E} = \vec{E}^e + \vec{E}^m = \frac{1}{j\omega\epsilon} rot\, rot\vec{A}^e - rot\vec{A}^m \tag{D.143}$$

$$\vec{H} = \vec{H}^e + \vec{H}^m = rot\vec{A}^e + \frac{1}{j\omega\mu} rot\, rot\vec{A}^m \tag{D.144}$$

D.2.2.1 Calculation of the Electric and Magnetic Field

Assumption: The medium is linear, homogeneous, isotropic, and lossless, and the field region is source-free.

Partial Problem I: $\vec{A}^e \neq \vec{0}$, $\vec{A}^m = \vec{0}$
With $div\, rot \equiv 0$, the condition $div\vec{B}^e = 0$ (D.137) can be fulfilled by defining

$$\vec{H}^e(\vec{r}) = rot\, \vec{A}^e(\vec{r}) \tag{D.145}$$

With (D.140), it follows that

$$rot\, \vec{E}^e = -j\omega\mu\, \vec{H}^e = -j\omega\mu\, rot\, \vec{A}^e \tag{D.146}$$

or

$$rot\left(\vec{E}^e + j\omega\mu\, \vec{A}^e \right) = \vec{0} \tag{D.147}$$

With $rot\,grad \equiv \vec{0}$, a scalar potential ϕ^e can be defined, which reads

$$-grad\ \phi^e = \vec{E}^e + j\omega\mu\ \vec{A}^e \tag{D.148}$$

Inserting this into (D.135) results (with $\vec{J}^e = 0$) in

$$\vec{E}^e = \frac{1}{j\omega\epsilon}\,rot\,\vec{H}^e = \frac{1}{j\omega\epsilon}\,rot\,rot\vec{A}^e = \frac{1}{j\omega\epsilon}\left(grad\,div\vec{A}^e - \vec{\nabla}^2\vec{A}^e\right) = -grad\phi^e - j\omega\mu\vec{A}^e \tag{D.149}$$

where ϕ^e denotes a scalar potential. After sorting, we obtain

$$\vec{\nabla}^2\ \vec{A}^e + k^2\ \vec{A}^e = grad\ \left(\ j\omega\epsilon\ \phi^e + div\ \vec{A}^e\ \right) \tag{D.150}$$

With the Lorentz standardization,

$$div\ \vec{A}^e = -j\omega\epsilon\ \phi^e \tag{D.151}$$

the electric vector potential $\vec{A}^e$ is uniquely determined by specifying the divergence of this vector potential. With (D.151), (D.150) reads

$$\vec{\nabla}^2\vec{A}^e + k^2\vec{A}^e = \vec{0} \tag{D.152}$$

When $\vec{A}^e$ is known from (D.152), the field components $\vec{E}^e$ and $\vec{H}^e$ can be derived from (D.145) and (D.149):

$$\vec{E}^e = \frac{1}{j\omega\epsilon}\ rot\ rot\ \vec{A}^e \tag{D.153}$$

$$\vec{H}^e = rot\ \vec{A}^e \tag{D.154}$$

Partial Problem II: $\vec{A}^m \neq \vec{0},\ \vec{A}^e = \vec{0}$:
In analogy to Partial Problem I, the electric field $\vec{E}^m$ can be derived from a magnetic vector potential $\vec{A}^m$,

$$\vec{E}^m(\vec{r}) = -\,rot\ \vec{A}^m(\vec{r}) \tag{D.155}$$

Substituted in (D.139), we obtain

$$rot\ \left(\ \vec{H}^m + j\omega\epsilon\ \vec{A}^m\ \right) = 0 \tag{D.156}$$

The term in parentheses can be described by a gradient field:

$$-\ grad\ \phi^m = \vec{H}^m + j\omega\epsilon\ \vec{A}^m \tag{D.157}$$

From (D.140), with (D.155), (D.157), and (D.149), we obtain:

$$\vec{\nabla}^2\vec{A}^m + k^2\vec{A}^m = grad\ \left(\ j\omega\mu\ \phi^m + div\ \vec{A}^m\ \right) \tag{D.158}$$

Recalling the Lorentz standardization,

$$div\ \vec{A}^m = -j\omega\mu\ \phi^m \tag{D.159}$$

again a decoupling of ϕ^m and $\vec{A}^m$ is achieved in analogy to Partial Problem I:

$$\vec{\nabla}^2\vec{A}^m + k^2\vec{A}^m = \vec{0} \tag{D.160}$$

Then the fields $\vec{E}^m$ and $\vec{H}^m$ are obtained from the vector potential $\vec{A}^m$ as follows [(D.155), (D.156)]:

$$\vec{E}^m = -rot\ \vec{A}^m \tag{D.161}$$

$$\vec{H}^m = \frac{1}{j\omega\epsilon}\, rot\ rot\ \vec{A}^m \tag{D.162}$$

D.2.2.2 Derivation of the Electromagnetic Field Components

In Section D.2.2.1, the electric and magnetic fields $\vec{E}$ and $\vec{H}$ have been derived from the vector potentials $\vec{A}^e$ and $\vec{A}^m$. We want to investigate special solutions now, for which the vector potentials have only a single component. We first suppose that this coincides with the propagation of the mode, which should be the z direction. Then, we will see that the solution of Maxwell equations will lead to E and H modes on standard waveguides. The TEM mode on transmission lines is included. We discuss the following cases: (1) $\vec{A}^e = \vec{e}_z\Psi^e$ and $\vec{A}^m = \vec{0}$, and (2) $\vec{A}^e = \vec{0}$ and $\vec{A}^m = \vec{e}_z\Psi^m$.

(1) $\vec{A}^e = \vec{e}_z\Psi^e$, $\vec{A}^m = \vec{0}$:
For the scalar potential Ψ^e, it follows directly from (D.152):

$$\Delta\psi^e + k^2\psi^e = 0 \tag{D.163}$$

The electrical and magnetic fields $\vec{E}^e$ and $\vec{H}^e$ can be calculated from (D.153) and (D.154). With $\vec{A}^e = \vec{e}_z\Psi^e$, it follows for $rot\ \vec{A}^e$ and $rot\ rot\ \vec{A}^e$ in Cartesian coordinate system:

$$rot\vec{A}^e = rot\ \vec{e}_z\,\psi^e = \begin{vmatrix} \vec{e}_x & \vec{e}_y & \vec{e}_z \\ \frac{\partial}{\partial x} & \frac{\partial}{\partial y} & \frac{\partial}{\partial z} \\ 0 & 0 & \psi^e \end{vmatrix} = \vec{e}_x\frac{\partial\psi^e}{\partial y} - \vec{e}_y\frac{\partial\psi^e}{\partial x} \tag{D.164}$$

$$rot\,rot\vec{A}^e = rot\left(\vec{e}_x\frac{\partial\psi^e}{\partial y} - \vec{e}_y\frac{\partial\psi^e}{\partial x}\right) = \begin{vmatrix} \vec{e}_x & \vec{e}_y & \vec{e}_z \\ \frac{\partial}{\partial x} & \frac{\partial}{\partial y} & \frac{\partial}{\partial z} \\ \frac{\partial\psi^e}{\partial y} & -\frac{\partial\psi^e}{\partial x} & 0 \end{vmatrix} \tag{D.165}$$

$$= \vec{e}_x\frac{\partial^2\psi^e}{\partial x\,\partial z} + \vec{e}_y\frac{\partial^2\psi^e}{\partial y\,\partial z} + \vec{e}_z\left(-\frac{\partial^2\psi^e}{\partial x^2} - \frac{\partial^2\psi^e}{\partial y^2}\right)$$

Recalling (D.163) and $\Delta = \partial^2/\partial x^2 + \partial^2/\partial y^2 + \partial^2/\partial z^2$, the last term in the parentheses can be written as $\partial^2 \psi^e/\partial z^2 + k^2 \psi^e$. Thus, the field components of $\vec{E}$ and $\vec{H}$ are calculated from (D.153) and (D.154) to be

$$E_x^e = \frac{1}{j\omega\epsilon}\frac{\partial^2 \psi^e}{\partial x\,\partial z} \qquad H_x^e = \frac{\partial \psi^e}{\partial y}$$

$$E_y^e = \frac{1}{j\omega\epsilon}\frac{\partial^2 \psi^e}{\partial y\,\partial z} \qquad H_y^e = -\frac{\partial \psi^e}{\partial x} \tag{D.166}$$

$$E_z^e = \frac{1}{j\omega\epsilon}\left(\frac{\partial^2}{\partial z^2} + k^2\right)\psi^e \qquad H_z^e = 0$$

The set of equations (D.166) describes an E mode ($E_z \neq 0$, $H_z = 0$) with respect to the z direction (TM_z). With the propagation term $e^{-jk_z z}$, the derivation to z delivers $\partial/\partial z = -jk_z$. Then (D.166) reads

$$E_x^e = \frac{k_z}{\omega\epsilon} H_y^e \qquad H_x^e = \frac{\partial \psi^e}{\partial y}$$

$$E_y^e = -\frac{k_z}{\omega\epsilon} H_x^e \qquad H_y^e = -\frac{\partial \psi^e}{\partial x} \tag{D.167}$$

$$E_z^e = \frac{1}{j\omega\epsilon}\left(k^2 - k_z^2\right)\psi^e \qquad H_z^e = 0$$

With the wave impedance of E modes (D.127), we obtain

$$E_x^e = Z_{FE} H_y^e \qquad H_x^e = \frac{\partial \psi^e}{\partial y}$$

$$E_y^e = -Z_{FE} H_x^e \qquad H_y^e = -\frac{\partial \psi^e}{\partial x} \tag{D.168}$$

$$E_z^e = \frac{k_t^2}{j\omega\varepsilon}\psi^e \qquad H_z^e = 0$$

where $Z_F = k_z/\omega\epsilon$.

(2) $\vec{A}^m = \vec{e}_z \Psi^m$, $\vec{A}^e = 0$:

With $\vec{A}^e = 0$, the electrical and magnetic field is calculated from (D.161) and (D.162), using (D.166) and (D.167):

$$E_x^m = -\frac{\partial \Psi^m}{\partial y} \qquad H_x^m = \frac{1}{j\omega\mu}\frac{\partial^2 \Psi^m}{\partial x\,\partial z}$$

$$E_y^m = \frac{\partial \Psi^m}{\partial x} \qquad H_y^m = \frac{1}{j\omega\mu}\frac{\partial^2 \Psi^m}{\partial y\,\partial z} \tag{D.169}$$

$$E_z^m = 0 \qquad H_z^m = \frac{1}{j\omega\mu}\left(\frac{\partial^2}{\partial z^2} + k^2\right)\Psi^m$$

The set of equations (D.169) describes an H mode ($H_z \neq 0$, $E_z = 0$) with respect to the z direction (TE_z). With the propagation term $e^{-jk_z z}$, the derivation to z delivers $\partial/\partial z = -jk_z$. Then (D.169) reads

$$
\begin{aligned}
E_x^m &= -\frac{\partial \Psi^m}{\partial y} & H_x^m &= -\frac{k_z}{\omega\mu} E_y^m \\
E_y^m &= \frac{\partial \Psi^m}{\partial x} & H_y^m &= \frac{k_z}{\omega\mu} E_x^m \\
E_z^m &= 0 & H_z^m &= \frac{1}{j\omega\mu}\left(k^2 - k_z^2\right)\Psi^m
\end{aligned}
\tag{D.170}
$$

With the wave impedance of H modes (D.132), we obtain

$$
\begin{aligned}
E_x^m &= -\frac{\partial \Psi^m}{\partial y} & H_x^m &= -\frac{1}{Z_F} E_y^m \\
E_y^m &= \frac{\partial \Psi^m}{\partial x} & H_y^m &= \frac{1}{Z_F} E_x^m \\
E_z^m &= 0 & H_z^m &= \frac{k_t^2}{j\omega\mu}\Psi^m
\end{aligned}
\tag{D.171}
$$

where $Z_F = \omega\mu/k_z$.

D.3 Longitudinal Section Modes

In Section D.2.2.2, we have chosen the vector potential $\vec{A}^e$ and $\vec{A}^m$ in the direction of mode propagation, which was defined as the z direction. The Maxwell equations deliver $\boldsymbol{E}$ (or $\boldsymbol{TM}$) and $\boldsymbol{H}$ (or $\boldsymbol{TE}$) modes, respectively, as solutions. The investigated waveguiding structures were supposed to be homogeneously filled with dielectric material over the cross-section. In waveguiding structures with inhomogeneous distributions of dielectric material, hybrid modes commonly exist as solutions of Maxwell equations with both longitudinal electric and magnetic field components. To fulfill the boundary conditions at the dielectric interface, both $(E)_z$ [or $(TM)_z$] and $(H)_z$ [or $(TE)_z$] modes are commonly needed. Thus, the discussed field problem can be solved in a classical way, as follows. For both field regions, the wave equations for the scalar potential functions Ψ^e and Ψ^m of the respective vector potentials in z direction are valid. Then the potential functions are defined, which fulfill the boundary conditions $E_z = 0$ ($\Psi^e = 0$) and $\partial H_z/\partial n = 0$ ($\partial\Psi^m/\partial n = 0$) on the infinitely conductive metal walls. Satisfying the continuity of the tangential electric and magnetic fields at the dielectric interfaces, an eigenvalue equation for the propagation coefficient k_z can be obtained. Often, it is more convenient to treat hybrid mode analysis with vector potentials perpendicular to the mode propagation direction; that is, normal to the dielectric interface. Such mode solutions are then termed longitudinal section modes.

D.3.1 Vector Potential in x Direction

First we derive the electric and magnetic fields $\vec{E}^e$ and $\vec{H}^e$ from an electric vector potential $\vec{A}^e = \Psi^e \vec{e}_x$ in x direction (Figure D.7). The derivation is similar to that described in Section D.2.2.2. Therefore, we confine ourselves to the final result.

$(E\ Mode)_x$: $\vec{A}^e = \Psi^e \vec{e}_x$

$$E_x^e = \frac{1}{j\omega\varepsilon}\left(\frac{\partial^2}{\partial x^2} + k^2\right)\psi^e \qquad H_x^e = 0$$

$$E_y^e = \frac{1}{j\omega\varepsilon}\frac{\partial^2 \Psi^e}{\partial x \partial y} \qquad H_y^e = \frac{\partial \psi^e}{\partial z} \qquad \text{(D.172)}$$

$$E_z^e = \frac{1}{j\omega\varepsilon}\frac{\partial^2 \psi^e}{\partial x \partial z} \qquad H_z^e = -\frac{\partial \psi^e}{\partial y}$$

The $(E)_x$ mode is also termed the longitudinal section electric mode [LSE_x, $(TM)_x$] with respect to the x direction ($E_x \neq 0$, $H_x = 0$).

$(H\ Mode)_x$: $\vec{A}^m = \Psi^m \vec{e}_x$

$$E_x^m = 0 \qquad H_x^m = \frac{1}{j\omega\mu}\left(\frac{\partial^2}{\partial x^2} + k^2\right)\psi^m$$

$$E_y^m = -\frac{\partial \psi^m}{\partial z} \qquad H_y^m = \frac{1}{j\omega\mu}\frac{\partial^2 \Psi^m}{\partial x \partial y} \qquad \text{(D.173)}$$

$$E_z^m = \frac{\partial \psi^m}{\partial y} \qquad H_z^m = \frac{1}{j\omega\mu}\frac{\partial^2 \psi^m}{\partial x \partial z}$$

The $(H)_x$ mode is also termed the longitudinal section magnetic mode [LSH_x, $(TE)_x$] with respect to the x direction ($H_x \neq 0$, $E_x = 0$).

D.3.2 Vector Potential in y Direction

For completion we also summarize the electromagnetic field components of longitudinal section modes with respect to the y direction:

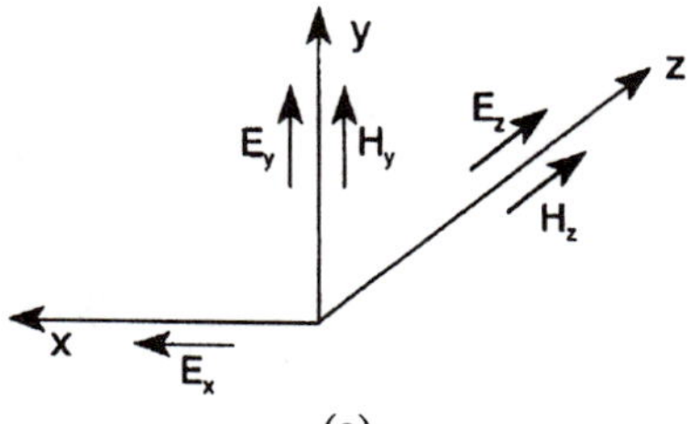

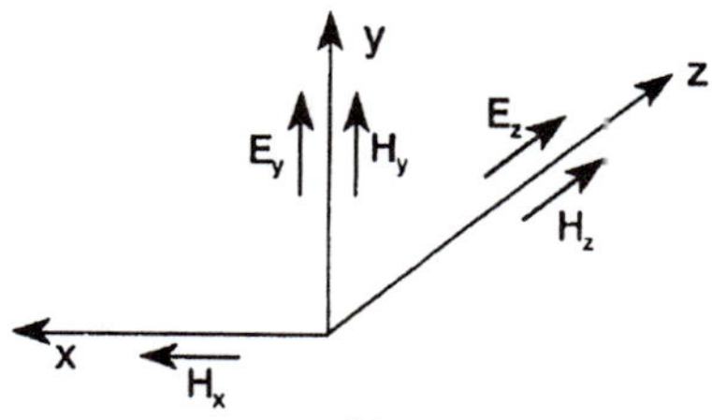

Figure D.7 Field components of (a) LSE_x and (b) LSH_x mode.

$(E\ Mode)_y$: $\vec{A}^e = \Psi^e \vec{e}_y$

$$E_x^e = \frac{1}{j\omega\epsilon}\frac{\partial^2\psi^e}{\partial x\partial y} \qquad H_x^e = -\frac{\partial\psi^e}{\partial z}$$

$$E_y^e = \frac{1}{j\omega\epsilon}\left(\frac{\partial^2}{\partial y^2} + k^2\right)\psi^e \qquad H_y^e = 0 \tag{D.174}$$

$$E_z^e = \frac{1}{j\omega\epsilon}\frac{\partial^2\psi^e}{\partial y\partial z} \qquad H_z^e = \frac{\partial\psi^e}{\partial x}$$

The set of equations denotes a longitudinal section electric mode $[LSE_y,\ (TM)_y]$ with respect to the y direction $(E_y \neq 0,\ H_y = 0)$.

$(H\ Mode)_y$: $\vec{A}^m = \Psi^m \vec{e}_y$

$$E_x^m = \frac{\partial\psi^m}{\partial z} \qquad H_x^m = \frac{1}{j\omega\mu}\frac{\partial^2\psi^m}{\partial x\partial y}$$

$$E_y^m = 0 \qquad H_y^m = \frac{1}{j\omega\mu}\left(\frac{\partial^2}{\partial y^2} + k^2\right)\psi^m \tag{D.175}$$

$$E_z^m = -\frac{\partial\psi^m}{\partial x} \qquad H_z^m = \frac{1}{j\omega\mu}\frac{\partial^2\psi^m}{\partial y\partial z}$$

Similarly, the set of equations denotes a longitudinal section magnetic mode $[LSH_y,\ (TE)_y]$ with respect to the y direction $(H_y \neq 0,\ E_y = 0)$.

D.4 Classification of Field Modes in Waveguiding Structures

TEM, E, and H modes are uniquely characterized by the divergence and curl properties of their transversal electric fields. In the following section, we will derive the respective relations, and discuss some transversal electrical mode patterns to comprehend their physical meaning. We make the assumptions that waveguiding structure is lossless, and the medium is isotropic, linear, and time-independent. The source-free Maxwell equations in complex form read:

$$\vec{\nabla} \times \vec{H} = j\omega\vec{D} \quad (\vec{J} \equiv \vec{0}) \tag{D.176}$$

$$\vec{\nabla} \times \vec{E} = -j\omega\vec{B} \tag{D.177}$$

Introducing transversal and longitudinal components for the electric and magnetic field strength $\vec{E}$ and $\vec{H}$ ($\vec{E} = \vec{E}_t + \vec{E}_z$, $\vec{H} = \vec{H}_t + \vec{H}_z$), and for the Nabla operator $\vec{\nabla}$ ($\vec{\nabla} = \vec{\nabla}_t + \vec{\nabla}_z$), results in a relationship between the transversal and longitudinal electric and magnetic fields. From (D.176), we get

$$j\omega\varepsilon\vec{E}_t = \vec{\nabla}_t \times \vec{H}_z + \vec{\nabla}_z \times \vec{H}_t \tag{D.178}$$

$$j\omega\varepsilon\vec{E}_z = \vec{\nabla}_t \times \vec{H}_t \tag{D.179}$$

and from (D.177)

$$-j\omega\mu\vec{H}_t = \vec{\nabla}_t \times \vec{E}_z + \vec{\nabla}_z \times \vec{E}_t \tag{D.180}$$

$$-j\omega\mu\vec{H}_z = \vec{\nabla}_t \times \vec{E}_t \tag{D.181}$$

D.4.1 TEM Mode

For a TEM mode, $E_z \equiv 0$, $H_z \equiv 0$ holds. From (D.179) and (D.181), we obtain directly

$$\vec{\nabla}_t \times \vec{H}_t = \vec{0} \qquad (\vec{E}_z \equiv \vec{0}) \tag{D.182}$$

$$\vec{\nabla}_t \times \vec{E}_t = \vec{0} \qquad (\vec{H}_z \equiv \vec{0}) \tag{D.183}$$

that is, the transversal electric and the transversal magnetic fields are curl-free in the cross-sectional plane. The circuital integral for $\vec{E}$ and $\vec{H}$ vanishes in the transversal plane. Therefore, the fields can be deduced from a potential field. Application of the transversal curl operator upon (D.180) leads to

$$-j\omega\mu\vec{\nabla}_t \times \vec{H}_t = \vec{\nabla}_z\left(\vec{\nabla}_t \cdot \vec{E}_t\right) \tag{D.184}$$

and with (D.182), this results in

$$\vec{\nabla} \cdot \vec{E}_t = 0 \tag{D.185}$$

In the case of a TEM mode, this means that the transversal curl and the transversal divergence are zero; or, in other words, the transversal vector field $\vec{E}_t$ is irrotational and solenoidal in the cross-section plane. With $\vec{\nabla} \times \vec{\nabla}\phi \equiv \vec{0}$, a curl-free vector field can be represented as the gradient of a scalar potential ϕ in any case:

$$\vec{E}_t = -\vec{\nabla}_t\,\phi \tag{D.186}$$

Considering that the fields are source-free, we derive with (D.185) the equation

$$\vec{\nabla}_t^2 \, \phi = 0 \tag{D.187}$$

which is valid within the field region. On the conductors, the boundary conditions also must be fulfilled; that is, the tangential components of $\vec{E}_t$ vanish on the conductor surface:

$$grad_t \, \phi \times \vec{n} = \vec{0}, \qquad \vec{E}_t \times \vec{n} = \vec{0} \tag{D.188}$$

This means that the vector $grad_t \, \phi$ points into the direction of the normal vector $\vec{n}$. Therefore, the contour of the conductor represents an equipotential line,

$$\phi = const \qquad \text{(on contour of conductor)} \tag{D.189}$$

D.4.2 *E* Modes

For an E mode, $E_z \neq 0$, $H_z \equiv 0$ is valid. From (D.179) and (D.181), we get directly:

$$\vec{\nabla}_t \times \vec{H}_t \neq \vec{0} \qquad (\vec{E}_z \neq \vec{0}) \tag{D.190}$$

$$\vec{\nabla}_t \times \vec{E}_t = \vec{0} \qquad (\vec{H}_z \equiv \vec{0}) \tag{D.191}$$

Therefore, the electric field of an E mode is curl-free within the cross-section plane. The divergence properties can be deduced from (D.190). The transversal fields of the electric and magnetic fields are coupled via the relationship

$$\vec{H}_t = \frac{1}{Z_F} (\vec{e}_z \times \vec{E}_t) \tag{D.192}$$

where Z_F is the wave impedance. The vectors $\vec{E}_t$, $\vec{H}_t$, and $\vec{e}_z$ are perpendicular to each other and form a "right-hand" orthogonal system. With (D.192), it follows from (D.190) that

$$\vec{\nabla}_t \cdot \vec{E}_t \neq 0 \tag{D.193}$$

Thus, we can conclude that the electric transversal field of E modes is irrotational, but that its divergence does not vanish.

D.4.3 *H* Modes

For an H mode, $E_z \equiv 0$, $H_z \neq 0$. is valid. From (D.179) and (D.181), we get directly:

$$\vec{\nabla}_t \times \vec{H}_t = \vec{0} \qquad (\vec{E}_z \equiv \vec{0}) \tag{D.194}$$

$$\vec{\nabla}_t \times \vec{E}_t \neq \vec{0} \qquad (\vec{H}_z \neq \vec{0}) \tag{D.195}$$

Therefore, the electric field of H modes is not curl-free in the cross-section plane. From (D.194), we derive with (D.180) under application of the transversal curl operator:

$$\vec{\nabla}_t \cdot \vec{E}_t = 0 \tag{D.196}$$

Thus, we have seen that the electric transversal field of H modes is solenoidal, but its curl does not vanish.

The following examples regarding the transversal electric field of some waveguide modes should illustrate the physical meaning of the derived mathematical relations. Figure D.8 shows the transversal electrical field distribution of the TEM mode in a coaxial transmission line. Testing the curl of the mode under analysis, we define a closed path C with path elements in parallel to the conductor walls and electric field lines.

The curl is defined as the limit value of the closed line integral divided by the area enclosed within C:

$$|\vec{\nabla}_t \times \vec{E}_t| = \lim_{\Delta A \to 0} \frac{\oint \vec{E}_t \cdot d\vec{s}}{\Delta A} \tag{D.197}$$

As can be seen in Figure D.8, along the path elements C_1 and C_3, $\vec{E}_t$ and $d\vec{s}$ are perpendicular to each other, so that this part gives no contribution to the line integral. Along path elements C_2 and C_4, $\vec{E}_t$ and $d\vec{s}$ are parallel. However, because the direction of $d\vec{s}$ is different along C_1 and C_3, both line integral contributions compensate to zero. Thus, the TEM field is indeed irrotational.

The divergence of the electric field $\vec{E}$ is defined by the equation

$$\vec{\nabla} \cdot \vec{E} = \lim_{\Delta V \to 0} \frac{\oint \vec{E} \cdot d\vec{A}}{\Delta V} \tag{D.198}$$

Reducing (D.198) to the two-dimensional (transversal) field problem, we can write

$$\vec{\nabla}_t \cdot \vec{E}_t = \lim_{\Delta A \to 0} \frac{\oint \vec{E}_t \cdot \vec{n}\, ds}{\Delta A} \tag{D.199}$$

where $\vec{n}$ is the normal vector perpendicular to the surface contour. Without loss of generality, we again define a surface with edges in parallel to the electric field lines and the contour of the conducting walls. As we can see from Figure D.8, all electric field lines that enter the surface are also leaving it. This property denotes a solenoidal field. Thus, we can summarize that the TEM mode comprises solenoidal and irrotational transversal electric fields.

Figure D.9 shows the transversal electric field distribution of the fundamental mode in rectangular waveguide, the H_{10} or TE_{10} mode. The electric field in the cross-section is independent of the height. Along the broadside, the field is sinusoidal. We again test the curl and divergence of the mode by defining a convenient closed path C, and area with contours parallel to the field lines and conducting walls. Because of the change of electric field density, the line integral does not vanish, so that the transverse field is not irrotational. However, the field is solenoidal, because all field lines that enter the defined area also leave it, although the field line density varies. Thus, we can conclude that *H* (or *TE*) modes in rectangular hollow waveguides are solenoidal but not irrotational.

Figure D.10 illustrates the transversal electric field of the H_{01} (or TE_{01}) mode. The closed line integral does not vanish, because of different values of the electric field strength along C_1 and C_3. Because all field lines are entering and leaving the defined area, the divergence is zero. Thus, *H* (or *TE*) modes in circular hollow waveguides are also solenoidal but not irrotational.

Figure D.11 shows the transverse electrical field of the E_{01} (or TM_{01}) mode. The electric field lines are coming perpendicular from the conductive walls, concentrating in the center, bending in a longitudinal direction, and then spreading and ending again perpendicular on the waveguide wall. Regarding the closed path in the cross-section, we find that the amplitude of the transverse component of the electric field is no longer constant along C_1 and C_3, respectively, but are the same, so that the path integral vanishes. Regarding the divergence, we must recognize that since the lines are bending from the transverse into longitudinal direction, not all lines that enter the area will also leave it. Therefore, the divergence will be nonzero. We can summarize that the transverse electrical field of *E* (or *TM*) modes in circular hollow waveguides are irrotational but not solenoidal.

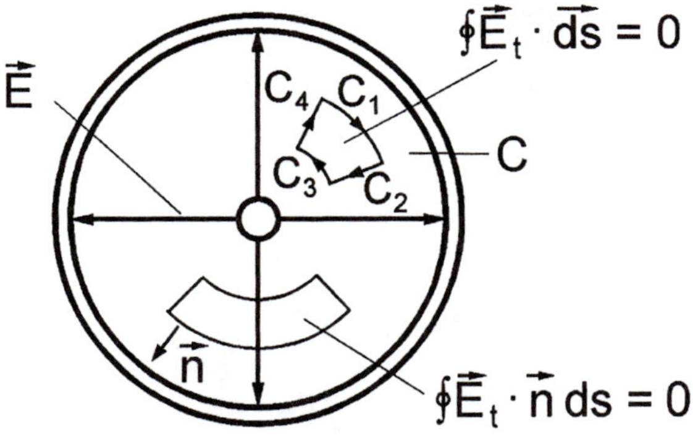

Figure D.8 Transversal electric field of a TEM mode in a coaxial transmission line.

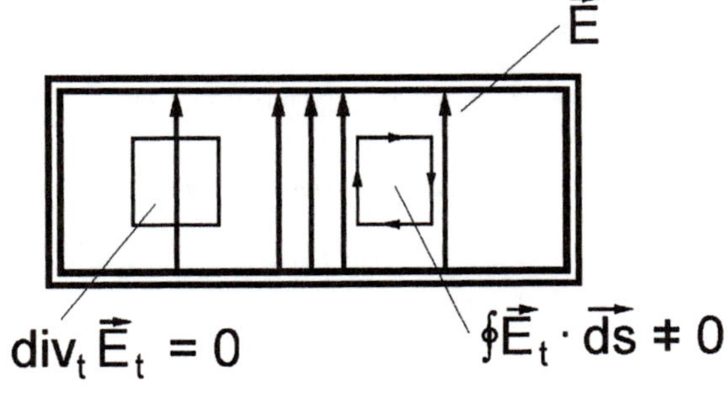

Figure D.9 Transversal electric field of H_{10} mode (TE_{10}) in a rectangular waveguide.

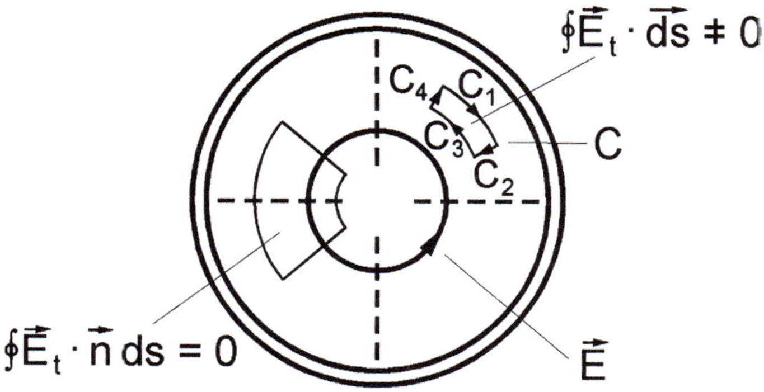

Figure D.10 Transversal electric field of H_{01} mode (TE_{01}) in a circular waveguide.

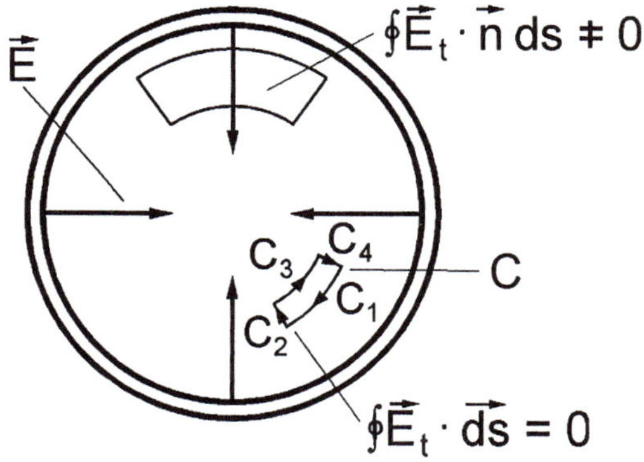

Figure D.11 Transversal electrical field of E_{01} mode (TM_{01}) in a circular waveguide.

References

[1] Wikipedia, *The Free Encyclopedia*, http://en.wikipedia.org.

[2] Maxwell, J. C., "A Dynamical Theory of the Electromagnetic Field," *Philosophical Transactions of the Royal Society of London*, 155, 1865, pp. 459–512.

[3] Maxwell, J. C., *A Treatise on Electricity and Magnetism*, 3rd ed., Vols. 1–2 (1891) (reprinted: New York: Dover, 1954).

[4] Mirshekar-Syahkal, D., *Spectral Domain Method for Microwave Integrated Circuits*, New York: John Wiley & Sons, 1990.

[5] Brand, H., *Schaltungslehre linearer Mikrowellennetze*, Stuttgart, Germany: S. Hirzel Verlag, 1970.

Appendix E

Definition of Insertion Loss

The insertion loss (IL) is defined as the reciprocal of the ratio of the power delivered to the load with and without insertion of a device between source and load, and is usually expressed in decibels [1]:

$$\text{Insertion Loss } (dB) = 10\log_{10} \frac{\text{power absorbed in the load without inserted device}}{\text{power absorbed in the load with inserted device}} \tag{E.1}$$

The measured values of the insertion loss depend on the actual values of the generator reflection coefficient Γ_G and load reflection coefficient Γ_L. This shall be exemplified by comparing the two circuits shown in Figures E.1 and E.2.

Figure E.1 shows a two-port device, interconnected between a generator with generator wave b_G and generator reflection coefficient Γ_G, and load with the load reflection coefficient Γ_L (network I). The device is characterized by the scattering matrix $\boldsymbol{S}$. Γ_1 is the two-port input reflection coefficient, and Γ_2 the load reflection coefficient. P_1^{abs} denotes the power which enters port 1, and P_2^{abs} is absorbed by the load.

Figure E.2(a) illustrates the connection of generator and load via an ideal transmission line with zero input and output eigenreflection and perfect transmission coefficients ($S_{12} = S_{21} = 1$). The network corresponds to a direct connection, as illustrated by Figure E.2(b). Following the definition (E.1), we can write

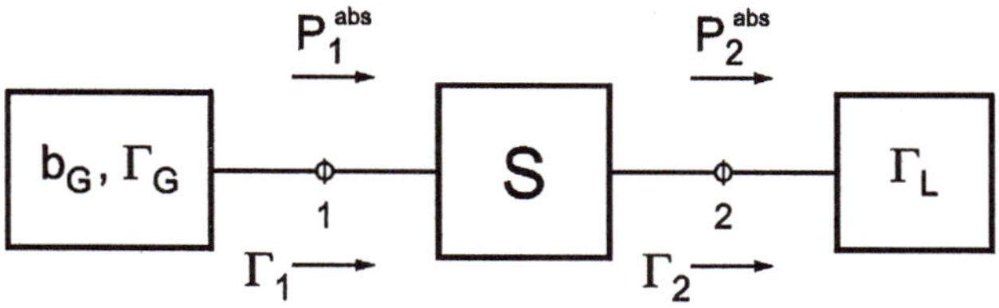

Figure E.1 Two-port device inserted between load and generator (network I).

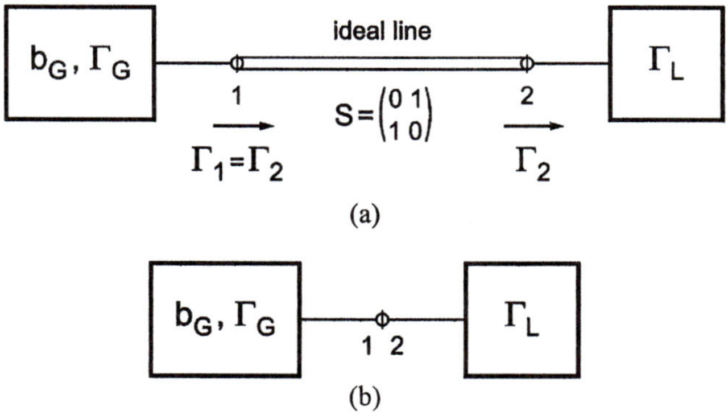

Figure E.2 Source and load connected via an ideal transmission line (direct connection, reference network II).

$$\mathrm{IL}_{21}/\mathrm{dB} = 10 \log_{10} \frac{P_1^{abs}(\Gamma_1 = \Gamma_2)}{P_2^{abs}} \tag{E.2}$$

P_2^{abs} of network I is calculated as

$$P_2^{abs} = \frac{1}{2}|b_G|^2 \frac{|S_{21}|^2 \left(1 - |\Gamma_2|^2\right)}{|1 - S_{11}\Gamma_G - S_{22}\Gamma_2 + \Gamma_G\Gamma_2 \det S|^2} \tag{E.3}$$

Regarding the reference network II, the power absorbed by the load $P_1^{abs}(\Gamma_1 = \Gamma_2)$, reads (note that $\Gamma_1^I \neq \Gamma_1^{II}$)

$$P_1^{abs}(\Gamma_1 = \Gamma_2) = \frac{1}{2}|b_G|^2 \frac{1 - |\Gamma_2|^2}{|1 - \Gamma_G\Gamma_1|^2} \tag{E.4}$$

Thus, the insertion loss is given as

$$\mathrm{IL}_{21}/\mathrm{dB} = 10 \log_{10} \frac{|1 - S_{11}\Gamma_G - S_{22}\Gamma_2 + \Gamma_G\Gamma_2 \det S|^2}{|S_{21}|^2 \; |1 - \Gamma_G\Gamma_1|^2} \tag{E.5}$$

Under the condition

$$\Gamma_G = \Gamma_2 = 0 \tag{E.6}$$

the insertion loss simply reads

$$\mathrm{IL}_{21}/\mathrm{dB} = -20 \log_{10} |S_{21}| \tag{E.7}$$

Reference

[1] Meinke-Gundlach, *Taschenbuch der Hochfrequenztechnik*, K. Lange and K.-H. Löcherer (eds.), 4th ed., Studienausgabe, Band 1, Grundlagen: Springer-Verlag, 1986.

Appendix F

Table of Quasi-Static Line Parameters After Hammerstad

No.	w/h	Z_0				$\epsilon_{eff}(f=0)$		
		$\epsilon_r = 1$	2.32	9.8	11.9	$\epsilon_r = 2.32$	9.8	11.9
1	20.00	16.14	10.91	5.41	4.92	2.19	8.88	10.76
2	19.90	16.21	10.96	5.44	4.94	2.19	8.88	10.75
3	19.80	16.28	11.01	5.46	4.96	2.19	8.87	10.75
4	19.70	16.35	11.06	5.49	4.99	2.19	8.87	10.75
5	19.60	16.42	11.11	5.52	5.01	2.19	8.87	10.74
6	19.50	16.50	11.16	5.54	5.03	2.19	8.86	10.74
7	19.40	16.57	11.21	5.57	5.06	2.19	8.86	10.73
8	19.30	16.65	11.26	5.59	5.08	2.18	8.86	10.73
9	19.20	16.72	11.32	5.62	5.11	2.18	8.86	10.73
10	19.10	16.80	11.37	5.65	5.13	2.18	8.85	10.72
11	19.00	16.88	11.42	5.67	5.16	2.18	8.85	10.72
12	18.90	16.96	11.48	5.70	5.18	2.18	8.85	10.71
13	18.80	17.04	11.53	5.73	5.21	2.18	8.84	10.71
14	18.70	17.12	11.59	5.76	5.23	2.18	8.84	10.71
15	18.60	17.20	11.64	5.79	5.26	2.18	8.84	10.70
16	18.50	17.28	11.70	5.81	5.28	2.18	8.83	10.70
17	18.40	17.36	11.76	5.84	5.31	2.18	8.83	10.69
18	18.30	17.44	11.81	5.87	5.33	2.18	8.83	10.69
19	18.20	17.53	11.87	5.90	5.36	2.18	8.82	10.69
20	18.10	17.61	11.93	5.93	5.39	2.18	8.82	10.68
21	18.00	17.70	11.99	5.96	5.42	2.18	8.82	10.68
22	17.90	17.78	12.05	5.99	5.44	2.18	8.81	10.67
23	17.80	17.87	12.11	6.02	5.47	2.18	8.81	10.67
24	17.70	17.96	12.17	6.05	5.50	2.18	8.80	10.66
25	17.60	18.04	12.23	6.08	5.53	2.18	8.80	10.66
26	17.50	18.13	12.29	6.11	5.56	2.18	8.80	10.65
27	17.40	18.22	12.36	6.15	5.58	2.18	8.79	10.65
28	17.30	18.32	12.42	6.18	5.61	2.18	8.79	10.65
29	17.20	18.41	12.48	6.21	5.64	2.17	8.79	10.64
30	17.10	18.50	12.55	6.24	5.67	2.17	8.78	10.64
31	17.00	18.60	12.61	6.28	5.70	2.17	8.78	10.63
32	16.90	18.69	12.68	6.31	5.73	2.17	8.77	10.63
33	16.80	18.79	12.75	6.34	5.76	2.17	8.77	10.62
34	16.70	18.89	12.81	6.38	5.80	2.17	8.77	10.62
35	16.60	18.98	12.88	6.41	5.83	2.17	8.76	10.61
36	16.50	19.08	12.95	6.45	5.86	2.17	8.76	10.61
37	16.40	19.18	13.02	6.48	5.89	2.17	8.75	10.60
38	16.30	19.28	13.09	6.52	5.92	2.17	8.75	10.60
39	16.20	19.39	13.16	6.56	5.96	2.17	8.75	10.59

No.	*w/h*	Z_0				$\epsilon_{eff}(f=0)$		
		$\epsilon_r = 1$	2.32	9.8	11.9	$\epsilon_r = 2.32$	9.8	11.9
40	16.10	19.49	13.24	6.59	5.99	2.17	8.74	10.59
41	16.00	19.60	13.31	6.63	6.02	2.17	8.74	10.58
42	15.90	19.70	13.38	6.67	6.06	2.17	8.73	10.58
43	15.80	19.81	13.46	6.70	6.09	2.17	8.73	10.57
44	15.70	19.92	13.53	6.74	6.13	2.17	8.73	10.57
45	15.60	20.03	13.61	6.78	6.16	2.17	8.72	10.56
46	15.50	20.14	13.69	6.82	6.20	2.16	8.72	10.55
47	15.40	20.25	13.77	6.86	6.23	2.16	8.71	10.55
48	15.30	20.36	13.85	6.90	6.27	2.16	8.71	10.54
49	15.20	20.48	13.93	6.94	6.31	2.16	8.70	10.54
50	15.10	20.59	14.01	6.98	6.35	2.16	8.70	10.53
51	15.00	20.71	14.09	7.02	6.38	2.16	8.69	10.53
52	14.90	20.83	14.17	7.07	6.42	2.16	8.69	10.52
53	14.80	20.95	14.25	7.11	6.46	2.16	8.68	10.52
54	14.70	21.07	14.34	7.15	6.50	2.16	8.68	10.51
55	14.60	21.19	14.43	7.20	6.54	2.16	8.68	10.50
56	14.50	21.32	14.51	7.24	6.58	2.16	8.67	10.50
57	14.40	21.44	14.60	7.28	6.62	2.16	8.67	10.49
58	14.30	21.57	14.69	7.33	6.66	2.16	8.66	10.49
59	14.20	21.70	14.78	7.38	6.70	2.16	8.66	10.48
60	14.10	21.83	14.87	7.42	6.75	2.15	8.65	10.47
61	14.00	21.96	14.96	7.47	6.79	2.15	8.65	10.47
62	13.90	22.10	15.06	7.52	6.83	2.15	8.64	10.46
63	13.80	22.23	15.15	7.57	6.88	2.15	8.64	10.45
64	13.70	22.37	15.25	7.61	6.92	2.15	8.63	10.45
65	13.60	22.51	15.35	7.66	6.97	2.15	8.62	10.44
66	13.50	22.65	15.45	7.71	7.01	2.15	8.62	10.43
67	13.40	22.79	15.55	7.77	7.06	2.15	8.61	10.43
68	13.30	22.94	15.65	7.82	7.11	2.15	8.61	10.42
69	13.20	23.08	15.75	7.87	7.15	2.15	8.60	10.41
70	13.10	23.23	15.85	7.92	7.20	2.15	8.60	10.41
71	13.00	23.38	15.96	7.98	7.25	2.15	8.59	10.40
72	12.90	23.53	16.07	8.03	7.30	2.15	8.59	10.39
73	12.80	23.69	16.17	8.09	7.35	2.14	8.58	10.39
74	12.70	23.84	16.28	8.14	7.40	2.14	8.57	10.38
75	12.60	24.00	16.40	8.20	7.45	2.14	8.57	10.37
76	12.50	24.16	16.51	8.26	7.51	2.14	8.56	10.36
77	12.40	24.33	16.62	8.32	7.56	2.14	8.56	10.36
78	12.30	24.49	16.74	8.38	7.61	2.14	8.55	10.35
79	12.20	24.66	16.86	8.44	7.67	2.14	8.54	10.34

No.	w/h	Z_0				$\epsilon_{eff}(f = 0)$		
		$\epsilon_r = 1$	2.32	9.8	11.9	$\epsilon_r = 2.32$	9.8	11.9
80	12.10	24.83	16.98	8.50	7.72	2.14	8.54	10.33
81	12.00	25.00	17.10	8.56	7.78	2.14	8.53	10.33
82	11.90	25.17	17.22	8.62	7.84	2.14	8.53	10.32
83	11.80	25.35	17.35	8.69	7.90	2.14	8.52	10.31
84	11.70	25.53	17.47	8.75	7.95	2.13	8.51	10.30
85	11.60	25.71	17.60	8.82	8.01	2.13	8.51	10.29
86	11.50	25.90	17.73	8.88	8.08	2.13	8.50	10.28
87	11.40	26.09	17.87	8.95	8.14	2.13	8.49	10.28
88	11.30	26.28	18.00	9.02	8.20	2.13	8.49	10.27
89	11.20	26.47	18.14	9.09	8.26	2.13	8.48	10.26
90	11.10	26.67	18.28	9.16	8.33	2.13	8.47	10.25
91	11.00	26.87	18.42	9.23	8.39	2.13	8.46	10.24
92	10.90	27.07	18.56	9.31	8.46	2.13	8.46	10.23
93	10.80	27.27	18.71	9.38	8.53	2.13	8.45	10.22
94	10.70	27.48	18.85	9.46	8.60	2.12	8.44	10.21
95	10.60	27.69	19.00	9.54	8.67	2.12	8.44	10.21
96	10.50	27.91	19.16	9.61	8.74	2.12	8.43	10.20
97	10.40	28.13	19.31	9.69	8.81	2.12	8.42	10.19
98	10.30	28.35	19.47	9.77	8.89	2.12	8.41	10.18
99	10.20	28.58	19.63	9.86	8.96	2.12	8.40	10.17
100	10.10	28.81	19.79	9.94	9.04	2.12	8.40	10.16
101	10.00	29.04	19.96	10.03	9.12	2.12	8.39	10.15
102	9.90	29.28	20.13	10.11	9.20	2.12	8.38	10.14
103	9.80	29.52	20.30	10.20	9.28	2.11	8.37	10.13
104	9.70	29.77	20.48	10.29	9.36	2.11	8.36	10.12
105	9.60	30.02	20.65	10.38	9.44	2.11	8.36	10.11
106	9.50	30.27	20.83	10.48	9.53	2.11	8.35	10.10
107	9.40	30.53	21.02	10.57	9.61	2.11	8.34	10.09
108	9.30	30.79	21.21	10.67	9.70	2.11	8.33	10.07
109	9.20	31.06	21.40	10.77	9.79	2.11	8.32	10.06
110	9.10	31.33	21.59	10.87	9.88	2.11	8.31	10.05
111	9.00	31.61	21.79	10.97	9.98	2.10	8.30	10.04
112	8.90	31.89	21.99	11.08	10.07	2.10	8.29	10.03
113	8.80	32.18	22.20	11.18	10.17	2.10	8.28	10.02
114	8.70	32.48	22.41	11.29	10.27	2.10	8.27	10.01
115	8.60	32.78	22.62	11.40	10.37	2.10	8.26	9.99
116	8.50	33.08	22.84	11.51	10.47	2.10	8.26	9.98
117	8.40	33.39	23.06	11.63	10.58	2.10	8.25	9.97
118	8.30	33.71	23.29	11.75	10.68	2.09	8.23	9.96
119	8.20	34.03	23.52	11.87	10.79	2.09	8.22	9.94

No.	*w/h*	Z_0				$\epsilon_{eff}(f=0)$		
		$\epsilon_r = 1$	2.32	9.8	11.9	$\epsilon_r = 2.32$	9.8	11.9
120	8.10	34.36	23.76	11.99	10.90	2.09	8.21	9.93
121	8.00	34.70	24.00	12.11	11.02	2.09	8.20	9.92
122	7.90	35.04	24.25	12.24	11.13	2.09	8.19	9.90
123	7.80	35.39	24.50	12.37	11.25	2.09	8.18	9.89
124	7.70	35.75	24.76	12.51	11.38	2.09	8.17	9.88
125	7.60	36.12	25.02	12.64	11.50	2.08	8.16	9.86
126	7.50	36.49	25.29	12.78	11.63	2.08	8.15	9.85
127	7.40	36.87	25.56	12.93	11.76	2.08	8.14	9.84
128	7.30	37.26	25.84	13.07	11.89	2.08	8.13	9.82
129	7.20	37.66	26.13	13.22	12.03	2.08	8.11	9.81
130	7.10	38.07	26.43	13.37	12.17	2.08	8.10	9.79
131	7.00	38.48	26.73	13.53	12.31	2.07	8.09	9.78
132	6.90	38.91	27.03	13.69	12.45	2.07	8.08	9.76
133	6.80	39.34	27.35	13.86	12.60	2.07	8.06	9.74
134	6.70	39.79	27.67	14.02	12.76	2.07	8.05	9.73
135	6.60	40.25	28.00	14.20	12.91	2.07	8.04	9.71
136	6.50	40.72	28.34	14.37	13.08	2.06	8.02	9.70
137	6.40	41.19	28.69	14.55	13.24	2.06	8.01	9.68
138	6.30	41.69	29.04	14.74	13.41	2.06	8.00	9.66
139	6.20	42.19	29.41	14.93	13.59	2.06	7.98	9.64
140	6.10	42.71	29.79	15.13	13.76	2.06	7.97	9.63
141	6.00	43.24	30.17	15.33	13.95	2.05	7.95	9.61
142	5.90	43.78	30.57	15.54	14.14	2.05	7.94	9.59
143	5.80	44.34	30.97	15.75	14.33	2.05	7.92	9.57
144	5.70	44.91	31.39	15.97	14.53	2.05	7.91	9.55
145	5.60	45.50	31.82	16.19	14.74	2.04	7.89	9.53
146	5.50	46.10	32.26	16.43	14.95	2.04	7.88	9.51
147	5.40	46.73	32.71	16.66	15.16	2.04	7.86	9.49
148	5.30	47.37	33.18	16.91	15.39	2.04	7.85	9.47
149	5.20	48.03	33.66	17.16	15.62	2.04	7.83	9.45
150	5.10	48.70	34.16	17.43	15.86	2.03	7.81	9.43
151	5.00	49.40	34.67	17.70	16.10	2.03	7.79	9.41
152	4.90	50.12	35.20	17.97	16.36	2.03	7.78	9.39
153	4.80	50.86	35.75	18.26	16.62	2.02	7.76	9.37
154	4.70	51.63	36.31	18.56	16.89	2.02	7.74	9.34
155	4.60	52.42	36.89	18.87	17.17	2.02	7.72	9.32
156	4.50	53.24	37.49	19.18	17.46	2.02	7.70	9.30
157	4.40	54.08	38.11	19.51	17.76	2.01	7.68	9.27
158	4.30	54.95	38.76	19.85	18.07	2.01	7.66	9.25
159	4.20	55.86	39.42	20.21	18.39	2.01	7.64	9.22

No.	w/h	Z_0				$\epsilon_{eff}(f=0)$		
		$\epsilon_r = 1$	2.32	9.8	11.9	$\epsilon_r = 2.32$	9.8	11.9
160	4.10	56.79	40.11	20.57	18.73	2.00	7.62	9.19
161	4.00	57.76	40.83	20.95	19.08	2.00	7.60	9.17
162	3.90	58.77	41.57	21.35	19.44	2.00	7.58	9.14
163	3.80	59.81	42.35	21.76	19.81	1.99	7.55	9.11
164	3.70	60.89	43.15	22.19	20.20	1.99	7.53	9.08
165	3.60	62.02	43.98	22.63	20.61	1.99	7.51	9.06
166	3.50	63.19	44.85	23.10	21.03	1.98	7.48	9.03
167	3.40	64.40	45.76	23.58	21.47	1.98	7.46	9.00
168	3.30	65.67	46.71	24.09	21.94	1.98	7.43	8.96
169	3.20	67.00	47.70	24.62	22.42	1.97	7.41	8.93
170	3.10	68.38	48.73	25.17	22.92	1.97	7.38	8.90
171	3.00	69.83	49.81	25.75	23.45	1.97	7.35	8.86
172	2.90	71.34	50.95	26.36	24.01	1.96	7.33	8.83
173	2.80	72.93	52.14	27.00	24.59	1.96	7.30	8.79
174	2.70	74.60	53.39	27.67	25.21	1.95	7.27	8.76
175	2.60	76.35	54.70	28.38	25.85	1.95	7.24	8.72
176	2.50	78.19	56.09	29.13	26.54	1.94	7.21	8.68
177	2.40	80.13	57.55	29.92	27.26	1.94	7.17	8.64
178	2.30	82.18	59.10	30.75	28.02	1.93	7.14	8.60
179	2.20	84.35	60.74	31.64	28.84	1.93	7.11	8.56
180	2.10	86.65	62.49	32.59	29.70	1.92	7.07	8.51
181	2.00	89.09	64.34	33.59	30.62	1.92	7.03	8.47
182	1.90	91.69	66.32	34.67	31.60	1.91	7.00	8.42
183	1.80	94.47	68.43	35.82	32.65	1.91	6.96	8.37
184	1.70	97.44	70.70	37.05	33.78	1.90	6.92	8.32
185	1.60	100.63	73.14	38.39	35.00	1.89	6.87	8.27
186	1.50	104.08	75.77	39.83	36.32	1.89	6.83	8.21
187	1.40	107.80	78.63	41.39	37.75	1.88	6.78	8.16
188	1.30	111.85	81.74	43.10	39.31	1.87	6.73	8.10
189	1.20	116.27	85.14	44.97	41.02	1.86	6.68	8.04
190	1.10	121.13	88.89	47.03	42.90	1.86	6.63	7.97
191	1.00	126.51	93.04	49.32	45.00	1.85	6.58	7.90
192	0.90	132.52	97.69	51.89	47.34	1.84	6.52	7.83
193	0.80	139.30	102.94	54.79	50.00	1.83	6.46	7.76
194	0.70	147.05	108.94	58.11	53.03	1.82	6.40	7.69
195	0.60	156.07	115.93	61.97	56.57	1.81	6.34	7.61
196	0.50	166.82	124.25	66.58	60.79	1.80	6.28	7.53
197	0.40	180.04	134.51	72.27	65.99	1.79	6.21	7.44
198	0.30	197.17	147.80	79.64	72.74	1.78	6.13	7.35
199	0.20	221.41	166.62	90.09	82.29	1.77	6.04	7.24
200	0.10	262.94	198.87	107.99	98.68	1.75	5.93	7.10

Appendix G

Circuit Theory

A circuit is an interconnection of n-port components that interact only at the ports. Each n-port is described by n^2 complex parameters. The circuit behavior can be predicted solely from knowledge of these parameters. Circuit analysis can be based on the voltage-current description, or make use of the S matrix concept.

G.1 Circuit Parameters Interrelating Voltage and Current

The electrical properties of an n-port can be described by various circuit parameters interrelating voltage V and current I at the ports of an n-port, such as H, Y, and Z. For a two-port (Figure G.1), the following relations are valid:

H *Parameter*

The H parameters, also termed hybrid coefficients, relate the input voltage and output current, respectively, to the input current and output voltage:

$$\begin{aligned} V_1 &= H_{11}\, I_1 + H_{12}\, V_2 \\ I_2 &= H_{21}\, I_1 + H_{22}\, V_2 \end{aligned} \tag{G.1}$$

The H matrix is often used for the characterization of transistor devices [1]. In particular

$$H_{21} = I_2 / I_1 \big|_{V_2 = 0} \tag{G.2}$$

denotes the current gain. H_{21} is measured with short-circuited termination at port 2.

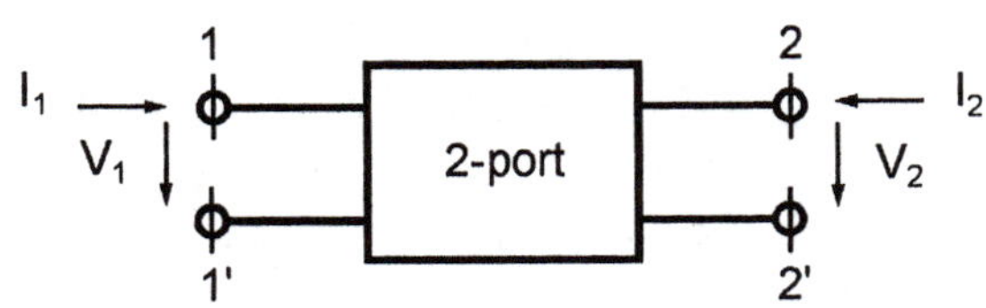

Figure G.1 Two-port representation and definition of voltage and current.

Y *Parameter*

The Y parameter relates the port currents to the port voltages,

$$\begin{aligned} I_1 &= Y_{11} V_1 + Y_{12} V_2 \\ I_2 &= Y_{21} V_1 + Y_{22} V_2 \end{aligned} \tag{G.3}$$

In matrix form, (G.3) reads

$$\begin{pmatrix} I_1 \\ I_2 \end{pmatrix} = \begin{pmatrix} Y_{11} & Y_{12} \\ Y_{21} & Y_{22} \end{pmatrix} \begin{pmatrix} V_1 \\ V_2 \end{pmatrix} \tag{G.4}$$

or

$$(I) = (Y)\,(V) \tag{G.5}$$

Z *Parameter*

The Z parameter relates the port voltages to the port currents,

$$\begin{aligned} V_1 &= Z_{11} I_1 + Z_{12} I_2 \\ V_2 &= Z_{21} I_1 + Z_{22} I_2 \end{aligned} \tag{G.6}$$

In matrix form, (G.6) reads,

$$\begin{pmatrix} V_1 \\ V_2 \end{pmatrix} = \begin{pmatrix} Z_{11} & Z_{12} \\ Z_{21} & Z_{22} \end{pmatrix} \begin{pmatrix} I_1 \\ I_2 \end{pmatrix} \tag{G.7}$$

or

$$(V) = (Z)\,(I) \tag{G.8}$$

The Z and Y matrices are the inverses of each other:

$$(Y) = (Z)^{-1} \tag{G.9}$$

Note that both the Z and Y matrices relate the total port voltages and currents. Equation (G.6) yields a guidance to determine the impedance coefficients. The Z_{ij} parameters can be found as

$$Z_{ij} = \left.\frac{V_i}{I_j}\right|_{I_k = 0 \text{ for } k \neq j} \tag{G.10}$$

which states that Z_{ij} can be found by driving port j with the current I_j, open-circuiting all other ports (so $I_k = 0$ for $k \neq j$), and measuring the open-circuit voltage at port i. Thus, Z_{ii} is the input impedance seen looking into port i when all other ports are open-circuited, and Z_{ij} is the transfer impedance between port i and j when all other ports are open-circuited. Similarly, from (G.3), Y_{ij} can be found as

$$Y_{ij} = \left.\frac{I_i}{V_j}\right|_{V_k = 0 \text{ for } k \neq j} \tag{G.11}$$

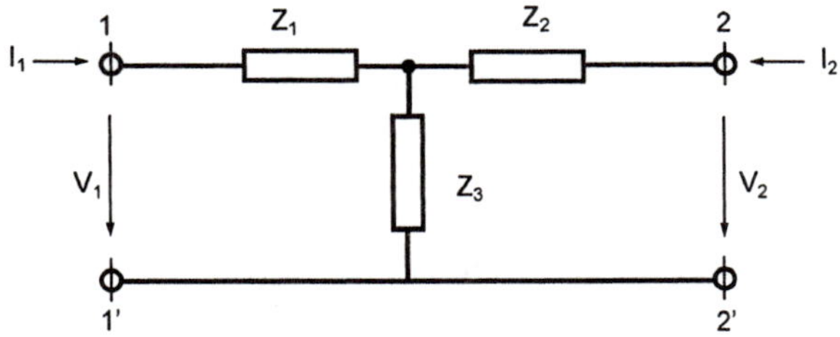

Figure G.2 Calculation of the Z parameters of a T-network.

which states that Y_{ij} can be determined by driving port j with the voltage V_j, short-circuiting all other ports (so $V_k = 0$ for $k \neq j$), and measuring the short-circuit current at port i.

Problem: Find the **Z** parameters of the two-port T-network shown in Figure G.2.

Solution: From (G.6), Z_{11} can be found as the input impedance of port 1 (terminal plane 1-1′) when port 2 (terminal plane 2-2′) is open-circuited:

$$Z_{11} = \left.\frac{V_1}{I_1}\right|_{I_2 = 0} = Z_1 + Z_3 \tag{G.12}$$

The transfer impedance Z_{12} can be found by measuring the open-circuit voltage at port 1 when a current I_2 is applied at port 2,

$$Z_{12} = \left.\frac{V_1}{I_2}\right|_{I_1 = 0} = \frac{V_2 \dfrac{Z_3}{Z_2 + Z_3}}{I_2} = Z_3 \tag{G.13}$$

Similarly, Z_{21} is found as

$$Z_{21} = \left.\frac{V_2}{I_1}\right|_{I_2 = 0} = \frac{V_1 \dfrac{Z_3}{Z_1 + Z_3}}{I_1} = Z_3 \tag{G.14}$$

With $Z_{12} = Z_{21}$, the circuit is reciprocal. Finally,

$$Z_{22} = \left.\frac{V_2}{I_2}\right|_{I_1 = 0} = Z_2 + Z_3 \tag{G.15}$$

Thus, the impedance matrix reads

$$Z = \begin{pmatrix} Z_1 + Z_3 & Z_3 \\ Z_3 & Z_2 + Z_3 \end{pmatrix} \tag{G.16}$$

G.2 S Matrix Concept

At microwave and millimeter-wave frequencies *H*, *Y*, and *Z*, parameters cannot be measured accurately because of the following reasons: (1) Microwave measuring instruments cannot measure directly the total voltage and the total current in the connection lines of an *n*-port, (2) accurate shorts and opens are difficult to achieve at microwave frequencies, and (3) active devices (transistors) become frequently unstable under short- or open-circuited conditions. Therefore, it is necessary to introduce new *n*-port parameters. Instead of port voltages and port currents, the operating conditions at an *n*-port at microwave frequencies are characterized by forward and backward waves (traveling waves, power waves) in the input and output ports. Their quantities (complex wave amplitudes) are related by the scattering parameters (*S* parameters). *S* parameters are the relationship between incident waves and reflected or transmitted waves. The experimental determination of the scattering parameters needs only resistive loading, rather than short or open circuits.

The total voltage V and current I at a terminal plane is given by

$$V = V^+ + V^- \quad \text{(G.17)}$$

$$I = I^+ + I^- \quad \text{(G.18)}$$

where V^+ and I^+ are incident, and V^- and I^- are reflected waves (Figure G.3). Normalized voltages and currents are defined as follows:

$$v = \frac{V}{\sqrt{Z_N}} \qquad [v] = \left[\frac{V}{\sqrt{\Omega}}\right] = \left[\sqrt{W}\right] \quad \text{(G.19)}$$

$$i = I\sqrt{Z_N} \qquad [i] = \left[A\sqrt{\Omega}\right] = \left[\sqrt{W}\right] \quad \text{(G.20)}$$

It is seen from (G.19) and (G.20) that both the normalized voltage and current have the same dimension. Z_N is an arbitrary normalization impedance. However, in a

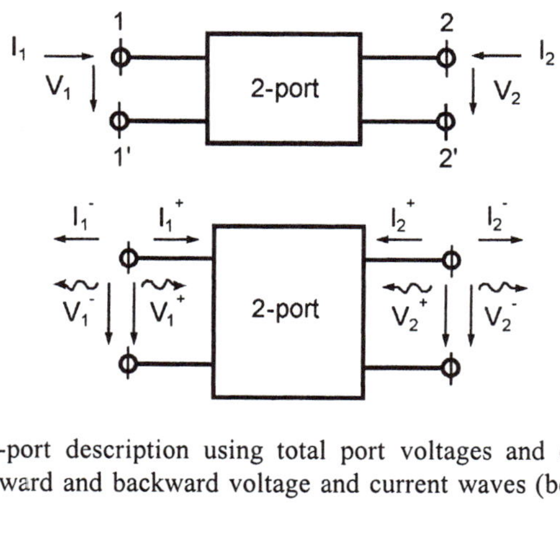

Figure G.3 Two-port description using total port voltages and currents (above), and corresponding forward and backward voltage and current waves (below).

microwave network comprised of interconnection transmission lines with a given characteristic impedance Z_0, it is useful to specify $Z_N = Z_0$. In the following, we assume transmission lines connected to the ports with a characteristic impedance Z_0, which is commonly 50Ω. Introducing incident and reflected power waves a and b, respectively, (G.19) and (G.20) are written as

$$v = \frac{V^+}{\sqrt{Z_0}} + \frac{V^-}{\sqrt{Z_0}} = a + b \tag{G.21}$$

$$i = I^+\sqrt{Z_0} - I^-\sqrt{Z_0} = a - b \tag{G.22}$$

Then from (G.21) and (G.22), the waves a and b are calculated as

$$a = \frac{V^+}{\sqrt{Z_0}} = I^+\sqrt{Z_0} \tag{G.23}$$

$$b = \frac{V^-}{\sqrt{Z_0}} = I^-\sqrt{Z_0} \tag{G.24}$$

From (G.23), we get

$$a^2 = \frac{V^+}{\sqrt{Z_0}} \cdot I^+\sqrt{Z_0} = V^+I^+ \tag{G.25}$$

Thus, the power wave amplitudes also can be written in the form

$$a = \sqrt{V^+I^+} \qquad [a] = \left[\sqrt{W}\right] \tag{G.26}$$

$$b = \sqrt{V^-I^-} \qquad [b] = \left[\sqrt{W}\right] \tag{G.27}$$

From (G.21) and (G.22), we obtain

$$a = \frac{1}{2}(v + i) \tag{G.28}$$

$$b = \frac{1}{2}(v - i) \tag{G.29}$$

The scattering matrix S relates the incident and reflected power waves:

$$\begin{aligned} b_1 &= S_{11}a_1 + S_{12}a_2 \\ b_2 &= S_{21}a_1 + S_{22}a_2 \end{aligned} \tag{G.30}$$

In matrix form, (G.30) reads

$$\begin{pmatrix} b_1 \\ b_2 \end{pmatrix} = \begin{pmatrix} S_{11} & S_{12} \\ S_{21} & S_{22} \end{pmatrix} \begin{pmatrix} a_1 \\ a_2 \end{pmatrix} \tag{G.31}$$

or

$$(b) = (S)\,(a) \tag{G.32}$$

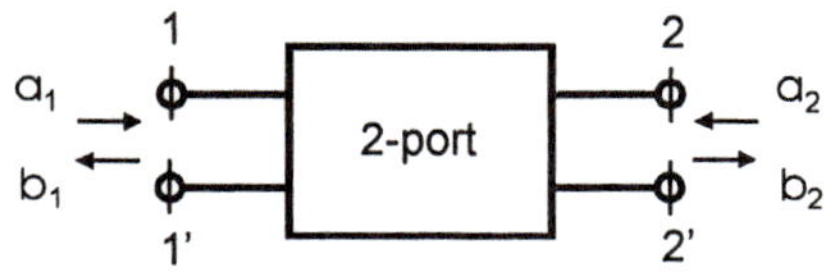

Figure G.4 Two-pole port (two-terminal pair) representation of a two-port and power wave definition.

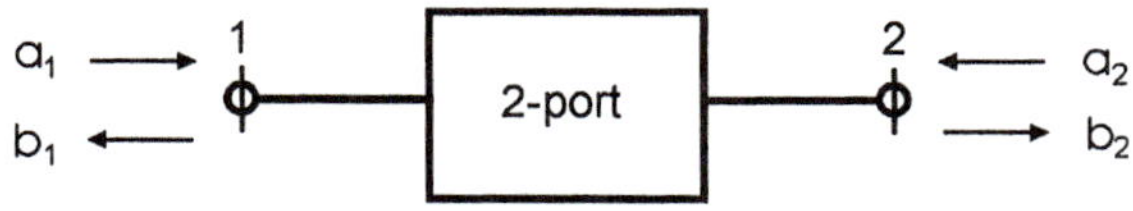

Figure G.5 Single-pole port representation of a two-port and power wave definition.

The scattering coefficients can be measured directly with a vector network analyzer. A specific element of the S matrix can be determined as

$$S_{ij} = \left.\frac{b_i}{a_j}\right|_{a_k = 0 \text{ for } k \neq j} \tag{G.33}$$

Equation (G.33) says that S_{ij} is found by driving port j with an incident wave of amplitude $a_j = V_j^+ / \sqrt{Z_0}$, and measuring the reflected wave amplitude $b_i = V_i^- / \sqrt{Z_0}$ coming out of port i. The incident waves on all ports except the jth port are set to zero, which means that all ports should be terminated in matched loads to avoid reflections. Thus, S_{ii} is the reflection coefficient looking into port i when all other ports are terminated in matched loads, and S_{ij} is the transmission coefficient from port j to port i, when all other ports are terminated in matched loads.

Figures G.4 and G.5 show the single- and two-pole representations of a two-port with defined incident and reflected power waves.

Problem: Find the S parameters of the 3-dB attenuator circuit shown in Figure G.6 [2].

Solution: From (G.30), S_{11} can be found as the reflection coefficient seen at port 1 when port 2 is terminated in a matched load ($Z_0 = 50\Omega$). The input eigenreflection coefficient reads

$$S_{11} = \left.\frac{b_1}{a_1}\right|_{a_2 = 0} = \left.\frac{V_1^- / \sqrt{Z_0}}{V_1^+ / \sqrt{Z_0}}\right|_{\substack{V_2^+ = 0 \\ Z_{L_2} = Z_0}} = \Gamma_{11'}\Big|_{\substack{a_2 = 0 \\ Z_{L_2} = Z_0}}$$

$$= \left.\frac{Z_{11'} - Z_0}{Z_{11'} + Z_0}\right|_{Z_{L_2} = Z_0} \tag{G.34}$$

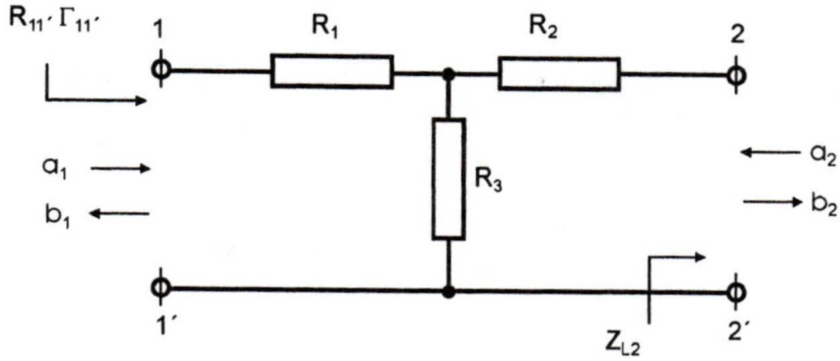

Figure G.6 A matched 3-dB attenuator with 50Ω input and output transmission lines ($R_1 = R_2 = 8.56\Omega$, $R_3 = 141.8\Omega$).

where $\Gamma_{11'}$ is the input reflection coefficient at port 1, and $Z_{11'}$, the input impedance. Z_{L_2} is the load at port 2. $Z_{11'}$ is calculated as

$$Z_{11'} = R_1 + \frac{R_3 \cdot (R_2 + Z_0)}{R_2 + R_3 + Z_0} = \left(8.56 + \frac{141.8(8.56 + 50)}{8.56 + 141.8 + 50}\right)\Omega = 50\,\Omega$$

Thus, from (G.34), and because of electrical symmetry of the network, $S_{11} = S_{22} = 0$. S_{21} can be found by applying an incident wave at port 1, a_1, and measuring the outcoming wave at port 2, b_2. This is equivalent to the transmission coefficient from port 1 to port 2.

$$S_{21} = \left.\frac{b_2}{a_1}\right|_{a_2=0} = \left.\frac{V_2^-}{V_1^+}\right|_{\substack{V_2^+=0 \\ Z_{L_2}=Z_0}} \tag{G.35}$$

From the fact that $S_{11} = S_{22} = 0$, we know that $V_1^- = 0$ when port 2 is terminated in $Z_0 = 50\Omega$ and that $V_2^+ = 0$ (see Figure G.7). In this case, $V_1^+ = V_1$ and $V_2^- = V_2$. S_{21} can then be written as

$$S_{21} = \left.\frac{V_2}{V_1}\right|_{Z_{L_2}=Z_0} \tag{G.36}$$

By applying a voltage S_{21} at port 1, this is divided twice, resulting in voltage drop

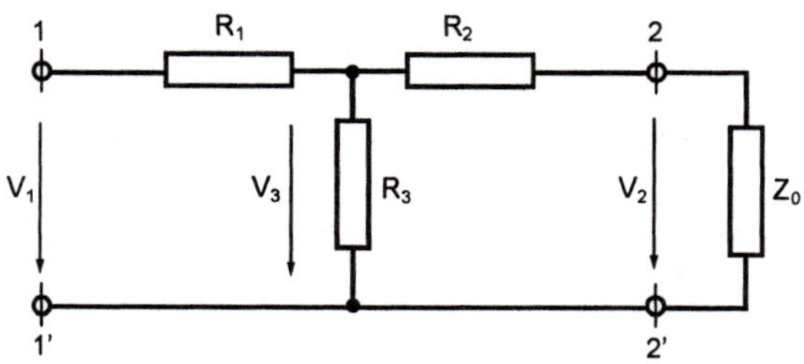

Figure G.7 Calculation of S_{21}.

V_3, and then in V_2 across the 50Ω load resistor at port 2. With the effective resistance R_3' comprising the shunted resistances R_3 and $R_2 + Z_0$,

$$R_3' = \frac{R_3(R_2 + Z_0)}{R_2 + R_3 + Z_0} = 41.44\,\Omega$$

the voltage ratio V_2/V_1 is calculated as

$$\frac{V_2}{V_1} = \frac{V_2}{V_3}\frac{V_3}{V_1} = \frac{Z_0}{R_2 + Z_0} \cdot \frac{R_3'}{R_1 + R_3'} = 0.853 \times 0.828 = 0.707$$

Thus, with (G.36), it follows

$$S_{12} = S_{21} = 0.707$$

The input power reads

$$P_1^+ = \frac{1}{2}|a_1|^2 = \frac{1}{2}|V_1^+ I_1^+| = \frac{1}{2}\frac{|V_1^+|^2}{Z_0} \tag{G.37}$$

then the output power is

$$P_2^- = \frac{1}{2}|b_2|^2 = \frac{1}{2}|V_2^- I_2^-| = \frac{1}{2}\frac{|V_2^-|^2}{Z_0} \tag{G.38}$$

With

$$V_2^- = S_{21} V_1^+ = \frac{1}{\sqrt{2}} V_1^+ \tag{G.39}$$

it follows

$$P_2^- = \frac{1}{2}\frac{|S_{21}|^2 |V_1^+|^2}{Z_0} = \frac{1}{2} P_1^+ \tag{G.40}$$

which is one-half (−3 dB) of the input power.

References

[1] Brand, H., *Schaltungslehre linearer Mikrowellennetze*, Stuttgart, Germany: S. Hirzel Verlag, 1970.

[2] Pozar, D. M., *Microwave Engineering*, Addison-Wesley Series in Electrical and Computer Engineering. Addison-Wesley Publishing Company, Inc., 1990.

Appendix H

Orthogonal Series Expansion Method

The method of orthogonal series expansion is applied to calculate the field distribution of the quasi-TEM mode in a shielded microstrip line [1], as shown in Figure H.1. It is assumed that the electromagnetic field within the shielded field region (Figure H.1) can be described by a series presentation of fields in subregions fulfilling the continuity condition at the interfaces. The quasi-TEM mode is an even mode. Even modes have the property that the electrical field distribution is symmetrical with respect to the x-z plane at $y = 0$. Thus, the mathematical problem can be simplified by treating only one-half of the microstrip structure, as shown in Figure H.2. The symmetry plane at $y = 0$ represents a magnetic wall or an ideal electrical open.

The modes in microstrip lines are hybrid modes with both longitudinal electric and magnetic field components. These fields can be described by the superposition of E (TM) modes (with longitudinal electric field component) and H (TE) modes (with longitudinal magnetic field component). As pointed out in Section D.2.2, the electric and magnetic fields read (D.143) and (D.144)

$$\vec{E} = \vec{E}^e + \vec{E}^m = \frac{1}{j\omega\epsilon} rot\, rot\, \vec{A}^e - rot\, \vec{A}^m$$

$$\vec{H} = \vec{H}^e + \vec{H}^m = rot\, \vec{A}^e + \frac{1}{j\omega\mu} rot\, rot\, \vec{A}^m$$

with

$$\vec{A}^e = \vec{e}_z\, \psi^e = \psi^e(x,y)\, e^{-jk_z z}\, \vec{e}_z$$

$$\vec{A}^m = \vec{e}_z\, \psi^m = \psi^m(x,y)\, e^{-jk_z z}\, \vec{e}_z$$

The equations are valid for the field components $\vec{E}$ and $\vec{H}$ in the subregions 1 to 5. $\vec{e}_z$ is the unit vector in the z-direction and k_z is the propagation coefficient in the z-direction. In the subregions, a series expansion can be made for the scalar potentials ψ^e and ψ^m, which fulfils the boundary conditions on the electric and magnetic walls. We consider first subregion 1.

a) Magnetic scalar potential ψ_1^m *in subregion 1:* In the x-direction, the condition $\partial H_z / \partial x \big|_{x=b,\, x=c} = 0$ must hold. Furthermore, $H_z = 0\big|_{y=0}$ must be fulfilled on the magnetic wall. The following series expansion for the magnetic scalar potential ψ_1^m in subregion 1 fulfills the conditions on the top metal wall ($x = c$), the magnetic wall ($y = 0$), and on the upper side of the strip ($x = b$):

$$\psi_1^m = \sum_{n=1}^{N+1} A_n \cos\left[\alpha_n^{(1)}(x - c)\right] \sin \beta_n^{(1)} y \tag{H.1}$$

where A_n ($n = 1, 2, ..., N + 1$) are the expansion coefficients, and $\alpha_n^{(1)}$ and $\beta_n^{(1)}$ are the transverse wavenumbers in the x- and y-directions of subregion 1. The first

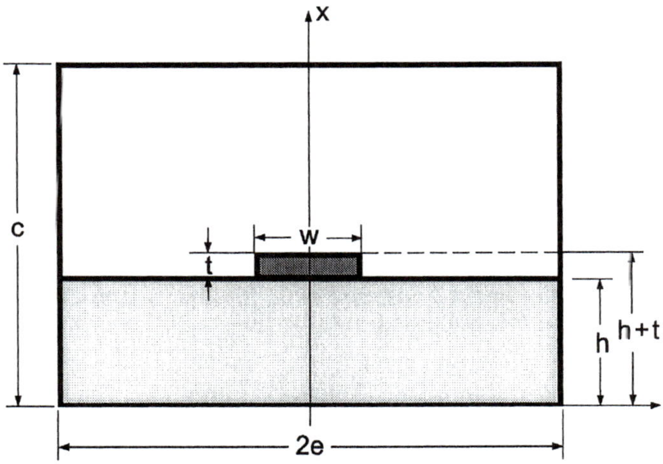

Figure H.1 Cross-section of the shielded microstrip transmission line.

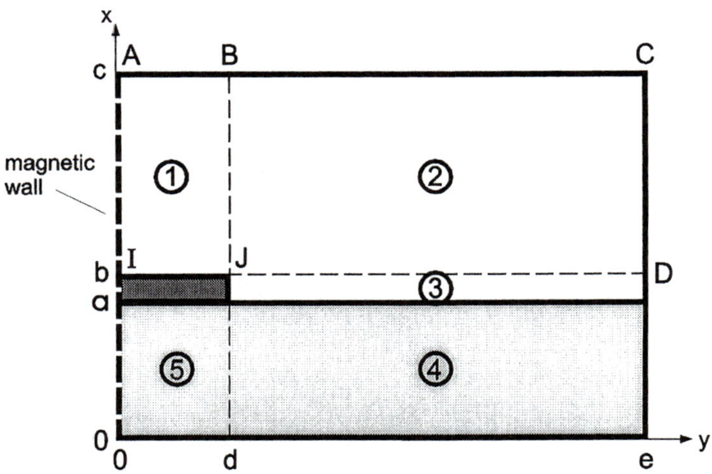

Figure H.2 Subdivision of the field region in the transversal plane.

expansion term ($n = 1$) can be identified as the quasi-TEM mode. Thus, with $N+1$ expansion terms, N higher-order H (TE) modes are being taken into account. The wavenumber $\alpha_n^{(1)}$ reads

$$\alpha_n^{(1)} = \frac{(n-1)\pi}{c-b} \tag{H.2}$$

and ensures that the normal magnetic component vanishes on the metallic strip at $x = b$. From the fundamental wavenumber relation $k_x^2 + k_y^2 + k_z^2 = k^2$, the following equation is obtained for β_n in region 1:

$$\alpha_n^{(1)^2} + \beta_n^{(1)^2} + k_z^2 = k_0^2 \tag{H.3}$$

or

$$\beta_n^{(1)} = \sqrt{k_0^2 - k_z^2 - \alpha_n^{(1)^2}} \tag{H.4}$$

b) Electric scalar potential ψ_1^e *in subregion 1:* In the x-direction, the condition $E_z = 0|_{x=b,c}$ must hold. Furthermore, $\partial E_z/\partial y|_{y=0} = 0$ must be fulfilled. The following series expansion for the electric scalar potential $\psi_1^{(e)}$ in subregion 1 fulfills the conditions:

$$\psi_1^{(e)} = \sum_{n=1}^{N} B_n \sin\left[\alpha_{n+1}^{(1)}(x-c)\right]\cos\beta_{n+1}^{(1)} y \tag{H.5}$$

where the sin-term directly forces the zero electric field on the conductive surface, and the cos-term satisfies the magnetic wall condition. B_n ($n = 1, 2, \ldots, N$) are the expansion coefficients, and α_{n+1} and β_{n+1} are the transverse wavenumbers in the x- and y-directions. The series expansion of the scalar potential function includes N higher-order E (TM) modes. The wavenumber $\alpha_{n+1}^{(1)}$ reads

$$\alpha_{n+1}^{(1)} = \frac{(n+1-1)\pi}{c-b} = \frac{n\pi}{c-b} \tag{H.6}$$

and takes care of the boundary at $x = b$.

c) Magnetic scalar potential ψ_2^m *in subregion 2:* In the x-direction, the condition $\partial H_z/\partial x|_{x=b,y=d;\,x=c} = 0$ must hold, and in the y-direction, $\partial H_z/\partial y|_{x=e} = 0$ must hold. The following series expansion for the magnetic scalar potential $\psi_2^{(m)}$ in subregion 2 fulfills the conditions:

$$\psi_2^m = \sum_{n=1}^{N+1} \bar{C}_n \cos\left[\alpha_n^{(1)}(x-c)\right]\cos\beta_n^{(1)}(y-e) + \tilde{C}_n \cos\left[\alpha_n^{(2)}(x-c)\right]\cos\beta_n^{(2)}(y-e) \tag{H.7}$$

where $\bar{C}_n$ and $\tilde{C}_n$ ($n = 1, 2, \ldots, N+1$) are the expansion coefficients, and $\alpha_n^{(2)}$ and

$\beta_n^{(2)}$ are the transverse wavenumbers in the x- and y-directions of subregion 2. The first expansion term can again be identified as the quasi-TEM mode. The wavenumbers $\alpha_n^{(2)}$ and $\beta_n^{(2)}$ read

$$\alpha_n^{(2)} = \frac{(n - 1)\pi}{c - b} \tag{H.8}$$

$$\beta_n^{(2)} = \sqrt{k_0^2 - k_z^2 - \alpha_n^{(2)^2}} \tag{H.9}$$

It should be noted that the expansion coefficients $\overline{C}_n$ describe tangential magnetic field components on the contour BCDJ, and the expansion coefficients $\tilde{C}_n$ describe tangential magnetic field components on the contour JBCD.

d) Electric scalar potential ψ_2^e *in subregion 2:* In the x-direction, the condition $E_z = 0|_{x=b,c}$ must hold. Furthermore, $\partial E_z / \partial y|_{y=b} = 0$ is valid. The following series expansion for the electric scalar potential $\psi_2^{(e)}$ in subregion 2 fulfills the conditions:

$$\psi_2^{(e)} = \sum_{n=1}^{N} \overline{D}_n \sin\left[\alpha_{n+1}^{(1)} (x - c)\right] \sin\beta_{n+1}^{(1)} (y - e) + \tilde{D}_n \sin\left[\alpha_{n+1}^{(2)} (x - c)\right] \sin\beta_{n+1}^{(2)} (y - e) \tag{H.10}$$

where $\overline{D}_n$ and $\tilde{D}_n$ $(n = 1, 2, \ldots, N)$ are the expansion coefficients, and $\alpha_{n+1}^{(2)}$ and $\beta_{n+1}^{(2)}$ are the transverse wavenumbers in the x- and y-directions of subregion 2. The scalar potential functions in regions 3, 4, and 5 can be obtained in a similar way.

Determination of the unknown expansion coefficients: The tangential field components across the interfaces between the subregions must be matched. Regarding, for example, the interface of regions 1 and 2 along contour $\overline{JB}$ $(y = d,\ b \le x \le c)$, it reads

$$E_{tang}^{(1)} = E_{tang}^{(2)}, \quad H_{tang}^{(1)} = H_{tang}^{(2)} \tag{H.11}$$

or for the respective Cartesian field components,

$$E_x^{(1)} = E_x^{(2)},\ E_z^{(1)} = E_z^{(2)},\ H_x^{(1)} = H_x^{(2)},\ H_z^{(1)} = H_z^{(2)} \tag{H.12}$$

For example, from the continuity condition of the H_z-component, it follows that with $\psi_1^m|_{y=d} = \psi_2^m|_{y=d}$

$$\psi_1^m|_{y=d} = \sum_{n=1}^{N+1} A_n \cos\left[\alpha_n^{(1)} (x - c)\right] \sin \beta_n^{(1)} d \tag{H.13}$$

$$\psi_2^{(m)}\Big|_{y=d} = \sum_{n=1}^{N+1} \overline{C}_n \cos\left[\alpha_n^{(1)}(x-c)\right]\cos\beta_n^{(1)}(d-e) + \tilde{C}_n \cos\left[\alpha_n^{(2)}(x-c)\right]\cos\beta_n^{(2)}(d-e) \quad \text{(H.14)}$$

Then,

$$\sum_{n=1}^{N+1} A_n \cos\left[\alpha_n^{(1)}(x-c)\right]\sin\beta_n^{(1)}d = \sum_{n=1}^{N+1} \overline{C}_n \cos\left[\alpha_n^{(1)}(x-c)\right]\cos\beta_n^{(1)}(d-e) + \tilde{C}_n \cos\left[\alpha_n^{(2)}(x-c)\right]\cos\beta_n^{(2)}(d-e) \quad \text{(H.15)}$$

Fulfilling the continuity for E_x, we need $E_x = -\partial\psi^m/\partial y$ (D.169) at the interface:

$$-\frac{\partial\psi_1^m}{\partial y}\Bigg|_{y=d} = \sum_{n=1}^{N+1} \beta_n^{(1)} A_n \cos\left[\alpha_n^{(1)}(x-c)\right]\cos\beta_n^{(1)}d \quad \text{(H.16)}$$

$$-\frac{\partial\psi_2^m}{\partial y}\Bigg|_{y=d} = \sum_{n=1}^{N+1} \beta_n^{(1)} \overline{C}_n \cos\left[\alpha_n^{(1)}(x-c)\right]\sin\beta_n^{(1)}(d-e) + \beta_n^{(2)} \tilde{C}_n \cos\left[\alpha_n^{(2)}(x-c)\right]\sin\beta_n^{(2)}(d-e) \quad \text{(H.17)}$$

Then with $\partial\psi_1^m/\partial y\big|_{y=d} = \partial\psi_2^m/\partial y\big|_{y=d}$, we obtain from (H.16) and (H.17)

$$\sum_{n=1}^{N+1} \beta_n^{(1)} A_n \cos\left[\alpha_n^{(1)}(x-c)\right]\cos\beta_n^{(1)}d = \sum_{n=1}^{N+1} \beta_n^{(1)} \overline{C}_n \cos\left[\alpha_n^{(1)}(x-c)\right]\sin\beta_n^{(1)}(d-e) + \beta_n^{(2)} \tilde{C}_n \cos\left[\alpha_n^{(2)}(x-c)\right]\sin\beta_n^{(2)}(d-e) \quad \text{(H.18)}$$

Homogeneous set of equations for the unknown expansion coefficients: The amplitude A_m can be obtained as follows. The fields are expanded with a suitable set of orthogonal functions. It is appropriate to choose the same functions used in the series of the scalar potentials. Multiplication of the continuity equation by $\cos\alpha_M^{(1)}(x-c)$, and integration with respect to x from points I and A (Figure H.2), yields:

$$A_m = \frac{2}{(c-b)\sin\beta_M^{(1)}}\left\{\sum_{n=1}^{N+1}\cos\beta_n^{(1)}\overline{C}_n\int_{x=b}^{c}\cos\left[\alpha_n^{(1)}(x-c)\right]\cos\alpha_M^{(1)}(x-c)\,dx + \sum_{n=1}^{N+1}\cos\beta_n^{(2)}(d-e)\,\tilde{C}_n\int_{x=b}^{c}\cos\beta_n^{(2)}(d-e)\cos\left[\alpha_M^{(1)}(x-c)\right]dx\right\} \quad \text{(H.19)}$$

Because expansion terms of region 1 have been used, it is also said that the field in region 2 has been expanded with the eigenfunctions of region 1. When we let $M = 1, 2, \ldots, N$, we get a homogeneous set of equations for the expansion coefficients. For a nontrivial solution, the determinant of the set of equations must be zero.

Reference

[1] Kowalski, G., and R. Pregla, "Dispersion Characteristics of Shielded Microstrips with Finite Thickness," *Arch. Elek. Übertragung*, Vol. 25, 1971, pp. 193–196.

Appendix I

Impedance Matrix of Step Discontinuity

Figure I.1 shows the top view of the microstrip discontinuity under study and Figure I.2 shows the equivalent circuit of a microstrip step discontinuity with an input line of characteristic impedance Z_{01} and output line of characteristic impedance Z_{02}. I_1 and V_1 are the current and voltage at port 1, and I_2 and V_2 are the current and voltage at port 2. The impedance matrix reads

$$\begin{pmatrix} V_1 \\ V_2 \end{pmatrix} = \begin{pmatrix} Z_{11} & Z_{12} \\ Z_{21} & Z_{22} \end{pmatrix} \begin{pmatrix} I_1 \\ I_2 \end{pmatrix} \tag{I.1}$$

Referring to the equivalent T-network (Section G.1), the impedance matrix is obtained as (G.16)

$$\begin{pmatrix} V_1 \\ V_2 \end{pmatrix} = \begin{pmatrix} Z_1 + Z_3 & Z_3 \\ Z_3 & Z_2 + Z_3 \end{pmatrix} \begin{pmatrix} I_1 \\ I_2 \end{pmatrix}$$

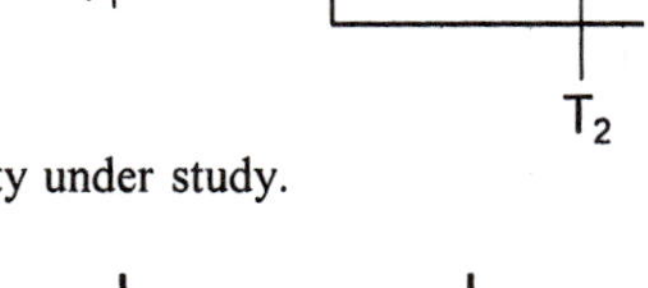

Figure I.1 Step discontinuity under study.

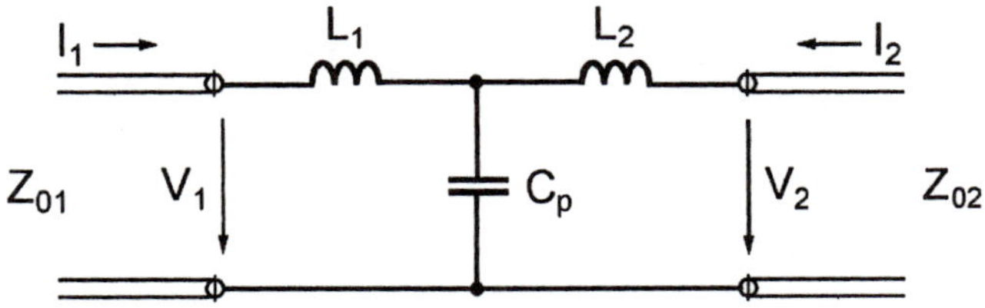

Figure I.2 Equivalent T-network of step discontinuity.

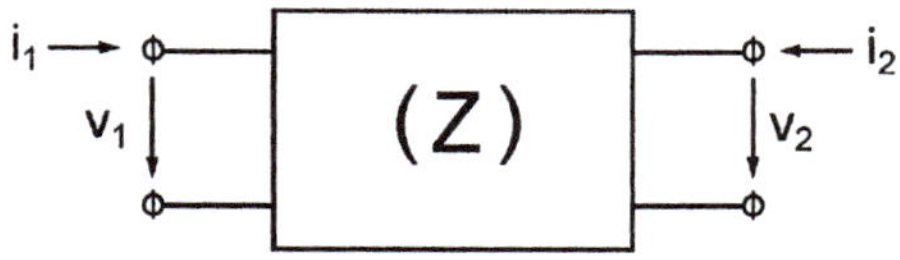

Figure I.3 Two-port representation of a step discontinuity with normalized impedance matrix.

With the given reactive elements, the Z matrix reads

$$(Z) = \frac{1}{j\omega C}\begin{pmatrix} 1-(\omega/\omega_0)^2 & 1 \\ 1 & 1-(\omega/\omega_0)^2 \end{pmatrix}\begin{pmatrix} I_1 \\ I_2 \end{pmatrix} \tag{I.2}$$

where $\omega_0 = 1/\sqrt{LC}$. Using normalized currents and voltages (Section G.2), and taking into account the different characteristic impedances at ports 1 and 2,

$$v_1 = \frac{V_1}{\sqrt{Z_{01}}}, \quad v_2 = \frac{V_2}{\sqrt{Z_{02}}}, \qquad i_1 = I_1\sqrt{Z_{01}}, \qquad i_2 = I_2\sqrt{Z_{02}} \tag{I.3}$$

equation (I.1) reads

$$v_1 = z_{11}i_1 + z_{12}i_2 \tag{I.4}$$
$$v_2 = z_{21}i_1 + z_{22}i_2 \tag{I.5}$$

or in matrix form,

$$\begin{pmatrix} v_1 \\ v_2 \end{pmatrix} = \begin{pmatrix} z_{11} & z_{12} \\ z_{21} & z_{22} \end{pmatrix}\begin{pmatrix} i_1 \\ i_2 \end{pmatrix}$$

$$= \begin{pmatrix} Z_{11}/Z_{01} & Z_{12}/\sqrt{Z_{01}Z_{02}} \\ Z_{21}/\sqrt{Z_{01}Z_{02}} & Z_{22}/Z_{02} \end{pmatrix}\begin{pmatrix} i_1 \\ i_2 \end{pmatrix} \tag{I.6}$$

Figure I.3 shows the simple representation of the two-port of the step discontinuity. The impedance matrix can be calculated from the (measured) S matrix [1],

$$Z = (E + S)\cdot(E - S)^{-1} \tag{I.7}$$

with the unit vector E.

Because the characteristic impedances of the ports are different, it is useful to normalize the matrix elements to a single reference impedance. In this case, the ideal transformer must be included to adequately describe the step discontinuity. In the following section, we will normalize all impedance parameters with respect to the input impedance Z_{01}.

In (I.4), only the last term $z_{12} i_2$ needs to be considered:

$$z_{12} i_2 = \frac{Z_{12}}{\sqrt{Z_{01} Z_{02}}} I_2 \sqrt{Z_{02}} = \frac{Z_{12}}{\sqrt{Z_{01} Z_{02}}} I_2 \sqrt{Z_{01}} \sqrt{\frac{Z_{02}}{Z_{01}}} \tag{I.8}$$

$$= \frac{Z_{12}}{\sqrt{Z_{01} Z_{02}}} \sqrt{\frac{Z_{02}}{Z_{01}}} I_2 \sqrt{Z_{01}} = z_{12}\, n\, i_2'$$

$$= \frac{Z_{12}}{Z_{01}} i_2' = z_{12}' i_2' \tag{I.9}$$

Primed quantities are normalized to the input impedance Z_{01}. With

$$i_2 = I_2 \sqrt{Z_{02}}, \qquad i_2' = I_2 \sqrt{Z_{01}} \tag{I.10}$$

the current i_2 is referred to the current i_2' as

$$i_2' = \sqrt{Z_{01}/Z_{02}}\, i_2 = i_2/n \tag{I.11}$$

Thus, with (I.9), (I.4) can be written as

$$v_1 = z_{11} i_1 + z_{12}' i_2'$$

$$= z_{11} i_1 + n\, z_{12}\, i_2' \tag{I.12}$$

Equation (I.5) reads in denormalized form

$$\frac{V_2}{\sqrt{Z_{02}}} = \frac{Z_{21}}{\sqrt{Z_{01} Z_{02}}} I_1 \sqrt{Z_{01}} + \frac{Z_{22}}{Z_{02}} I_2 \sqrt{Z_{02}} \tag{I 13}$$

The first term is modified as follows:

$$\frac{V_2}{\sqrt{Z_{02}}} = \frac{V_2}{\sqrt{Z_{01}}}\sqrt{\frac{Z_{01}}{Z_{02}}} = v_2'/n = v_2 \tag{I.14}$$

or

$$v_2' = n\, v_2 \tag{I.15}$$

The second term is rewritten as

$$\frac{Z_{21}}{\sqrt{Z_{01}Z_{02}}} = \frac{Z_{21}}{Z_{01}}\,\frac{Z_{01}}{\sqrt{Z_{01}Z_{02}}} = \frac{Z_{21}}{Z_{01}}\sqrt{\frac{Z_{01}}{Z_{02}}}$$

$$= z_{21}'/n = z_{21} \tag{I.16}$$

Thus,

$$z_{21'} = n\, z_{21} = \frac{Z_{21}}{\sqrt{Z_{01}Z_{02}}}\sqrt{\frac{Z_{02}}{Z_{01}}} = \frac{Z_{21}}{Z_{01}} \tag{I.17}$$

The impedance z_{22} is reformulated as

$$z_{22} = \frac{Z_{22}}{Z_{02}} = \frac{Z_{22}}{Z_{01}}\,\frac{Z_{01}}{Z_{02}} = z_{22}'/n^2 \tag{I.18}$$

or

$$z_{22}' = z_{22}\, n^2 \tag{I.19}$$

With (I.11), (I.15), (I.16), and (I.19), (I.5) reads

$$v_2'/n = z_{21}'/n\ i_1 + z_{22}'/n^2\ n\, i_2' \tag{I.20}$$

Multiplying (I.20) by n yields

$$\begin{aligned} v_2' &= z_{21}'\, i_1 + z_{22}'\, i_2' \\ &= n\, z_{21}\, i_1 + n^2\, z_{22}\, i_2' \end{aligned} \tag{I.21}$$

Figure I.4 shows the equivalent network of the step discontinuity with all impedance parameters normalized to the input characteristic impedance Z_{01}.

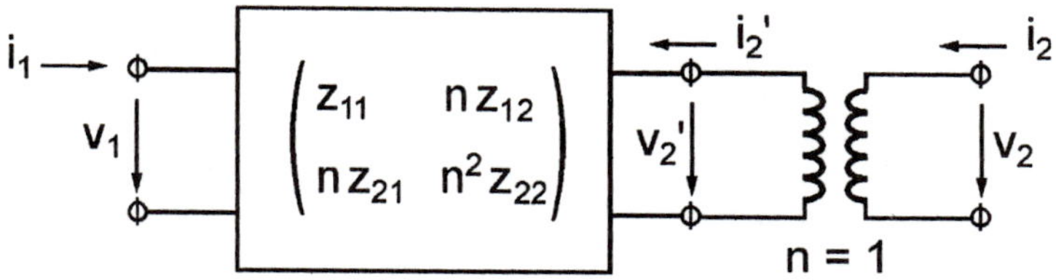

Figure I.4 Equivalent circuit of a step discontinuity with all impedance parameters referenced to the input characteristic impedance [2–4].

References

[1] Brand, H., *Schaltungslehre linearer Mikrowellennetze*, Stuttgart: S. Hirzel Verlag, 1970.

[2] Kompa, G., *Die frequenzabhängigen Übertragungseigenschaften von Streifenleitungs-Wellenwiderstandssprüngen, Filtern und Stichleitungen in Mikrostrip-Technik*, Doctoral Thesis, Aachen: RWTH, 1975.

[3] Kompa, G., "S-Matrix Computation of Microstrip Discontinuities with a Planar Waveguide Model," *Arch. Elektr. Übertr.*, Vol. 30, 1976, pp. 58–64.

[4] Kompa, G., "Design of Stepped Microstrip Components," *The Radio and Electronic Engineer*, Vol. 48, No. 1/2, January/February 1978, pp. 53–63.

Appendix J

Coupled Microstrip Lines

In this appendix, the basic design parameters of coupled microstrip lines are outlined. We assume two uniform coupled microstrip lines of the same width in one plane, as illustrated in Figure 2.23. Asymmetric coupled lines in an inhomogeneous medium are analyzed in [1]. First, we will analyze the quasi-static frequency region, and subsequently dispersion.

J.1 Quasi-Static Coupled Line Parameters

It was discussed in Section 2.8 that two fundamental modes with zero cut-off frequency, the even and odd modes, can propagate along the coupled lines. As illustrated in Figure 2.25, the electromagnetic field distributions show strong differences. With growing slot width *s*, the coupling of the two lines becomes weaker, and the distribution of the electromagnetic field approaches that of two isolated single microstrip lines. When the distance *s* is continuously reduced, the even mode degenerates to a microstrip mode with a doubled microstrip width. A similar limit consideration is not possible for the odd mode. Its existence depends on the different potential of the coupled lines. Therefore, a limit value $s \to 0$ for the odd mode parameters, such as effective dielectric constant and characteristic impedance, cannot be found. This circumstance hampers the derivation of design formulas from numerical results based on full-wave analysis.

Figure J.1 shows the principal dependence of the quasi-static effective dielectric constant of the even mode, $\epsilon_{eff_e}(f=0)$, and odd mode, $\epsilon_{eff_o}(f=0)$, respectively, as a function of the normalized coupling slot s/h [2]. From the plot, it becomes apparent that

$$\epsilon_{eff_e}(f=0) \ge \epsilon_{eff}(f=0) \ge \epsilon_{eff_o}(f=0) \tag{J.1}$$

For the characteristic impedances, the following inequality is valid [3]:

$$Z_{0_e}(\epsilon_r, f=0) \ge Z_0(\epsilon_r, f=0) \ge Z_{0_o}(\epsilon_r, f=0) \tag{J.2}$$

In the limit case $s/h \to \infty$, both strip conductors merge to a single microstrip line with characteristic impedance $Z_0(2w/h)$. The capacitance per unit length has doubled, and the inductance per unit length halved with respect to the single line. Therefore, the limit value for the even characteristic impedance is obtained as $Z_{0e}(s/h \to \infty) = 2Z_0(2w/h)$ (compare Figure 3.2). An odd mode does not exist, so that

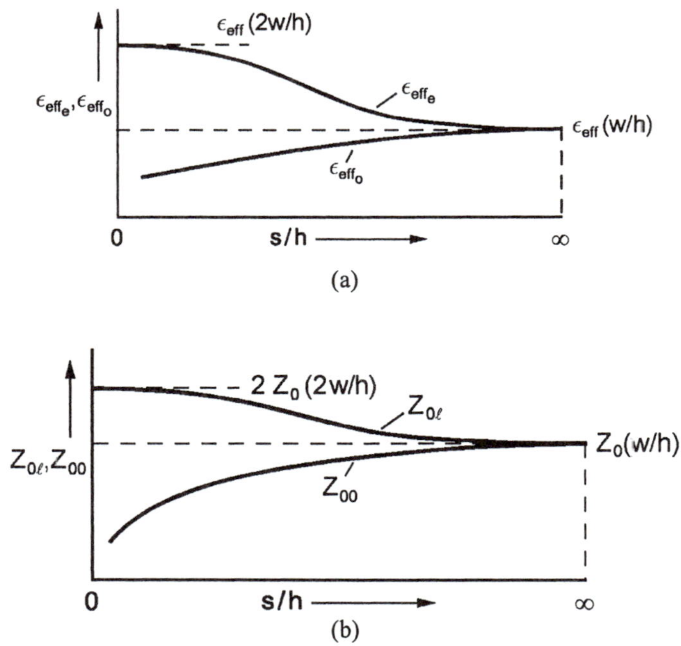

Figure J.1 (a) Principal dependence of effective dielectric constant and (b) characteristic impedance of even and odd mode versus normalized slot width *s/h*.

a limit value for the odd characteristic impedance cannot be defined. An analytical relation for the dielectric constant $\epsilon_{eff_e}(f=0)$ and characteristic impedance $Z_{0_e}(f=0)$ of the even mode must include the following limit values. For simpler writing, we omit the zero-frequency indication for quasi-static validity.

$$\epsilon_{eff_e}(w/h)\Big|_{s/h\to 0} = \epsilon_{eff}(2w/h) \tag{J.3}$$

$$\epsilon_{eff_e}(w/h)\Big|_{s/h\to \infty} = \epsilon_{eff}(w/h) \tag{J.4}$$

$$Z_{0_e}(w/h)\Big|_{s/h\to 0} = 2\,Z_0(2w/h) \tag{J.5}$$

$$Z_{0_e}(w/h)\Big|_{s/h\to \infty} = Z_0\,(w/h) \tag{J.6}$$

Regarding the fundamental odd mode, there is a much stronger electromagnetic air-field in a very narrow coupling slot compared to an isolated microstrip line. This implies a correspondingly lower value of effective dielectric constant ϵ_{eff_o}. With increasing slot width, the field energy is increasingly concentrated in the dielectric substrate, so that ϵ_{eff_o} is assumed to increase monotonously,

$$d\,[\epsilon_{eff_o}(f=0)]\,/\,d(s/h) \geq 0 \tag{J.7}$$

with the limit value for ϵ_{eff_o} and Z_{0_o}, respectively,

$$\epsilon_{eff_o}(w/h)\Big|_{s/h \to \infty} = \epsilon_{eff}(w/h) \tag{J.8}$$

$$Z_{0_o}(w/h)\Big|_{s/h \to \infty} = Z_0(w/h) \tag{J.9}$$

Analysis of coupled microstrip lines has been carried out by several authors. In 1968, Bryant and Weiss [4] published a notable work about an electrostatic solution of a coupled pair of microstrip lines of the same widths. Characteristic impedance and effective dielectric constant are graphically presented as a function of the w/h ratio ($0 \leq w/h \leq 2.0$), with the normalized coupling slot s/h ($s/h = 0.3, 0.6, 1.0$) as a parameter. However, curves are only given for three values of dielectric constant ($\epsilon_r = 1.0, 9.0, 16.0$). A look-up table for $\epsilon_r = 10.0$ is also included. In [5], a method is proposed to find interpolation design curves for any other substrate. An early computer program for the calculation of coupled line parameters is described in [6]. Based on the results of Bryant and Weiss [4, 6], Figure J.2 shows a diagram in which the characteristic impedance of the even mode $Z_{0_e}(f=0)$ is plotted against that of the odd mode $Z_{0_o}(f=0)$ with the normalized width w/h and coupling slot s/h [7]. Figure J.3 shows the corresponding effective dielectric constants [7].

Akhtarzad et al. [8] published closed-form expressions in 1975. The relative error compared with the results in [4] has been found to be approximately 10% [9]. A respective design example of a microstrip coupler is described in [10].

Pregla and Kowalski [11] developed an algorithm for the calculation of the propagation coefficient and characteristic impedance of both single and coupled microstrip lines. Although the programming is easy, the involved improper integral needs computation time that becomes severe with multirepetitive optimization.

Garg and Bahl [9] published in 1979 semiempirical design equations. They decomposed the total capacitance of coupled lines in terms of the main capacitance and fringing capacitances on the outside edges of the coupled lines, and fringing capacitances in the slot region. The capacitances are described by analytical functions. With respect to the published results in [4], a maximum error of 3% is claimed by the authors for the characteristic impedances Z_{0_e} and Z_{0_o} within the region

$$\epsilon_r \geq 1,\ 0.2 \leq w/h \leq 2,\ 0.05 \leq s/h \leq 2$$

The maximum error for $\epsilon_{eff_e}(0)$ and $\epsilon_{eff_o}(0)$ is obtained as 4%.

Substantial progress in calculation accuracy was achieved by Hammerstad and Jensen in 1980 [12]. The formulas of the effective dielectric constant of even and odd modes are similarly composed as (3.121) for a single microstrip line with modified function parameters:

$$\epsilon_{eff_m}(u, g, \epsilon_r) = \frac{\epsilon_r + 1}{2} + \frac{\epsilon_r - 1}{2}\,F_m(u, g, \epsilon_r) \tag{J.10}$$

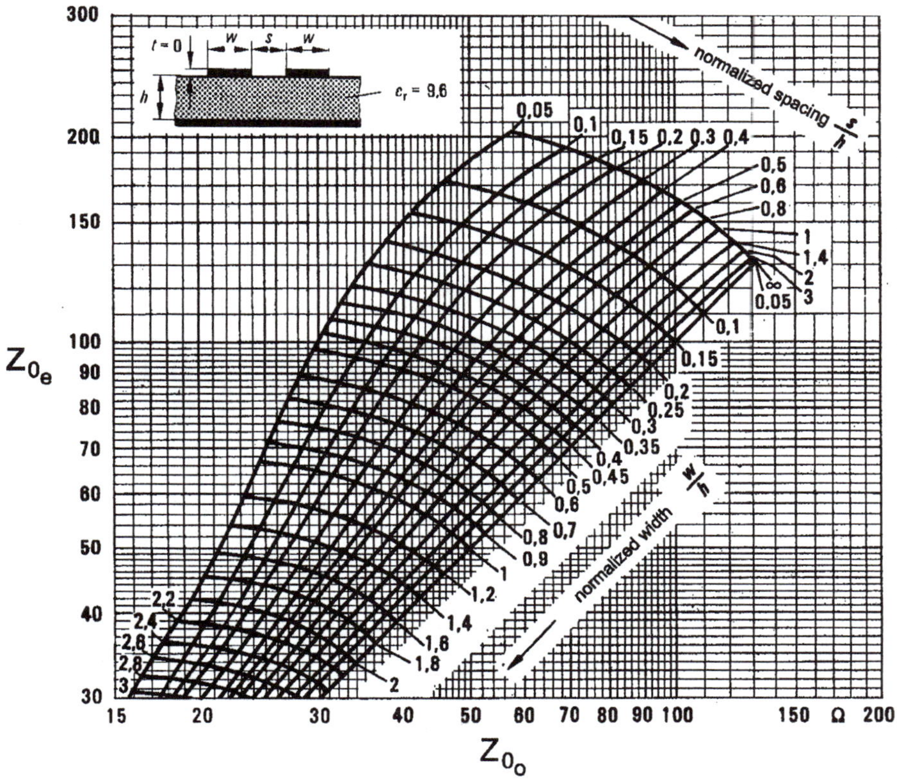

Figure J.2 Quasi-static characteristic impedance of the even and odd mode of coupled microstrip line with normalized microstrip width *w/h* and coupling slot *s/h* as parameters. (*From:* [7]. © 1983 Springer-Verlag. Reprinted with permission.)

where the index m is set to e for the even mode and o for the odd mode. The function $F_{e,o}$ has been found to be

$$F_e(u,g,\epsilon_r) = \left[1 + \frac{10}{\mu(u,g)}\right]^{-a(\mu)b(\epsilon_r)} \tag{J.11}$$

$$F_o(u,g,\epsilon_r) = f_o(u,g,\epsilon_r)\left(1 + \frac{10}{u}\right)^{-a(u)b(\epsilon_r)} \tag{J.12}$$

where

$$\mu(u,g) = g\exp(-g) + u\,\frac{20 + g^2}{10 + g^2} \tag{J.13}$$

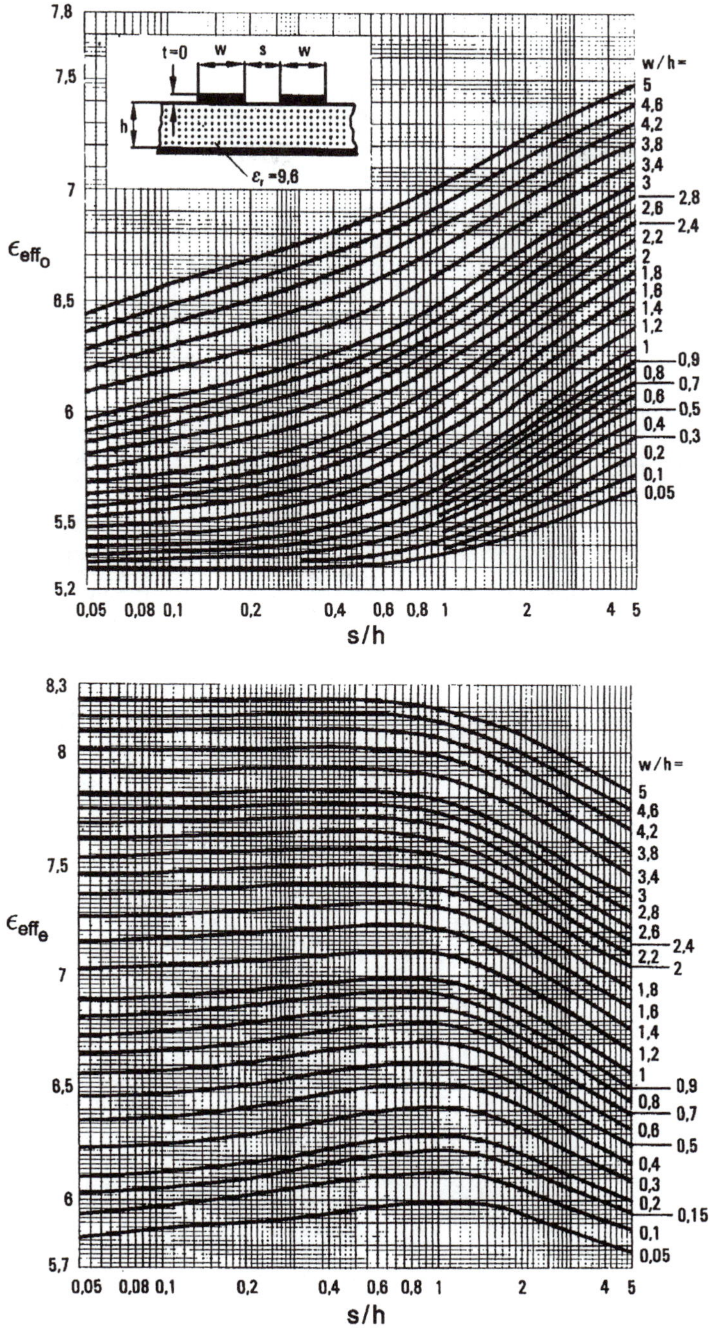

Figure J.3 Effective dielectric constant of the even and odd mode of coupled microstrip line as a function of normalized coupling slot with *w/h* as a parameter. (*From:* [7]. © 1983 Springer-Verlag. Reprinted with permission.)

$$f_o(u,g,\epsilon_r) = f_{o1}(g,\epsilon_r)\exp\left[p(g)\ln u + q(g)\sin\left(\pi\frac{\ln u}{\ln 10}\right)\right] \quad (J.14)$$

$$p(g) = \exp\left(-0.745\,g^{0.295}\right)/\cosh\left(g^{0.68}\right) \quad (J.15)$$

$$q(g) = \exp(-1.366 - g) \quad (J.16)$$

$$u = w/h, \qquad g = s/h \quad (J.17)$$

$$f_{01}(g,\epsilon_r) = 1 - \exp\left\{-0.179\,g^{0.15} - \frac{0.328\,g^{r(g,\epsilon_r)}}{\ln\left[\exp(1) + (g/7)^{2.8}\right]}\right\} \quad (J.18)$$

$$r(g,\epsilon_r) = 1 + 0.15\left\{1 - \frac{\exp\left[1 - (\epsilon_r - 1)^2/8.2\right]}{1 + g^{-6}}\right\} \quad (J.19)$$

The exponential terms $b(\epsilon_r)$ and $a(u)$ in (J.12), and in $a(\mu)$ in (J.11), respectively, are already known from (3.121), and are rewritten for completeness,

$$a(u) = 1 + \frac{1}{49}\ln\left[\frac{u^4 + (u/52)^2}{u^4 + 0.432}\right] + \frac{1}{18.7}\ln\left[1 + \left(\frac{u}{18.1}\right)^3\right] \quad (J.20)$$

$$b(\epsilon_r) = 0.564\left(\frac{\epsilon_r - 0.9}{\epsilon_r + 3.0}\right)^{0.053} \quad (J.21)$$

The characteristic impedances read

$$Z_{0_m}(\epsilon_r = 1,u,g) = Z_0(\epsilon_r = 1,u)/\left[1 - Z_0(\epsilon_r = 1,u)\,\phi_m(u,g)/\eta_0\right] \quad (J.22)$$

where

$$\phi_e(u,g) = \varphi(u)/\{\psi(g)\{\alpha(g)\,u^{m(g)} + [1 - \alpha(g)]\,u^{-m(g)}\}\} \quad (J.23)$$

$$\varphi(u) = 0.864\,u^{0.172} \quad (J.24)$$

$$\psi(g) = 1 + \frac{g}{1.45} + \frac{g^{2.09}}{3.95} \quad (J.25)$$

$$\alpha(g) = 0.5\exp(-g) \quad (J.26)$$

$$m(g) = 0.2175 + \left[4.113 + \left(\frac{20.36}{g}\right)^6\right]^{-0.251} + \frac{1}{323}\ln\frac{g^{10}}{1 + \left(\frac{g}{13.8}\right)^{10}} \quad (J.27)$$

$$\phi_o(u,g) = \phi_e(u,g) - \frac{\theta(g)}{\psi(g)} \exp\left[\beta(g)\, u^{-n(g)} \ln u\right] \tag{J.28}$$

$$\theta(g) = 1.729 + 1.175 \ln\left(1 + \frac{0.627}{g + 0.327 g^{2.17}}\right) \tag{J.29}$$

$$\beta(g) = 0.2306 + \frac{1}{301.8} \ln \frac{g^{10}}{1 + \left(\frac{g}{3.73}\right)^{10}} + \frac{1}{5.3} \ln\left(1 + 0.646 g^{1.175}\right) \tag{J.30}$$

$$n(g) = \left\{\frac{1}{17.7} + \exp\left[-6.424 - 0.76 \ln g - \left(\frac{g}{0.23}\right)^5\right]\right\} \ln \frac{10 + 68.3 g^2}{1 + 32.5 g^{3.093}} \tag{J.31}$$

For the single microstrip line, the following relation was used in (J.22) to calculate the air-filled characteristic impedance with a maximum error of 0.3% for $u = w/h > 0.06$ (note that the more accurate relation (3.120) [12] was found at a later date):

$$Z_0(\epsilon_r = 1, u) = \eta_0 / \left(u + 1.98\, u^{0.172}\right) \tag{J.32}$$

The errors in the even and odd mode characteristic impedances were claimed to be less than 0.8%, and less than 0.6% for the effective dielectric constants.

Based on theoretical results [13], Kirschning investigated the validity of the relations proposed by Hammerstad in the following parameter space:

$$w/h = 0.1\,(0.1)\,1;\ 1.5;\ 2\,(2)\,10$$
$$s/h = 0.1\,(0.1)\,1;\ 1.5;\ 2\,(2)\,6$$
$$\epsilon_r = 2.35,\ 3.75,\ 6.4,\ 9.7,\ 12\,(2)\,18 \tag{J.33}$$

where the number in parentheses denotes the increment of the respective variable. It turned out that re-optimization of the quasi-static effective dielectric constant of the even mode did not show any improvement. Using the notation after Kirschning [2], the final relation for $\epsilon_{eff_e}(f = 0)$ reads

$$\epsilon_{eff_e}(f = 0) = \frac{\epsilon_r + 1}{2} + \frac{\epsilon_r - 1}{2} \left(1 + 10/P_7\right)^{-P_8 P_9} \tag{J.34}$$

where

$$P_7 = g_n \exp(-g_n) + u \frac{20 + g_n^2}{10 + g_n^2} \tag{J.35}$$

$$P_8 = 1 + \frac{1}{49} \ln \left[\frac{P_7^4 + (P_7/52)^2}{P_7^4 + 0.432} \right] + \frac{1}{18.7} \ln \left[1 + \left(\frac{P_7}{18.1} \right)^3 \right] \quad (J.36)$$

$$P_9 = 0.564 \left(\frac{\epsilon_r - 0.9}{\epsilon_r + 3.0} \right)^{0.053} \quad (J.37)$$

$$u = w/h, \qquad g_n = g/h \quad (J.38)$$

A maximum error of 1.3% was found for the characteristic impedance $Z_{0_e}(f = 0)$ within a reduced value interval of the dielectric constant $\epsilon_r = 2.35, 6.4, 9.7$ of the even mode. An essential improvement could be obtained by additional optimization of the parameters for the characteristic impedance $Z_{0_e}(f = 0)$ of the even mode with remaining maximum error of 0.6%:

$$Z_{0_e}(\epsilon_r = 1, f = 0) = Z_0(\epsilon_r = 1, f = 0) \Big/ \left[1 - Z_0(\epsilon_r = 1, f = 0) R_{21}/\eta_0 \right] \quad (J.39)$$

where

$$R_{18} = 0.8695\, u^{0.194} \quad (J.40)$$

$$R_{19} = 1 + \frac{g_n}{1.33} + \frac{g_n^{2.31}}{5.29} \quad (J.41)$$

$$R_{20} = 0.1975 + \left[16.6 + (8.4/g_n)^6 \right]^{-0.387} + \frac{1}{241} \ln \left[\frac{g_n^{10}}{1 + (g_n/3.4)^{10}} \right] \quad (J.42)$$

$$R_{21} = \frac{2 R_{18}}{R_{19} \left\{ \exp(-g_n)\, u^{R_{20}} + \left[2 - \exp(-g_n) \right] u^{-R_{20}} \right\}} \quad (J.43)$$

$$u = w/h, \qquad g_n = g/h, \qquad \eta_0 = 376.73\,\Omega \quad (J.44)$$

As pointed out in [1], Equation (J.10) for the effective dielectric constant of the odd mode can be applied to alumina substrate, showing good agreement with theoretical results. However, regarding lower values of dielectric constant (for example, $\epsilon_r = 2.35$), a maximum error of approximately 4% was observed for $w/h = 8$ and $s/h \to 10$. In this parameter region, the plotted function $\epsilon_{eff_o} = f(s/h, w/h = const.)$ is no longer monotonously increasing with growing normalized slot value s/h, which affects assumption (J.7).

Because of the discussed problems, Kirschning has fitted numerical data to a new functional expression:

$$\epsilon_{eff_o}(f = 0) = \epsilon_{eff}(f = 0) - \left[\epsilon_{eff}(f = 0) - \frac{\epsilon_r + 1}{2} - P_{11} \right] \exp(-P_{12} g_n^{P_{13}}) \quad (J.45)$$

where

$$P_{10} = \frac{0.747\,\epsilon_r}{0.15 + \epsilon_r} \tag{J.46}$$

$$P_{11} = 0.7287\left[\epsilon_{eff}(f=0) - \frac{\epsilon_r + 1}{2}\right]\left[1 - \exp(-0.179u)\right] \tag{J.47}$$

$$P_{12} = P_{10} - (P_{10} - 0.207)\exp(-0.414u) \tag{J.48}$$

$$P_{13} = 0.593 + 0.694\exp(-0.562u) \tag{J.49}$$

Compared with theoretical results, the error for the parameter values in (J.33) turned out to be between -0.5% and $+0.45\%$.

Equation (J.22) for the characteristic impedance of the odd mode was re-optimized and delivered the following result:

$$Z_{0_e}(\epsilon_r = 1, f=0) = Z_0(\epsilon_r = 1, f=0)\,/\left[1 - Z_0(\epsilon_r = 1, f=0)\,R_{27}/\,\eta_0\right] \tag{J.50}$$

where

$$R_{22} = 1.794 + 1.14\ln\left(1 + \frac{0.638}{g_n + 0.517\,g_n^{2.34}}\right) \tag{J.51}$$

$$R_{23} = 0.2305 + \frac{1}{281.3}\ln\left[\frac{g_n^{10}}{1 + (g_n/5.8)^{10}}\right] + \frac{1}{5.1}\ln\left(1 + 0.598\,g_n^{1.154}\right) \tag{J.52}$$

$$R_{24} = \left(10 + 190\,g_n^2\right)/\left(1 + 82.3\,g_n^3\right) \tag{J.53}$$

$$R_{27} = R_{21} - \frac{R_{22}}{R_{19}}\exp\left[R_{23}\,u^{-R_{26}}\ln(u)\right] \tag{J.54}$$

with R_{19} in (J.41) and R_{21} in (J.43),

$$R_{25} = \exp\left[-6.5 - 0.95\ln(g_n) - g_n/0.15)^5\right] \tag{J.55}$$

$$u = w/h,\; g_n = g/h,\; \eta_0 = 376.73\,\Omega \tag{J.56}$$

The error related to theoretical calculations lies in the range from -0.6% to $+0.5\%$.

J.2 Dispersive Coupled Line Parameters

Getsinger [14] has adapted the dispersion relation of a single microstrip line (4.8) to coupled lines. However, as stated in [12], the results are asymptotically wrong for extreme values of gap width. Conversely, it is ascertained in [1] that rather good agreement with theoretical results is given for alumina, which was the substrate material primarily under investigation.

The functional structure proposed by Getsinger was adopted by Kirschning to describe the frequency-dependent effective constants $\epsilon_{eff_e}(f)$ and $\epsilon_{eff_0}(f)$, introducing the already-derived relations for the limit conditions (J.3), (J.4), and J.8). Thus, it is defined

$$\epsilon_{eff_{e,o}}(f_n) = \epsilon_r - \frac{\epsilon_r - \epsilon_{eff_{e,o}}(f=0)}{1 + P_{e,o}(f_n)} \tag{J.57}$$

The following expression has been found for $P_e(f_n)$:

$$P_e(f_n) = P_3 P_4 \left[\left(0.1844\, P_{16} + P_5 P_6\right) f_n \right]^{1.5763} \tag{J.58}$$

where P_3, P_4, P_5, and P_6 are defined in (4.15), and

$$P_{14} = 0.334 \exp\left[-3.3\,(\epsilon_r/15)^3\right] + 0.746 \tag{J.59}$$

$$P_{15} = P_{14} \exp\left[-(f_n/18)^{0.368}\right] \tag{J.60}$$

$$P_{16} = 1 + 4.069\, P_{15}\, g_n^{0.479} \exp\left[-1.347\, g_n^{0.595} - 0.17\, g_n^{2.5}\right] \tag{J.61}$$

$$g_n = g/h, \qquad f_n = (f/\text{GHz}) \cdot (h/\text{mm}) \tag{J.62}$$

Verification of (J.56) has been done for a substrate height $h = 1$ mm and for the following parameter set:

$$w/h = 0.1\,(0.1)\,1;\ 1.5;\ 2\,(2)\,10 \tag{J.63}$$

$$s/h = 0.1\,(0.1)\,1;\ 1.5;\ 2\,(2)\,6 \tag{J.64}$$

$$\epsilon_r = 2.35,\ 3.75,\ 6.4,\ 9.7,\ 12\,(2)\,18 \tag{J.65}$$

$$f/\text{GHz} = 2\,(2)\,10;\ 12.5\,(2.5)\,30 \tag{J.66}$$

A maximum relative error of 1.4% was found for the given parameter combinations. It has been observed that maximum errors occur with narrow widths $w/h \leq 1$.

Because no limit condition exists for the effective dielectric constant for the odd mode in the case of an infinitely narrow coupling slot, the analytical description of dispersion needs higher complexity of fitting functions.

$$P_o(f_n) = P_3 P_4 \left[\left(0.1844 + P_5 P_6\right) P_{24}\, f_n \right]^{1.5763} \tag{J.67}$$

where

$$R_{17} = 0.7168 \left[1 + \frac{1.076}{1 + 0.0576\,(\epsilon_r - 1)} \right] \tag{J.68}$$

$$P_{18} = P_{17} - \frac{2.486}{\pi}\left[\exp(-f_n/20)^{1.424}\right]\arctan\left[2.481\,(\epsilon_r/8)^{0.946}\right] \tag{J.69}$$

$$P_{19} = 0.242\,(\epsilon_r - 1)^{0.55} \tag{J.70}$$

$$P_{20} = \frac{2}{\pi}\left[\exp(-0.34 f_n) - 1\right]\arctan\left[1.263\,(u/3)^{1.629}\right] \tag{J.71}$$

$$P_{21} = P_{18} + \frac{1 - P_{18}}{1 + 1.183\,u^{1.376}} \tag{J.72}$$

$$P_{22} = 1.695\,P_{19}/\left(0.414 + 1.605\,P_{19}\right) \tag{J 73}$$

$$P_{23} = 0.8928 + 0.1072\left\{1 - \exp\left[-0.42\,(f_n/20)^{3.215}\right]\right\} \tag{J.74}$$

$$P_{24} = abs\left\{1 - 0.8928\,\frac{P_{21}}{P_{23}}\,(1 + P_{20})\exp\left[-P_{22}\,g_n^{1.092}\right]\right\} \tag{J.75}$$

$$u = w/h,\; g_n = g/h,\; \eta_0 = 376.73\,\Omega \tag{J.76}$$

The attained accuracy is again 1.4% for the given parameter combinations in (J.63) to (J.66), however, with a slightly reduced upper frequency of 25 GHz (h = 1 mm). The same error is also mantained in a broader frequency range $f \le 30$ GHz with a reduced upper value of dielectric constant ($\epsilon_r \le 12.9$).

As pointed out in [1], the description of the frequency-dependent characteristic impedances of the even and odd modes, based on the P-I definition with convenient accuracy, needs a rather large number of modeling terms. Similar to the approach of the even effective dielectric constant, the analytical limit values for narrow and wide coupling slots are included in the dispersion formula. The frequency-dependent characteristic impedance of the even mode $Z_{0_e}(\epsilon_r, f)$ reads

$$Z_{0_e}(\epsilon_r, f) = Z_{0_e}(\epsilon_r, f=0)\left\{\frac{0.9408\left[\epsilon_{eff}(f_n)\right]^{\tilde{R}_8} - 0.9603}{(0.9408 - R_9)\left[\epsilon_{eff}(f_n = 0)\right]^{\tilde{R}_8} - 0.9603}\right\}^{R_{17}} \tag{J.77}$$

where

$$R_{28} = 0.893\left[1 - \frac{0.3}{1 + 0.7\,(\epsilon_r - 1)}\right] \tag{J.78}$$

$$R_{29} = 2.121 \frac{(f_n/20)^{4.91}}{1 + R_{28}(f_n/20)^{4.91}} g_n^{0.902} \exp(-2.87 g_n) \tag{J.79}$$

$$R_{30} = 1 + 0.038(\epsilon_r/8)^{5.1} \tag{J.80}$$

$$R_{31} = 1 + 1.203 \frac{(\epsilon_r/15)^4}{1 + (\epsilon_r/15)^4} \tag{J.81}$$

$$R_{32} = \frac{1.887\, g_n^{R_{31}} \exp(-1.5\, g_n^{0.84})}{1 + 0.41(f_n/15)^3 u^{2/R_{30}} / (0.125 + u^{1.626/R_{30}})} \tag{J.82}$$

$$R_{33} = R_{32}\left[1 + \frac{9}{1 + 0.403(\epsilon_r - 1)^2}\right] \tag{J.83}$$

$$R_{34} = 0.394\left\{1 - \exp\left[-1.47(u/7)^{0.672}\right]\right\}\left\{1 - \exp\left[-4.25(f_n/20)^{1.87}\right]\right\} \tag{J.84}$$

$$R_{35} = \frac{0.61\left\{1 - \exp\left[-2.13(u/8)^{1.593}\right]\right\}}{1 + 6.544\, g_n^{4.17}} \tag{J.85}$$

$$R_{36} = \frac{0.21\, g_n^4}{(1 + 0.018\, g_n^{4.9})(1 + 0.1\, u^2)\left[1 + (f_n/24)^3\right]} \tag{J.86}$$

$$R_{37} = R_{36}\left[0.09 + \frac{1}{1 + 0.1(\epsilon_r - 1)^{2.7}}\right] \tag{J.87}$$

$\tilde{R}_8$ is a modified version of R_8,

$$\tilde{R}_8 = R_8 - R_{29} + R_{33} - R_{34} + R_{35} + R_{37} \tag{J.88}$$

and $\tilde{R}_4$ reads

$$\tilde{R}_4 = 0.016 + (0.0514\, R_{38}\, \epsilon_r)^{4.524}$$

where

$$R_{38} = abs\left[1 - 42.54\, g_n^{0.133} \exp(-0.812\, g_n) \frac{u^{2.5}}{1 + 0.033\, u^{2.5}}\right] \tag{J.89}$$

Within the line parameter space (h = 1 mm)

$$0.1 \le w/h \le 10$$
$$0.1 \le s/h \le 10$$
$$\epsilon_r \le 12.9$$
$$f/\mathrm{GHz} \le 30 \tag{J.90}$$

a maximum error of 2% is obtained. This accuracy is also attained for $\epsilon_r \le 18$ for the reduced frequency range $f/\mathrm{GHz} \le 25$.

For the characteristic impedance of the odd mode, the following relation has been proposed in [1]:

$$Z_{0_o}(\epsilon_r, f) = Z_0(\epsilon_r, f=0) + \frac{Z_{0_o}(\epsilon_r, f=0)\left[\dfrac{\epsilon_{eff_o}(f_n)}{\epsilon_{eff_o}(f=0)}\right]^{R_{39}} - R_{40}\, Z_0(\epsilon_r, f_n)}{1 + R_{41} + R_{42}\,(0.46\, g_n)^{2.2}} \tag{J.91}$$

where

$$R_{39} = \frac{0.925\,(f_n/R_{43})^{1.536}}{1 + 0.3\,(f_n/30)^{1.536}} \tag{J.92}$$

$$R_{40} = 1 + \frac{0.005 f_n R_{44}}{\left[1 + 0.812\,(f_n/15)^{1.9}\right](1 + 0.025\,u^2)} \tag{J.93}$$

$$R_{29} = 2.121\,\frac{(f_n/20)^{4.91}}{1 + R_{28}\,(f_n/20)^{4.91}}\, g_n^{0.902} \exp(-2.87 g_n) \tag{J.94}$$

$$R_{41} = \frac{2.506\, R_{45}\, u^{0.894}}{3.575 + u^{0.894}}\left[\frac{f_n(1 + 1.3u)}{99.25}\right]^{4.29} \tag{J.95}$$

$$R_{42} = \frac{0.3 f_n^2}{10 + f_n^2}\left\{1 + \frac{0.7\,(\epsilon_r - 1)^2}{0.3\left[5 + (\epsilon_r - 1)^2\right]}\right\} \tag{J.96}$$

$$R_{43} = 30 - R_{46} - 22.2\,\frac{\left[(\epsilon_r - 1)/13\right]^{12}}{1 + \left[(\epsilon_r - 1)/13\right]^{12}} \tag{J.97}$$

$$R_{44} = 0.4\, g_n^{0.84}\left[1 + \frac{2.5\,(\epsilon_r - 1)^{1.5}}{5 + (\epsilon_r - 1)^{1.5}}\right] \tag{J.98}$$

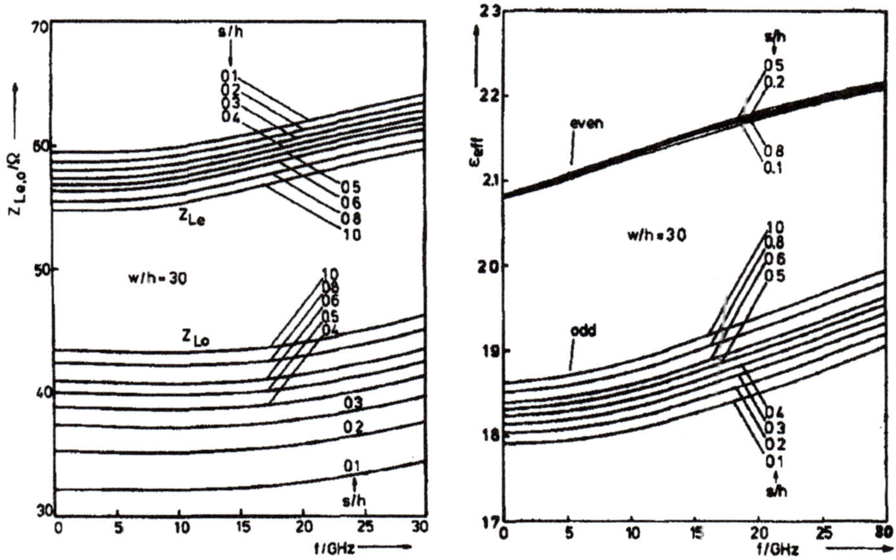

Figure J.4 The frequency-dependent even- and odd-mode characteristic impedances and effective dielectric constants of coupled microstrip lines on a plastic substrate (RT-Duroid 5870, ε_r = 2.35, h = 0.79 mm, w/h = 3.0). (*From:* [3]. © 1984 IEEE. Reprinted with permission.)

$$R_{45} = \frac{0.149(\epsilon_r - 1)^3}{94.5 + 0.038(\epsilon_r - 1)^3} \tag{J.99}$$

$$R_{46} = \frac{15.16}{1 + 0.196(\epsilon_r - 1)^2} \tag{J.100}$$

For the line parameter space defined in (J.90,) a maximum error of 2.5% is claimed for a frequency range $f \le 25$ GHz. Regarding dielectric constant values in the range $\epsilon_r \le 18$, the upper frequency is 20 GHz for the same relative error. Figure J.4 shows an example of the frequency-dependent characteristic impedances and effective dielectric constants of the even and odd modes of coupled microstrip lines on RT-Duroid 5870 ($\epsilon_r = 2.35$, $h = 0.79$ mm, $w/h = 3.0$) [3]. It can be seen that the odd mode parameters are more sensitively dependent on the given range of normalized spacing s/h of coupled lines. In particular, the effective dielectric constant is nearly independent on the s/h parameter value for the even mode.

References

[1] Tripathi, V. K., "Asymmetric Coupled Transmission Lines in an Inhomogeneous Medium," *IEEE Trans.*, Vol. MTT-23, 1975, pp. 734–739.

[2] Kirschning, M., *Entwicklung von Näherungsmodellen für den rechnergestützten Entwurf von hybriden und monolithischen Schaltungen in Mikrostreifenleitungstechnik*, Doctoral Thesis, University of Duisburg, 1984.

[3] Kirschning, M., and R. H. Jansen, "Accurate Wide-Range Design Equations for the Frequency-Dependent Characteristic of Parallel Coupled Microstrip Lines," *IEEE Trans.*, Vol. MTT-32, 1984, pp. 83–90.

[4] Bryant, T. G., and J. A. Weiss, "Parameters of Microstrip Transmission Lines and of Coupled Pairs of Microstrip Lines," *IEEE Trans.*, Vol. MTT-16, 1968, pp. 1021–1027.

[5] Shamasundara, S. D., and N. Singh, "Design of Coupled Microstrip Lines," *IEEE Trans.*, Vol. MTT-25, 1977, pp. 232–233.

[6] Bryant, T. G., and J. A. Weiss, "MSTRIP (Parameters of Microstrip); Computer Program Description," *IEEE Trans.*, Vol. MTT-16, 1971, pp. 418–419.

[7] Hoffmann, R. K., *Integrierte Mikrowellenschaltungen*, Berlin: Springer-Verlag, 1983.

[8] Akhtarzad, S., T. R. Rowbotham, and P. B. Johns, "The Design of Coupled Microstrip Lines," *IEEE Trans.*, Vol. MTT-23, 1975, pp. 486–492.

[9] Garg, R., and I. J. Bahl, "Characteristics of Coupled Microstrip-Lines," *IEEE Trans.*, Vol. MTT-27, 1979, pp. 700–705.

[10] Edwards, T., *Foundations for Microstrip Circuit Design*, 2nd ed., New York: John Wiley & Sons, 1992.

[11] Pregla, R., and G. Kowalski, "Simple Formulas for the Determination of the Characteristic Constants of Microstrips," *Arch. Elektr. Übertr.*, Vol. 28, 1974, pp. 339–340.

[12] Hammerstad, E., and O. Jensen, " Accurate Models for Microstrip Computer-Aided Design," *IEEE MTT-S Int. Microwave Symp. Digest*, 1980, pp. 407–409.

[13] Jansen, R. H., "Unified User-Oriented Computation of Shielded, Covered and Open Planar Microwave and Millimeter-Wave Transmission Line Characteristics," *IEE J. Microwaves, Opt. & Acoust.*, 3, 1979, pp. 14–22.

[14] Getsinger, W. J., "Dispersion of Parallel Coupled Microstrip," *IEEE Trans.*, Vol. MTT-21, 1973, pp. 145–146.

About the Author

Günter Kompa received Dipl.-Ing. and Dr.-Ing. degrees in electrical communication from the Technical University of Aachen, Aachen, Germany, in 1970 and 1975, respectively.

In 1976, he joined Endress & Hauser Company, Germany, focusing on microwave and laser radar research and development. In 1980, he became the head of the department of research and development of sensors for level control.

For his publication "Design of Stepped Microstrip Components" (*The Radio and Electronic Engineer*, Vol. 48, No. 1/2, January/February 1978, pp. 53–63), he won the 1978 Heinrich-Hertz Award, offered by the Institute of Radio and Electronic Engineers, London.

Since 1984, he has been a professor heading the Department of Microwave Engineering at the University of Kassel, Germany.

From 1990 to 2002, Dr. Kompa was a referee for the German Ministry of Research and Technology BMBF (Bundesministerium für Bildung, Wissenschaft, Forschung and Technologie), and, from 1997 to 2001, he was a referee of the German Academic Exchange Service (DAAD). He established the international masters course "Electrical Communication Engineering (ECE)" at the University of Kassel in 1997, which was among the first established international postgraduate degree courses in Germany funded by the government. With the newly established accreditation agencies, it achieved accreditation in 2003.

In 1990, Dr. Kompa organized the International IEEE Workshop on "Measurement Techniques for Microwave Device Characterization and Modeling" in Stuttgart, Germany; and in 1997 the International IEEE Workshop on "Experimentally Based FET Device Modeling and Related Nonlinear Circuit Design" at the University of Kassel, Germany, supported by the DFG (German Research Coucil), the European Research Office (ERO, London), the European Office of Aerospace Research and Development (EOARD, London), and the Office of Naval Research–Europe (ONR-EUR, London).

He was a speaker in 1992 at the International IEEE Workshop "Integrated Nonlinear Microwave and Millimeter-Wave Circuits (INMIC)" in Duisburg, Germany, and in 1993 at the workshop "Advanced Microwave Devices, Characterization and Modeling" on the occasion of the 23rd European Microwave Conference in Madrid, Spain.

Dr. Kompa was invited for the inaugural address of the International Conference on "Modern Design, Manufacturing and Measurement" at the Tsinghua University, Beijing, China, in 1993.

He was coorganizer of the DFG doctorate program "Devices and Materials for Microsystem Engineering" at the University of Kassel for six years, starting in 1995.

Dr. Kompa is a member of the professional societies of IEEE and VDE. He was on the honorary international advisory board for the Asia-Pacific Microwave Conference at Singapore in 1999.

His current research interests cover mainly circuit and system design of ultra-wideband (UWB) microwave and laser radars, multitone characterization and modeling of high-power devices and amplifiers, and power amplifier linearization.

He is author and coauthor of more than 150 technical papers in the fields of microstrip technique, laser radar technology, microwave and millimeter-wave measurement, and circuit design at microwave and millimeter-wave frequencies for communications and sensor application. He has filed numerous patents in Europe, the United States, and Japan.

He has been offering lectures on high-frequency technology, with emphasis on microwaves and millimeter waves, designs of linear and nonlinear microstrip circuits, propagation on waveguiding structures, and high-frequency sensor techniques.

Index

Recent Titles in the Artech House Microwave Library

Active Filters for Integrated-Circuit Applications, Fred H. Irons

Advanced Techniques in RF Power Amplifier Design, Steve C. Cripps

Automated Smith Chart, Version 4.0: Software and User's Manual, Leonard M. Schwab

Behavioral Modeling of Nonlinear RF and Microwave Devices, Thomas R. Turlington

Broadband Microwave Amplifiers, Bal S. Virdee, Avtar S. Virdee, and Ben Y. Banyamin

Computer-Aided Analysis of Nonlinear Microwave Circuits, Paulo J. C. Rodrigues

Design of FET Frequency Multipliers and Harmonic Oscillators, Edmar Camargo

Design of Linear RF Outphasing Power Amplifiers, Xuejun Zhang, Lawrence E. Larson, and Peter M. Asbeck

Design of RF and Microwave Amplifiers and Oscillators, Pieter L. D. Abrie

Digital Filter Design Solutions, Jolyon M. De Freitas

Distortion in RF Power Amplifiers, Joel Vuolevi and Timo Rahkonen

EMPLAN: Electromagnetic Analysis of Printed Structures in Planarly Layered Media, Software and User's Manual, Noyan Kinayman and M. I. Aksun

FAST: Fast Amplifier Synthesis Tool—Software and User's Guide, Dale D. Henkes

Feedforward Linear Power Amplifiers, Nick Pothecary

Fundamentals of Nonlinear Behavioral Modeling for RF and Microwave Design, John Wood and David E. Root, editors

Generalized Filter Design by Computer Optimization, Djuradj Budimir

High-Linearity RF Amplifier Design, Peter B. Kenington

High-Speed Circuit Board Signal Integrity, Stephen C. Thierauf

Intermodulation Distortion in Microwave and Wireless Circuits, José Carlos Pedro and Nuno Borges Carvalho

Lumped Elements for RF and Microwave Circuits, Inder Bahl

Microwave Circuit Modeling Using Electromagnetic Field Simulation, Daniel G. Swanson, Jr. and Wolfgang J. R. Hoefer

Microwave Component Mechanics, Harri Eskelinen and Pekka Eskelinen

Microwave Engineers' Handbook, Two Volumes, Theodore Saad, editor

Microwave Filters, Impedance-Matching Networks, and Coupling Structures, George L. Matthaei, Leo Young, and E.M.T. Jones

Microwave Materials and Fabrication Techniques, Second Edition, Thomas S. Laverghetta

Microwave Mixers, Second Edition, Stephen A. Maas

Microwave Radio Transmission Design Guide, Trevor Manning

Microwaves and Wireless Simplified, Third Edition, Thomas S. Laverghetta

Modern Microwave Circuits, Noyan Kinayman and M. I. Aksun

Modern Microwave Measurements and Techniques, Second Edition, Thomas S. Laverghetta

Neural Networks for RF and Microwave Design, Q. J. Zhang and K. C. Gupta

Noise in Linear and Nonlinear Circuits, Stephen A. Maas

Nonlinear Microwave and RF Circuits, Second Edition, Stephen A. Maas

QMATCH: Lumped-Element Impedance Matching, Software and User's Guide, Pieter L. D. Abrie

Practical Analog and Digital Filter Design, Les Thede

Practical Microstrip Design and Applications, Günter Kompa

Practical RF Circuit Design for Modern Wireless Systems, Volume I: Passive Circuits and Systems, Les Besser and Rowan Gilmore

Practical RF Circuit Design for Modern Wireless Systems, Volume II: Active Circuits and Systems, Rowan Gilmore and Les Besser

Production Testing of RF and System-on-a-Chip Devices for Wireless Communications, Keith B. Schaub and Joe Kelly

Radio Frequency Integrated Circuit Design, John Rogers and Calvin Plett

RF Design Guide: Systems, Circuits, and Equations, Peter Vizmuller

RF Measurements of Die and Packages, Scott A. Wartenberg

The RF and Microwave Circuit Design Handbook, Stephen A. Maas

RF and Microwave Coupled-Line Circuits, Rajesh Mongia, Inder Bahl, and Prakash Bhartia

RF and Microwave Oscillator Design, Michal Odyniec, editor

RF Power Amplifiers for Wireless Communications, Steve C. Cripps

RF Systems, Components, and Circuits Handbook, Ferril A. Losee

Stability Analysis of Nonlinear Microwave Circuits, Almudena Suárez and Raymond Quéré

TRAVIS 2.0: Transmission Line Visualization Software and User's Guide, Version 2.0, Robert G. Kaires and Barton T. Hickman

Understanding Microwave Heating Cavities, Tse V. Chow Ting Chan and Howard C. Reader